STRUCTURE AND PROPERTIES OF FAT CRYSTAL NETWORKS

SECOND EDITION

STRUCTURE AND PROPERTIES OF FAT CRYSTAL NETWORKS

SECOND EDITION

Alejandro G. Marangoni
Leendert H. Wesdorp

CRC Press
Taylor & Francis Group
Boca Raton   London   New York

CRC Press is an imprint of the
Taylor & Francis Group, an **informa** business

CRC Press
Taylor & Francis Group
6000 Broken Sound Parkway NW, Suite 300
Boca Raton, FL 33487-2742

First issued in paperback 2023

Version Date: 20120803

ISBN 13: 978-1-03-265220-7 (pbk)
ISBN 13: 978-1-4398-8762-2 (hbk)
ISBN 13: 978-0-429-11115-0 (ebk)

DOI: 10.1201/b12883

Contents

Introduction

Humans have used fats and oils in their diets since prehistoric times. Fats are important because they are a rich and concentrated source of calories—they are the most concentrated caloric source in nature (9 kcal/g), provide essential fatty acids, and carry many important water-insoluble micronutrients, such as vitamins A, E, K, phytosterols, beta-carotene, and lutein among many others. Fats also provide organoleptic characteristics to foods, which enhances their flavor, texture, lubricity, and satiety, making them more desirable by the consumer. Due to the usefulness of fats and oils as food ingredients, research is always oriented toward furthering our knowledge of their physical and chemical characteristics, both in the bulk and dispersed states.

Fats and oils are a subgroup of lipids. Lipids are organic compounds with the common characteristic of being soluble in nonpolar organic solvents such as isobutanol and hexane and generally insoluble in water. The terms "fat" and "oil" are used interchangeably. The use of each term is based on the physical state of the material at room temperature, which will be dependent on geographical latitude and altitude and the time of the year. In our view, a fat or oil should be defined as a TAG mixture that is solid or liquid at 25°C, respectively, in order to avoid ambiguities. Fats are not usually 100% solid, but rather mixtures of hard crystalline solids intimately associated with liquid oil, in the range of 10%–90% (Bailey, 1950).

Chemically, fats are mixtures of usually over 95% of triacylglycerol (TAGs) molecules and 1%–5% minor components, including phospholipids, glycolipids, free fatty acids, monoacyglycerols (MAGs), diacylglycerols (DAGs), tocopherols, and phytosterols to name a few. TAG molecules consist of a glycerol backbone with three fatty acids esterified to the three alcohol groups at specific locations referred to as sn-1, sn-2, and sn-3, as seen in Figure I.1.

These fatty acids can differ based on chain length, saturation, branching, and the presence of *trans* or *cis* double bonds. Methods to isolate TAGs and to determine their fatty acid composition and positional distribution within a TAG are well established (Christie and Han, 2010; The Lipid Library at http://lipidlibrary.aocs.org).

TAGs are usually classified into three groups with respect to the fatty acids that are present. Monoacid TAGs have the same fatty acids present in positions sn-1, sn-2, and sn-3, such as tristearin and tripalmitin. TAG molecules with two or three different types of fatty acids are called diacid or triacid TAGs, both of which are referred to as mixed-acid TAGs. Diacid TAGs can be further placed into subgroups based on which fatty acid occupies position sn-2, making the TAG molecule thus symmetric or asymmetric. Fats, in turn, can be composed of only one type of TAG molecule (not commonly) or they can be a combination of many different TAG molecules. For example, milk fat contains a still unknown number of TAGs, definitely greater than 200 molecular species (Gresti et al., 1993; Jensen, 1995), while 80% of all TAGs in high oleic sunflower oil are triolein molecules (Gunstone, 2004).

The most influential structural characteristics in a TAG molecule that affect their physical properties (e.g., melting and crystallization behavior, solid fat content,

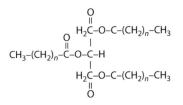

FIGURE I.1   The structure of a typical saturated triacylglycerol molecule.

polymorphism, nano- and microstructure, mechanical properties, oil binding capacity) include the length of the fatty acid chains; the number, position, and configuration of double bonds (saturated vs. unsaturated or *cis* vs. *trans*); and, finally, the stereospecific position of the fatty acid on the glycerol backbone (Small, 1986). The most common TAGs contain fatty acids with chain lengths between 4 and 22 carbon atoms with 0–6 double bonds (Gunstone, 2004). A characteristic of having a fat composed of a mixture of different TAG molecules is that the material does not display a unique melting point, but rather a melting range defined by chemical nature and molecular interactions of this complex mixture. This is due both to the great variety of molecular species present and because of the complex phase behavior between and among TAG species (Bailey, 1950; Rossel, 1967; Timms, 1984; Small, 1986). Moreover, TAGs can crystallize into different solid state structures, depending on crystallization conditions (temperature, shear, time), giving rise to polymorphism (Clarkson and Malkin, 1934; Chapman, 1962; Sato, 2001). Understanding the way that TAGs arrange and interact is critical if an understanding of the macroscopic functionality of fats is sought.

An interesting aspect of fats is that they behave like elastic solids until a deforming stress exceeds a certain value (yield value, yield force, or yield stress), at which point the product starts to flow like a viscous fluid (Haighton, 1959; deMan and deMan, 2002). This phenomenon, termed plasticity, arises from the fact that the crystallized material forms a fat crystal network that entraps liquid oil.

Many of the sensory attributes of fat and fat-structured food products, like spreadability, mouthfeel, texture, and flavor, are strongly influenced by the physical characteristics of the fat crystal network. The mechanical and organoleptic properties, as well as the stability (shelf life) of materials such as chocolate, butter, margarine, and spreads, are, to a great extent, determined by crystallization of the material. This is why we need to understand fat crystallization and its relationship to structure: in order to control and engineer material properties. In food technology, the concept of "structure" relates to the organization of a number of similar or dissimilar elements, their binding into a unit, and the interrelationships between the individual elements or their groupings (Raeuber and Nikolaus, 1980). The concept of structure has, therefore, organizational, constructive, and relative meanings. The organizational structure is derived from the combined action of structural elements and their groupings. The constructive structure has building and hierarchical aspects, and the relative structure refers to the level of interest in a study. In food systems, structuring plays an important role in determining bulk material properties. Fundamentally, the microstructural level, which is established by supramolecular assemblies or phase discontinuities and can vary from

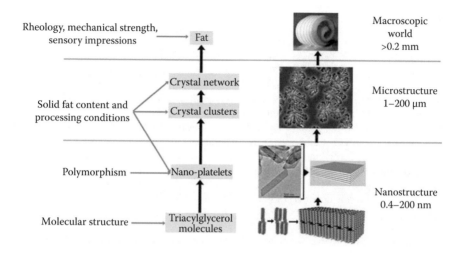

FIGURE I.2 Structural hierarchy in a fat crystal network—from molecules to material.

0.5 to 200 μm, has an enormous influence on the macroscopic properties of the network (Marangoni et al., 2012). It is crucial to understand the structural organization present in a material and the macroscopic properties because it can guide the development of specific characteristics in manufactured materials. In a fat crystal network, a structural hierarchy has been recognized to describe the formation and interaction of the elements that affect the final properties of a three-dimensional fat crystal network and therefore the foods structured by such network. Figure I.2 depicts the proposed structural hierarchy in fat crystal networks. It summarizes our current knowledge of the structure of edible fats, from TAG molecules to progressively larger supramolecular assemblies until the macroscopic world is reached. This view summarizes the work of many researchers in the field over the past 50 years, such as van den Tempel (1961), deMan and Beers (1987), Heertje (1993), and Marangoni et al. (2012). The complete view of this structural hierarchy was not possible until the recent characterization of the nanoscale in fats (Acevedo and Marangoni, 2010a,b).

In this second edition, we have revisited concepts and approaches used in the study of fat crystal networks. New developments have been included, in particular intermolecular interactions. Analytical methods used in the study of fat crystal networks have been thoroughly updated, while challenges related to a better understanding of phase equilibria in fats remain unchanged. This book can be used by the student interested in furthering their fundamental knowledge of fat structure and functionality or by the student interested in properly characterizing their material. It is important to understand the structural organization present in a material and relate it to macroscopic properties because it can guide efforts in replacing ingredients, optimizing functionality, and improving health in a rational fashion. Ultimately, this will also result in a more focused, less frustrating, and less expensive endeavor. In the words of Heerje and Leunis (1997), "Product attributes such as spreadability, hardness and work softening are determined at least partly by the shape and size of the individual fat crystals and the way in which these crystals interact to form clusters, agglomerates and networks."

REFERENCES

Acevedo, N.C. and A.G. Marangoni. 2010a. Characterization of the nanoscale in triacylglyc-erol crystal networks. *Cryst. Growth Des.* 10: 3327–3333.

Acevedo, N.C. and A.G. Marangoni. 2010b. Towards nanoscale engineering of triacylglycerol crystal networks. *Cryst. Growth Des.* 10: 3334–3339.

Bailey, A.E. 1950. *Bailey's Industrial Oil and Fat Products*. New York: Interscience Publishers.

Chapman, D. 1962. The polymorphism of glycerides. *Chem. Rev.* 62: 433–456.

Christie, W.W. and X. Han. 2010. *Lipid Analysis: Isolation, Separation, Identification and Lipidomic Analysis*. Cambridge, U.K.: Woodhead Publishing Ltd.

Clarkson, C.E., and T. Malkin. 1934. Alternation in long-chain compounds. Part II. An X-ray and thermal investigation of the triglycerides. *J. Chem. Soc.*, 666–671.

Gresti, J., M. Bugaut, C. Maniongui, and J. Bezard. 1993. Composition of molecular species of triacylglycerols in bovine milk fat. *J. Dairy Sci.* 76: 1850–1869.

Gunstone, F.D. 2004. *The Chemistry of Oils and Fats*. Coventry, U.K.: Blackwell Publishing Ltd.

Haighton, A.J. 1959. The measurement of the hardness of margarine and fats with cone pen-etrometers. *J. Am. Oil Chem. Soc.* 36: 345–348.

Heertje, I. 1993. Microstructure studies in fat research. *Food Microstruct.* 12: 77–94.

Heertje, I. and M. Leunis. 1997. Measurement of shape and size of fat crystals by electron microscopy. *Lebensm. Wiss. Technol.* 30: 141–146.

Jensen, R.G. 1995. *Handbook of Milk Composition*. New York: Academic Press.

deMan, J.M., and A.M. Beers. 1987. Fat crystal networks: Structure and rheological proper-ties. *J. Text. Stud.* 18: 303-318.

deMan, J.M., and L. deMan. 2002. Texture of fats. In *Physical Properties of Lipids*, eds. A.G. Marangoni and S.S. Narine. New York: Marcel Dekker, pp. 191–217.

Marangoni, A.G., N. Acevedo, F. Maleky, E. Co, F. Peyronel, G. Mazzanti, B. Quinn, and D. Pink. 2012. Structure and functionality of edible fats. *Soft Matter*. DOI: 10.1039/C1SM06234D.

Raeuber, H.J., and H. Nikolaus. 1980. Structure of foods. *J. Text. Stud.* 11: 187–198.

Rossel, J.B. 1967. Phase diagrams of triglyceride systems. *Adv. Lipid Res.* 5: 353–408.

Sato, K. 2001. Crystallization behavior of fats and lipids—A review. *Chem. Eng. Sci.* 56: 2255–2265.

Small, D.M. 1986. *The Physical Chemistry of Lipids*. New York: Plenum Press.

van den Tempel, M. 1961. Mechanical properties of plastic disperse systems at very small deformations. *J. Colloid Sci.* 16: 284–296.

Timms, R.E. 1984. Phase behavior of fats and their mixtures. *Prog. Lipid Res.* 23: 1–38.

1 Crystallography and Polymorphism

1.1 CRYSTAL LATTICES

Early in the history of modern science it was suggested that the regular external form of crystals (their morphology) implied an internal regularity. Visual inspection of many different crystals led to the realization that they all corresponded to one of only seven regular shapes called the *seven crystal systems*. Which system a crystal belongs to is determined by measuring the angles between its faces, and deciding how many axes are needed to define the principal features of its shape. Figure 1.1 depicts the three crystallographic axes (a, b, c) and angles (α, β, γ) that define the shape of a crystal. Notice how x, y, z do not necessarily correspond to a, b, c, except for a rectangular shape.

For example, if three equivalent, mutually perpendicular axes are required to define the crystal shape, then the crystal belongs to the *cubic system*. If one axis perpendicular to two that make an obtuse angle is required, then the crystal belongs to the *monoclinic system*. The characteristics of the seven crystal systems are summarized in Table 1.1. We will return to this table shortly after we discuss lattices. At that point, the significance of the extra column in Table 1.1 will become clear.

The difficulty with a visual classification procedure is that crystal faces grow at different rates, and so the appearance of the crystal may be distorted. Moreover, in some cases there may be accidental equivalences of axes. Therefore, in order to classify crystals unambiguously, we must do so on the basis of the *internal* symmetry elements they possess.

1.2 LATTICES AND UNIT CELLS

A crystal is an orderly array of symmetrically arranged particles. In order to be precise about the internal organization of the crystal, we need to distinguish the individual units from which the crystal is built. We introduce the following definitions:

1. *The asymmetric unit* is the particle (ion, molecule) from which the crystal is built.
2. *The space lattice* is a three-dimensional, infinite array of points, each of which is surrounded in an identical way by its neighbors. That is, the space lattice defines the basic structure of the crystal. In some cases there may be an asymmetric unit at each lattice point, but that is not necessary. For instance, each lattice point might be at the center of a cluster of three asymmetric units. The space lattice is, in effect, the abstract scaffolding for the crystal structure.

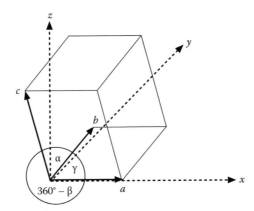

FIGURE 1.1   Diagram depicting the three crystallographic axes a, b, c, the corresponding crystallographic angles α, β, γ within the context of a set of Cartesian coordinates and a putative unit cell.

TABLE 1.1

The Seven Crystal Systems and Associated Lattice Types and Their Structural Characteristics

System	Axes	Angles	Lattice Type
Cubic	$a = b = c$	$\alpha = \beta = \gamma = 90°$	P, F, I
Rhombohedral (trigonal)	$a = b = c$	$\alpha = \beta = \gamma < 120° \neq 90°$	P
Tetragonal	$a = b \neq c$	$\alpha = \beta = \gamma = 90°$	P, I
Hexagonal	$a = b \neq c$	$\alpha = \beta = 90°, \gamma = 120°$	P
Orthorhombic	$a \neq b \neq c$	$\alpha = \beta = \gamma = 90°$	P, C, I, F
Monoclinic	$a \neq b \neq c$	$\alpha = \gamma = 90°, \beta \neq 90°$	P,C
Triclinic	$a \neq b \neq c$	$\alpha \neq \beta \neq \gamma$	P

P, primitive; I, body centered; F, face centered; C, base centered.

---

3. *The crystal structure* is obtained by associating with each lattice point an assembly of asymmetric units in a symmetrical arrangement which is identical for each lattice point.
4. *The unit cell* is the fundamental unit from which the entire crystal may be constructed purely by translational displacements (like a brick in a wall). The unit cell contains all the symmetry elements of the crystal.

A unit cell must possess the overall symmetry of the crystal. It follows that we should expect to be able to account for the external morphology of a crystal in terms of the symmetry of its unit cell. The morphology, however, as we mentioned before, also depends on the relative rates of growth of the different crystal faces, but the underlying unit cell structure is uniform. Therefore, if we can identify the symmetry elements of the unit cell, we shall have an unambiguous classification of the crystal system.

In three dimensions there are 14 types of unit cells that can stack together and give rise to a space lattice: the Bravais lattices. The Bravais lattices fall into seven groups, corresponding precisely to the seven crystal systems (Table 1.1). In other words, the seven crystal systems correspond to the existence of seven regular internal shapes that may be packed together to fill all space. Furthermore, the occurrence of the crystal classes reflects the presence of the corresponding symmetry elements in the unit cells. Therefore, a triclinic crystal morphology indicates the presence of a unit cell with triclinic symmetry, and so on. A very detailed visual rendition of the different crystal systems can be found at http://www.youtube.com/watch?v=PWQ89UoxOK8

In addition to the aforementioned seven crystal systems, we must also consider the way the atoms are distributed within the unit cell that makes up the crystal lattice. Those with lattice points only at the corners of the unit cell are called primitive (P); body-centered (I) unit cells have lattice points in the center of the unit cell, face-centered (F) unit cells have lattice points in the center of their faces, and base-centered (C) unit cells have lattice points only at two of the four faces of the unit cell (Figure 1.2). Thus, considering 7 several crystal systems and the types of lattices present in each family, we find 14 types of lattices in nature.

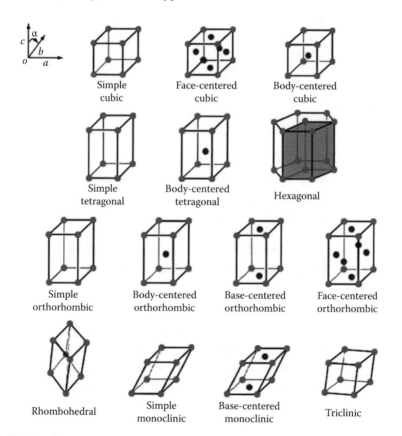

FIGURE 1.2   The fourteen Bravais lattices.

1.3   MILLER INDICES

Even in a rectangular lattice, a large number of planes can be identified. Consider a three-dimensional rectangular lattice formed from a unit cell of sides a, b (Figure 1.3). Four sets of planes have been drawn through these lattice points. These planes can be identified according to the projected distances along the a, b, and c axes that a plane covers between two consecutive lattice points. For example, the four sets in the illustration can be denoted respectively, in a clockwise fashion starting from the upper left hand corner, as $(1a, 1b)$, $(1a, \infty b)$, $(-1a, 1b)$, and $(\infty a, 1b)$. If we agree to quote distances along the axes in terms of the lengths of the unit cells, these planes can be specified more simply as $(1, 1)$, $(1, \infty)$, $(-1, 1)$, and $(\infty, 1)$. The lattice in Figure 1.3 corresponds to the top view of a three-dimensional rectangular lattice in which the unit cell has a length "c" in the z-direction. Thus, all four sets of planes intersect the z axis at infinity, and therefore, the complete labels are $(1, 1, \infty)$, $(1, \infty, \infty)$, $(-1, 1, \infty)$, and $(\infty, 1, \infty)$. The appearance of ∞ is inconvenient, and a way of eliminating it is to deal with the reciprocal of the indices. The, so-called, Miller indices are the reciprocals of the numbers in the brackets, with fractions cleared. For example, the $(1a, 1b, \infty c)$ plane, abbreviated as the $(1, 1, \infty)$ plane, becomes (110) in the Miller system. This label is used to refer to the complete set of equally spaced planes parallel to this one. Similarly, the $(1, \infty, \infty)$ plane becomes (100). Negative indices are written with a bar over the number. The three indices in the Miller system are denoted as (hkl).

One should remember that the unit cell need not be rectangular (see Figure 1.1); the procedure works equally well when axes are not perpendicular. For that reason, the axes are called a, b, c rather than x, y, z. As well, the smaller the value of h in (hkl), the more nearly parallel the plane to the a axis (the same applies to k and the b axis and l and the c axis). When $h = 0$, the planes intersect the a axis at infinity, and so $(0, k, l)$ are parallel to the a axis. Similarly, $(h0l)$ and $(hk0)$ are parallel to the b and c axis, respectively.

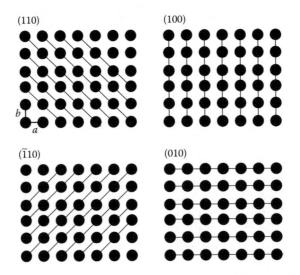

FIGURE 1.3   Examples of some of the many planes present within a crystal and their corresponding Miller indices.

The Miller indices are very useful for calculating the separation between planes (*d*). From the extension of the Pythagorean theorem to three dimensions, it is possible to show that for a general orthorhombic lattice (a rectangular, or orthogonal, lattice based on a unit cell with different lengths of crystallographic axes),

$$\frac{1}{d_{hkl}^2} = \left(\frac{h}{a}\right)^2 + \left(\frac{k}{b}\right)^2 + \left(\frac{l}{c}\right)^2 \tag{1.1}$$

This equation simplifies for the case of a cubic lattice (*a* = *b* = *c*) to

$$\frac{1}{d_{hkl}^2} = \frac{(h^2 + k^2 + l^2)}{a^2} \tag{1.2}$$

Equations 1.1 and 1.2 only apply to the case where $\alpha = \beta = \gamma = 90°$ (cubic, orthogonal, and tetragonal unit cells). More complex equations are required for the other unit cell geometries, where estimates of the crystallographic angles are required. One thing, though, should be clear at this point: the prediction of the characteristic spacing, *d*, *for a specific crystallographic plane*, requires knowledge of the dimensions and shape of the unit cell. This information is usually not available in the area of fat crystallography.

1.4 POWDER X-RAY DIFFRACTION AND BRAGG'S LAW

Since x-rays have wavelengths comparable to atomic spacings, they are diffracted when passed through a crystal. Diffraction arises as a result of interference between waves. Where their amplitudes are in-phase, the waves augment each other and the intensity is enhanced; where their amplitudes are out-of-phase they cancel, and the intensity is decreased.

Consider a stack of reflecting lattice planes (Figure 1.4). The path length difference between the two rays is 2*x*, which equals 2*d*sinθ, where *d* is the layer spacing and θ is the glancing angle. For many glancing angles, the path-length difference is not an integral number of wavelengths, thus resulting in destructive interference. However, when the path-length difference is an integral number of wavelengths (2*x* = *n*λ), the reflected waves are in-phase and interfere constructively. It follows that a bright reflection should be observed when the glancing angles fulfill the Bragg condition:

$$n\lambda = 2d \sin \theta \tag{1.3}$$

Usually, all reflections are considered to be first-order (*n* = 1), with higher order reflections (*n* = 2, *n* = 3, …) incorporated into *d*. Thus, Bragg's law can be rewritten as

$$\lambda = 2d_{hkl} \sin \theta \tag{1.4}$$

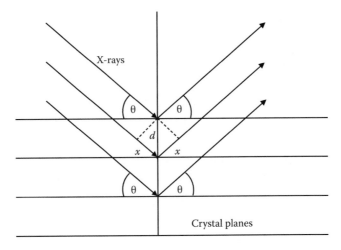

FIGURE 1.4 Geometry of the refection of x-rays from crystal planes used in the derivation of Bragg's law.

In this case, an nth order reflection is considered to arise from the (nh, nk, nl) planes. For example, $d_{220} = 1/2d_{110}$. In general, increasing the indexes uniformly by a factor of n decreases the separation distance by the same factor (Figure 1.5).

In most fats and oils work, single crystal XRD is not carried out, but rather powder XRD. Fats are polycrystalline materials, a randomly oriented heap of tiny crystals. Some of the crystals will, however, be in the correct orientation to satisfy Bragg's law, even for monochromatic radiation. In principle, each set of (hkl) planes gives rise to a diffraction cone, because some of the randomly orientated crystallites

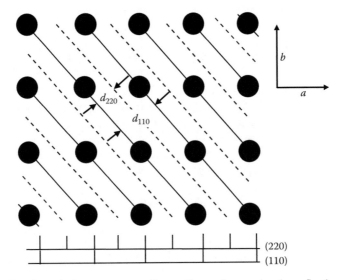

FIGURE 1.5 Crystal planes corresponding to first and second order reflections and their characteristic d-spacing.

can diffract the incident beam. This powder technique is equivalent to rotating a single crystal over all possible orientations relative to the incident x-ray beam. In fats and oils crystallography, the major application of powder XRD is to qualitatively fingerprint crystal forms since many such patterns have been recorded. The powder technique is useful for a qualitative analysis of the sample, and for an initial determination of the dimensions and symmetry of the unit cell, but it cannot provide the detailed information about the electron density distribution that is available from single-crystal methods. Having said this, it is possible to predict the single crystal structure of a fat by making assumptions about its structure. For example, Peschar et al. (2004) predicted the single crystal structure of cocoa butter using powder diffraction techniques, by assuming it was mostly composed of SOS (stearic-oleic-stearic) TAGs. This type of analysis requires specialized know-how that is beyond the scope of this chapter.

1.5   TYPICAL POWDER XRD SETUP

Figure 1.6 depicts a possible configuration for a powder XRD experiment. In this particular setup, the sample is positioned in the middle of the camera and the walls of the camera are lined with photographic film. The incoming x-ray beam will be diffracted by the crystal planes, when Bragg's law is obeyed, at angle θ relative to the through beam (Figure 1.6). From knowledge of the distance between the camera and the sample (r), as well as the distance between the through beam and the particular reflection (the length of the arc, l), it is possible to determine the angle α, which is equal to 2θ using

$$\alpha = 360° \frac{l}{2\pi r} \qquad (1.5)$$

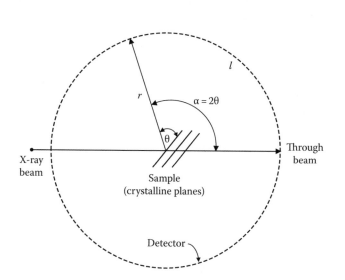

FIGURE 1.6   A powder x-ray diffraction experimental setup.

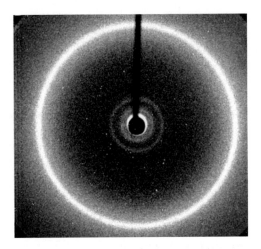

FIGURE 1.7 Debye rings for cocoa butter crystallized at 15°C collected using a two-dimensional CCD detector.

Since $\alpha = 2\theta$, we can determine the spacing between crystal planes corresponding to that particular reflection using Bragg's law,

$$d_{hkl} = \frac{\lambda}{2\sin\theta} \tag{1.6}$$

A densitometric scan of the film will provide an intensity-scattering angle profile for the crystalline material being analyzed. This profile is called the powder spectrum.

Many other setups are possible. Many modern machines have two-dimensional detectors (CCD detectors) which can detect the position and intensity of all reflections simultaneously, as shown in Figure 1.7, which is a powder spectrum of cocoa butter crystallized at 15°C. X-rays are usually transmitted and scattered through the sample in this setup and is thus referred to as x-ray scattering. These detectors are well-suited for the study of the dynamics of crystallization processes. The detector could also be a scintillation counter mounted on a motorized goniometer which travels and collects intensity data at the different scattering angles, as shown in Figure 1.8 for form V of cocoa butter at room temperature. X-rays are usually reflected from the sample in this setup and the technique is referred to as x-ray diffraction instead. The resolution obtained with such setup is very high; however, it takes a long period of time to collect a spectrum which includes both wide angle and small angle data (step sizes are in the order of 0.01°). Thus, the dynamics of crystallization processes are difficult to study. A densitometric scan of Figure 1.7 would result in a spectrum such as the one shown in Figure 1.8. A line detector, on the other hand, is a fixed detector (does not move) which allows the recording of diffracted light intensity for all diffraction angles simultaneously. Of course, there are ranges of usefulness, and therefore a couple of these can be placed so as to cover the required range of 2θ. There a so many types of detectors (or cameras as they are usually referred to), that

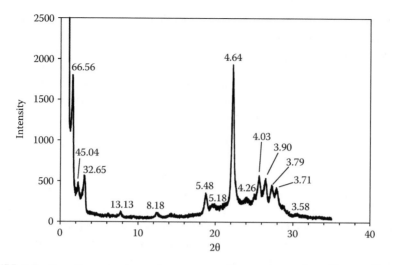

FIGURE 1.8 One-dimensional powder spectrum of cocoa butter form V, crystallized at 22°C for 20 days. The numbers correspond to the d-spacings in angstroms.

it would be difficult to discuss them all in this short chapter. In fat crystallography, the most important consideration is whether both wide and small angle data needs to be recorded and whether time resolution is required. If there are requirements for monitoring changes in solid state structure as a function of time, highly sensitive two-dimensional detectors and/or fixed line-detectors will be required. Don't buy the wrong instrument for your application!

1.6 INDEXING REFLECTIONS

From knowledge of the of the angle 2θ, values of d can be determined. The dimensions of the unit cell (a, b, and c) can be determined by identifying the d-spacings that correspond to the (100), (010), and (001) planes, respectively. For example, assuming a cubic unit cell, one can calculate d-spacings corresponding to all hkl planes using Equation 1.7, and compare them to experimentally determined values. However, the reader probably immediately noticed the requirement of knowledge of the type of unit cell present and the fact that this example focused conveniently only on a cubic unit cell. The situation even for a general orthorhombic case where $a \neq b \neq c$ would be much more complicated. Suffice to say that there are more sophisticated techniques to do this.

$$d_{hkl} = \frac{a}{(h^2 + k^2 + l^2)^{1/2}} \qquad (1.7)$$

Some types of unit cells give characteristic and easily recognizable patterns of diffraction lines. Thus, it is possible to generate characteristic spectra for known unit cell types and experimental results can be compared to these.

Of note is that not all (*hkl*) planes will produce a diffraction line. Consider the (100), (110), and (111) planes of primitive, face-centered, and body-centered cubic lattices. In a face-centered lattice, the (100) and (110) planes sandwich planes sharing the same lattice points. The same is true for the (100) and (110) planes of a body-centered lattice. Reflections from these planes will therefore destructively interfere with reflections from the intervening planes. It follows that only in crystals with a structure based on the primitive lattice will all three of these planes produce diffraction lines.

1.7 CRYSTALLOGRAPHIC STRUCTURE OF FATS

Most of our knowledge of the crystalline structure of fats comes from x-ray diffraction studies. Fat crystal structure has profound effects on the physical properties (consistency, melting point) of the fat. The adopted crystal structure depends on the specific type of TAG molecular species present, with characteristic fatty acid composition and distribution, TAG purity, and crystallization conditions (temperature, rate of cooling, shear, seeds, and solvent).

When a triacylglycerol molecule crystallizes, the molecule adopts an asymmetric tuning fork configuration, or chair (Figure 1.9). This figure corresponds to the single crystal structure of tricaprin published by Jensen and Mabis in 1966. Indicated on the figure are the *b* and *c* axes of the unit cell. This view of the crystal lattice is along the *b–c* plane. In TAG crystallography, the *c* axis refers to the long axis of the unit cell, while the *b* and *a* axes refer to the short axes of the unit cell. In this representation, the other short axis *a* would be protruding behind the *a–b* plane. Also notice in this figure that the acyl chains at positions sn-1 (I) and sn-2 (II) are arranged opposite to each other, while the acyl chain at position sn-3 is parallel to the sn-1 chain. One can imagine how important the configuration of the glycerol moiety must be in order to accommodate such packing. Several of these "chairs" will then align side-by-side to maximize van der Waal's interactions (Figure 1.10). The single crystal structure of tricaprin is shown here again, with proper bond angles and lengths. The triclinic unit cell is also indicated here as a box around the molecular pair.

In general, long spacings mirror the distance between methyl end groups for triacylglycerols and are usually comprised of two (Jensen and Mabis, 1966), three (Goto et al., 1992), or four (van Langevelde et al., 2000; Sato et al., 2001) chain lengths of the fatty acid constituents of the TAG (Figure 1.11). Long spacings for triacylglycerols increase with increasing fatty acyl chain length and decrease with increasing tilt angle. In crystallographic terms, these reflections correspond to separation distances (*d*-spacings) between (001) planes (d_{001}).

1.7.1 SINGLE CRYSTAL STRUCTURES

Very few single crystal structures have been determined for triacylglycerols. Vand and Bell (1951) first attempted to determine the crystal structure of trilaurin. They came very close, being able to define the packing of the hydrocarbon chains, as well as the dimensions and space group of the unit cell. The first full crystal structure of a triacylglycerol is attributed to L.H. Jensen and A.J. Mabis in 1963. They determined the

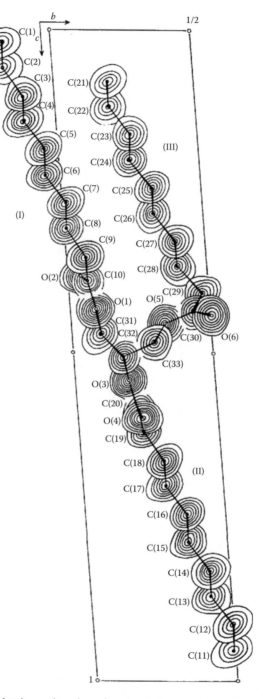

FIGURE 1.9 Electron density sections through carbon and oxygen atoms from a single crystal of tricaprin parallel to the *b–c* plane. (From Jensen, L.H. and Mabis, A.J., *Acta Crystallogr.,* 21, 770, 1966.)

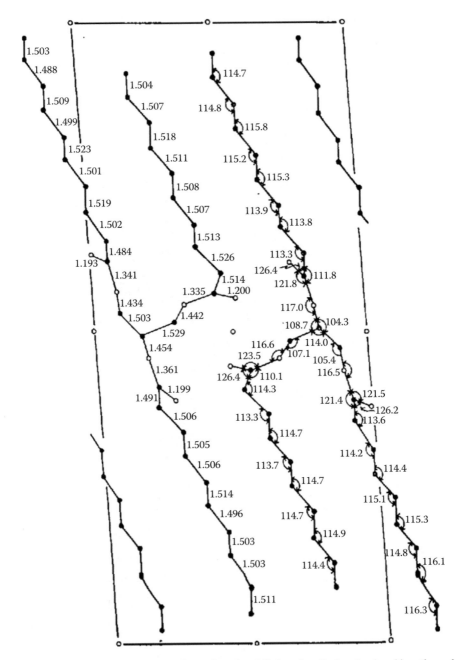

FIGURE 1.10 Molecular outline for a tricaprin triclinic unit cell, showing bond lengths and angles not involving hydrogen atoms. (From Jensen, L.H. and Mabis, A.J., *Acta Crystallogr.*, 21, 770, 1966.)

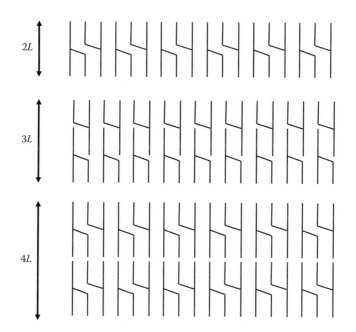

FIGURE 1.11 Stacking possibilities for TAG within lamellae in fat crystal. The letter "*L*" denotes a characteristic fatty acid chain length.

structure of the β-form of tricaprin. The refinement of their structure appeared in 1966. The crystal structure of the β-form of 1,2-dipalmitoyl-3-acetyl-sn-glycerol was reported by Goto et al. (1992). Of note is also the reported structure for a brominated triacylglycerol, namely, 2-11-bromoundecanoyl-1,1'-dicaprin (Doyne and Gordon, 1968).

The structure of the β' form was not reported until the year 2000, mainly due to difficulties in growing a large enough crystal for x-ray diffraction studies. The single crystal structures of the β' forms of 1,3-dicapryl-2-lauryl-sn-glycerol (van Langevelde et al., 2000) and 1,2-dipalmitoyl-3-myristoyl-sn-glycerol (Sato et al., 2001) have been determined experimentally. Surprisingly, both these structures are of the 4*L* type (Figure 1.12).

1.7.2 POLYMORPHISM

Polymorphic forms of fats are solid phases of the same chemical composition which upon melting yield identical liquid phases (like diamond and carbon). The first report of fat polymorphism was that of Duffy (1853) (Figure 1.13). Duffy identified three melting points for mutton and beef stearine, vegetable tallow, palmitine from palm oil, margarine from butter, and margarine from human fat, depending on crystallization conditions. The author did not attribute the differences in melting point to polymorphism at the time. It took another 81 (!) years for the definitive work in this area to be published. Clarkson and Malkin published the first powder x-ray diffraction studies on TAG polymorphism. The work is still first class even by today's standards. These authors recorded characteristic x-ray spectra and melting points for three

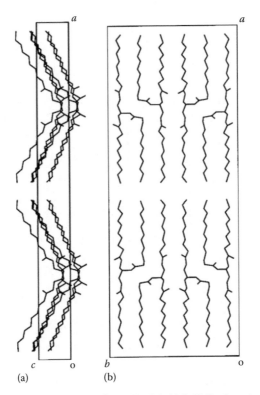

FIGURE 1.12   (a) Packing diagram of β′ CLC (10:0-12:0-10:0) viewed perpendicular to the *a–c* plane showing the bending of the molecules. (b) Packing diagram of β′ CLC viewed perpendicular to the *a–b* plane showing the chain packing. (From van Langevelde, A. et al., *Acta Crystallogr.*, B56, 1103, 2000.)

polymorphic forms of a series of homogenous TAGs, ranging from tricaprin (C10:0) to tristearin (C18:0). They clearly showed the long spacing and melting point dependence on carbon chain length, and identified the characteristic short spacings associated with the different TAGs studied (Figure 1.14). In a series of remarkable papers, they also characterized the polymorphism of unsymmetrical mixed TAGs (Carter and Malkin, 1939), monoglycerides (Malkin and El Shurbagy, 1936), diglycerides (Malkin et al., 1937), and TAGs containing erucic, brassidic, oleic, and elaidic acids (Carter and Malkin, 1947). However, the nomenclature used by Malkin to classify the different polymorphs in TAGs was not completely correct. He characterized the different forms as "glass," alpha, and beta, when in reality it was later shown that there was no glassy state in TAGs (Ferguson and Lutton, 1941; Lutton, 1945, 1950). Lutton (1945) helped clarify this issue and his nomenclature is the one used ever since in the field (Figure 1.15). There are usually three major polymorphic forms in fats (or TAGs), namely, the alpha (α), beta prime (β′), and beta (β) forms, in order of increasing stability, melting point, and density. Finally, the seminal review in the area of polymorphism, summarizing what is known with certainty in this area was written by Chapman (1962).

	Formulæ	Solidifying Points	Melting Points		
			1	**2**	**3**
Mutton-stearine	$C_{34}H_{34}O_4$	51·7°	52·0°	64·2°?	69·7°
Beef-stearine	$C_{34}H_{34}O_4$	50·5	51·0	63·0	67·0
Substance from vegetable tallow	—	45·0	45·6	62·0	64·5
Palmitine from palm oil	$C_{32}H_{32}O_4$	45·5	46·0	61·7	62·8
Margarine from butter	$C_{34}H_{34}O_4$	40·0	40·5	51·0	52·6
Margarine (?) from human fat	—	43·5	44·2	54·5	56·0
Cocinine	$C_{22}H_{22}O_4$	29·3	33·5		
Elaidine[‡]	$C_{36}H_{34}O_4$	23·7 / 28·0	38·0		
Stearic acid	$C_{34}H_{34}O_4$	65·8	68·0		
Palmitic acid	$C_{32}H_{32}O_4$	59·0	61·0		
Margaric acid from butter	$C_{34}H_{34}O_4$	50·5	52·3		
Stearic ether	$C_{38}H_{38}O_4$	33·0	33·7		
Cerotic ether[§]	$C_{58}H_{58}O_4$	60·0	60·3		
Cerotin (alcohol)	$C_{54}H_{56}O_2$	81·0	81·0		
Cerotene	$C_{54}H_{54}$	57·0	57·8°		
Chinese wax	$C_{108}H_{108}O_4$	80·5	81·0		
Paraffin	$(CH)_{2n}$	43·5	43·5		

FIGURE 1.13 Melting points for different natural fats and fatty substances demonstrating the existence of different melting points for the same material. (From Duffy, P., *J. Chem. Soc.*, 5, 197, 1853.)

As mentioned earlier, powder x-ray diffraction was, and still is, the definitive tool to characterize TAG and fat polymorphism. The characteristic spectra for the alpha (a), beta prime (b), and beta (c) forms of cocoa butter are shown in Figure 1.16, using a two-dimensional camera. Powder x-ray spectra of the three polymorphic forms present in fully hydrogenated canola oil, analyzed using one-dimensional detector mounted on a goniometer in a theta-theta configuration, are shown in Figure 1.18. This figure is quite revealing. One can clearly notice the short spacings in the wide angle x-ray scattering (WAXS) region which define the type of polymorph present, as discovered by Malkin in 1934. Moreover, one can also determine the size of the TAG lamellae present from the long spacings in the small angle x-ray scattering (SAXS) region. Notice how the *d*-spacings corresponding to the 001 plane are larger for the alpha form than for the beta prime and beta forms. This is due to the increased tilting of the chains in the more stable forms, which leads to an effective reduction in the *d*-spacing of the 001 plane. The orientation of the acyl chains within the unit cell can be characterized by their angle of tilt (τ), which corresponds to the angle between the crystallographic axis, *c*, and the *a*–*b* plane. This angle of tilt was determined for the beta form of trilaurin to be 61°35′ by Vand and Bell (1951). We can continue on this line of thought with an interesting example. If we assume that the tilt angle of acyl chains in the alpha polymorph is zero, this would correspond to the maximum length of an all-extended configuration of a 2L TAG polytype. Changes in the long spacing of the 2L alpha

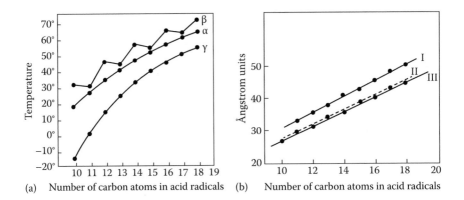

Long spacings of the triglycerides (Å.U.).

No. of C atoms in acid	10	11	12	13	14	15	16	17	18
Stable β-form	26.8	29.6	31.2	34.1	35.8	38.9	40.6	43.5	45
Unstable α-form*	—	33.0	35.6	37.7	41.2	42.9	45.6	48.5	50.6
Vitreous form	No long spacings.								

* As a rule only two orders appear for this form, and consequently the data are not quite so accurate as those for the β-forms. We consider the error to be less than ±0.3 Å.U.

Side spacings of the triglycerides (Å.U.).

β-Form, even acids	3.7*	3.9*	4.6	5.3
β-Form, odd acids	3.65*	4.0*	4.6	5.3
α-Form, and glass		4.2		

FIGURE 1.14 Chain-length dependence of single-acid TAG melting points and long spacings for their respective polymorphic forms. (a) Melting points of the triglycerides; (b) Long spacings of the triglycerides determined from small-angle powder x-ray diffraction. Characteristic reflections in the small and wide x-ray scattering regions are also listed at the bottom for the different polymorphic forms identified. (From Clarkson, C.E. and Malkin, T., *J. Chem. Soc.*, 670, 1934).

	This Laboratory				Malkin		
M.p., (°C)	**Main Short Spacings**	**Long Spacings**	**Name of Form**	**M.p., (°C)**	**Short Spacings**	**Long Spacings**	**Name of Form**
54	4.14 Å.	50.6 Å.	Alpha	54.5	4.2 Å. diffuse	.....	Gamma (Glass)
64	3.78, 4.18*	46.8	Beta prime[a]	65.0	4.2	50.6 Å.	Alpha
73.1	3.68, 3.84, 4.61, 5.24	45.15	Beta	71.5	3.7, 3.9, 4.6, 5.3	45	Beta

[a] Beta prime for mixed glycerides according to Malkin—approx. 3.8, 4.2.[2]
The situation is analogous for other homologs.

FIGURE 1.15 Table I from a well known paper by Lutton (1945) showing discrepancies between Malkin's classification of the different polymorphs present in TAGs and fats and his classification. Lutton's nomenclature prevailed.

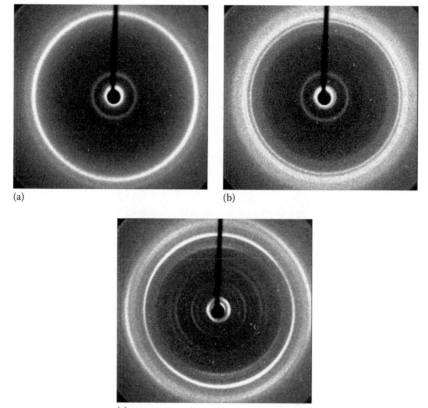

(a)   (b)

(c)

FIGURE 1.16   Powder x-ray diffraction Debye rings for cocoa butter crystallized at (a) −5°C for 5 days (alpha form), (b) 15°C for 7 days (beta prime form), and (c) 26°C for 28 days (beta form).

form upon transformation to the 2L beta form for tristearin, triplamitin, trimyristin, and trilaurin (and others) are reported in Table 3 of the review by Ferguson and Lutton (1941), taken from the work of Malkin. It would be very straightforward to show that the angle of tilt can then be determined using the following simple equation:

$$\tau = 90° - \cos^{-1}\left(\frac{d_{001}(\beta \text{ or } \beta')}{d_{001}(\alpha)} \right)$$ (1.8)

Plugging the values reported in Ferguson and Lutton's review into this equation, we obtain tilt angle values of 62.7°, 62.9°, 60.33°, and 61.21° for tristearin, triplamitin, trimyristin, and trilaurin, respectively. The value of 61.21° for trilaurin obtained using this approximation compares well with the value of 61.58° reported by Vand and Bell (1951) using single crystal data. It would seem that the angle of tilt for all beta polymorphic forms is the same, with an average value for the series analyzed earlier of 61.8°. If we analyze the spectra in Figure 1.17 for fully-hydrogenated canola oil,

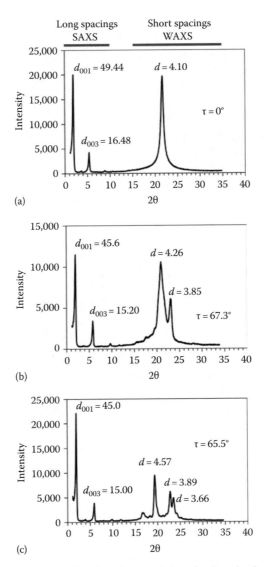

FIGURE 1.17 Powder spectra showing characteristic reflections in the small angle (long spacings) and wide angle (short spacings) x-ray scattering regions for the three polymorphs present in fully hydrogenated canola oil, namely, the alpha (a), beta prime (b), and beta (c) forms. Also indicated are the angles of tilt for the chains within the crystal(τ).

we obtain angles of tilt for the beta prime and beta forms of 67.3° and 65.5°, respectively. These values are very similar to the values we determined for pure TAGs. It is interesting to notice that the angle of tilt for the beta prime form is very similar to that of the beta form. Another interesting deduction from this analysis, and the fact that the c axis is very different in length from the a and b axes is that for this to be true, then the unit cell of the alpha form must be tetragonal, hexagonal, or orthorhombic (see Table 1.1 for reassurance).

1.7.2.1 Energetics of Crystallization as It Relates to Polymorphism

The alpha form is metastable and will thus transform into more stable forms. Two crystalline forms are called "monotropic" when one is more stable than the other, and recrystallization will take place in the direction of the more stable form only. Transformation of one polymorphic form into another is possible with or without melting of the solid (melt-mediated vs. solid state). Figure 1.18 summarizes the possible dynamics of fat crystallization and recrystallization in terms of the polymorphism of the solid state. Notice how the three polymorphic forms can form directly from the melt. Also evident is the irreversibility of the α to β′ to β transformation (monotropism).

Figure 1.19 summarizes many of the principles governing the metastable crystallization of fats. Triacylglycerols are relatively large (small) molecules. It takes a relatively long time for the long hydrocarbon chains to align and form a stable crystal form. The driving force for nucleation (and thus crystallization) is the degree of undercooling (or supercooling). When a melt is cooled below its melting point, a driving force for crystallization is established, which is proportional to the extent of this undercooling. Below its melting point, a material is in a metastable region where even though the material is undercooled, it does not crystallize. In this region, molecules are trying to adopt stable packing arrangements for the formation of a crystalline solid and decrease this driving force (the chemical potential between the solid

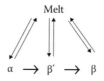

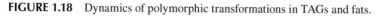

FIGURE 1.18 Dynamics of polymorphic transformations in TAGs and fats.

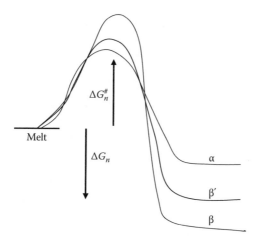

FIGURE 1.19 Diagram depicting the free energy of activation for the formation of a particular polymorph as well as the free energy difference between products and reactants (thermodynamic stability).

and the melt). Given sufficiently high degrees of supercooling, molecules do not have sufficient time to pack in the most thermodynamically stable configuration and thus tend to crystallize in metastable forms. The key concept here is that the metastable forms have a lower activation free energy on nucleation ($\Delta G^{\#}$) than the more stable forms (Figure 1.19). Thus, the metastable forms will form more readily. This is a purely kinetic effect. Also notice how the free energy of formation (ΔG) of the metastable forms is lower than that of the stable forms, since the free energy of the metastable form is the highest. Thus, even though the metastable forms have a lower thermodynamic stability, they tend to form more readily due to the kinetics of the process. This explains why metastable forms form first, even though their melting point is lower than that of the more stable forms.

1.7.2.2 Subcells and Subcell Packing

Even though after Chapman's (1962) review, it would have seemed we knew everything there was to know about polymorphism, we still did not have a good idea of which crystallographic planes gave rise to the "short spacings" observed for TAGs and natural fats (DeSouza et al., 1990). It is really remarkable that both pure homogenous TAGs and natural fats give rise to similar diffraction patterns in the WAXS region, since natural fats are mixtures of sometimes hundreds of different TAGs. One remarkable paper, in my view, was again that of Vand and Bell (1951). Not only did these authors nearly solve the single crystal structure of trilaurin over a decade earlier than Jensen and Mabis' first report in 1963, but also managed to index the reflections of their trilaurin crystal (Figure 1.20). Scanning down the first column, which reports the relative intensities for different reflections, we notice the strongest reflections for this crystal are at 4.55, 3.85, and 3.72 Å, which correspond very closely to the values of the beta polymorph reflections determined by Clarkson and Malkin in 1934. What is also very interesting here is that these reflections do not correspond to a single plane in the crystal but to a combination of up to four different planes with similar diffraction properties. This is probably the reason why it is very difficult and never unambiguous to index reflections in TAGs and natural fats. For natural fats the situation is ever worse since they are mixtures of many different crystals, so one has to contend with similarly diffracting planes and a mixed bag of crystals. One may argue that indexing natural fats is therefore a meaningless exercise.

In an attempt to provide some crystallographic meaning to the characteristic WAXS reflections—the short spacings—researchers started to make use of the concept of a subcell, within the unit cell (Vand, 1951). In order to visualize the packing of the acyl chains within the unit cell, a section perpendicular to the chain axis is required. This section forms a *plane lattice* defined by subcell lattice parameters a_o, b_o, and a crystallographic angle γ_o (Figure 1.21). This figure corresponds to the triclinic subcell of trilaurin determined by Vand and Bell (1951). Methyl and carboxyl groups are not included in the subcell lattice. Ethylene is the smallest unit within the hydrocarbon chain which constitutes a three-dimensional entity. Diagrams like the one shown in Figure 1.22 depict the way in which ethylene groups within the long hydrocarbon chains are arranged relative to each other, shown here for the three major polymorphs found in TAGs. As pointed out by Abrahamson et al. (1978) in their review on lateral packing of hydrocarbon chains, the data upon which this

Intensity	Spacing (A.)	Main-Cell Index*	Subcell Index
2	15·85	002	—
20	10·49	003	—
1	8·30	$10\bar{3}$	—
4	7·83	004	—
1	6·18	104, $20\bar{1}$	—
1	5·81	$20\bar{3}$, $10\bar{5}$, $20\bar{2}$	—
20	5·36	010, $01\bar{1}$, 202	—
20	5·20	006, $01\bar{2}$, 011	—
6	4·92	$20\bar{4}$, $10\bar{6}$, $01\bar{3}$	—
1	4·80	012, $1\bar{1}3$	—
400	4·55	$20\bar{5}$, 106, $11\bar{2}$, $11\bar{3}$	010
20	4·38	204	—
1	4·22	$20\bar{6}$	—
20	4·09	$30\bar{1}$, 300	—
200	3·853	$\bar{2}1\bar{3}$, $10\bar{8}$, $20\bar{7}$	$\bar{1}10$
300	3·728	302	100
40 D	3·593	$30\bar{5}$, 303	—
10 D	3·398	$10\bar{9}$, 304	—
2	3·289	—	—
20 D	3·144	$20\bar{9}$, $1.0.\bar{10}$	—
4 D	2·974	$40\bar{2}$, $40\bar{3}$, 306, $40\bar{4}$	—
1	2·832	—	—
1	2·743	403	—
20	2·571	405, $\bar{1}2\bar{5}$, 411	$\bar{1}20$, 110
2	2·416	223	020
4	2·363	$\bar{3}.1.10$	$\bar{1}11$
2	2·291	$1.0.\bar{12}$, $4.0.\bar{10}$	001
4 D	2·242	$1.0.\bar{14}$, 0.1.12	011
2	2·186	$4.0.\bar{11}$	$10\bar{1}$
2	2·143	$\bar{2}27$	—
2	2·080	$4.0.\bar{12}$	—
2	2·059	$\bar{2}28$	$\bar{1}21$
1	1·978	$2.1.\bar{15}$	$01\bar{1}$
4	1·930	$\bar{4}2\bar{8}$, 604	$\bar{2}20$, 200
2	1·777	—	021, $11\bar{1}$, $\bar{2}21$
2	1·670	—	120
2	1·632	—	$20\bar{1}$, 111

D = diffuse line

* Not exhaustive; indices of reflexions known to be weak from single-crystal photographs are omitted

FIGURE 1.20 Table 1 from Vand and Bell (1951) showing the set of indexed reflections for a trilaurin crystal.

$$a_0 = \frac{1}{a_s^*} \csc \gamma_s^* = 4.078 \text{ A.;}$$

$$b_0 = \frac{1}{b_s^*} \csc \gamma_s^* = 5.209 \text{ A.;}$$

$$\gamma_0 = 180° - \gamma_s^* = 114° \; 23'.$$

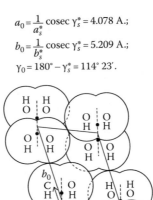

FIGURE 1.21   Triclinic acyl chain subcell chain packing in a trilaurin crystal as determined by Vand and Bell (1951).

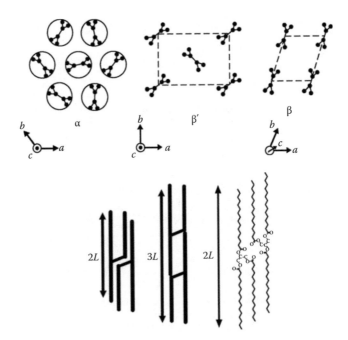

FIGURE 1.22   Putative TAG acyl chain subcell packing which defines the type of polymorphism in all TAGs and fats.

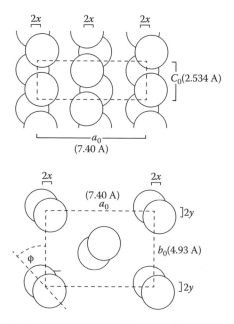

FIGURE 1.23 Orthorhombic perpendicular chain subcell packing for long chain paraffin hydrocarbons as determined by Bunn (1939).

extrapolation was based was that of Vand and Bell (1951) for the triclinic subcell packing of trilaurin and that of Bunn (1939) for orthorhombic packing of long chain paraffins (not TAGs), as shown in Figure 1.23. Moreover, there is no evidence anywhere for hexagonal subcell packing for an alpha TAG polymorph. Work by Muller in 1930 and earlier demonstrated disorder in the orientation of ethylene groups within long acyl chains of paraffin waxes close to their melting points. This is the closest we have got to describing acyl chain packing in TAGs for the alpha form.

Single crystal studies are definitive in their ability to answer these questions. Vand and Bell (1951), Jensen and Mabis (1966), and Doyne and Gordon (1968) reported triclinic unit cells and subcells for their TAGs in the beta polymorphic form. It would seem that the beta form in a triclinic crystal. However, the beta form of 1,2 dipalmitoyl-3-acetyl-sn-glycerol reported by Goto et al. (1992) has a monoclinic unit cell and a triclinic subcell. Regardless of this, the subcell packing for this beta form is also triclinic. The situation is more interesting for the beta prime form. Van Langevelde et al. (2000) reported an Iba2 space group for their single crystal structure of CLC (10:0-12:0-10:0), which belongs to the orthorhombic crystal system. They also reported that the subcell was orthorhombic as well. It would seem that we were justified to base our classification system on a single piece of work on long-chain paraffins from 1939. However, the second single crystal structure for the beta prime form of PPM (16:0-16:0-14:0) reported by Sato et al. (2001) was monoclinic space group C2 as reported previously by Birker et al. (1991) for LML (12:0-14:0-12:0). However, the subcell structure reported in Sato's work was of a hybrid type (Abrahamson et al., 1978). A hybrid subcell displays characteristics of many different types of subcells (Figure 1.24). Moreover, to complicate things further, his hybrid subcell, HS3, had

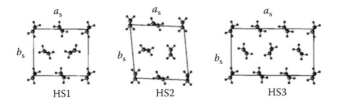

FIGURE 1.24 A hybrid subcell (HS) arrangement was detected in the β' polymorphic form of PPM (16:0-16:0-14:0). The specific type detected was HS3, which is a combination of monoclinic, triclinic, and orthorhombic subcell. (From Sato, K. et al., *J. Lipid Res.*, 42, 338, 2001.)

never been observed before. Only HS1 and HS2 have been previously reported. So, here we have the case where the unit cell of the beta prime form of a TAG is not orthorhombic and the subcell packing is hybrid. Moreover, there is no definitive evidence that the subcell of an alpha form is packed in a hexagonal fashion. It would seem that the use of subcell structure to define different polymorphs present in fats, other than for the beta form, is not warranted yet. Future work needs to focus on this important issue for the beta prime, and more importantly, for the alpha form. Single crystal work is definitely the most direct route, but judicious use of complementary *techniques* and modelling (Birker et al., 1991; Van Langevelde et al., 1999; Van de Streek et al., 1999) could lead to significant advances in the near future.

REFERENCES

Abrahamsson, S., B. Dahlén, H. Löfgren, and I. Pascher. 1978. Lateral packing of hydrocarbon chains. *Prog. Chem. Fats Other Lipids*. 16: 125–143.

Birker, P.J.M.W.L., S. de Jong, E.C. Roijers, and T.C. van Soest. 1991. Structural investigations of β' triacylglycerols: An x-ray diffraction and microscopic study of twinned β' crystals. *J. Am. Oil Chem. Soc.* 68: 895–906.

Bunn, C.W. 1939. The crystal structure of long-chain normal paraffin hydrocarbons. The "shape" of the methylene group. *Trans. Faraday Soc.* 35: 482–491.

Carter, M.G.R. and T. Malkin. 1939. An x-ray examination of the glycerides. Part VII. Unsymmetrical mixed triglycerides. *J. Chem. Soc.* 1518–1521.

Carter, M.G.R. and T. Malkin. 1947. An x-ray examination of the glycerides. Part VIII. Glycerides of erucic, brassidic, oleic, and elaidic acids. *J. Chem. Soc.* 554–558.

Chapman, D. 1962. The polymorphism of glycerides. *Chem. Rev.* 62: 433–456.

Clarkson, C.E. and T. Malkin. 1934. Alternation in long chain compounds. Part II. An x-ray and thermal investigation of the triglycerides. *J. Chem. Soc.* 666–671.

DeSouza, V., J.M. deMan, and L. deMan. 1990. Short spacings and polymorphic forms of natural and commercial solid fats: A review. *J. Am. Oil Chem. Soc.* 67: 835–843.

Doyne, T.H. and J.T. Gordon. 1968. The crystal structure of a diacid triglyceride. *J. Am. Oil Chem. Soc.* 45: 333–334.

Duffy, P. 1853. On certain isomeric transformations of fats. *J. Chem. Soc.* 5: 197–210.

Ferguson, R.H. and E.S. Lutton. 1941. The polymorphic forms or phases of triglyceride fats. *Chem. Rev.* 29: 355–384.

Goto, M., D.R. Kodali, D.M. Small, K. Honda, K. Kozawa, and T. Uchida. 1992. Single crystal structure of a mixed-chain triacylglycerol: 1,2-Dipalmitoyl-3-acetyl-sn-glycerol. *Proc. Natl. Acad. Sci. USA* 89: 8083–8086.

Jensen, L.H. and A.J. Mabis. 1963. Crystal structure of β-tricaprin. *Nature* 197: 681–682.

Jensen, L.H. and A.J. Mabis. 1966. Refinement of the structure of β-tricaprin. *Acta Crystallogr.* 21: 770–781.

Lutton, E.S. 1945. The polymorphism of tristearin and some of its homologs. *J. Am. Oil Chem. Soc.* 67: 524–527.

Lutton, E.S. 1950. Review of polymorphism of saturated even glycerides. *J. Am. Oil Chem. Soc.* 27: 276–281.

Malkin, T. and M.R. El Shurbagy. 1936. An x-ray and thermal examination of glycerides. Part II. The α-monoglycerides. *J. Chem. Soc.* 1628–1634.

Malkin, T., M.R. El Shurbagy, and M.L. Meara. 1937. An x-ray and thermal examination of glycerides. Part III. The αα′-diglycerides. *J. Chem. Soc.* 1409–1413.

Muller, A. 1930. The crystal structure of the normal paraffins at temperatures ranging from that of liquid air to the melting points. *Proc. Roy. Soc.* 127: 417–430.

Peschar, R., M.M. Pop, D.J.A. De Ridder, J.B. van Mechelen, R.A.J. Driessen, and H. Schenk. 2004. Crystal structures of 1,3-distearoyl-2-oleoylglycerol and cocoa butter in the β(V) phase reveal the driving force behind the occurrence of fat bloom on chocolate. *J. Phys. Chem. B.* 108: 15450–15453.

Sato, K., M. Goto, J. Yano, K. Honda, D.R. Kodali, and D.M. Small. 2001. Atomic resolution structure analysis of β′ polymorph crystal of a triacylglycerol: 1,2-Dipalmitoyl-3-myristoyl-sn-glycerol. *J. Lipid Res.* 42: 338–345.

Van Langevelde, A., K. van Malssen, R. Driessen, K. Goubitz, F. Hollander, R. Peschar, P. Zwart, and H. Schenk. 2000. Structure of $C_nC_{n+2}C_n$-type (n = even) β′-triacylglycerols. *Acta Crystallogr.* B56: 1103–1111.

Van Langevelde, A., K. van Malssen, E. Sonnenveld, R. Peschar, and H. Schenk. 1999. Crystal packing of a homologous series β′-stable triacylglycerols. *J. Am. Oil Chem. Soc.* 76: 603–609.

Van de Streek, J., P. Verwer, R. de Gelder, and F. Hollander. 1999. Structural analogy between β′ triacylglyerols and n-alkanes. Toward the crystal structure of β′ p.p + 2.p Triacylglycerols. *J. Am. Oil Chem. Soc.* 76: 1333–1341.

Vand, V. 1951. Method for determining the signs of the structure factors of long-chain compounds. *Acta Crystallogr.* 4: 104–105.

Vand, V. and I.P. Bell. 1951. A direct determination of the crystal structure of the β form of trilaurin. *Acta Crystallogr.* 4: 465–469.

2 Nucleation and Crystalline Growth Kinetics

2.1 INTRODUCTION TO CRYSTALLIZATION

The crystallization of multicomponent triacylglycerol (TAG) mixtures, for example, edible fats, is of paramount importance in the manufacture of products such as chocolate, ice cream, and butter (Garside, 1978; Hartel, 2001). The crystallization behavior and structure of these materials is extremely sensitive to heat and mass transfer conditions and compositional changes (Chapman, 1962; Timms, 1984; Kellens et al., 1992; Herrera et al., 1999; Sato, 1999, 2001; Herrera, 2002; Martini et al., 2002a,b; Mazzanti et al., 2003, 2004, 2005, 2007, 2009; Sato and Ueno, 2005) and will ultimately affect mechanical strength, flow behavior, and sensory texture (Herrera and Hartel, 2000a,b,c; Campos et al., 2002; Campos and Marangoni, 2010; Brunello et al., 2003; Dibildox Alvarado et al., 2004). Of particular interest is the nucleation behavior of these systems, since important structural features are a direct consequence of nucleation behavior, such as crystallite number, size, and morphology, as well as the spatial distribution of mass (Marangoni and McGauley, 2003). In what follows, we will review theoretical formalisms used to characterize both the nucleation and crystallization behavior of edible fats. An excellent and widely read review on fat crystallization was written in 1988 by Boistelle (1988) and Kashchiev's book on nucleation is a must read for more advanced students of the art (Kashchiev, 2000). Philippe Rousset (2002) wrote an excellent review on modeling crystallization kinetics of triacylglycerols and is a highly recommended material. The most current critical review of the art has recently been written by Keshra Sangwal (2012), who summarizes most serious efforts in modeling crystallization processes in triacylglycerols. The purpose of this chapter is to offer the reader this author's view on how to model the kinetics of crystallization of fats from a basic knowledge of the kinetic theory of gases. The approach is rather phenomenological and highly practical, however stressing fundamental understanding of what is being done, with the undergraduate and early graduate students in mind.

2.1.1 NUCLEATION OVERVIEW

Fat is a multicomponent mixture of TAGs. When the temperature of fat is decreased below the melting temperature of the highest melting TAG in the mixture, the melt becomes supersaturated in this particular TAG species. This so-called undercooling or supercooling below the melting temperature of the crystallizing species is equivalent to a solution supersaturation, and represents the thermodynamic driving force for the change in state, from liquid to solid.

Fats usually have to be undercooled by at least 5°C–10°C before they begin to crystallize. For a few degrees below the melting point, the melt exists in a metastable region. In this region, molecules begin to aggregate into tiny clusters called embryos. At these low degrees of undercooling, embryos continuously form and breakdown, but do not persist to form stable nuclei. The energy of interaction between TAG molecules has to be greater than the thermal energy of the molecules in the melt ($k_B T$) so as to overcome Brownian effects. For these flexible molecules, it is not sufficient to interact; molecules have to adopt a specific conformation in order to form a stable nucleus. The adoption of this more stable conformation is relatively slow, thus explaining the existence of a metastable region. As the undercooling is increased (i.e., at lower temperatures), stable nuclei of a specific critical size are formed. The Gibbs free energy change associated with the formation of a crystal nucleus (ΔG_n) includes contributions from both surface (positive) and volume changes (negative) and is defined by

$$\Delta G_n = A_n \delta - V_n \frac{\Delta \mu}{V_m^s} \tag{2.1}$$

where
  A_n is the surface area of a nucleus
  δ is the surface free energy per unit area
  V_n is the volume of a nucleus
  $\Delta \mu$ is the chemical potential difference between solid and liquid (related to the degree of supersaturation)
  V_m^s is the molar volume of the solid

The formation of a crystal leads to the creation of a solid–liquid interface, resulting in a positive contribution to the free energy of nucleation. On the other hand, the formation of a crystal also leads to a decrease in the chemical potential difference between the solid and the liquid. Thus, at a critical nucleus size (referred to as a critical radius [r_c] in this discussion), a maximum in the free energy of nucleation profile will be observed, where an increase proportional to A_n is exactly balanced by a decrease proportional to V_n. Beyond r_c, the free energy of nucleation will continuously decrease. In order to minimize their free energy, clusters smaller than r_c will breakdown and those larger than r_c will continue to grow.

The three most common types of nucleation include primary homogeneous, primary heterogeneous, and secondary nucleation. With significant undercooling, primary homogenous nucleation may occur in a melt. Primary homogenous nucleation occurs in pure solutions, in the absence of foreign interfaces. In reality, however, nucleation is usually catalyzed by the presence of foreign particles or interfaces. Primary heterogeneous nucleation requires a significantly lower supersaturation than homogeneous nucleation because foreign surfaces reduce the effective surface free energy by decreasing the crystal–melt interfacial tension (δ in Equation 2.1). "Catalytically active impurities," which can include surface edges and/or foreign particle surfaces, create order in small regions of the melt, thus serving as templates for nucleation. Secondary nucleation, the process where new crystal nuclei form on contact with existing crystals, or crystal fragments, will proceed once some primary nucleation has taken place. The number of nuclei needed to induce crystallization in a bulk system is very low. In contrast, each droplet in an emulsified system must contain a nucleus for crystallization to take place. Since the likelihood of each globule containing a nucleus or catalytic impurity is low, emulsified systems require much more undercooling to induce crystallization compared to bulk systems.

2.1.2 QUANTIFICATION OF THE DRIVING FORCE FOR CRYSTALLIZATION

The degree of supersaturation of a particular ith TAG in a melt can be quantified by the supersaturation ratio (β_i) where

$$\beta_i = \frac{[c_i]}{[c_i^*]} \tag{2.2}$$

where $[c_i]$ is the concentration of a particular ith TAG species in the melt under supersaturation conditions, while $[c_i^*]$ corresponds to the equilibrium solubility (concentration) of the particular ith TAG species in the melt (the "solubility limit") under the same temperature and pressure conditions. It is usual to express the driving force for crystallization as the natural logarithm of β (ln β), since this quantity is directly proportional to the chemical potential difference ($\Delta\mu$) between the ith crystallized TAG species in the solid state and the ith TAG species in solution in the melt:

$$\Delta\mu_i = RT \ln \beta_i \tag{2.3}$$

where

    R is the universal gas constant ($R = 8.314\,J/mol\,K$)
    T is temperature at which the crystallization is taking place [K] (recall this is an isothermal treatment)

This quantity, ln β_i, is what is referred to as the "degree of supersaturation."

In a melt, however, this driving force for crystallization is more conveniently expressed as a function of the degree of undercooling (ΔT) rather than the degree of supersaturation, and thus

$$\Delta\mu_i = \Delta H_{m,i} \frac{(T_{m,i} - T)}{T_{m,i}} \tag{2.4}$$

where

    $\Delta H_{m,i}$ is the enthalpy of melting [J/mol] of the ith TAG species in solution with other solid and/or liquid TAGs or in a neat state (no solvent or cosolvent)
    $T_{m,i}$ is the melting temperature of the ith TAG species [K] in solution with other solid and/or liquid TAGs or in a neat state
    T is the temperature at which crystallization is taking place [K]

The degree of supersaturation and the degree of undercooling are just two ways of describing the same field. Combining Equations 2.3 and 2.4 yields the result

$$\ln \beta = \alpha \Delta T \tag{2.5}$$

where

$$\alpha = \frac{\Delta H_m}{RTT_m} \tag{2.6}$$

When using these equations, please keep in mind what experiment you are carrying out—either freezing point depression or solubility at different temperatures.

A convenient way of calculating the change in chemical potential upon dilution from the neat state (without solvent or cosolvent), basically a freezing point depression, can be done by using

$$\Delta \mu_i = \Delta H_{m,i}^0 \frac{(T_{m,i}^0 - T_{m,i}^{sol})}{T_{m,i}^0} \tag{2.7}$$

where

$\Delta H_{m,i}^0$ is the melting enthalpy of the neat (without solvent or cosolvent) pure solid component

$T_{m,i}^0$ is the melting temperature of the neat pure solid component

$T_{m,i}^{sol}$ is the melting temperature of the solid in solution, at a particular concentration

Table 2.1 shows a typical calculation of the degree of supersaturation and chemical potential difference for mixtures of fully hydrogenated canola oil and high-oleic sunflower oil crystallized for 1 day at 30°C.

TABLE 2.1

Enthalpy of Melting (ΔH_m), Peak Melting Temperature (T_m), Chemical Potential Difference ($\Delta \mu$), and Degree of Supersaturation (ln β) for Mixtures of FHCO and HOSO Crystallized for 1 Day at 40°C (313.15 K) under Static, Isothermal Conditions

FHCO (%)	20	30	40	50	60	70	80	90	100
ΔH_m (kJ/mol)	32.5	61.4	73.4	95.4	106.8	120.3	129.6	148.4	164.9
T_m (K)	333.3	336.5	337.4	338.8	340.7	342.0	344.2	344.4	344.9
$\Delta \mu$ (kJ/mol)	10.0	11.4	11.8	12.5	13.3	13.9	14.9	15.0	15.2
ln β	3.84	4.37	4.53	4.80	5.11	5.34	5.72	5.76	5.84

The molecular weight of FHCO used was 891.48 g/mol. Chemical potential differences were calculated using Equation 2.4.

FHCO, fully hydrogenated canola oil; HOSO, high oleic sunflower oil.

When making comparisons between the crystallization behavior of two fats, it is imperative to keep in mind that comparisons should be made at similar degrees of supersaturation (or undercooling). If fats have very different melting points and enthalpies of melting, it is obvious and trivial that one should notice differences in their crystallization kinetics and eventual crystal size, rheological properties, and solid fat content (SFC). Although the crystallization behavior of different fats at the same temperature is of industrial importance, fundamental crystallization research, on the other hand, *must* thrive to make comparisons between systems under similar supersaturation conditions.

2.1.3 Better Understanding the Chemical Potential

In this discussion, we have used the term "chemical potential" freely and without in-depth discussion. In this field, it is very important to have a good understanding of what is chemical potential actually is and its relationship to supersaturation and undercooling. In the vignette that follows, we will derive the expression for a chemical potential from first principles and will show how it is related to the supersaturation ratio. The readers who do not feel the need to further discuss chemical potential can safely skip the entire section.

Consider the basic relationship between Gibbs free energy (G), enthalpy (H), and entropy (S) in a closed system:

$$G = H - TS \qquad (2.8)$$

Taking the differential of this relationship yields

$$\partial G = \partial H - T\partial S - S\partial T \qquad (2.9)$$

Considering that if the enthalpy is related to the internal energy (E) of a system and the pressure (P)–volume (V) work carried out by the system by $H = E + PV$, then $\partial H = \partial E + P\partial V + V\partial P$. Substituting this last expression for ∂H in Equation 2.9 yields

$$\partial G = \partial E + P\partial V + V\partial P - T\partial S - S\partial T \qquad (2.10)$$

The change in internal energy of a system ($-\partial E$) is related to the heat (∂q) absorbed or generated by the system and the work carried out by the system ($-\partial w$) by $\partial E = \partial q - \partial w$. Considering only PV work where $\partial w = P\partial V$, and for the case of a reversible process where $\partial q_{rev} = \partial S/T$, the change in internal energy of the system can be expressed as $\partial E = T\partial S - P\partial V$. Substituting this last expression for the change in internal energy of the system into Equation 2.10 and canceling out terms results in the important expression for the change in free energy of a system as a function pressure, volume, and temperature:

$$\partial G = V\partial P - S\partial T \qquad (2.11)$$

For an isothermal process, $\partial T = 0$ and thus Equation 2.11 further simplifies to

$$\partial G = V \partial P \tag{2.12}$$

As is usually done in thermodynamics, we need to obtain an expression for the change in free energy (ΔG) in a system when going from a "standard state" to a particular state we are interested in. In order to do this we need to integrate Equation 2.12 from the standard state, of free energy G^0, to the particular ith state we are interested in, with free energy G_i. Considering that according to the ideal gas law $V = nRT/P$, the required integration has the form

$$\int_{G^0}^{G_i} \partial G = nRT \int_{P^0}^{P_i} \frac{\partial P}{P} \tag{2.13}$$

where P^0 is the pressure of the standard state has been agreed upon to be 1 atm. Considering that $P^0 = 1$ atm, the result of the integration yields

$$G_i - G^0 = nRT \ln P_i \tag{2.14}$$

Let's continue this discussion by considering our states of interest to be the under-cooled or supersaturated state with a free energy G_{SS}, and the equilibrium solubility state, in the absence of any supersaturation effects, G_{eq}, at the equilibrium melting temperature of the solid, then

$$G_{SS} - G^0 = nRT \ln P_{SS}$$
$$G_{eq} - G^0 = nRT \ln P_{eq} \tag{2.15}$$

Since the standard states of both systems are the same, we can define a change in free energy from an equilibrium solubility state to a supersaturate state as

$$\Delta G = (G_{SS} - G^0) - (G_{eq} - G^0) = nRT \ln P_{SS} - nRT \ln P_{eq}$$
$$\Delta G = G_{SS} - G_{eq} = nRT \ln \frac{P_{SS}}{P_{eq}} \tag{2.16}$$

The change in free energy *per mol* of material for a change from the supersaturated state to the equilibrium solubility state corresponds to the chemical potential change ($\Delta\mu$) for the state change:

$$\Delta\mu = \frac{\Delta G}{n} = RT \ln \frac{P_{SS}}{P_{eq}} \tag{2.17}$$

One most invoke one more law before this treatment is over and that is Henry's law, which states that the partial pressure (P_i) of a solute in the gas phase above an infinitely dilute solution of such solute, of concentration c_i [mol/L], can be predicted from $P_i = k_H c_i$, where k_H is the Henry's law constant, which is a function of both temperature and the type of solvent. For a dilute solution, the concentration of the solute is approximately proportional to its mole fraction, x_i, and Henry's law can be written as $P_i = k_H x_i$. This can be compared to Raoult's law, $P_i = P^0 x_i$ where P^0 is the pressure of the pure solvent in the absence of any other dissolved species. At first sight, Raoult's law appears to be a special case of Henry's law where $k_H = P^0$. This is true for pairs of closely related substances, such as is the case of a TAG crystallizing in a TAG solvent, which obey Raoult's law over the entire composition range. Such mixtures are called "ideal mixtures," which actually is the case for TAG mixtures. However, in general, both laws are limit laws, and they apply at opposite ends of the compositional range. The vapor pressure of the component in large excess, such as the solvent in a dilute solution, is proportional to its *mole* fraction, and the constant of proportionality is the vapor pressure of the pure substance (Raoult's law). The vapor pressure of the solute is also proportional to the solute's mole fraction, but the constant of proportionality is different and must be determined experimentally (Henry's law). In mathematical terms, for Raoult's law,

$$\lim_{x \to 1} \left(\frac{P}{x} \right) = P^0$$

while for Henry's law,

$$\lim_{x \to 0} \left(\frac{P}{x} \right) = k_H$$

Raoult's law can also be related to nongas solutes.

Substituting the partial pressures of the supersaturated states and equilibrium states in Equation 2.16, and crossing out k_H or P^0 from the expression, yields

$$\Delta\mu = RT \ln \frac{[c_{ss}]}{[c_{eq}]} \qquad (2.18)$$

Comparison of Equations 2.18 and 2.2 indicates that the ratio $[c_{ss}]/[c_{eq}]$ corresponds to the supersaturation ratio (β) and that Equations 2.18 and 2.3 are thus identical.

An alternative insight into the chemical potential can be achieved by considering the famous relationship between the standard state Gibbs free energy (ΔG^0) and the equilibrium constant for a process at equilibrium,

$$\Delta G^0 = -RT \ln K_{eq} \qquad (2.19)$$

This free energy change arises from changes in the standard state enthalpy and entropy of the system, namely, $\Delta G^0 = \Delta H^0 - T\Delta S^0$. Combining these two expressions yields an expression for the relationship between the equilibrium constant and the enthalpic and entropic changes taking place in the system under standard state conditions:

$$\ln K_{eq} = -\frac{\Delta H^0}{RT} + \frac{\Delta S^0}{R} \qquad (2.20)$$

Upon differentiation of this equation as a function of temperature, one obtains the familiar van't Hoff relationship for the temperature dependence of the equilibrium constant of a reaction:

$$\frac{\partial \ln K_{eq}}{\partial T} = \frac{\Delta H^0}{RT^2} \qquad (2.21)$$

Definite integration of Equation 2.21 with temperature boundaries, T_1 and T_2,

$$\int_{T_1}^{T_2} \partial \ln K_{eq} = \int_{T_1}^{T_2} \frac{\Delta H^0}{RT^2} \partial T \qquad (2.22)$$

yields

$$\ln K_{eq} = \left[-\frac{\Delta H^0}{RT} \right]_{T_1}^{T_2} = -\frac{\Delta H^0}{R}\left(\frac{1}{T_2} - \frac{1}{T_1} \right) = \frac{\Delta H^0}{R}\frac{(T_2 - T_1)}{T_1 T_2} \qquad (2.23)$$

Let us consider using this equation for the description of the crystallization transition of a solution of a molten solid TAG A in a liquid TAG B upon cooling from T_m, to a temperature T, below T_m. For this case, the equilibrium constant of the reaction is

$$K_{eq} = \frac{[c_{i,l}^T]}{[c_{i,l}^{T_m}]}$$

where $c_{i,l}$ is the concentration of TAG A dissolved in TAG B at either of the two temperatures. Obviously, the concentration of TAG A in TAG B at T_m, where TAG A is completely molten and dissolved in TAG B, is much higher than at T.

Introducing this expression into Equation 2.23 for $T_1 = T_m$ and $T_2 = T$, yields

$$\ln \frac{[c_{i,l}^T]}{[c_{i,l}^{T_m}]} = -\frac{\Delta H_m^0}{RT}\frac{(T_m - T)}{T_m} \qquad (2.24)$$

and by multiplying both sides by RT and (-1) and inverting the logarithmic term to get rid of the negative sign, yields

$$RT \ln \frac{[c_{i,l}^{T_m}]}{[c_{i,l}^{T}]} = \Delta H_m^0 \frac{(T_m - T)}{T_m} \qquad (2.25)$$

Close inspection of Equation 2.2 reveals that the ratio $K_{eq} = [c_{i,l}^{T_m}]/[c_{i,l}^{T}]$ is merely the supersaturation ratio β, which makes Equation 2.25 identical to Equations 2.3 and 2.4.

2.2 CRYSTALLIZATION KINETICS

2.2.1 NUCLEATION

2.2.1.1 Isothermal Steady-State Nucleation Theory

The conventional Gibbs–Thomson formulation expresses the overall free energy change resulting from the formation of a spherical nucleus as the sum of a surface term and a volume term:

$$\Delta G_n = 4\pi r^2 \delta - \frac{4}{3}\pi r^3 \frac{\Delta\mu}{V_m^s} \qquad (2.26)$$

where
ΔG_n is free energy change associated with the formation of spherical nucleus [J]
r is crystal radius [m]
δ is the solid–liquid surface free energy per unit area [J/m²], or crystal–melt interfacial tension
$\Delta\mu$ is the chemical potential difference between the liquid and the solid [J/mol], while V_m^s is molar volume of the solid [m³/mol]

The creation of a solid–liquid interface requires energy, leading to an increase in the free energy of the system; however, the creation of a nucleus also causes a decrease in the free energy of the system, driven by a decrease in the supersaturation of the system upon crystallization. Thus, at a certain critical radius, a maximum in the $\Delta G_n - r$ profile will be observed (Figure 2.1). As discussed earlier, this ΔG_n maximum is associated with the critical radius of the nucleus (r_c).

Envision a particular ith TAG component in a fat mixture. If the fat melt is cooled instantaneously below the melting point of this ith component, it will crystallize until its saturation concentration in the melt is reached and thermodynamic equilibrium reestablished. At this point, the chemical potential of the pure ith solid, $\mu_i^*(s)$, equals the chemical potential of that same ith component dissolved in the melt, $\mu_i(l)$, along with other components:

$$\mu_i^*(s) = \mu_i(l) \qquad (2.27)$$

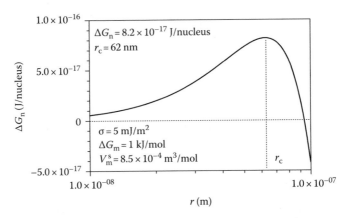

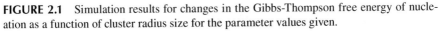

FIGURE 2.1 Simulation results for changes in the Gibbs-Thompson free energy of nucleation as a function of cluster radius size for the parameter values given.

The chemical potential of the ith component in the melt equals

$$\mu_i(1) = \mu_i^*(1) + RT \ln a_i(1) \tag{2.28}$$

where
  $\mu_i^*(1)$ is the chemical potential of a pure liquid of the ith component
  $a_i(1)$ corresponds to the activity of the ith component in the melt
  $a_i(1) = \gamma_i x_i(1)$, where γ_i is the activity coefficient and $x_i(1)$ is the concentration

Thus we can write

$$\mu_i^*(s) = \mu_i^*(1) + RT \ln a_i(1) \tag{2.29}$$

and therefore

$$\Delta\mu = \mu_i^*(s) - \mu_i^*(1) = RT \ln a_i(1) \tag{2.30}$$

If the solid phase was composed of a mixture of fats, rather than a single component, then we would have to take into account the concentration (activity) of the ith component in the solid as well. Thus, after the material has crystallized and equilibrium has been reestablished, the chemical potential of the solid and the liquid would be equal:

$$\mu_i^*(s) + RT \ln a_i(1) = \mu_i^*(1) + RT \ln a_i(1) \tag{2.31}$$

The chemical potential difference between the pure ith solid and the pure ith liquid could then be expressed as a function of the concentration (activity) of the ith component in the solid and the liquid:

$$\Delta\mu = \mu_i^*(s) - \mu_i^*(1) = RT \ln a_i(1) - RT \ln a_i(s)$$

$$= RT \ln \frac{a_i(1)}{a_i(s)} \tag{2.32}$$

The change in nucleation free energy as a function of nucleus size is given by

$$\frac{\partial \Delta G_n}{\partial r} = 8\pi r \delta - 4\pi r^2 \frac{\Delta \mu}{V_m^s} \tag{2.33}$$

The formation of a stable crystal nucleus takes place above a critical radius (r_c). This radius corresponds to a maximum in the free energy versus nuclei radius profile. At this point,

$$\frac{\partial \Delta G_n}{\partial r} = 0 \tag{2.34}$$

Equation 2.34 can then be rearranged to

$$8\pi r_c \delta = 4\pi r_c^2 \frac{\Delta \mu}{V_m^s} \tag{2.35}$$

and an expression for this critical radius obtained:

$$r_c = \frac{2\delta V_m^s}{\Delta \mu} \tag{2.36}$$

Substituting r_c for r in the Gibbs–Thompson model,

$$\Delta G_n^{r_c} = 4\pi \left(\frac{2\delta V_m^s}{\Delta \mu} \right)^2 \delta - \frac{4}{3} \pi \left(\frac{2\delta V_m^s}{\Delta \mu} \right)^3 \frac{\Delta \mu}{V_m^s} \tag{2.37}$$

and after rearrangement, leads to a useful expression

$$\Delta G_n^{r_c} = \frac{16\pi\delta^3 (V_m^s)^2}{(\Delta \mu)^2} - \frac{32\pi\delta^3 (V_m^s)^3 \Delta \mu}{3(\Delta \mu)^3 V_m^s} = \frac{16\pi\delta^3 (V_m^s)^2}{(\Delta \mu)^2} - \frac{32\pi\delta^3 (V_m^s)^2}{3(\Delta \mu)^2}$$

$$= \frac{16\pi\delta^3 (V_m^s)^2}{3(\Delta \mu)^2} \tag{2.38}$$

As shown before several times, $\Delta \mu_i = \Delta H_m \dfrac{\Delta T}{T_m}$, and Equation 2.38 can be written as

$$\Delta G_n^{r_c} = \frac{16\pi\delta^3 (V_m^s)^2 T_m^2}{3\Delta H_m^2 (\Delta T)^2} \tag{2.39}$$

This free energy of nucleation corresponds to an activation free energy of nucleation per nucleus—the free energy barrier required to form a nucleus of critical radius r_c.

2.2.1.2 Theory of Reaction Rates

Absolute reaction rate theory is a collision theory that assumes that chemical activation occurs through collisions between molecules. The central postulate of this theory is that the rate of a chemical reaction is given by the rate of passage of the activated complex through the transition state.

This theory is based on two assumptions: a dynamical bottleneck assumption and an equilibrium assumption. The first asserts that the rate of a reaction is controlled by the decomposition of an activated transition-state complex, and the second one asserts that an equilibrium exists between reactants (A and B) and the transition-state complex, $C^{\#}$, namely, $A + B \rightleftharpoons C^{\#} \rightarrow$ products.

It is therefore possible to define an equilibrium constant for the conversion of reactants in the ground state to an activated complex in the transition state. For the previous reaction,

$$K_{eq}^{\#} = \frac{[C^{\#}]}{[A][B]} \tag{2.40}$$

At equilibrium, the rate of the forward reaction (v_1) is equal to the rate of the reverse reaction (v_{-1}), $v_1 = v_{-1}$. Therefore, for the reaction $A + B \rightleftharpoons C^{\#}$ at equilibrium,

$$k_1[A][B] = k_{-1}[C^{\#}] \tag{2.41}$$

and therefore

$$K_{eq} = \frac{[\text{Products}]}{[\text{Reactants}]} = \frac{[C^{\#}]}{[A][B]} = \frac{k_1}{k_{-1}} \tag{2.42}$$

Since $\Delta G^0 = -RT \ln K$, the temperature dependence of the logarithm of the equilibrium constant for the standard state equals

$$\frac{\partial \ln K}{\partial T} = \frac{\Delta G^0}{RT^2} \tag{2.43}$$

and since $\ln K = \ln k_1 - \ln k_{-1}$, then

$$\frac{\partial \ln k_1}{\partial T} - \frac{\partial \ln k_{-1}}{\partial T} = \frac{\Delta G^0}{RT^2} \tag{2.44}$$

The change in the molar standard state free energy of a system undergoing a chemical reaction from reactants to products (ΔG^0) is equal to the free energy required for reactants to be converted to products minus the free energy required for products to be converted to reactants. Moreover, the energy required for reactants to be converted to products is equal to the difference in energy between the ground state and transition state of the reactants $(\Delta G_1^{\#})$, while the energy required for products to be converted to reactants is equal to the difference in energy between the ground state and transition state of the products $(\Delta G_{-1}^{\#})$ (Figure 2.2).

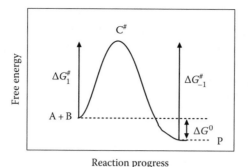

FIGURE 2.2   Changes in free energy as a function of the progress of a chemical reaction.

Therefore, the change in the molar free energy of a system undergoing a chemical reaction from reactants to products can be expressed as $\Delta G^0 = G_{\text{products}} - G_{\text{reactants}} = \Delta G_1^{\#} - \Delta G_{-1}^{\#}$.

The equation can therefore be expressed as two separate differential equations corresponding to the forward and reverse reactions:

$$\frac{\partial \ln k_1}{\partial T} = \frac{\Delta G_1^{\#}}{RT^2} + A \quad \text{and} \quad \frac{\partial \ln k_{-1}}{\partial T} = \frac{\Delta G_{-1}^{\#}}{RT^2} + A \qquad (2.45)$$

Arrhenius determined that for many reactions, $A = 0$. Therefore, indefinite integration of Equation 2.45 for either the forward or reverse reactions,

$$\int \partial \ln k_r = \frac{\Delta G^{\#}}{R} \int \frac{\partial T}{T^2} \qquad (2.46)$$

yields the expression

$$\ln k_r = \ln c - \frac{\Delta G^{\#}}{RT} \quad \text{or} \quad k_r = c e^{-(\Delta G^{\#}/RT)} \qquad (2.47)$$

where $\Delta G^{\#}$ is the molar free energy of activation. By using statistical thermodynamic arguments, it is possible to show that the preexponential factor c equals

$$c = \kappa \upsilon \approx \frac{k_B T}{h} \qquad (2.48)$$

where
   κ is the transmission coefficient (dimensionless)
   υ is the frequency [s^{-1}] of the normal mode oscillation of the transition-state complex along the reaction coordinate—more rigorously the average frequency of barrier crossing

The transmission coefficient, which can differ dramatically from unity, includes many correction factors, including tunneling, barrier recrossing correction, and solvent frictional effects. This expression can also equal $k_B T/h$, where k_B is Boltzman's constant [J/K], while h is Planck's constant [J s].

Thus an expression for the rate of a first order chemical reaction involving N molecules can have the form

$$v = k_r N = \frac{N k_B T}{h} e^{-\Delta G^{\#}/RT} \tag{2.49}$$

where
v is the rate of the reaction [s^{-1}]
k is the first order reaction rate constant [s^{-1}]
N is the number of molecules participating in the reaction

2.2.1.3 Determination of the Free Energy of Nucleation for an Isothermal Process

The assumption of this approach is that the activation free energy of nucleation ($\Delta G_n^{\#}$) to be the molar equivalent of the free energy of nucleation for a nucleus of critical radius r_c ($\Delta G_n^{r_c}$). In this treatment, it is thus necessary to transform $\Delta G_n^{r_c}$ [J/nucleus] to a free energy of nucleation per total amount of nucleating material [J/molecule or J/mol]. This is achieved by dividing $\Delta G_n^{r_c}$ by the number of molecules per nucleus ($N_{m/n}$):

$$\Delta G_n^{\#} = \frac{\Delta G_n^{r_c}}{N_{m/n}} \tag{2.50}$$

This free energy of activation for nucleation term can now be used in Equation 2.49, where v now denotes the rate of nucleation [s^{-1}], and N the number of molecules in the melt participating in the nucleation process.

The rate of nucleation is approximated by the inverse of the induction time of crystallization (τ):

$$v \sim \frac{1}{\tau} \tag{2.51}$$

Thus, Equation 2.49 can be expressed as

$$\frac{1}{\tau} \sim \frac{N k_B T}{h} e^{-\Delta G_n^{\#}/k_B T} \tag{2.52}$$

and

$$\frac{1}{\tau T} \sim \frac{N k_B}{h} e^{-\Delta G_n^{\#}/k_B T} \tag{2.53}$$

Taking the natural logarithm on both sides results in

$$\ln(\tau T) \sim -\ln \frac{Nk_B}{h} + \frac{\Delta G_n^{\#}}{k_B T} \tag{2.54}$$

Introducing Equation 2.39 into Equation 2.54 yields

$$\ln(\tau T) \sim -\ln \frac{Nk_B}{h} + \frac{\Delta G_n^{\#}}{k_B T} = -\ln \frac{Nk_B}{h} + \frac{\left(16\pi\delta^3 (V_m^s)^2 T_f^2 / 3\Delta H_f^2 (\Delta T)^2\right)}{k_B T} \tag{2.55}$$

After rearrangement, this equation can be expressed as

$$\ln(\tau T) \sim -\ln \frac{Nk_B}{h} + \frac{16\pi\delta^3 (V_m^s)^2 T_f^2}{3k_B \Delta H_f^2} \left(\frac{1}{T(\Delta T)^2} \right) \tag{2.56}$$

Thus, a plot of $\ln \tau T$ versus $1/T(\Delta T)^2$ should yield a straight line with a slope, m [K^3], which

equals:

$$m = \frac{16\pi\delta^3 (V_m^s)^2 T_f^2}{3k_B \Delta H_f^2} \tag{2.57}$$

Since

$$\Delta G_n^{\#} = \frac{16\pi\delta^3 (V_m^s)^2 T_f^2}{3\Delta H_f^2 (\Delta T)^2} \tag{2.58}$$

the free energy of nucleation can then be determined from

$$\Delta G_n^{\#} = \frac{mk_B}{(\Delta T)^2} \tag{2.59}$$

Figure 2.3 depicts a typical determination of the free energy of activation for nucleation for palm oil. Two distinct linear regions can be appreciated above and below 27°C, probably corresponding to the nucleation behavior of two different fractions or polymorphic phases. Figure 2.4 shows predicted changes in the free energy of nucleation as a function of temperature from parameters derived experimentally for each one of these distinct phases.

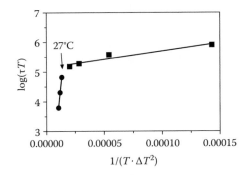

FIGURE 2.3   A plot of log (ΔT) versus $1/(T \cdot \Delta T^2)$ yields two distinct linear regions slopes that permit the calculation of activation free energies of nucleation of each region. Induction times of crystallization (τ, s) were determined using a cloud point analyzer.

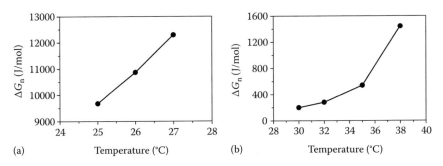

FIGURE 2.4   Activation free energies of nucleation (J/mol) versus temperature (°C) calculated from the slopes of the linear regions in Figure 2.3, for the fractions crystallizing in the range of (a) 25°C–27°C and (b) 30°C–38°C.

Another useful application of this nucleation analysis lies in the determination of the crystal–melt interfacial tension (δ). If reliable calorimetric data for the enthalpy and temperature of fusion are available, and an estimate of the molar volume can be obtained, the solid–melt interfacial tension can be directly calculated from

$$\delta = \left(\frac{3 m k_B \Delta H_f^2}{16 \pi (V_m^s)^2 T_f^2} \right)^{1/3} \tag{2.60}$$

2.2.1.4   Estimates of ΔH_f and V_m^s

Estimates of the enthalpy of fusion should be obtained experimentally using differential scanning calorimetry. It is not possible to predict a priori the enthalpy of fusion of a complex mixture of triacylglycerols and minor polar components that have been crystallized, most probably, under dynamic conditions. Knowledge of the complex phase behavior of TAGs crystallized under nonequilibrium, dynamic conditions does not exist yet.

The density of a solid TAG mixture, crystallized under relevant conditions, should also be determined experimentally by picnometry. The molar volume of a fat can then be calculated from density data using

$$\text{Molar volume (m}^3/\text{mol)} = \frac{\text{MW (g/mol)}}{\text{Density (g/mL)}}(1 \cdot 10^{-6}\,\text{m}^3/\text{mL}) \qquad (2.61)$$

Experimentally determined melting enthalpies and temperatures, as well as densities, for different pure TAG polymorphs are listed in Appendix Va and Table 10.3, respectively, of Donald Small's (1986) book. Suffice to say, good experimental data are required for these types of analyses. The nature of the material (multicomponent system, presence of impurities, metastable nucleation, etc.) though, makes it quite difficult to obtain good data.

2.2.1.5   Metastability and Free Energy of Nucleation

Even at moderate degrees of undercooling, triacylglycerols do not nucleate in their most stable polymorphic form, but rather nucleate in a metastable form. This can be explained by close inspection of Equation 2.58. Being closer in nature to the liquid state, the α form of TAGs has a lower surface free energy, or crystal–melt interfacial tension. This, in turn will lead to a lower free energy of nucleation, and thus a higher nucleation rate. The crystal–melt surface free energy is the main factor responsible for the metastable nucleation behavior of triacylglycerols and fats. This effect must be enormous. For an α-form at a particular temperature, the degree of undercooling (ΔT), melting enthalpy (ΔH_f), and melting temperature (T_f) are lower, and the molar volume (V_m) is higher (lower density), than for more stable crystal forms. This should result in a higher activation free energy for the nucleation of an α-form than for more stable crystal forms. However, the opposite is true. The activation free energy of nucleation is actually lower for an α-form, resulting in a higher rate of nucleation, than for more stable crystal forms. Close inspection of Equation 2.58 reveals that for this to be the case, the surface free energy term (δ) must be much lower for the α-form than for more stable crystalline phases. Future studies should pay more attention to the effects of δ on crystallization behavior.

2.2.2   Isothermal Crystal Growth—The Avrami Model

The Avrami model (Avrami, 1939, 1940, 1941) can be used to quantify crystallization kinetics and provides an indication of the nature of the crystal growth process. Applied to the study of fat crystallization, the Avrami equation has the following form:

$$\frac{\text{SFC}}{\text{SFC}_{\max}} = 1 - e^{-kt^n} \qquad (2.62)$$

where
  n is the Avrami exponent [dimensionless]
  k is the Avrami constant [t^{-n}]
  SFC corresponds to the solid fat content at a particular time [%]
  $\text{SFC}_{\max}$ corresponds to the maximum SFC achieved at a particular temperature [%]

This model was developed to describe the kinetics of liquid–solid phase transitions in metals and its principles were first applied to polymer crystallization in the 1950s (Christian, 1965; Sharples, 1966). The Avrami model is the most commonly used model used in the study of fat crystallization. The equation describes an event in which there is an initial lag-period, where crystallization occurs very slowly, followed by a subsequent rapid increase in crystal mass. This model takes into account that crystallization occurs by both nucleation and crystal growth and is based on the assumptions of isothermal transformation conditions, spatially random nucleation, and linear growth kinetics in which the growth rate of the new phase depends only on temperature and not on time. It is also assumed that the density of the growing bodies is constant.

The Avrami parameters provide information on the nature of the crystallization process. The constant (k) represents a crystallization rate constant. It is primarily a function of the crystallization temperature, generally obeys an Arrhenius-type temperature dependency, and takes both nucleation and crystal growth rates into account. Half-times of crystallization, $t_{1/2}$, reflect the magnitudes of k and n according to

$$t_{1/2} = \left(\frac{\ln 2}{k} \right)^{1/n} \tag{2.63}$$

The Avrami exponent, n, sometimes referred to as an index of crystallization, indicates the crystal growth mechanism. This parameter is a combined function of the time dependence of nucleation and the number of dimensions in which growth takes place. Nucleation is either instantaneous, with nuclei appearing all at once early on in the process, or sporadic, with the number of nuclei increasing linearly with time. Growth either occurs as rods, disks, or spheres in one, two, or three dimensions, respectively. Table 2.2 shows the value of the Avrami exponent, n, expected for various types of nucleation and growth.

Although n should be an integer, fractional values are usually obtained, even in cases where the model fits the data quite well. Deviations from integer values for n have been explained as simultaneous development of two (or more) types of crystals, or similar crystals from different types of nuclei (sporadic vs. instantaneous).

TABLE 2.2

Values for the Avrami Exponent (n) for Different Types of Nucleation and Growth

N	Type of Crystal Growth and Nucleation Expected
$3 + 1 = 4$	Spherical growth from sporadic nuclei
$3 + 0 = 3$	Spherical growth from instantaneous nuclei
$2 + 1 = 3$	Platelike growth from sporadic nuclei
$2 + 0 = 2$	Platelike growth from instantaneous nuclei
$1 + 1 = 2$	Rod/needle/fiber-like growth from sporadic nuclei
$1 + 0 = 1$	Rod/needle/fiber-like growth from instantaneous nuclei

FIGURE 2.5 Microstructure of milk fat crystallized isothermally at 5°C. The Avrami index for the crystallization process leading to this microstructure was $n = 1$.

Deviations may also occur in cases where spherical crystals arise from initially rod- or platelike nuclei. In these situations, n is continually changing. In some cases of metals and alloys in which growth is diffusion controlled, fractional exponents often correlate with specific growth mechanisms which can be confirmed microscopically.

In the case of fats, we have adopted the philosophy of Christian (1965), where specific Avrami exponents are associated with certain growth modes determined by microscopy. For example, for milk fat crystallized at high degrees of supercooling, we would expect a lower free energy of nucleation and a higher rate of nucleation, thus leading to the formation of a large number of nuclei at the onset of the crystallization process (instantaneous nucleation). The rate of crystal growth would also be quite high. This in turn would lead to a more one-dimensional growth. The end result would be a granular microstructure composed of a large number of small crystals (Figure 2.5).

On the other hand, for milk fat crystallized at low degrees of supercooling, we would expect a higher free energy of nucleation and a lower rate of nucleation, thus leading to the formation of a small number of nuclei, possibly in a sporadic fashion (sporadic nucleation). The rate of crystal growth would also be lower. This in turn would lead to a more multidimensional growth. The end result would be a "clustered" microstructure composed of a small number of large crystals (Figure 2.6).

2.2.2.1 Derivation of the Model
Abbreviations

A	is the area through which diffusion takes place [m²]
A_g	is the area of the crystal involved in growth [m²]
c	is the concentration of supersaturated material [M]
c^*	is the equilibrium saturation concentration of the material [M]

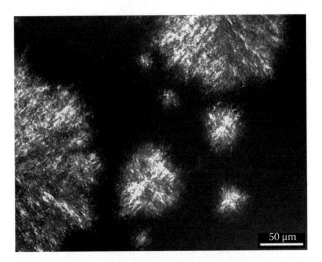

50 μm

FIGURE 2.6 Microstructure of milk fat crystallized isothermally at 25°C. The Avrami index for the crystallization process leading to this microstructure was $n = 4$.

$(c - c^*)$	is the concentration difference between the amount of supersaturated material at a particular time and the equilibrium saturation concentration of material at a particular temperature [M]
g	is the growth rate constant for the crystal's radius, length or height [s^{-1}]
h	is the crystal height [m]
j	is the nucleation rate constant [m^{-3}/s]
k	is the Avrami constant [s^{-n}]
K	is the mass transfer, or diffusion (D), coefficient [m^2/s]
k_g	is the rate constant for single crystal growth (combined specific mass transfer coefficient and surface reaction rate constant) [m/s]
κ	is the surface reaction rate constant [m^2/s]
l	is the crystal length [m]
m_c	is the mass of a single crystal at a particular time [kg]
m_s	is the total mass of solids present in the system at a particular time [kg]
m_{max}	is the maximum total mass of solids present in the system at infinite time [kg]
$(m_{max} - m_s)$	is the mass of supercooled material that has not crystallized yet [kg]
m_T	is the total amount of mass in the system (solids plus liquid) [kg]
MW	is the molecular weight of the diffusing molecules [kg/mol]
N_c	is the total number of crystals in system
Q	is the number of moles of material [mol]
r	is the crystal radius [m]
ρ	is the crystallite density [m^{-3}]
t	is the crystallization time [s]
V_T	is the volume of the system [m^3]
ΔC	is the concentration difference across boundary layer [M]
Δx	is the thickness of layer across which diffusion takes place—the boundary layer [m]

The starting point of this treatment is the empirical chemical diffusion equation (Fick's first law in one dimension):

$$\frac{\partial Q}{\partial t} = KA\left(\frac{\Delta C}{\Delta x}\right)$$

(2.64)

This Equation 2.64 describes the transport of mass per unit area across a concentration gradient. Since the increase in mass of a crystal is a function of the amount of mass that diffuses to the crystal surface, we can write

$$\frac{\partial m_c}{\partial t} = \frac{\partial Q}{\partial t} MW$$

(2.65)

The chemical diffusion equation describing the growth of a single crystal can therefore be written as

$$\frac{\partial m_c}{\partial t} = k_g A_g (c - c^*)$$

(2.66)

where

$$k_g = \left(\frac{1}{K} + \frac{1}{\kappa}\right)^{-1} (\Delta x)^{-1}$$

(2.67)

and

$$(\Delta C)MW = (c - c^*)$$

(2.68)

The rate constant for single crystal growth (k_g) accounts for the possibility of diffusion control at small relative velocities and surface reaction control at high relative velocities. K varies with solution properties like viscosity and agitation, but κ does not. As well, K varies little with temperature, while κ can change dramatically upon cooling.

The total increase in solids in the system is the product of the increase in mass of a single crystal times the number of growing crystals in the system:

$$\frac{\partial m_s}{\partial t} = N_c k_g A_g (c - c^*)$$

(2.69)

and since

$$(c - c^*) = \frac{(m_{max} - m_s)}{V_T}$$

(2.70)

Equation 2.69 can be thus be expressed as

$$\frac{\partial m_s}{\partial t} = N_c k_g A_g \frac{(m_{max} - m_s)}{V_T} \tag{2.71}$$

The number of crystals per unit volume in the system is defined as

$$\rho_c = \frac{N_c}{V_T} \tag{2.72}$$

Considering Equation 2.72, and after variable separation, Equation 2.71 can be rearranged to

$$\frac{\partial m_s}{(m_{max} - m_s)} = k_g \rho_c A_g \partial t \tag{2.73}$$

This is the basic equation that can be used to derive the final form of the Avrami equation for different growth geometries and types of nucleation.

2.2.2.1.1 Spherical Growth with Instantaneous Nucleation

The surface area of a spherical crystal involved in crystal growth is

$$A_g = 4\pi r^2 \tag{2.74}$$

The linear growth rate of the crystal radius in time is expressed as

$$r = gt \tag{2.75}$$

Introducing Equations 2.74 and 2.75 into Equation 2.73 results in

$$\frac{\partial m_s}{(m_{max} - m_s)} = k_g \rho_c 4\pi g^2 t^2 \partial t \tag{2.76}$$

Integration of Expression 2.76 for the boundary conditions $m_s = 0$ at $t = 0$, and m_s at t,

$$\int_0^{m_s} \frac{\partial m_s}{(m_{max} - m_s)} = 4\pi k_g \rho_c g^2 \int_0^t t^2 \partial t \tag{2.77}$$

results in the equation

$$\ln\left(\frac{m_{max}}{m_{max} - m_s}\right) = \frac{4}{3}\pi k_g \rho_c g^2 t^3 \tag{2.78}$$

This expression can be transformed and rearranged to

$$\frac{m_s}{m_{max}} = 1 - e^{-(4/3)\pi k_g \rho_c g^2 t^3} \tag{2.79}$$

Moreover, the mass fraction (m_s/m_{max}) can be expressed as the ratio of solid fat contents:

$$\frac{V_s}{V_{max}} = \frac{\dfrac{V_s}{V_T}}{\dfrac{V_{max}}{V_T}} = \frac{\dfrac{m_s/\rho_s}{m_T/\rho_{s+l}}}{\dfrac{m_{max}/\rho_s}{m_T/\rho_{s+l}}} = \frac{\dfrac{m_s}{m_T}}{\dfrac{m_{max}}{m_T}} = \frac{SFC}{SFC_{max}} \tag{2.80}$$

Thus, the Avrami equation describing the growth of a spherical crystal under conditions of instantaneous nucleation can be expressed as

$$\frac{SFC}{SFC_{max}} = 1 - e^{-k_A t^3} \tag{2.81}$$

where

$$k_A = \frac{4}{3}\pi k_g \rho_c g^2 \tag{2.82}$$

2.2.2.1.2 Spherical Growth with Sporadic Nucleation

The treatment starts with Equation 2.76, shown here again for the sake of clarity,

$$\frac{\partial m_s}{(m_{max} - m_s)} = k_g \rho_c 4\pi g^2 t^2 \partial t \tag{2.76}$$

For sporadic nucleation, the change in the number of nuclei as a function of time is given by

$$\rho_c = \frac{N_c}{V_T} = jt \tag{2.83}$$

Introducing Equation 2.83 into Equation 2.76 results in

$$\frac{\partial m_s}{(m_{max} - m_s)} = k_g j 4\pi g^2 t^3 \partial t \tag{2.84}$$

Integration of Expression 2.84 for the boundary conditions $m_s = 0$ at $t = 0$ and m_s at t,

$$\int_0^{m_s} \frac{\partial m_s}{(m_{max} - m_s)} = 4\pi k_g j g^2 \int_0^t t^3 \partial t \tag{2.85}$$

results in the equation

$$\ln\left(\frac{m_{max}}{m_{max} - m_s}\right) = \pi k_g j g^2 t^4 \tag{2.86}$$

that can be transformed and rearranged to

$$\frac{m_s}{m_{max}} = 1 - e^{-\pi k_g j g^2 t^4} \tag{2.87}$$

Thus, the Avrami equation describing the growth of a spherical crystal under conditions of sporadic nucleation can be expressed as

$$\frac{SFC}{SFC_{max}} = 1 - e^{-kt^4} \tag{2.88}$$

where

$$k_A = \pi k_g j g^2 \tag{2.89}$$

2.2.2.1.3 Platelike Growth with Instantaneous Nucleation

Consider a rectangular plate growing in the X–Y plane, but not in the Z-plane. The area involved in crystal growth is therefore

$$A_g = 4lh \tag{2.90}$$

where the dimension l is increasing linearly in time according to

$$l = gt \tag{2.91}$$

Introducing Equations 2.90 and 2.91 into 2.73, results in

$$\frac{\partial m_s}{(m_{max} - m_s)} = k_g \rho_c 4 g t h \partial t \tag{2.92}$$

Integration of Expression 2.92 for the boundary conditions $m_s = 0$ at $t = 0$, and m_s at t,

$$\int_0^{m_s} \frac{\partial m_s}{(m_{max} - m_s)} = 4hk_g\rho_c g \int_0^t t\partial t \tag{2.93}$$

results in the equation

$$\ln\left(\frac{m_{max}}{m_{max} - m_s}\right) = 2hk_g\rho_c gt^2 \tag{2.94}$$

that can be transformed and rearranged to

$$\frac{m_s}{m_{max}} = 1 - e^{-2hk_g\rho_c gt^2} \tag{2.95}$$

Thus, the Avrami equation describing the growth of a platelike crystal under conditions of instantaneous nucleation can be expressed as

$$\frac{SFC}{SFC_{max}} = 1 - e^{-k_A t^2} \tag{2.96}$$

where

$$k_A = 2hk_g\rho_c g \tag{2.97a}$$

2.2.2.1.4 Platelike Growth with Sporadic Nucleation

The treatment starts with Equation 2.92, shown here again for the sake of clarity,

$$\frac{\partial m_s}{(m_{max} - m_s)} = 4hk_g\rho_c gt\partial t \tag{2.92}$$

For sporadic nucleation, the change in the number of nuclei as a function of time is given Equation 2.83, shown here again for the sake of clarity,

$$\rho_c = \frac{N_c}{V_T} = jt \tag{2.83}$$

Introducing Equation 2.83 into Equation 2.92 results in

$$\frac{\partial m_s}{(m_{max} - m_s)} = 4hk_g jgt^2\partial t \tag{2.97b}$$

Integration of Expression 2.97b for the boundary conditions $m_s = 0$ at $t = 0$, and m_s at t,

$$\int_0^{m_s} \frac{\partial m_s}{(m_{max} - m_s)} = 4hk_g jg \int_0^t t^2 \partial t \qquad (2.98)$$

results in the equation

$$\ln\left(\frac{m_{max}}{m_{max} - m_s}\right) = \frac{4}{3} hk_g jgt^3 \qquad (2.99)$$

that can be transformed and rearranged to

$$\frac{m_s}{m_{max}} = 1 - e^{-(4/3)hk_g jgt^3} \qquad (2.100)$$

Thus, the Avrami equation describing the growth of a platelike crystal in two dimensions under conditions of sporadic nucleation can be expressed as

$$\frac{SFC}{SFC_{max}} = 1 - e^{-kt^3} \qquad (2.101)$$

where

$$k_A = \frac{4}{3} hk_g jg \qquad (2.102)$$

2.2.2.1.5 Rodlike Growth with Instantaneous Nucleation

Consider a cylinder growing in length, but not in cross section. The area involved in crystal growth is therefore

$$A_g = 2\pi r^2 \qquad (2.103)$$

Introducing Equation 2.103 into Equation 2.73 results in

$$\frac{\partial m_s}{(m_{max} - m_s)} = k_g \rho_c 2\pi r^2 \partial t \qquad (2.104)$$

Integration of Expression 2.104 for the boundary conditions $m_s = 0$ at $t = 0$, and m_s at t,

$$\int_0^{m_s} \frac{\partial m_s}{(m_{max} - m_s)} = 2\pi r^2 k_g \rho_c \int_0^t \partial t \qquad (2.105)$$

results in the equation

$$\ln\left(\frac{m_{\max}}{m_{\max} - m_s}\right) = 2\pi r^2 k_g \rho_c t \tag{2.106}$$

that can be transformed and rearranged to

$$\frac{m_s}{m_{\max}} = 1 - e^{-2\pi r^2 k_g \rho_c t} \tag{2.107}$$

Thus, the Avrami equation describing the growth of a rodlike crystal in one dimension under conditions of instantaneous nucleation can be expressed as

$$\frac{SFC}{SFC_{\max}} = 1 - e^{-k_A t} \tag{2.108}$$

where

$$k_A = 2\pi r^2 k_g \rho_c \tag{2.109}$$

2.2.2.1.6 Rodlike Growth with Sporadic Nucleation

The treatment starts with Equation 2.104, shown here again for the sake of clarity,

$$\frac{\partial m_s}{(m_{\max} - m_s)} = 2\pi r^2 k_g \rho_c \partial t \tag{2.104}$$

For sporadic nucleation, the change in the number of nuclei as a function of time is given by Equation 2.83, shown here again for the sake of clarity,

$$\rho_c = \frac{N_c}{V_T} = jt \tag{2.83}$$

Introducing Equation 2.83 into Equation 2.104 results in

$$\frac{\partial m_s}{(m_{\max} - m_s)} = 2\pi r^2 k_g jt \partial t \tag{2.110}$$

Integration of Expression 2.110 for the boundary conditions $m_s = 0$ at $t = 0$, and m_s at t,

$$\int_0^{m_s} \frac{\partial m_s}{(m_{\max} - m_s)} = 2\pi r^2 k_g j \int_0^t t \partial t \tag{2.111}$$

results in the equation

$$\ln\left(\frac{m_{max}}{m_{max}-m_s}\right)=\pi r^2 k_g j t^2 \tag{2.112}$$

that can be transformed and rearranged to

$$\frac{m_s}{m_{max}}=1-e^{-\pi r^2 k_g j t^2} \tag{2.113}$$

Thus, the Avrami equation describing the growth of a rodlike crystal in one dimension under conditions of sporadic nucleation can be expressed as

$$\frac{SFC}{SFC_{max}}=1-e^{-k_A t^2} \tag{2.114}$$

where

$$k_A=\pi r^2 k_g j \tag{2.115}$$

2.2.2.2 Use of the Model

The Avrami model has the general form

$$\frac{SFC}{SFC_{max}}=1-e^{-k_A t^n} \tag{2.116}$$

This model can be linearized by a double logarithmic transformation after rearrangement to yield

$$\ln\left(1-\frac{SFC}{SFC_{max}}\right)=-k_A t^n \tag{2.117}$$

and

$$\ln\left(-\ln\left(1-\frac{SFC}{SFC_{max}}\right)\right)=\ln\left(k_A\right)+n\ln\left(t\right) \tag{2.118}$$

Thus, a plot of $\ln(-\ln(1-SFC/SFC_{max}))$ versus $\ln t$ should yield a straight line with slope $= n$ and y-intercept $= \ln k_A$. Figure 2.7a shows the evolution of SFC as a function of time for the isothermal crystallization of cocoa butter at 32°C. Transformation of the SFC-t data according to Equation 2.105 should yield a straight line when plotted. However, a straight line is not obtained, as usually is the case (Figure 2.7b). Thus, only the linear region of the transformed data is used to calculate the slope and y-intercept (Figure 2.7c).

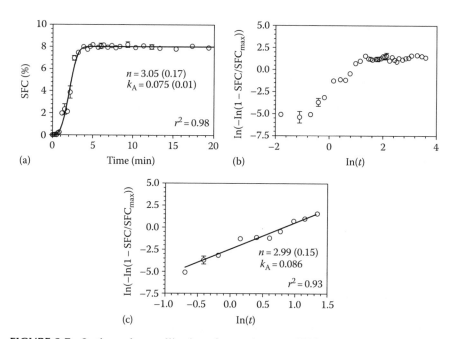

FIGURE 2.7   Isothermal crystallization of cocoa butter at 32°C. (a) Nonlinear regression of the Avrami model to the crystallization data. (b) Double logarithmic transformation for linearization purposes of the same data set. (c) Linear regression analysis of the double logarithmic linearization of the data in (b) focusing only the linear region.

Alternatively, SFC versus time data can be fitted to the Avrami model in its nonlinear form, using standard curve-fitting (nonlinear regression) routines available in most modern graphical programs, and estimates of k_A and n obtained. Model fits to the data and estimates of the kinetic parameters are shown in Figure 2.7a. As can be appreciated, results obtained by these two methods are usually comparable.

Sometimes, multiple-step growth curves are observed in crystallization experiments. Multiple steps may be due to fractionation or polymorphic transformations taking place during crystallization. As an example, we have milk fat crystallized at 10°C (Figure 2.8). The entire process could be modeled using a multicomponent Avrami model of the form

$$\mathrm{SFC} = \mathrm{SFC}_0 + \sum_{i=1}^{n} \mathrm{SFC}_{\mathrm{max},i}(1 - e^{-k_i t^{n_i}}) \tag{2.119}$$

where SFC_0 is the initial SFC, usually zero.

Fitting a single-component Avrami model to the data results in a completely different set of kinetic parameters (Figure 2.9). In this case, the single-component analysis represents an average of the two events.

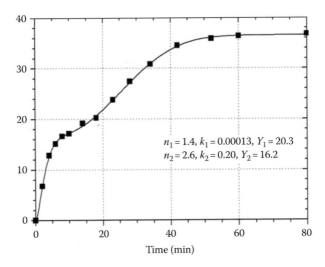

FIGURE 2.8 SFC evolution in time during the isothermal crystallization of milk fat at 10°C. Shown here is a double-Avrami fit to the experimental data.

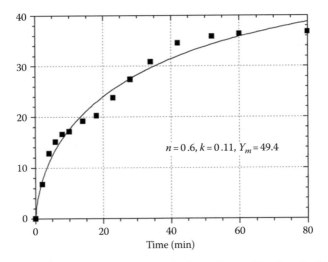

FIGURE 2.9 SFC evolution in time during the isothermal crystallization of milk fat at 10°C. Shown here is a single-Avrami fit to the experimental data.

The Avrami model does not implicitly take induction times into consideration. If induction times (τ) are known, these can be incorporated into the Avrami model as

$$\frac{SFC}{SFC_{max}} = 1 - e^{-k_A(t-\tau)^n} \tag{2.120}$$

This is particularly important point since a process with a long induction time could be mistakenly interpreted as a crystallization process with an unusually high induction time. In my experience, Avrami exponents higher than ~5 are most probably artifactual. Artifacts will be introduced, for example, when sample cooling is inefficient. This is usually not a problem with small samples (i.e., less than 200 mg). However, a "dry bath" cooling system where heat is exchanged from metal through air to a sample in a plastic or glass containers is not appropriate for fat crystallization studies. Even 2–4 g samples of fat will not be cooled fast enough for proper kinetic studies to be carried out (please refer to experimental methods section for further examples).

As a final note, I would like to comment on the determination of the energy of activation from the temperature dependence of the Avrami constant using the Arrhenius model

$$\ln k_A = \ln A - \frac{E_a}{RT} \tag{2.121}$$

A plot of $\ln k_A$ versus $1/T$ should yield (and does yield) a straight line with positive slope $= -E_a/R$. This, however, is only correct if all the k_A's have the same units, that is, if the Avrami exponent (n) is the same at all temperatures (not likely). Remember that the Avrami constant has units of $(time)^{-n}$; thus unless the Avrami exponent remains constant for all temperatures used, an Arrhenius analysis using Avrami constants is not strictly valid.

2.3 ISOTHERMAL CRYSTALLIZATION KINETICS AND MICROSTRUCTURE

2.3.1 RELATIONSHIP BETWEEN ISOTHERMAL NUCLEATION KINETICS AND THE FRACTAL DIMENSION OF A FRACTAL CLUSTER

In a study of the relationship between crystallization behavior and microstructure in cocoa butter (Marangoni and McGauley, 2003), we discovered a strong effect of temperature on both crystallization kinetics and microstructure, as characterized by the box-counting fractal dimension. We thus plotted the values of the box-counting dimension as a function of the inverse of the induction time, in semilogarithmic coordinates (Figure 2.10a), and as a function of the Avrami exponent (Figure 2.10b). Some very interesting relationships became apparent. The inverse of the induction time of a process is proportional to the rate of that process. For nucleation, the inverse of the induction time (τ) is proportional to the nucleation rate (J), namely, $J \sim 1/\tau$. Moreover, as shown previously, the natural logarithm of the nucleation rate is proportional to the activation free energy of nucleation ($\Delta G^{\#}$). Taking the natural logarithm to both sides of Equation 2.49, we obtain

$$\ln J = \alpha - \frac{\Delta G^{\#}}{k_B T} \tag{2.122}$$

where
k_B is the Boltzmann constant
T is the absolute temperature

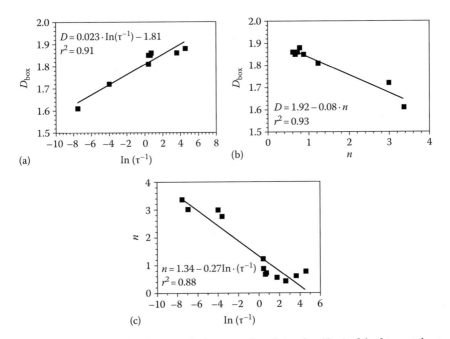

FIGURE 2.10 Relationship between the box-counting dimension (D_{box}) of the fat crystal network microstructure and natural logarithm of the inverse of the induction time (a). Relationship between the fractal dimension and the Avrami exponent (b). Relationship between the Avrami exponent and the and natural logarithm of the inverse of the induction time (c).

The parameter α corresponds to

$$\alpha = \ln\left(\frac{NkT}{h}\right) \tag{2.123}$$

where
   h is Planck's constant
   N is the number of molecules participating in the crystallization process

Thus, our results suggested that there is direct relationship between the nucleation rate and the resulting microstructure of the fat crystal network of the form

$$D - D^* = \beta \ln J \tag{2.124}$$

where
   β is a constant characteristic of a particular system
   D^* is the box-counting fractal dimension of a microstructure arising from a crystallization process with a nucleation rate of unity ($J = 1$)
   D is the box-counting dimension of the microstructure viewed by polarized light microscopy (PLM)

The box-counting dimension is very sensitive to the degree of fill of the embedding space. Thus, it stands to reason that a higher nucleation rate should lead to the creation of a microstructure that fills space more homogeneously and without voids—a large number of small crystals.

This equation could also be expressed as a function of the activation free energy of nucleation:

$$D - D^* = \kappa - \beta \frac{\Delta G^{\#}}{kT} \tag{2.125}$$

where $\kappa = \alpha\beta$.

These results suggest that the microstructure is partly a consequence of the energetics of nucleation. These results agree in principle with the work of Rousset (2002) on the modeling of microstructural growth in POP (1-palmitoyl, 2-oleyl, 3-palmitate), one of the main triacylglycerols in cocoa butter. In Figure 2.8 of his work, Rousset demonstrates how different nucleation regimes and rates lead to the development of different microstructures. The microstructures shown would have different fractal dimensions.

A statistically significant relationship ($p < 0.001$) also existed between the box-counting fractal dimension and the Avrami exponent (Figure 2.10b). This fractal dimension was inversely related to the Avrami exponent. The Avrami exponent is a function of the type of nucleation (instantaneous or sporadic) and the dimensionality of growth, and thus it is not surprising that it is strongly correlated to the final microstructure of the system. We also found a correlation between the nucleation rate and the Avrami exponent (Figure 2.10c). This is not surprising either since the Avrami exponent contains information about the type of nucleation process. Interestingly, it is possible to define an Avrami exponent at a nucleation rate of unity ($J = 1$).

In what follows, we will derive the relationship in Equation 2.125 from first principles.

The mass of an object created in a sequence of random steps will grow in time in a linear fashion:

$$M \sim t \tag{2.126}$$

since the amount of mass accrued during time t is proportional to the number of random steps, $j(t)$, performed by the diffusing triaclyglycerol molecule, $j(t) \sim t$.

For a fractal object,

$$M \sim \xi^D \tag{2.127}$$

where
   ξ is the diameter of the cluster
   D is the fractal dimension

Equating Equation 2.126 to Equation 2.127, we obtain

$$\xi \sim t^{1/D} \tag{2.128}$$

or

$$\xi = kt^{1/D} \tag{2.129}$$

where k is a rate constant for growth. Thus, if an object obeys the linear mass deposition behavior, and its growth is fractal in nature, a log-log plot of ξ versus t will be linear with a slope (s) equal to $1/D$. For the case where the growth of the cluster is Euclidean, the fractal dimension would equal the Euclidean dimension (d), and its value would be 2 or 3 depending on the type of crystal growth mechanism. The slope of the log-log plot of ξ versus t would then the 1/2 or 1/3.

For a fractal object structured as a particle cluster, number of particles in the cluster scale as a function of cluster diameter as

$$N = c\xi^{D} \tag{2.130}$$

where
   N is the number of particles within a fractal cluster
   c is a system-specific constant
   ξ is the diameter of the fractal cluster

Combining Equations 2.129 and 2.130, we obtain

$$N = ck^{D}t \tag{2.131}$$

The differential of N with time ($J = \partial N/\partial t$) has thus the form

$$J = ck^{D} \tag{2.132}$$

Taking the natural log of Expression 2.132 yields

$$\ln J = \ln c + D \ln k \tag{2.133}$$

Upon rearrangement, Equation 2.133 can be expressed as Equation 2.122, where

$$\beta = \frac{1}{\ln k} \tag{2.134}$$

and

$$D^* = -\frac{\ln c}{\ln k} \tag{2.135}$$

This treatment was derived and is based on experiments carried out on a microscope slide (microscopy) and for small amounts of crystallizing fat. These systems are not heat-transfer limited, and thus represent an idealized situation, and will be useful in comparative studies between fats in the laboratory.

2.3.2   RELATIONSHIP BETWEEN FRACTAL CLUSTER SIZE AND THE ISOTHERMAL FREE ENERGY OF NUCLEATION

Spherulites are nearly spherical, densely branched, polycrystalline clusters formed under highly nonequilibrium conditions (Keith and Padden, 1963; Magill, 2001). They are observed in a wide range of materials, from polymers and metallic glasses to volcanic rocks and edible fats. Understanding the factors that influence the microstructure of these materials is important since their mechanical strength and failure characteristics strongly depend on microstructure.

Spherulites arise when crystal growth rates are large compared to diffusion rates in the melt (Keith and Padden, 1963; Goldenfeld, 1987; Magill, 2001). Spherulitic melt crystallization is observed in multicomponent systems and/or in systems with small self-diffusion coefficients and low crystallization rates, and is characterized by fibrous crystal habits and profuse noncrystallographic branching.

Work by Liu and coworkers (Liu and Sawant, 2002a,b; Wang et al., 2006) has demonstrated the fractal nature of supramolecular materials structured as fibrous networks of interconnected nanosized crystalline fibers or particles. Using a modified Avrami equation, they have shown that the crystalline growth kinetics of several of these materials is consistent with that of fractal objects. Moreover, they also demonstrate that the structure and branching growth of their organogel networks is reminiscent to those of a Cayley fractal tree. They then proceed to show how the growth of the network is a function of the degree of supersaturation of the system in a manner consistent with nucleation and growth processes rather than diffusion-limited processes. The proposed mechanism for the formation of such fractal Cayley tree structures thus includes an initial nucleation event, followed by growth, branching, further growth, branching, and so forth. The branching process at the tip of the growing fibers was suggested to be noncrystallographic (crystallographic mismatch) in nature. These authors have also proposed that spherulitic growth arises due to surface integration kinetic effects rather than diffusional instabilities in protein crystals (Chow et al., 2002).

Recent work by Grasany and coworkers (Grasany et al., 2004, 2005, 2006) has focused on the development of a coarse-grained phase field theory of crystal nucleation and polycrystalline growth. This remarkable development has led to the proposal of a general mechanism of polycrystalline growth, and the successful prediction of the observed spherulite morphologies and categories, as well as their transformation kinetics. This work suggests that spherulitic growth arises due to diffusional instabilities, as well as three nucleation modes of new grains at the growth front. This growth front nucleation is a secondary nucleation process that arises due to three main reasons: (1) the presence of foreign particles (static heterogeneities), (2) trapping of orientational disorder due to a reduced rotational diffusion coefficient relative to the translational diffusion coefficient of the crystallizing molecules (dynamic heterogeneities), and (3) noncrystallographic branching.

The internal structure of a spherulite arises from a complex interplay of heat and mass transfer processes during its formation. It was only recently that Owen and Bergmann demonstrated fractal scaling behavior in polyhydroxybutyrate spherulites over a length scale ranging from 12 to 300 nm using ultrasmall angle x-ray scattering techniques (Owen and Bergmann, 2004). Considering that the lamellar periodicity of the polymer used in that study was ~6 nm, this suggests that the fractal distribution of mass was related to the spatial distribution of bundles, or clusters, of lamellae in the interior of relatively large spherulites (10–100 μm). The use of the concept and techniques of fractal geometry to describe the spherulitic/dendritic/branched texture of disequilibrium silicate minerals was first proposed by Fowler et al. (1989).

Fats are the main structural components in many food products such as margarine, chocolate, and butter. The sensory characteristics and mechanical properties of these materials are highly dependant on the structure of the underlying fat crystal network. This fat crystal network is built by the interaction of polycrystalline triglyceride clusters (spherulites), which provides firmness or solid-like behavior to the plastic fats. Work in our laboratory has focused on the development of a structural-mechanical theory of elasticity of polycrystalline networks (Narine and Marangoni, 1999a,b; Marangoni, 2000; Marangoni and Rogers, 2003; Awad et al., 2004). Even though the principles of fractal geometry were successfully used to develop our models, we had never directly demonstrated fractal scaling of the crystal cluster size with the volume fraction of solids until recently (Marangoni and Ollivon, 2007). Since our work had also suggested a relationship between nucleation energetics and the fractal dimension (Marangoni and McGauley, 2003; Batte and Marangoni, 2005), then it would be possible to predict the size of the polycrystalline clusters in fat crystal networks (or any polycrystalline material structured as a collection of clusters) from knowledge of their free energy of nucleation. This, in turn, would allow for the prediction of the elastic modulus and yield stress of such materials from knowledge of their crystallization behavior.

The size of a fractal cluster (ξ) is related to the volume fraction of solids (Φ) in a power law fashion as described by the classic fractal relationship

$$\xi \sim a\Phi^{1/D-d} \tag{2.136}$$

where
   a is the size of a primary particle/crystal that makes up the cluster
   D is the mass fractal dimension
   d is the Euclidean dimension in which the fractal object is embedded

For a square-embedding space lattice filled with square particles, this relationship *can* be an equality. However, this is usually not the case, and one should consider the preexponential factor to be just that—a constant (p).

Combining Equations 2.136 and 2.125, one obtains a relationship between the size of a fractal cluster and the activation free energy of nucleation,

$$\xi = p\Phi^{\dfrac{1}{Z-\beta\dfrac{\Delta G^{\#}}{RT}}} \tag{2.137}$$

where $Z = D^* - d + \kappa$.

A generic triglyceride mixture, refined and fractionated palm oil was used in this study. The fatty acid composition of the triglyceride mixture was 47.5% palmitic (16:0), 5.3% stearic (18:0), 37.6% oleic (18:1), and 7.5% linoleic (18:2) acids. The remainder of the mass was made up of small amounts of 12:0, 14:0, 18:3, and 20:0 acids. The softening point (melting range) of the sample was 25°C–28°C while the iodine value was in the range 43–48. Fats were heated at 80°C for 30 min to erase crystal memory and blended with hot colza oil (a diluent) in 10% (w/w) increments. Palm-colza oil blends were heated at 80°C for 30 min to erase crystal memory. All glassware was preheated to 80°C to avoid crystallization of the melted fats during sampling. A small droplet ($\sim$10 µL) of the melted fat was placed on a microscope glass slide (25 mm × 75 mm × 1 mm). A cover slip (22 mm × 22 mm) was then placed parallel to the plane of the slide and centered on the drop of sample to ensure that the sample thickness was uniform ($\sim$20 µm). Four slides of each sample were stored at 20°C at room temperature for 24 h. Samples were then observed under polarized light using an Nikon Eclipse 600 light microscope equipped with a 20× plan Achromat Ph1 objective lens. A neutral color balance filter was place in the light path and images were acquired using a Nikon Coolpix 950 digital camera at room temperature.

Images were processed using Adobe Photoshop 5.5 (Adobe Systems Inc., San Jose, CA). Original color images were first converted to 8 bit grayscale mode and their contrast and intensity adjusted using the "auto levels" function in Photoshop. Spherulites micrographs destined for further analysis were then thresholded using the standard "threshold" function in Photoshop. The mass fractal dimension (D) was estimated using the IP*Measure→Fractal Dimension plug-in filter for Photoshop (Reindeer Graphics, Image Processing Toolkit 5.0, http://www.reindeergraphics.com). This erosion-dilation measure (EDM) fractal dimension relates to the boundary irregularity or roughness of the object (Russ, 2002). Spherulite size was measured manually using Image J 1.34n (Wayne Rasband, National Institute of Health, http://rsb.info.nih.gov/ij/) and the average and standard deviation determined. This size corresponds to the average equatorial diameter of the spherulite. At least five micrographs were taken from each of the four different slides and between 5 and 50 features were used in the analysis. Palm-colza oil blends were heated at 80°C for 30 min to erase crystal memory.

X-ray glass capillary tubes (1.5 mm internal diameter) were preheated to 80°C to avoid crystallization of the melted fats during sample preparation and filled with the melt. Samples were allowed to crystallize at room temperature for 24 h. Powder x-ray diffraction (XRD) measurements were conducted using an in-house diffractometer with a Cu source (λ = 1.54 Å) operated at 40 kV and 20 mA, and position-sensitive detectors. Scattering intensity is shown as function of the scattering vector q, where $q = \dfrac{4\pi}{\lambda}\sin(\theta) = \dfrac{2\pi}{d}$, d (Å) corresponds to the interplanar distance, λ corresponds to the x-ray wavelength, and 2θ is the diffraction Bragg angle.

Most of the spherulites encountered in this study belong to category 1, that is, spherulites that grow radially from the nucleation site, branching intermittently to maintain a space-filling character. The sizes of triglyceride spherulites encountered in this study are in the range 10–100 μm, similar to that of polymer spherulites. Changes in spherulite size as a function of the mass fraction of the solid triglyceride fraction in liquid oil are shown in Figure 2.11a through d. Spherulite size increases as the mass fraction of solids decreases due to a reduction in nucleation events in favor of crystal growth. Powder x-ray diffraction studies on the different blends of solid triglyceride with liquid oil at small angles (Figure 2.11e) and wide angles (Figure 2.11f) indicate that the same polymorphic form is present in all samples. This demonstrates that changes in spherulite size as a function of dilution are not due to changes in the polymorphism of the material, but due to changes in the driving force, that is, the chemical potential, for crystallization. This is an important "check" when dealing with complex multicomponent mixtures of different molecular species.

Analysis of changes in spherulite diameter (ξ) as function of mass fraction of solids (Φ) revealed a power law dependence of the spherulite diameter to the solids' mass fraction (Figure 2.12c). Moreover, this power law dependence could be modeled using the classic fractal scaling relationship in two-dimensional space, $\xi \sim \Phi^{1/(D-2)}$, yielding a fractal dimension of 1.71. In order to confirm the validity of this fractal dimension value, we carried out image analysis on typical spherulites and obtained a fractal dimension of 1.73 (Figure 2.12a and b). The agreement between these two methods was encouraging and strongly supports the argument for a fractal distribution of mass within the interior of the spherulites. We then proceeded to fit the model (Equation 2.138) to the data (Figure 2.12d), with encouraging results—the fit was acceptable and the parameter estimates were reasonable and in the range expected for this particular solid triglyceride mixture. In this analysis, the free energy of nucleation was fixed as a constant.

In conclusion, here we show that triglyceride spherulites can be considered as polycrystalline fractal clusters, and that the size of these spherulites can be, in principle, predicted from knowledge of their crystallization behavior. Since the mechanical properties of fats are a function of the size of such polycrystalline clusters ($K = k_{L}L/\xi$, where K is the force constant of the material, L is the size of the system in the direction of the applied stress, ξ is the diameter of the fractal clusters, and k_{L} is the force constant of the links between fractal clusters), these findings provide support for the use of fractal structural-mechanical models to describe the rheological properties of plastic fats.

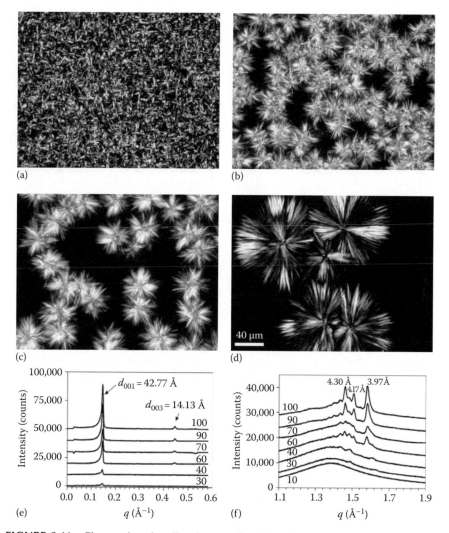

FIGURE 2.11   Changes in spherulite size as a function of the mass fraction of solids for 90%–100% solid triglyceride (a), 70%–80% solid triglyceride (b), 60% solid triglyceride (c), and 50% solid triglyceride (d). Powder x-ray diffraction spectra at small angles (e) and wide angles (f) for the different proportions of the solid triglyceride in oil.

2.3.3   FRACTAL GROWTH OF MILK FAT CRYSTALS IS UNAFFECTED BY MICROSTRUCTURAL CONFINEMENT

Fractal objects are best known as self-similar patterns that cannot be explained by classical geometry. Fractals are scale invariant, meaning that an object will look statistically similar at different length scales. For example, if a piece of a fractal object is cut out and magnified, the resulting object will, on average, look the same as the original one. This, so-called self-similarity is characteristic of fractal objects. To determine the fractal dimension of a growing object, it is necessary to measure

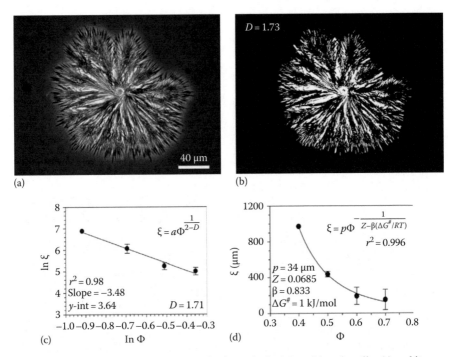

FIGURE 2.12 Phase contrast micrograph of a typical triglyceride spherulite (a) and its corresponding thresholded version (b). Changes in spherulite diameter (ξ) as a function of the mass fraction of solid triglyceride (Φ) obey a power-law relationship (c). Changes in spherulite diameter as a function of the mass fraction of solid triglyceride can be accurately predicted using the model developed in this study (d). Indicated are the goodness of fit (r^2) and model parameter estimates.

the variation in the object's volume (V) with its linear size (R), $V(R) \sim R^D$. Here, D is typically a noninteger number called the fractal dimension, which is smaller than the Euclidean dimension (d) of the space the fractal is embedded in, $D < d$. If the space is completely or homogeneously filled, then $D = d$. The volume term can be replaced with mass of that cluster (M), assuming, as it is usual in the field of mass fractals, that $M \sim V$ (Vicsek, 1992).

If a crystal cluster grows in a fractal fashion, then M will be proportional to time (t), that is, $M \sim t$. This linear mass accumulation arises if the amount of mass accrued in time is proportional to the number of random steps, $j(t)$, performed by the diffusing molecules and thus $j(t)$ is directly proportional to t (Rothschild, 1998). For fractal objects, cluster mass scales to cluster diameter (ξ) as $M \sim \xi^D$. Consequently, the diameter of the growing cluster scales in time as $\xi \sim t^{1/D}$. Thus, if an object obeys the linear mass deposition condition, and its growth is fractal in nature, a log-log plot of ξ versus t will be linear with a slope equal to $1/D$. For the case where the growth of the cluster is Euclidean, the fractal dimension would equal d, and its value would be 2 or 3 depending on the type of crystal growth mechanism (two- or three-dimensional). The slope of the log-log plot of ξ versus t for these two cases would then be 1/2 or 1/3.

For example, Rothschild (1998) studied the electrodeposition of silver. The radius (r) of the growing deposits was measured over time (t) and plotted on a log-log scale. The fractal dimension for silver agglomeration was then calculated from the inverse of the slope. The results of this graph method were in agreement with the theoretical value for the fractal dimension for silver agglomeration of $D = 2.53$.

Crystallization of materials under spatial confinement has been found to change its bulk properties. For example, materials confined to nanoscale dimensions display properties that differ from the corresponding unconfined case due to their reduced dimensionality and large interfacial effects (Alba Simionesco et al., 2003). Phase transition pressures and temperatures are shifted from their bulk values and new phases can appear due to surface forces. A study on polymer confinement found that crystallinity was greatly reduced and crystallization rate was much slower under confinement (Weimann et al., 1999). Moreover, heterogeneous nucleation on the surface governed the crystallization process.

The impetus for this study comes from the recent experimentally obtained relationship between the nucleation rate (J) and the box-counting fractal dimension (D) of cocoa butter crystal networks, as discussed earlier. The objectives of this case study were to determine whether the growth of fat crystals is fractal in nature, and whether the thickness of the crystallizing layer can affect the dimensionality of the growth process as well as final microstructure.

Anhydrous milk fat was crystallized under constant conditions using two thicknesses, 20 and 170 μm. The milk fat was heated to 80°C for 20 min before a drop was placed on a preheated (80°C) glass microscope slide and covered by preheated (80°C) glass coverslip. The slides were cooled under the microscope at 5°C/min to 21°C. Polarized micrographs were obtained for single milk fat crystal clusters and nucleation studies using a 40× objective lens and a 10× objective lens respectively on an Olympus BH microscope (Olympus, Tokyo, Japan). The microscope was equipped with a XC75 CCD camera and LG-3 capture board (Scion Corporation, Frederick, MD). Scion Image 1.62 was used to take PLM at specific time intervals, average each image over 16 frames, and subtract off the background. The images were manually thresholded using the same value for each image in that particular run. The threshold level was determined by the value that best distinguished the grayscale images from the background. Once thresholded, the images were inverted and the diameter was manually measured in ImageJ (Image processing and analysis in Java).

Phase transition analysis recorded the changing intensity of light in 2 s intervals of the crystallizing milk fat. The data storage began once the sample completed cooling from 60°C to 21°C at 5°C/min. The Automatic Petroleum Analyzer (Phase Technology, Richmond, Canada) was used for this purpose. SFC of the crystallizing milk fat was measured to verify the approximation of mass proportional to time. The SFC tubes were cooled with the equivalent cooling rate to milk fat on the glass slides from 80°C to 21°C, then measured every 30 s with pulsed nuclear magnetic resonance (pNMR) using a Bruker mq20 Series NMR Analyzer (Bruker, Milton, Ontario, Canada).

The first approximation to calculating the fractal growth of a milk fat crystal cluster assumes the mass of the growing cluster is proportional to the time ($M \sim t$). This was

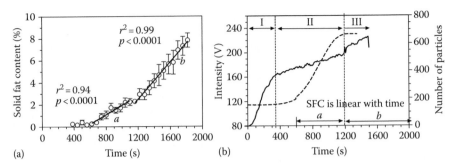

FIGURE 2.13 (a) Changes in solid fat content (SFC) with time for milk fat cooled to 21°C. (b) Changes in the number of particles (—) and scattered light intensity (---) over time for milk fat cooled to 21°C.

verified using pNMR of milk fat cooled under the same conditions as milk fat on a glass slide. The pNMR measurements, shown in Figure 2.13a, have a linear region between 600–1200 s (a) and 1200–1800 s (b) from when the milk fat was cooled to 21°C.

These two linear regions suggest that mass deposition is directly proportional to time after 600 s. However, this linearity may have existed at earlier times, in the region where the pNMR spectrometer was not sensitive enough to detect small amount of solids in the sample. Figure 2.13b shows the existence of three major growth regions. The first region (I) corresponds to the time period where nucleation took place. No SFC was detected in this region due to the low concentration of solids. The second region (II) shows a slower increase in the number of nuclei. Here, the energy in the system was mostly devoted to crystal growth. It can be noted that the time dependence of the SFC in Figure 2.13a and the time dependence of the scattered light intensity (Figure 2.13b) are almost identical. Thus, both these methods are more sensitive to crystal growth than nucleation events. The small but abrupt increase in the number of particles at the boundary of regions II and III corresponds to the time at which the transition between regions a and b took place (Figure 2.13a). This jump in the number of particles may have been caused by a polymorphic transformation in the material, or due to the crystallization of a particular fraction with different chemical composition, that is, a different phase. The scattered light intensity signal becomes saturated in region III; however, crystal growth is still occurring at this point, as shown in region b of the SFC curve. Mass deposition in this region is also linear in time. The sudden decrease in the number of particles observed in region III is due to the crystal impingement (e.g., they touch), where their merging leads to an effective reduction in the number of identifiable crystals in the sample. Figure 2.13 illustrates that the mass of the growing crystal clusters is proportional to crystallization time after 600 s at 21°C. The linear region used for calculating the slope in the following analyses went from 600 to 1200 s, representing region a of the SFC curve.

Polarized light micrographs of milk fat clusters crystallized at 20°C in 20 and 170 μm sample thicknesses are shown in Figure 2.14.

The 20 μm thick sample (Figure 2.14a) is more defined and more sharply focused than the 170 μm sample (Figure 2.14b). The thick sample appeared hazy due to the extra mass in the z-axis of the specimen. Material above and below the focal plane

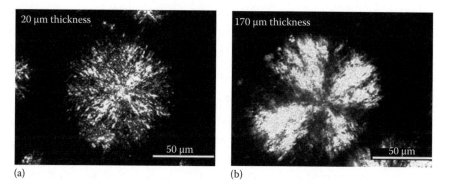

(a) (b)

FIGURE 2.14 Polarized light micrographs of 20 μm (a) and 170 μm (b) thick samples of milk fat.

causes haze due to the scattering of light from crystallites. The size of the crystal clusters was generally similar for thick and thin samples; however, there was much more movement and agglomeration processes taking place in the 170 μm samples.

The log-log plot of diameter of the cluster versus time for the thick and thin samples was linear in time from 170 s after reaching 21°C onward (Figure 2.15).

The slope of each line was calculated using linear regression in the region of time from 600 to 1200 s, since this satisfies the linear mass deposition condition $M \sim t$ (for all linear regressions, $0.96 < r^2 < 0.99$). However, the entire range was linear. The values obtained using the entire range were not significantly different from the 600–1200 s range. The dimensionality of growth is fractal in each graph, suggesting that milk fat crystals grow in a fractal fashion. The average fractal dimensions for the 20 and 170 μm thick samples were 1.49 ± 0.019 and 1.40 ± 0.106, respectively. There was more variability in the 170 μm samples, which was expected because of material movement and haze in the images. There were, however, no statistically significant differences between the fractal dimensions obtained at the two thickness ($p > 0.05$). The average fractal dimension was $D = 1.44$. Moreover, microstructural confinement did not affect the growth fractal dimension for milk fat crystal clusters.

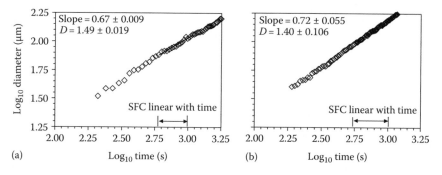

(a) (b)

FIGURE 2.15 Log–log plot of diameter of the cluster versus time for 20 μm (a) and a 170 μm (b) samples of milk fat. Values reported correspond to the mean of standard errors of three replicates.

To conclude, we have shown that milk fat crystal clusters grow in a fractal fashion, which was unaffected by microstructural confinement in the range 20–170 µm. This includes the usual range of sample thicknesses used in microscopy studies of crystallizing materials.

2.3.4 COMPARISON OF EXPERIMENTAL TECHNIQUES USED IN LIPID CRYSTALLIZATION STUDIES

The first step in the determination of the free energy of nucleation is the determination of the induction time of nucleation (τ). In order to determine the free energy of nucleation using standard models, the fat melt has to be cooled instantaneously to the specified crystallization temperature. If the fat starts crystallizing while still cooling, we are under dynamic crystallization condition (in contrast to static, isothermal conditions) and the standard Fisher–Turnbull treatment is not applicable. An important point to be made is whichever technique is used for the determination of the induction time (light scattering, NMR, turbidimetry, microscopy), it must be sensitive to the early stages of the crystallization process and hopefully as close as possible to the nucleation event. What follows is a short study aimed at addressing the appropriateness of commonly used analytical techniques in the determination of the induction time of nucleation taken from the work of Wright et al. (2000b).

Crystallization of fats is considered to encompass two distinct events: nucleation and crystal growth. While a stable nucleus must be formed before crystal growth can occur, these events are not mutually exclusive. Nucleation may take place while crystals grow on existing nuclei. In our investigations into the effects of minor components on milk fat crystallization, it became clear that minor components delayed crystallization of milk fat triacylglycerols. However, it was difficult to discern whether the effects were at the nucleation or crystal growth level. Distinguishing between nucleation and crystal growth constitutes a major challenge in lipid crystallization studies.

The shape of a crystallization curve can provide some insight into the mode of crystal growth. However, the nucleation step is more elusive because the methods typically used in these studies are relatively insensitive. pNMR, which measures SFC, and light scattering techniques, which measure absorbance or transmittance of light, are commonly used to monitor lipid crystallization. Anyone familiar with the pNMR method knows that, at times, small amounts of crystals are visible in the melt before any solids are detected. Clearly, at this stage, well beyond the induction time for nucleation, the pNMR signal is measuring crystal growth. Turbidimetry, while more sensitive than pNMR, for example, shorter induction times are obtained, also has its limitations. We found a very strong correlation between induction times determined by pNMR and turbidimetry for the three fat systems used in this study: milk fat, milk fat TAGs, and milk fat TAGs with 0.1% added milk fat–derived diacylglycerols. This suggested that increases in turbidity are also due to mass deposition of crystals and not only nucleation (Figure 2.16). However, here we can also appreciate how the slope of this relationship varies from system to system, thus suggesting that although proportional to each other, induction times of crystallization are not measuring exactly the same event.

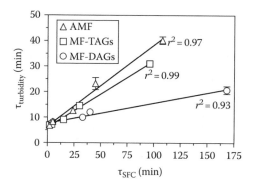

FIGURE 2.16 Induction time by turbidity ($\tau_{Turbidity}$) versus induction time by SFC (τ_{SFC}) for anhydrous milk fat (AMF), milk fat triacylglycerols (MF-TAGs), and MF-TAGs with 0.1% diacylglycerols added back (MF-DAGs) crystallized at 5°C, 10°C, 15°C, 20°C, 22.5°C, 25°C, and 27.5°C.

It would be beneficial to have a convenient way of unambiguously determining nucleation induction times when seeking to understand the effects of varying composition and processing conditions on nucleation, and it is essential if the induction times are used as a measure of the rate of nucleation for determination of activation energies of nucleation.

Induction times determined by pNMR, turbidity, and light scattering measurements were compared to those determined using PLM in conjunction with image analysis. Isothermal DSC was attempted as a fifth method for comparison. However, because of the inherent lack of sensitivity at the high cooling rates required to obtain isothermal crystallization conditions, it was abandoned.

For the PLM method at 22.5°C, AMF, MF-TAGs, and MF-DAGs were preheated to 80°C for 10 min before a drop of each was placed on a preheated (80°C) glass microscope slide and covered with a preheated (80°C) glass coverslip. The samples were imaged with a Zeiss polarized light microscope using a 10× objective and equipped with a CCD video camera. Temperature of the slide was maintained at 22.5°C. Crystallization was followed by capturing an image every 15 s for 30 min. The images were processed using Image Tool (The University of Texas Health Science Center, San Antonio, TX). A background subtraction was performed initially by subtracting the initial image (time = 0 s) for each of AMF, MF-TAGs, and MF-DAGs from every other image in the respective crystallization run. The images were manually thresholded, using the same value for every image in each of AMF, MF-TAGs, and MF-DAGs. The threshold level was that which was found to most accurately reflect the original grayscale images. Once the images were thresholded, the relative amounts of black and white pixels in each image were determined. The amount of black (representing crystal mass) was plotted as a function of crystallization time.

For the light scattering studies, a phase transition analyzer (Phase Technology, Richmond, British Columbia, Canada) was used. A 150 μL sample, preheated to 80°C for 30 min, was pipetted into the sample container of the analyzer, which was preheated and maintained at 75°C using a thermoelectric cooler. Thereafter, the

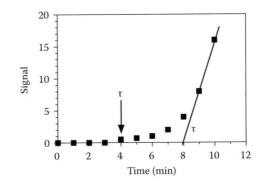

FIGURE 2.17 Determination of crystallization induction times by baseline deviation and linear extrapolation to the time axis.

sample was rapidly cooled from 75°C to 22.5°C at a controlled rate of 50°C/min. When the sample reached 22.5°C, it was held at this temperature and crystallization was continuously monitored using an optical scattering approach. In this setup, a beam of light impinges on the sample from above. A matrix of optical sensors, in tandem with a lens system, is also placed perpendicularly above the sample. When crystals start to appear in the sample, the incident beam is scattered by the solid–liquid phase boundaries and scattered light impinges via the lens onto the detectors. As more and more crystal mass develops, the signal output increases and is automatically recorded.

Crystallization curves for pNMR, turbidity, light scattering, and image analysis were normalized by dividing each value by the maximum crystallization value. The resulting fractional crystallization values were compared. Induction times were determined by extrapolating the linear portion of the crystallization curves to the time axis and by baseline deviation as shown in Figure 2.17.

Induction times were taken as the time from when the samples were placed at 22.5°C until crystallization began. Strictly speaking, however, the induction time should be time the melt spent in the metastable region, below its melting temperature. The induction time should be determined as the time from the moment that the sample reaches the set crystallization temperature to that time where the first crystal nuclei appear. In all experiments, however, it takes some time for isothermal conditions to be established. It is possible that differences in the cooling rates between the four methods could influence crystallization. This will be true regardless of how induction time is defined. Considering time zero as the point at which 22.5°C is reached ignores the fact that crystallization will occur between the melting temperature and 22.5°C. AMF, MF-TAG, and MF-DAG have Mettler dropping points of approximately 34°C. Therefore, during cooling below 34°C and at 22.5°C, the fats experience the same degree of supercooling. This eliminates concerns of having different thermodynamic factors at work in the three fats and makes the comparison of their crystallization behaviors much easier to define. However, there are very real concerns regarding differences in cooling rates between the different experimental methods. The cooling curves for the pNMR, turbidity, and light scattering are shown in Figure 2.18. A cooling curve was not determined for the microscopy experiment,

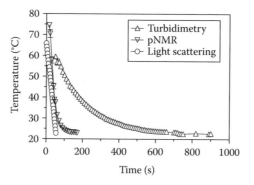

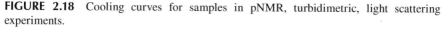

FIGURE 2.18 Cooling curves for samples in pNMR, turbidimetric, light scattering experiments.

TABLE 2.3
Rate Constants of Cooling Determined from the Initial Linear Decrease in Temperature to 34°C for pNMR, Turbidimetry, and Light Scattering Spectroscopy

Experimental Method	Cooling Rate Constant (°C/s)
pNMR	−1.060 ± 0.119
Turbidimetry	−0.156 ± 0.008
Light scattering	−0.774 ± 0.000

although the temperature was found to equilibrate at 22.5°C within approximately 30 s. The rate constants of cooling for the initial decrease in temperature to the dropping point (34°C) are shown in Table 2.3.

Figure 2.17 and Table 2.3 show that the cooling for the pNMR and light scattering are reasonably rapid and linear to 22.5°C. Similarly, the microscope slides reached the crystallization temperature very rapidly. This makes it easier to assign any differences observed between these experimental methods to the sensitivity of the methods, and not specifically to different cooling rates. Newtonian-type cooling was observed for the turbidity experiments. This meant that 22.5°C was not reached for nearly 15 min. This is longer than most of the induction times determined for turbidimetry (Table 2.4). Therefore, these induction times actually correspond to crystallization at temperatures higher than 22.5°C and we must be careful in drawing information from the results. In the cases where the crystallization temperature was attained very quickly and within approximately the same time frame (pNMR, light scattering, and microscopy), the comparison between the methods is more appropriate. Every effort should be made to achieve rapid and linear cooling during crystallization experiments to avoid such complications. In the case of turbidimetry, this resulted because of the poor heat transfer between the sample and the surrounding metal cell holder through which cooling water was circulated. More efficient heat transfer was achieved when the glass pNMR tubes were plunged into a water bath at

TABLE 2.4

Induction Times (Minutes) Determined by Linear Extrapolation and Baseline Deviation for AMF, MF-TAGs, and MF-DAGs Monitored by pNMR, Turbidimetry, Light Scattering Spectroscopy, and Polarized Light Microscopy Coupled to Image Analysis

	pNMR	Turbidimetry	Light Scattering	Microscopy
By extrapolation				
AMF	28.2 (4.2%)	14.9 (8.4%)	9.9 (0.5%)	3.0 (3.8%)
MF-TAGs	14.7 (6.2%)	12.3 (11.8%)	6.7 (1.1%)	1.5 (3.0%)
MF-DAGs	33.3 (4.9%)	12.9 (11.0%)	7.1 (0.7%)	5.3 (1.8%)
Baseline deviation				
AMF	21.7 (4.1%)	16.3 (8.1%)	7.5	3.0
MF-TAGs	11.0 (9.1%)	11.8 (3.7%)	5.0 (10.0%)	1.5
MF-DAGs	34.0 (5.9%)	12.7 (9.5%)	6.5	3.5

the crystallization temperature, in the chamber of the phase transition analyzer, and when the glass microscope slides were placed on a metal platform.

Figure 2.19 shows thresholded polarized light micrographs of MF-TAGs at various crystallization times at 22.5°C. Crystallization curves for AMF, MF-TAGs, and MF-DAGs by pNMR, turbidity, and PLM image analysis are shown in Figure 2.20.

MF-TAGs crystallized first, followed by AMF and MF-DAGs. MF-DAGs had the longest induction times determined by pNMR, while by turbidimetry and microscopy, AMF had the longest induction times. Crystallization curves for AMF, MF-TAGs, and MF-DAGs obtained from measurements of light scattering intensities are shown in Figure 2.21. In this case, MF-TAGs crystallized first and AMF had the longest induction times. The induction times for the crystallization curves are reported in Table 2.4.

Induction times determined as the time of deviation from the baseline were shorter than those calculated by extrapolation of the linearly increasing curves, although the same trends were observed. Also, Table 2.4 shows that while the relative trends were similar, there were large differences in the absolute value of onset times of crystallization between the four methods. Despite the fact that the absolute values for the induction times differed, removal of the minor components consistently decreased the induction times of the triacylglycerols. We can thus be certain that milk fat minor components exhibit an inhibitory effect on triacylglycerol crystallization.

Table 2.4 shows that the induction times determined by pNMR were the longest, while those determined by the image analysis technique were the shortest. Therefore, with the image analysis approach, we were able to detect some early crystallization events beyond the sensitivity of the other methods. The higher sensitivity demonstrated allowed for the detection of early crystallization events, possibly in the vicinity of the true nucleation events. If cooling rates could be controlled in a better fashion, turbidimetry could be a more sensitive technique for the study of

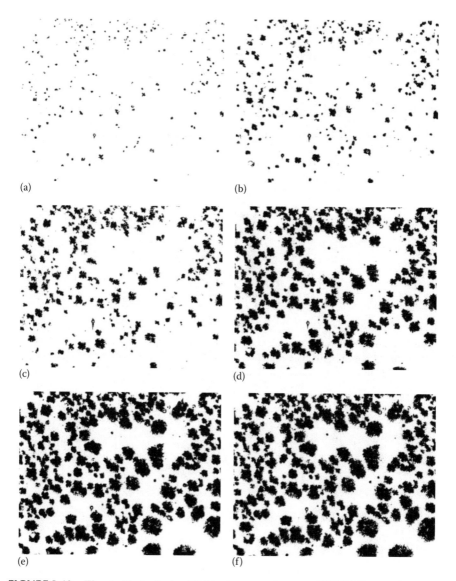

FIGURE 2.19 Thresholded polarized light microscope images of MF-TAGs at various crystallization times at 22.5°C: (a) 1 min, (b) 3 min, (c) 5 min, (d) 10 min, (e) 20 min, and (f) 28 min.

the early stages of a crystallization process than pNMR. Poor heat transfer was the major disadvantage with the turbidimetric experiments, although there are other inherent limitations. Light scattering spectroscopy proved to be a more sensitive method than turbidimetry and pNMR. In this method, the intensity of scattered light, rather than the attenuation of the signal intensity (I/I_0) is measured. The particular geometry of the sample cell and positioning of the detectors also maximizes the collection of the scattered light. This technique proved to be very convenient, user-friendly, and reproducible.

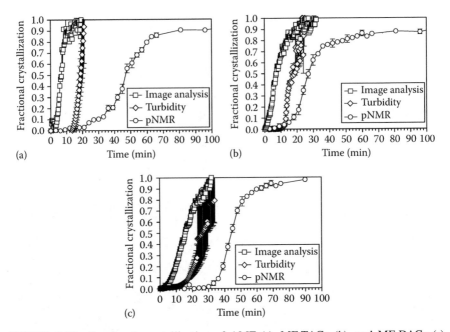

FIGURE 2.20 Fractional crystallization of AMF (a), MF-TAGs (b), and MF-DAGs (c) determined by pulsed NMR measurements of solid fat content, turbidity measurements, and polarized light microscopy coupled to image analysis at 22.5°C. Symbols in (a) and (b) represent the average and standard errors of three replicates.

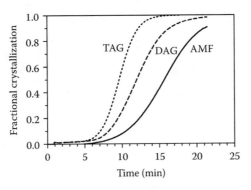

FIGURE 2.21 Fractional crystallization of anhydrous milk fat (AMF), milk fat triglycerides (TAG), and milk fat diglycerides with 0.1% added milk fat diacylglycerols (DAG), as determined by light scattering spectroscopy.

PLM has inherent advantages over turbidimetry. The PLM technique exploits the difference in refractive index of a beam of incident light polarized in two perpendicular directions. This phenomenon is known as birefringence. Anisotropic materials, such as a fat crystal, will display birefringence. Since fat crystals are birefringent, they will appear as sharp bright objects in a nonbirefringent, and therefore dark, background. The use of polarizers set at 90° removes most of the nonbirefringent background signal (colored melt and scattering impurities), thereby considerably increasing the signal-to-noise ratio. As well, since all of the transmitted light beam in the field of view is collected by the lenses and focused on the camera, signal intensity and therefore sensitivity are increased.

Ultimately, the experimental technique of choice will depend on the application. pNMR provides the best method to characterize the overall crystallization process. For this reason, it is suitable for use in the Avrami analysis. The water bath–based cooling used in the pNMR experiments also offered rapid cooling and accurate temperature control. Both turbidity and scattering intensity signals tend to become saturated prior to the completion of the crystallization process. Thus, it is not possible to obtain reliable data on the latter stages of crystallization. Although turbidity seems to offer the advantage of improved sensitivity, in our experience, there can be large errors associated with its measurement, concerns with poor reproducibility, and major challenges with temperature control. Light scattering improves on this because it measures reflectance of light as opposed to transmittance. It offers extreme sensitivity to early crystallization events, is easy to use, and requires only a small volume of sample. Cooling rates and temperature can also be accurately controlled in the instrument. Microscopy coupled with image analysis also proved to be sensitive and had good temperature control, although it was the most cumbersome technique. It does have the advantage that morphological information can be acquired simultaneously as the kinetics are quantified.

In many studies of fat crystallization, researchers have equated increases in the volume of crystallized material (V) to increases in absorbance (A) due to scattering at a particular wavelength, and thus

$$\frac{V - V_0}{V_m - V_0} = \frac{A - A_0}{A_m - A_0} \tag{2.138}$$

The first problem with this assumption is that A_m is usually taken as the off-scale turbidity value. This maximum absorbance does not correspond to the end of crystallization, or the maximum volume or mass of crystallized material achieved. It simply represents the point at which the crystallizing material becomes too opaque, and the amount of transmitted light becomes negligible. However, the crystallization process continues well after the turbidity values have gone off scale. Without a true maximum absorbance value, which corresponds to the end of the crystallization process, the Avrami equation cannot be used.

Secondly, the relationship between the amount of absorbed light due to scattering (A_s) and particle concentration is very complex (Campbell and Dwek, 1984). Consider the relationship between A_s and concentration:

$$A_s = -\ln\frac{I}{I_0} = \tau c l \tag{2.139}$$

In the Rayleigh regime, no multiple scattering (from the same particle) occurs, that is, particles in a dilute medium are smaller than $\sim\lambda/10$ and are considered isotropic point scatterers. As well, as for all scattering experiments, it is assumed that no absorption takes place. For this case

$$\tau = \frac{16\pi}{3} R_{90^\circ} \qquad (2.140)$$

where
   τ is a turbidity parameter analogous to an extinction coefficient
   R_{90° is the Rayleigh ratio at 90° and refers to primary scattering from unit volume
      of solution

For Rayleigh scattering,

$$R_{90^\circ} = Kc\overline{M}_w \qquad (2.141)$$

where
   K is a constant
   c is concentration (grams per unit volume)
   $\overline{M}_w$ is the weight-average molecular weight of the particle (grams per mole)

Substituting Equations 2.121 and 2.122 into Equation 2.120 yields the expression

$$A_s = \frac{16\pi}{3} Kc^2 \overline{M}_w \qquad (2.142)$$

Thus, A_s is not only a function of the amount of material present in the suspension, but also of its size. During crystallization, the number and the size of particles increase in time, making the equivalence of A_s with concentration somewhat dubious. Moreover, the absorbance due to turbidity increases as the square of the concentration of scattering material.

As particles become larger than $\lambda/10$, these assumptions do not apply, and we enter the MIE scattering regime, which complicates matters further. In this regime, a light wave can scatter from several points within the same particle and constructive/destructive interference from the interaction between those scattered waves can take place. For example, if we use 600 nm light, the limit of applicability of turbidity measurements would be for particles 60 nm or smaller, which is very close to the onset of nucleation.

Thirdly, an observed decrease in transmitted light could be due to light refraction, an apparent change in velocity of the transmitted light beam as it travels through the sample. Fat crystals are extremely birefringent, and light is transmitted through them. This could cause significant refraction of the incident light and lead to a drop in the intensity of the transmitted light. This increase in absorbance would then have nothing to do with the volume or mass of crystals present in the sample.

For these reasons, turbidity measurements should not be used in the quantitative kinetic characterization of crystallization processes, other than for the detection of induction times.

2.4 NONISOTHERMAL NUCLEATION OF FATS

2.4.1 ISOTHERMAL, NEAR-ISOTHERMAL, AND NONISOTHERMAL PROCESSES

In the industrial manufacture of food products, crystallization usually takes place under nonisothermal conditions, where temperature is changing as the material crystallizes in time. However, unlike for the case of isothermal nucleation and crystallization discussed earlier, very few theoretical tools exist to model the nucleation behavior of these complex organic mixtures under nonisothermal conditions.

Classic nucleation theory mainly addresses isothermal processes, where the temperature drops from the melting temperature to a set crystallization temperature is instantaneous (Figure 2.22a). These theories and models address situations where the temperature remains constant during the reaction and the temperature drop from the melting temperature, T_m, to the set crystallization temperature, T_{set}, is instantaneous (Figure 2.22a). As well, it is assumed that crystallization occurs only when T_{set} is reached and not prior to this. For the isothermal case, time zero is assumed to be the start of the experiment as the temperature at the beginning of the experiment is assumed to be T_{set}. This model is suited to the study of systems that are not heat-transfer-limited. Under isothermal conditions, the crystallization process can be characterized by an induction time (t_i), which is the time required for the appearance of the first solid nuclei at T_{set} under the influence of a thermal driving force. The induction time t_i is proportional to the degree of supersaturation (solutions), or degree of supercooling or undercooling (melts) (ΔT), which is the difference between the equilibrium melting temperature of the material (T_m) and the set crystallization temperature (T_{set}), that is, $T_m - T_{set}$.

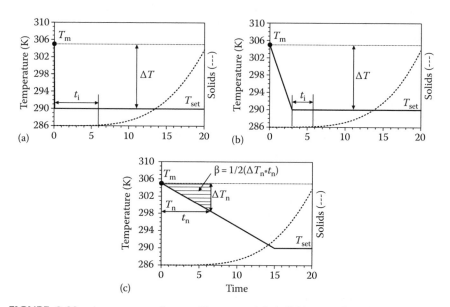

FIGURE 2.22 Appearance of crystalline material (solids) as a function of time for isothermal (a), near-isothermal (b), and non-isothermal (c) crystallization processes.

Experimental realities limit the speed at which a system can reach the set temperature. Limitations in heat transfer will result in a gradual reduction in temperature (Figure 2.22b) as opposed to an instantaneous drop observed in Figure 2.22a. These conditions can be considered "near-isothermal." It is assumed, in this case, that crystallization does not begin until after some time the set crystallization temperature has been reached. For a "near-isothermal" case, then, it is still possible to determine an induction time of nucleation and treat the crystallization process as if it were taking place under isothermal conditions. For this case, it is important to remember that time zero corresponds to the time when the system reaches T_{set}. The induction in this case is the time interval between the attainment of the set crystallization temperature and the time of the first appearance of solid nuclei.

In the industrial manufacture of crystalline products, crystallization takes place under nonisothermal conditions. Under these conditions, crystallization occurs prior to attaining the set crystallization temperature (T_{set}) (Figure 2.22c). For this situation, both the nonisothermal induction time of nucleation (t_n) and the undercooling at nucleation (ΔT_n) have different meanings than for the isothermal case. However, very few practical theoretical tools are available to model the nucleation behavior of these complex organic mixtures under nonisothermal conditions. Interested readers are advised to read the review by Sangwal (2012) on the current status of theory, as well work by Rybin (2003), McGraw and Hu (2003), Wyslouzil and Seinfeld (1992), and Kurasov (1999). The chapter on nucleation at variable supersaturation in Kashchiev's book is particularly enlightening (Kashchiev, 2000). In this section, however, a new approach to the modeling of non-isothermal nucleation of complex multicomponent molecular mixtures of triacylglycerols will be presented (Marangoni et al., 2005, 2006). Even though this approach is phenomenological in nature, it may represent a practical way to tackle the problem. The first challenge is the definition of the time-dependent exposure of the system to supercooling.

2.4.2 FORMULATION OF THE TIME-DEPENDENT SUPERCOOLING PARAMETER

To model nonisothermal nucleation, a new quantity or parameter that characterizes the driving force for nucleation has to be defined. This parameter will be shown to embody the dynamic of the system. It was previously mentioned that the induction time of nucleation and undercooling at nucleation have different meanings for the isothermal versus nonisothermal cases. The first step, therefore, in the formulation of a nonisothermal model, is the redefinition of these parameters. Upon examination of Figure 2.22c, one can notice that it is not the temperature differential, ΔT, that is the driving force for nucleation. Instead, it is the time the system has been exposed to a particular temperature differential. In other words, the supercooling of the system is a dynamic quantity as opposed to being a static quantity as for the isothermal case: ΔT is changing in time as the material crystallizes and changes composition. Thus, a supercooling-time exposure has to be defined. This corresponds to the exposure of the system to supercooling until the initiation of nucleation. This is calculated as the area under the supercooling-time trajectory from the time when the system crosses the melting temperature (T_m) to the time where the first crystal nuclei appear (t_n). The temperature at which the first crystal nuclei appears is called the nucleation temperature (T_n). Notice that $T_n \neq T_{set}$ and therefore, $[\Delta T = (T_m - T_{set})] \neq [\Delta T_n = (T_m - T_n)]$. If the cooling rate is assumed to be constant, the dynamic supercooling-time exposure (β) at the onset of nucleation can thus be defined as

$$\beta = \frac{1}{2} \Delta T_n t_n \tag{2.143}$$

Another important parameter in the characterization of a nonisothermal system is the cooling rate. A linear cooling rate is defined as

$$\phi = \frac{\Delta T}{\Delta t} \tag{2.144}$$

As previously discussed, at the crystallization temperature T_c, $t = t_n$. At the melting temperature T_m, $t_0 = 0$. Substituting these into Equation 2.2 gives

$$\phi = \frac{(T_m - T_n)}{t_n - 0} = \frac{\Delta T_n}{t_n} \tag{2.145}$$

Substituting $\Delta T_n / \phi$ for t_n in the expression for the supercooling-time exposure at nucleation (Equation 2.1) gives the following expression:

$$\beta = \frac{1}{2} \frac{(\Delta T_n)^2}{\phi} \tag{2.146}$$

The parameter β corresponds to the triangular area under the curve for the supercooling-time curve (Figure 2.22c). This parameter takes into consideration the amount of supercooling in time required for nucleation to start. It is important to realize that the parametrization of the data relative to temperature as well as time is necessary for a proper description of nucleation under nonisothermal conditions. The dynamic supercooling-time exposure β incorporates a thermodynamic component in the form of the supercooling at nucleation (ΔT_n) and a kinetic component in the form of the nonisothermal nucleation induction time (t_n).

Finally, the square root of β corresponds then to an effective supercooling experienced by the system at nucleation:

$$\sqrt{\beta} = \frac{\Delta T_n}{\sqrt{2\phi}} \tag{2.147}$$

This effective supercooling is a linear function of the chemical potential difference between the melt and the crystalline solid ($\Delta \mu$). This chemical potential difference is related to ΔT_n as $\Delta \mu_n = \frac{\Delta H_m}{T_m} \Delta T_n$, where ΔH_m is the enthalpy of melting of the crystalline material, T_m is the melting temperature of the crystalline material, while ΔT_n corresponds to the degree of undercooling at nucleation. Thus, the effective supercooling is related to chemical potential difference at nucleation as

$$\Delta\mu_n = \frac{\Delta H_m}{T_m}\sqrt{2\phi\beta} \qquad (2.148)$$

In this treatment, the relationship between the nucleation rate under nonisothermal conditions and the chemical potential difference at nucleation will be modeled in a statistical fashion using an exponential probability density function. This approach is in contrast to the Gibbs–Thompson approach, where an estimate of the interfacial energy of the crystal nucleus is required. The statistical approach presented here allows for the determination of a clustering energy of nucleation without the need to determine the crystal–melt interfacial energy, and only requires the determination of a supersaturation-related term, the dynamic supercooling-time exposure or effective supercooling at nucleation ($\sqrt{\beta}$). This statistical approach was chosen also due to the complexity of the systems studied—some natural fats contain over 300 different triglycerides, of which dozens may crystallize simultaneously and can form complex mixtures among themselves. Under these conditions, it is difficult obtain accurate estimates of either the chemical potential or the interfacial energy.

2.4.3   PROBABILISTIC APPROACH TO MODELING NONISOTHERMAL NUCLEATION KINETICS

Work from our group (Marangoni et al., 2005, 2006) has shown that the normalized nucleation rate (J/J_{max}) had an exponential dependence on $\sqrt{\beta}$, where J_{max} corresponds to the maximum nucleation rate. This observation raised the possibility of modeling nonisothermal nucleation kinetics statistically, in a similar fashion as for the kinetic theory of gases, using an exponential probability density function. In order to develop this argument logically, we must revisit the basic premises of kinetic theory.

The rate of a reaction (v) is a function of the concentration of molecules with sufficient energy to overcome an energy barrier to the particular reaction (N^*), and thus $v = k[N^*]$, where k is the rate constant for the reaction and N^* corresponds to the concentration of molecules in the activated state. In the kinetic theory of gases, the molecules in the activated state are those molecules with sufficient energy, and in the proper orientation, to undergo the chemical reaction. For the case of nucleation reactions, N^* would correspond to the concentration of molecules in the metastable state, just prior to the nucleation event. The proportion of molecules in the appropriate state to undergo a reaction (from energetic and conformational considerations) will be given by (N^*) = $p(x)(N_T)$, where N_T is the total concentration of reactant, and $p(x)$ corresponds to the probability density function (pdf), which describes the frequency distribution of the particular event.

Following this line of reasoning, the effective supercooling $\sqrt{\beta}$ was assumed to be distributed in an exponential fashion, with an exponential pdf, $p\left(\sqrt{\beta};k\right)$, of the form

$$p\left(\sqrt{\beta};k\right) = \begin{cases} ke^{-k\sqrt{\beta}}; \sqrt{\beta} \geq 0 \\ 0 \quad\quad ; \sqrt{\beta} < 0 \end{cases} \tag{2.149}$$

The parameter k is called the rate parameter. The rate parameter has to satisfy the condition $k > 0$. This pdf applies to values of the randomly distributed variable belonging to the set $\sqrt{\beta} \in [0;\infty)$. The scale parameter ($\mu$) is simply the inverse of the rate parameter and represents the mean, or expected value, of an exponentially distributed random variable, $E[\sqrt{\beta}] = \mu = 1/k$. Thus, this pdf is appropriate to model our situation where our random variable has to always be greater than zero, and the mean is fixed.

Exponential distributions are used to model memoryless Poisson, or stochastic, processes, which take place with constant probability per unit time or distance. This is the reason why exponential pdfs are extensively used to model random processes such as Brownian motion and diffusional processes. In the case of Brownian motion, the future position of a molecule is independent of its current position. In our case, we assume that our nucleation phase transition initiation event takes place with a constant probability per unit effective supercooling (β), possibly not unreasonable considering the constant cooling rates used.

Another interesting property of an exponential pdf is that among all continuous pdfs, with support $[0;\infty)$, the exponential pdf with $\mu = 1/k$ has the highest entropy. Many physical systems tend to move toward maximal entropy configurations over time (principle of maximum entropy).

Considering all of what was discussed earlier, the rate of the nucleation reaction (J) will thus be given by

$$J = k_p N_T k e^{-k\sqrt{\beta}} = J_{max} k e^{-k\sqrt{\beta}} \tag{2.150}$$

2.4.4 CLUSTERING ENERGY FOR NONISOTHERMAL NUCLEATION

By combining Equations 2.150 and 2.147, the relative nucleation rate can be expressed as an exponential function of the inverse of the square root of the cooling rate:

$$\frac{J}{J_{max}} = ke^{-k\frac{\Delta T_n}{\sqrt{2\phi}}} \tag{2.151}$$

Using this model, it is possible to determine the energy that needs to be lost from the system in order to initiate the nucleation process. Consider that from T_m to T_n, no phase change has taken place. Thus, up to this point, strictly specific heat Q_m has been removed from the system ($Q_m = C_p \Delta T$), where Q_m is the specific heat removed from the system upon cooling per gram of material, C_p is the specific heat [J/g] of the material. Substituting Q_m/C_p for ΔT in Equation 2.151 leads to the expression

$$\frac{J}{J_{max}} = ke^{-\frac{Q_m}{C_p\sqrt{2\phi}}} \tag{2.152}$$

where Q_m is proposed here to represent the specific energy [J/g] that needs to be lost by the system while travelling through the metastable region prior to the formation of the first stable nucleus. For the first stable nucleus to form, clusters have to reach, on average a size greater than a critical diameter, thus the name "clustering energy." Thus, from a nonlinear fit of J/J_{max} versus ϕ data, both the parameters k and Q_m can be determined. This quantity Q_m can then be multiplied by the average molecular weight of the triacylglycerols [MW, g/mol], to obtain the molar clustering energy of nucleation (Q_M) [J/mol].

2.4.5 SPECIAL CASE WHEN β IS VERY SMALL

As shown in Equation 2.150, the nonisothermal expression has the form

$$\frac{J}{J_{max}} = ke^{-k\sqrt{\beta}} \tag{2.153}$$

The first two terms of the Taylor expansion of this function are

$$\frac{J}{J_{max}} = ke^{-ka} - k^2 e^{-ka}\left(\sqrt{\beta} - a\right) \tag{2.154}$$

For the case where $a = 0$, this expression reduces to

$$\frac{J}{J_{max}} \approx k - k^2\sqrt{\beta} \tag{2.155}$$

Thus, for the case where the effective supercooling at nucleation $\left(\sqrt{\beta}\right)$ is in the vicinity of $a = 0$, the exponential pdf can be approximated by a linear function. This situation would apply to processes with very small metastable regions that nucleated at very small degrees of supersaturation, or undercooling ($\Delta T \approx 0$), and/or with very short induction times ($t_i \approx 0$).

In order to determine the nucleation clustering energy using this expression, we can use the same arguments as earlier. Replacing $\sqrt{\beta}$ with $\Delta T_n/\sqrt{2\phi}$ and ΔT_n with Q_m/C_p, we obtain

$$\frac{J}{J_{max}} = k - k^2\frac{\Delta T_n}{\sqrt{2\phi}} = k - k^2\frac{Q_m}{C_p\sqrt{2\phi}} \tag{2.156}$$

The slope of the J/J_{max} versus $\sqrt{\beta}$ plot corresponds to $k^2 Q_m/C_p$, while the y-intercept corresponds to k, and thus Q_m can be easily obtained.

2.4.6 Nonisothermal Nucleation of Five Commercial Fats—A Practical Example of This Approach

2.4.6.1 Materials and Methods Used

Fat samples: The fat samples used in this study included anhydrous milk fat (AMF), partially hydrogenated palm oil (PHPO), palm oil (PO), chemically interesterified and hydrogenated palm oil (IHPO), and a partially hydrogenated blend of palm oil and palm stearin (PH(PO/PS)). The fatty acid profile of these samples is shown in Table 2.5.

Chemical composition: The fatty acid composition was determined using gas chromatography. A column of 5 mm outer diameter, 3 mm internal diameter, and 1.5 m long was filled with 10% silar 9CP on chromosorb W, AW 80/100 mesh. This column was placed in a gas chromatography (GC) equipment (Shimadzu GC-8A, Kyoto, Japan). The chromatograph oven was set at 60°C and then a temperature ramp was programmed from 60°C to 210°C at 8°C/min. The detector and injector temperature was held at 230°C. Nitrogen was used as the carrier gas and both hydrogen and air were used to feed the flame ionization detector (FID). Before chromatographic analysis, the methyl esters of the TAG's fatty acids were generated; therefore, 50 mg of sample was placed in a vial and dissolved in 2 mL of iso-octane. 200 μL of 2 N KOH in MeOH was added. The mixtures were vortexed for 1 min, and after resting for 5 min, 2 drops of methyl orange were added. Finally, the sample was titrated with 2 N HCl until a pink endpoint was observed. 0.5 μL of the organic phase was injected to the chromatograph. The resulting peaks were integrated using

TABLE 2.5

Fatty Acid Composition (% w/w) of the Samples Used in This Study

Fatty Acids	Milk Fat	PHPO	Palm Oil	IHPO	PH(PO/PS)
C4:0	1.5 ± 0.1	0	0	0	0
C6:0	1.4 ± 0.1	0	0	0	0
C8:0	1.0 ± 0.1	0	0	0	0
C10:0	2.5 ± 0.1	0	0	0	0
C12:0	3.0 ± 0.1	0	0	0.3 ± 0.1	0.2 ± 0.1
C14:0	9.8 ± 0.2	0.9 ± 0.1	0.9 ± 0.1	0.9 ± 0.1	1.0 ± 0.1
C14:1	1.6 ± 0.1	0	0	0	0
C16:0	28.4 ± 0.7	42.0 ± 0.6	40.3 ± 0.5	39.3 ± 0.2	45.3 ± 0.8
C16:1	3.0 ± 0.1	0	0	0	0
C17:0	0.8 ± 0.1	0	0	0	0
C18:0	12.4 ± 0.1	9.9 ± 0.4	7.8 ± 0.7	19.4 ± 0.2	11.4 ± 0.2
C18:1	27.2 ± 0.4	42.6 ± 0.2	39.1 ± 0.1	30.9 ± 0.2	35.3 ± 0.3
C18:2	4.5 ± 0.3	3.6 ± 0.3	10.4 ± 0.1	7.7 ± 0.1	5.7 ± 0.2
C18:3 and C20:0	1.1 ± 0.1	1.1 ± 0.1	1.5 ± 0.1	1.4 ± 0.1	1.1 ± 0.1
C20:1	1.1 ± 0.1	0	0	0	0
C22:0 and C20:4	0.6 ± 0.1	0	0	0	0
Total	100	100	100	100	100

a Shimadzu integrator (C-R3A Chromatopac). Three determinations on each of three separate samples were carried out. The average and standard deviation are reported.

Differential scanning calorimetry: The thermal behavior of the samples was studied by means of a DSC2910 differential scanning calorimeter (DSC) (TA Instruments, Mississauga, Ontario, Canada). A 5–10 mg sample of melted fat was placed in an aluminum DSC pan and was heated from 20°C to 60°C, held at this temperature for 30 min and then cooled at different cooling rates (0.5°C, 1°C, 2°C, 3°C, 4°C, and 5°C/min) to the set temperature (T_{set}). This temperatures were set to 15°C, 25°C, 26°C, 30°C, and 20°C for AMF, PH(PO/PS), PHPO, IHPO, and PO, respectively. The peak crystallization temperature (T_n) was determined from these profiles. Two different procedures were used to determine the melting temperatures by means of DSC. First, the sample was crystallized as described before and kept at T_{set} for 5 min, followed by heating from T_{set} to 60°C at 5°C/min. The peak melting temperatures were determined from these melting profiles. The second procedure used was to quench the samples from 60°C to 20°C and then store them at 5°C for 1 month to ensure that the most stable polymorph was generated. Melting thermograms were obtained as before (heating from 10°C to 60°C at 5°C/min) to obtain the peak melting temperatures of the samples. Three determinations on each of three separate samples were carried out. The average and standard deviations are reported.

Induction times of crystallization (PTA): Induction times of crystallization were studied by means of a Fats & Oils Phase Transition Analyzer (PSA-70V-HT, Phase Technology, Richmond, British Columbia, Canada). The analyzer is a light turbidimeter that detects the appearance of the first crystals during the crystallization process of a fat sample. Samples were melted for 30–45 min at 80°C and then 150 μL were placed in the crystallization cell. The temperature of the cell was raised to 60°C, held at this temperature for 15 min, and then it was dropped at 0.5°C, 1°C, 2°C, 3°C, 4°C, and 5°C/min to the different T_{set} described in the DSC experiments. Both the crystallization temperatures and the induction times were calculated from these experiments as the time when the first crystals appear, which is evidenced by a deviation in the baseline of the laser signal. The induction time for nonisothermal nucleation (t_c) was calculated as the time at which the first crystals were detected minus the time required to reach the melting point temperature. Three determinations on each of three separate samples were carried out. The averages and standard deviations are reported.

This technique was also used to determine melting points. For this determination, melted samples (80°C for 30 min) were placed in the PTA cell at 60°C and then cooled to 10°C at 40°C/min. Upon reaching this temperature, samples were incubated at this temperature for 30 min and then heated at 1°C/min and 5°C/min until melted. The melting temperature was determined as the temperature at which the signal of the laser becomes constant and is equal to the baseline. As no significant differences were found between the different heating rates assayed using this method, the melting points reported in this study are the average of the two melting temperatures obtained at these heating rates and their standard deviations.

Polarized light microscopy: Triacylglycerol crystals are birefringent and their crystallization behavior and microstructure can therefore be conveniently characterized using PLM. For PLM studies, a small droplet (~10 μL) of melted fat (130°C for 10 min) was placed on a preheated (130°C) glass slide, and then a preheated (130°C) glass cover was placed over the sample to produce a uniform thick fat film. The slides were then cooled from 130°C to 30°C at 0.5°C/min, 1°C/min, and 5°C/min using a cold stage (Linkam LTS-350 hot/cold stage, Linkam Instruments, Surrey, United Kingdom). The microstructure of the fat samples was observed by using an Olympus BH microscope (Tokyo, Japan). Images were acquired with a Sony XC-75 CCD video camera (Sony Corporation, Tokyo, Japan), and an LG-3 PCI frame grabber using Scion Image (Scion Corporation, Frederick, MD). The crystallization processes were followed by capturing an image every 10 s. The images acquired were thresholded using the built-in automatic threshold function of Image J (National Institutes of Health, Bethesda, MD). The number of particles and the area fraction of the crystals are obtained by using the "Particle Analysis" function of Image J. In this study, we report the average and standard deviation of three replicates for particle size and number.

2.4.6.2 Results

The chemical composition of the five different systems studied is reported in the following.

The crystallization behavior of triacylglycerol and its structure are extremely sensitive to cooling rate (Campos et al., 2002). For example, an increase in cooling rate usually leads to a decrease in crystal size (Figure 2.23).

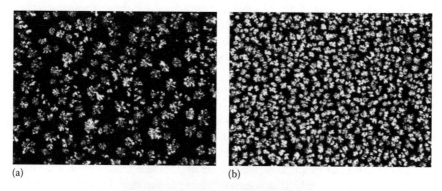

(a)   (b)

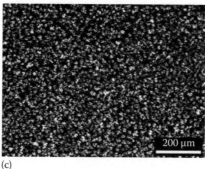

(c)

FIGURE 2.23 Polarized light micrographs of IHPO crystallized at 0.5 K/min (a), 1 K/min (b), and 5 K/min (c).

For all systems studied, an increase in cooling rate during crystallization leads to a decrease in the crystallization temperature (T_0 and T_c), and an increase in nucleation rate, resulting in a greater number of smaller crystals. These effects are summarized in Table 2.6. Moreover, these fats crystallized in the same polymorphic form—the material had the same melting temperature, enthalpy of melting and powder x-ray diffraction patterns (not shown), irrespective of the cooling rate experienced during crystallization.

We crystallized each of the three different fats at three different cooling rates and determined the number of crystals as a function of time (Figure 2.24a through c). The nucleation rate was determined by taking the first derivative of these profiles (Figure 2.24d through f). For the purpose of our analysis, we used the peak nucleation rate (J_p). In all subsequent discussions, we will refer to this peak nucleation rate (J_p) as the nucleation rate (J). This novel method of determining the nucleation rate of birefringent materials worked extremely well and we confirmed that an increase in cooling rate always lead to an increase in nucleation rate (Figure 2.24g). It is worth noticing in these figures that the number of nuclei does decrease at certain time periods. This could be due to a melt-mediated polymorphic transformation and/or stepwise crystallization. We concentrated on the initial increase in the number of crystals in our analysis.

Armed with this new approach and reliable estimates of the nucleation rate derived from image analysis of polarized light micrographs in time, we sought to relate the nucleation rate of the material to its supercooling-time exposure discussed earlier (β). The nucleation rate was found to be an exponential function of the square root of β and thus data from individual systems were fitted to the model to obtain the maximum nucleation rate (J_{max}). Nucleation rate data was then normalized using this value, and the relative nucleation rates (J/J_{max}) determined for the three systems at the different cooling rates. Figure 2.25 finally shows the excellent fit of the normalized nucleation rate to the model.

The microscopy method for determining the nucleation rate was very powerful, but is also very time-consuming. Instead, we decided to expand the study to more systems than the three discussed earlier and explore more cooling rates. To do so, we resorted to the use of turbidimetric measurements for the determination of the induction time of nucleation, which is inversely related to the rate of nucleation, and can be determined very fast.

2.4.6.2.1 Determination of the Melting Temperature of a Complex Fat Mixture

The melting point is defined as the temperature at which a substance changes from solid to liquid state. A pure crystalline solid will melt at a specific temperature, and thus, melting point is a characteristic of a substance that can be used to identify it. However, when dealing with natural fat systems, which are complex muticomponent mixtures of TAG molecules and other minor components, the melting point determination is not an easy task, especially if the fat tends to crystallize in different polymorphic forms as well. For this reason, it is very important to specify which techniques (capillary method, differential scanning calorimetry, Mettler dropping point, etc.), and crystallization conditions were used to determine the "melting point." For example, when using DSC, the peak temperature value obtained from the

TABLE 2.6

Effects of Cooling Rate on the Crystallization and Melting Behavior of Multicomponent Fats (PHPO, IHPO, PHPOPS)

Sample	φ (K/min)	T_c (°C)[a]	T_0 (°C)[b]	T_m (°C)[c]	ΔH (J/g)[d]	Particle Area[e] (μm²)	Particle Number[f]
PHPO	0.5	31.6 (0.26)	32.3 (0.55)	46.3 (0.14)	71.9 (5.68)	6.9 (0.18)	3320 (40)
	1.0	31.3 (0.20)	31.3 (0.42)	45.7 (0.72)	72.4 (6.68)	8.2 (0.22)	2970 (52)
	5.0	31.0 (0.11)	29.9 (0)	44.2 (0.17)	74.0 (6.68)	5.3 (0.27)	3650 (63)
IHPO	0.5	37.0 (0.81)	36.6 (0.11)	49.6 (0.98)	64.7 (2.16)	126.4 (3.69)	228 (6)
	1.0	36.5 (0.92)	36.2 (0.08)	49.5 (0.66)	65.0 (1.76)	74.7 (2.31)	443 (11)
	5.0	35.4 (0.61)	34.7 (0.08)	49.5 (0.78)	64.2 (0.25, 4)	36.3 (0.32)	925 (7)
PH(PO/PS)	0.5	31.1 (0.30)	32.1 (0.55)	46.5 (0.57)	53.4 (0.08, 4)	11.7 (1.01)	2260 (111)
	1.0	30.6 (0.15)	30.7 (0.06)	47.0 (0.03)	54.4 (0.54, 4)	6.7 (0.11)	3180 (23)
	5.0	30.4 (0.10)	29.9 (0)	46.7 (0.56)	53.4 (0.22, 4)	5.1 (0.04)	4000 (7)

Reported are the averages and standard deviations of *n* replicates.

PHPO, partially hydrogenated palm oil; IHPO, interesterified hydrogenated palm oil; PHPOPS, partially hydrogenated mixture of palm oil and palm stearin.

[a] Peak crystallization temperature was determined by differential scanning calorimetry (DSC).

[b] Crystallization onset temperature was determined by laser turbidimetry (PTA analysis).

[c] Peak melting temperature was determined by differential scanning calorimetry.

[d] Enthalpy of melting was determined by differential scanning calorimetry.

[e] Particle diameters were determined by polarized light microscopy and image analysis.

[f] The number of crystals was determined by polarized light microscopy and image analysis.

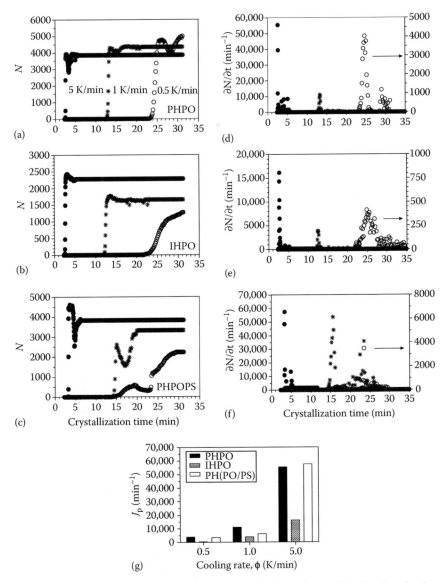

FIGURE 2.24 Changes in number of crystals as a function of crystallization time for three multicomponent fat systems (a–c), and the corresponding first derivatives of the patterns (d–f). Effect of cooling rate on the nucleation rate of the fats (g).

melting profile of a crystallized fat is usually a reasonable indicator of the average melting temperature of the sample. However, in most cases, the TAGs in the fat crystallize in different polymorphic forms (phases), and/or fractionation occurs. In these cases, several peaks are obtained, and thus, choosing the correct peak that represents the melting temperature of the samples is not straightforward. Figure 2.26 shows an example of the melting profiles of all samples used in this work when crystallized at 1°C/min. Figure 2.26a and b show the melting behavior of samples PH(PO/PS),

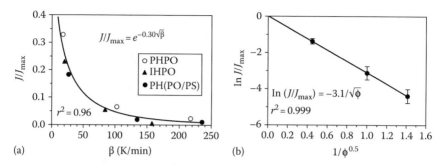

FIGURE 2.25   (a) Dependence of the relative nucleation rate on the supercooling-time potential of the different systems studied. (b) Dependence of the relative nucleation rate on cooling rate. A higher cooling rate leads to a larger value of β.

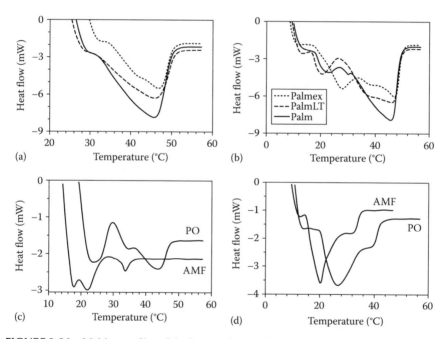

FIGURE 2.26   Melting profiles of the fat samples used in this study immediately after reaching the crystallization temperature (a and c) and after 1 month of storage at 5°C (b and d).

IHPO and PHPO 5 min after reaching T_{set} (A) and after 1 month storage at 5°C. Figure 2.26c and d show the same pattern but for AMF and PO. It is evident from this figure that not only the type of sample, but also the time spent at the set temperature affects the thermal behavior of the sample and therefore the peak melting temperature value obtained.

In the global analysis carried out in this work, we need an average melting temperature for the kinetic analysis. Thus, we have a challenge, since this temperature has to be related to the initial triacylglycerol fraction that nucleates. Moreover, the determination of this average/global melting temperature has to be

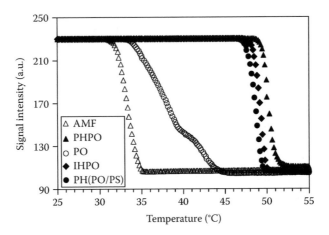

FIGURE 2.27 Phase transition analyzer (light scattering) profiles as a function of increasing temperature used to determine the melting point of the different fat systems used in this work.

reasonably reproducible. Figure 2.27 shows the melting profile of the five samples obtained using the phase transition analyzer (PTA). From this figure, we can notice that the determination of a global or average value for the melting point of the fat is much more straightforward (however, not necessarily more correct) than when using the DSC. Here, the melting temperature is determined as the temperature at which the signal intensity decreases to baseline and becomes constant. The melting points obtained were 49.8°C ± 0.9°C, 35.5°C ± 0.2°C, 49.9°C ± 0.2°C, 51.3°C ± 1.7°C, and 45.6°C ± 0.4°C for IHPO, AMF, PH(PO/PS), PHPO, and PO, respectively. This method enables the determination of melting points as the temperature when the last crystals in the fat melt away. We believe that this method offers a convenient way to determine an average value for the melting point of a fat. Certainly this is not the true melting point of the fat, since the fat is composed of hundreds of different triacylglycerol molecules, each with a unique melting point. We resort to using this global melting temperature in order to be able to carry out the analysis. Otherwise, this analysis would not be possible.

Table 2.7 shows the melting and crystallization temperatures obtained from DSC and PTA together with the induction times of crystallization determined by PTA. The DSC peak melting temperatures and the PTA end of melt temperatures reported in this table were determined from the melting profiles obtained after incubating the samples, 5 min (DSC) and 30 min (PTA) after reaching T_{set}. Thus, they represent the melting temperature of the solids that are formed at the onset of the crystallization process.

2.4.6.2.2 Nucleation Rate Determination

The nucleation rate was estimated from the inverse of the induction time of nucleation obtained from the PTA experiments (Table 2.7). This method of estimating the nucleation rate is convenient since it is experimentally accessible, however, it is only an approximation. A nucleation rate corresponds to the number of nuclei appearing per

TABLE 2.7

Comparison between Melting (T_m) and Crystallization Temperatures (T_c) Calculated from the Melting Profile from the DSC and the PTA

Cooling Rate (°C/min)	T_m (DSC) (°C)	T_c (DSC) (°C)	T_m (PTA) (°C)	T_c (PTA) (°C)	t_i (s)
AMF					
0.5	33.6 ± 0.1	17.8 ± 0.7	35.5 ± 0.2	21.4 ± 0.6	1734 ± 69
1	33.6 ± 0.2	16.2 ± 0.2		19.9 ± 0.4	956 ± 25
2	34.0 ± 0.2	15.1 ± 0.1		18.9 ± 0.3	508 ± 10
3	34.3 ± 0.1	15.6 ± 1.4		18.8 ± 0.2	343 ± 14
4	34.2 ± 0.3	14.7 ± 0.2		18.3 ± 0.5	264 ± 7
5	33.5 ± 0.1	14.4 ± 0.1		19.5 ± 0.1	195 ± 22
10				16.0 ± 0.2	119 ± 6
15				15.8 ± 0.9	81 ± 7
20				15.9 ± 1.0	60 ± 5
PHPO					
0.5	45.7 ± 0.4	27.3 ± 0.1	51.3 ± 1.7	34.6 ± 0.1	2050 ± 14
1	45.7 ± 0.1	27.7 ± 0.2		32.9 ± 1.7	1128 ± 106
2	46.0 ± 0.2	27.4 ± 0.1		30.6 ± 0.1	635 ± 8
3	45.7 ± 0.1	27.3 ± 0.1		30.4 ± 0.3	428 ± 8
4	45.6 ± 0.3	27.2 ± 0.1		30.1 ± 0.5	326 ± 8
5	45.6 ± 0.1	27.1 ± 0.1		31.2 ± 0.7	247 ± 10
10				29.0 ± 0.9	136 ± 7
15				28.4 ± 0.6	94 ± 12
20				28.7 ± 0.8	70 ± 3
PO					
0.5	43.7 ± 0.1	20.3 ± 0.01	45.6 ± 0.4	25.3 ± 1.0	2471 ± 169
1	43.0 ± 0.1	19.8 ± 0.1		24.0 ± 0.1	1323 ± 18
2	43.3 ± 0.1	19.5 ± 0.1		23.0 ± 0.3	691 ± 10
3	43.0 ± 0.3	19.5 ± 0.1		22.9 ± 0.8	466 ± 16
4	43.0 ± 0.3	19.4 ± 0.1		21.7 ± 0.8	359 ± 21
5	42.8 ± 0.5	19.4 ± 0.1		23.0 ± 0.8	281 ± 13
10				21.3 ± 0.3	149 ± 8
15				21.1 ± 1.5	100 ± 7
20				21.1 ± 1.2	75 ± 16
IHPO					
0.5	47.2 ± 0.5	31.9 ± 1.8	49.8 ± 0.9	37.4 ± 0.4	1527 ± 57
1	46.7 ± 0.1	32.0 ± 0.1		34.5 ± 0.2	937 ± 10
2	47.0 ± 0.5	31.3 ± 0.1		34.1 ± 0.2	483 ± 6
3	46.6 ± 0.4	29.8 ± 0.1		33.9 ± 0.1	325 ± 4
4	32.7 ± 0.6	30.1 ± 0.1		33.8 ± 0.3	230 ± 66
5	33.2 ± 0.2	30.1 ± 0.2		34.6 ± 1.1	187 ± 16

(continued)

TABLE 2.7 (continued)
Comparison between Melting (T_m) and Crystallization
Temperatures (T_c) Calculated from the Melting Profile
from the DSC and the PTA

Cooling Rate (°C/min)	T_m (DSC) (°C)	T_c (DSC) (°C)	T_m (PTA) (°C)	T_c (PTA) (°C)	t_i (s)
10				33.0 ± 0.3	103 ± 9
15				32.6 ± 0.7	70 ± 12
20				32.4 ± 0.2	54 ± 16
PH(PO/PS)					
0.5	45.7 ± 0.3	30.0 ± 0.1	49.9 ± 0.2	35.1 ± 0.3	1806 ± 32
1	46.0 ± 0.1	29.2 ± 0.5		32.1 ± 0.1	1089 ± 7
2	46.2 ± 0.2	28.7 ± 0.1		31.4 ± 0.3	569 ± 11
3	46.3 ± 0.1	28.4 ± 0.3		31.3 ± 0.4	380 ± 10
4	46.3 ± 0.1	28.3 ± 0.1		30.8 ± 0.7	279 ± 54
5	46.1 ± 0.2	27.6 ± 0.3		31.5 ± 0.9	226 ± 18
10				30.0 ± 0	122 ± 30
15				30.2 ± 0.5	80 ± 11
20				29.1 ± 0.1	64 ± 11

Induction times (t_i) calculated from PTA runs are also included. All values represent averages and standard deviations of three separate determinations on different samples.

DSC, differential scanning calorimeter; PTA, polymer thermal analyzer.

unit time, while this estimated nucleation rate is the inverse of an induction time. In our previous work (Marangoni et al., 2006), we determined the nucleation rate of fats using PLM by counting the number of reflections appearing per unit time. The method used in this study is different. Even though we only had three cooling rates in our previous experiment, we plotted the nucleation rate obtained using the PTA versus the nucleation rate obtained by PLM. The correlation coefficient (r^2) obtained was 0.82 and the slope 1.1 ± 0.2. Thus, agreement was reasonable, thus validating the approximation $J \sim 1/\tau$. This, however, has to be validated further in other systems under different conditions.

As shown in Table 2.6, crystallization temperatures obtained for the same sample at the same cooling rate were higher for PTA than for DSC, suggesting that the first technique was more sensitive than the second one at detecting the appearance of the first crystals (onset of crystallization). Due to the high sensitivity of the phase transition analyzer (Wright et al., 2000), we can consider that the induction times determined using this technique are reasonable estimates of the induction times of nucleation. For all the samples used in this study, the induction times were shorter when the samples were crystallized at higher cooling rates. This is an expected result since the higher the cooling rate, the shorter the time required to reach the crystallization temperature. Also, from Table 2.7, we can appreciate that when the cooling rate is high, samples crystallize at lower temperatures suggesting that the *time of*

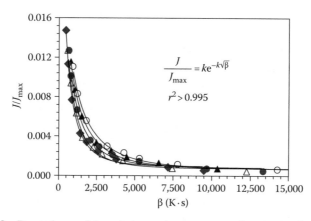

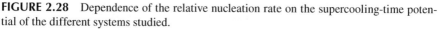

FIGURE 2.28 Dependence of the relative nucleation rate on the supercooling-time potential of the different systems studied.

exposure to supercooling is of key importance. Considering that the induction times calculated during a nonisothermal crystallization include the time needed to reach a specific crystallization temperature, it is necessary to find an appropriate parameter that describes this time–temperature dependence. The supercooling-time exposure (β) described earlier captures this dynamic reasonably well.

The exponential relationship between the relative nucleation rate and square root of β can be appreciated in Figure 2.28. The fit of the model to the data was excellent and statistically better ($p < 0.05$) than for an exponential relationship between the relative nucleation rate and β. Moreover, model fits were significantly different for the different systems studied ($p < 0.0001$). It is possible to determine the clustering energy of nucleation for nucleation using the approach developed in this study. Table 2.8 shows the values for k, J_{max}, and Q_m determined for each sample.

We can observe that the lower the Q_m value, the higher the J, as expected from kinetic theory, if Q_m represents the clustering energy of nucleation (possibly the energy of activation for the nucleation process). One interesting finding is that the J_{max} values were not statistically different from each other ($p > 0.05$), with an average value of 1.25 with a standard deviation of 0.02. Thus, it thus possible predict the nucleation rate, or induction time of nucleation, of triacylglycerols at different cooling rates, under nonisothermal conditions, by using

$$J\,[\mathrm{s}^{-1}] = \frac{1}{t_c} = 1.25ke^{-Q_m/Z\sqrt{\phi}} \tag{2.157}$$

The parameters required are listed in Table 2.8; the clustering energy of nucleation in Expression 2.157 should be used with units of kJ/mol.

Moreover, a reasonable approximation of the nucleation rate of fats as a function of cooling rate can be obtained using

$$J\,[\mathrm{s}^{-1}] = \frac{1}{t_c} = 0.0745e^{-0.86/\sqrt{\phi}} + 0.0008 \tag{2.158}$$

TABLE 2.8

Exponential Constant (k), Molar Nucleation Clustering Energy (Q_M), and Maximum Nucleation Rate (J_{max}) for the Systems Studied

System	k^a (K$^{-1/2}$s$^{-1/2}$)	Q_M^b (kJ/mol)	J_{max}^a (s^{-1})
AMF	0.0775ab (0.00480)	31.4	1.32a (0.079)
PHPO	0.0622bc (0.00339)	39.1	1.23a (0.056)
PO	0.0558c (0.00329)	43.6	1.20a (0.056)
IHPO	0.0848a (0.00510)	28.7	1.26a (0.070)
PH(PO/PS)	0.0701ac (0.00557)	34.6	1.23a (0.084)
Average ($n = 5$)	0.070 (0.0050)	39.3	1.25 (0.020)

a Values reported are the average and standard error ($n = 9$). Values followed by the same letter within a column are not significantly different from each other ($p > 0.05$).
b Molar nucleation clustering energy was calculated using an average TAG molecular weight of 800 g/mol.

It is worth pointing out that the entire kinetic characterization was carried out using simple light scattering device. The PTA analyzer was used to characterize the melting as well as the crystallization behavior. This procedure could thus be completely automated and a standard method developed for the characterization of the nonisothermal nucleation behavior of fats.

In summary, in this work, we have developed a new way of characterizing nucleation kinetics under nonisothermal conditions by parametrization of the data, considering both time and supercooling effects. A supercooling-time exposure parameter was defined and found to be related to the nucleation rate in a simple exponential decay fashion. The parametrization procedure used in the analysis of the nucleation kinetics of the fats crystallized at different cooling rates allowed for the determination of a cooling-rate independent clustering energy of nucleation.

REFERENCES

Alba Simionesco, C., E. Dumont, B. Frick, B. Geil, D. Morineau, V. Teboul, and Y. Xia. 2003. Confinement of molecular liquids: Consequences on thermodynamic, static and dynamical properties of benzene and toluene. *Eur. Phys. J.* 12: 19–28.
Avrami, M. 1939. Kinetics of phase change I. General Theory. *J. Chem. Phys.* 7: 1103–1112.
Avrami, M. 1940. Kinetics of phase change II. Transformation-time relations for random distribution of nuclei. *J. Chem. Phys.* 8: 212–224.

Avrami, M. 1941. Kinetics of phase change III. Granulation, phase change, and microstructure. *J. Chem. Phys.* 9: 177–184.

Awad, T., M.A. Rogers, and A.G. Marangoni. 2004. Scaling behavior of the elastic modulus in colloidal networks of fat crystals. *J. Phys. Chem. B* 108: 171–179.

Batte, H.D. and A.G. Marangoni. 2005. Fractal growth of milk fat crystals is unaffected by microstructural confinement. *Cryst. Growth Des.* 5: 1703–1705.

Boistelle, R. 1988. Fundamentals of Nucleation and Crystal Growth. In: *Crystallization and Polymorphism of Fats and Fatty Acids*, Garti, N., K. Sato, Eds. New York: Marcel Dekker, pp. 189–226.

Brunello, N., S.E. McGauley, and A.G. Marangoni. 2003. Mechanical properties of cocoa butter in relation to its crystallization behavior and microstructure. *Lebensmittel-Wissenschaft und-Technologie* 36: 525–532.

Campbell, I. and R.A. Dwek, R.A. 1984. *Biological Spectroscopy*. Menlo Park, CA: The Benjamin Cummings Publishing Company, Inc.

Campos, R. and A.G. Marangoni. 2010. Molecular composition dynamics and structure of cocoa butter. *Cryst. Growth Des.* 10: 205–217.

Campos, R., S.S. Narine, and A.G. Marangoni. 2002. Effect of cooling rate on the structure and mechanical properties of milk fat and lard. *Food Res. Int.* 35: 971–981.

Chapman, D. 1969. Polymorphism of glycerides. *Chem. Rev.* 62: 433–456.

Chow, P.S., X.Y. Liu, J. Zhang, and R.B.H. Tan. 2002. Spherulitic growth kinetics of protein crystals. *Appl. Phys. Lett.* 81: 1975–1977.

Christian, J.W. 1965. *The Theory of Transformations in Metals and Alloys: An Advanced Textbook in Physical Metallurgy*. London, U.K.: Pergamon Press, pp. 471–495.

Dibildox-Alvarado, E., J. Neves Rodrigues, L.A. Gioelli, J. Toro-Vazquez, and A.G. Marangoni. 2004. Effect of crystalline microstructure on oil migration in a semisolid fat matrix. *Cryst. Growth Des.* 4: 731–736.

Fowler, A., H.E. Stanley, and G. Daccord. 1989. Disequilibrium silicate mineral textures: Fractal and non-fractal features. *Nature* 341: 134–138.

Garside, J. 1978. General principles of crystallization. In: *Food Structure and Behavior*, Blanshard, J.M.W., P. Lillford, Eds. London, U.K.: Academic Press, pp. 35–49.

Goldenfeld, N. 1987. Theory of spherulitic crystallization. *J. Cryst. Growth* 84: 601–608.

Grasany, L., T. Pusztai, T. Borzsonyi, G. Toth, G. Tegze, J.A. Warren, and J.F. Douglas. 2006. Phase field theory of crystal nucleation and polycrystalline growth: A review. *J. Mater. Res.* 21: 309–319.

Grasany, L., T. Pusztai, T. Borzsonyi, J.A. Warren, and J.F. Douglas. 2004. A general mechanism of polycrystalline growth. *Nat. Mater.* 3: 645–650.

Grasany, L., T. Pusztai, T. Borzsonyi, J.A. Warren, and J.F. Douglas. 2005. Growth and form of spherulites. *Phys. Rev. E* 72: 011605.

Hartel, R.W. 2001. Nucleation. In: *Crystallization in Foods*, Hartel, R.W., Ed. Frederick, MD: Aspen Publishers Inc., pp. 145–188.

Herrera, M.L. 2002. Crystallization of hydrogenated sunflower oil. In: *Physical Properties of Lipids*, Marangoni A.G., S.S. Narine, Eds. New York: Marcel Dekker, pp. 449–477.

Herrera, M.L., M. Gatti, and R.W. Hartel. 1999. A kinetic analysis of crystallization of a milk fat model system. *Food Res. Int.* 32: 289–298.

Herrera, M.L. and R.W. Hartel. 2000a. Effect of processing conditions on crystallization kinetics of a milk fat model system. *J. Am. Oil Chem. Soc.* 77: 1177–1187.

Herrera, M.L. and R.W. Hartel. 2000b. Effect of processing conditions on physical properties of a milk fat model system: Rheology. *J. Am. Oil Chem. Soc.* 77: 1189–1195.

Herrera, M.L. and R.W. Hartel. 2000c. Effect of processing conditions on physical properties of a milk fat model system: Microstructure. *J. Am. Oil Chem. Soc.* 77: 1197–1204.

Kashchiev, D. 2000. Nucleation at variable supersaturation. In: *Nucleation: Basic Theory with Applications*, Kashchiev, D., Ed. Oxford, U.K.: Butterworth Heinemann, pp. 279–289.

Keith, H.D. and F.J. Padden. 1963. A phenomenological theory of spherulitic crystallization. *J. Appl. Phys.* 34: 2409–2421.

Kellens, M., W. Meeussen, and H. Reynaers. 1992. Study of the polymorphism and the crystallization kinetics of tripalmitin: A microscopic approach. *J. Am. Oil Chem. Soc.* 69: 906–911.

Kurasov, B.V. Multicomponent nonisothermal nucleation 1. Kinetic equation. arXiv: cond-mat/9909058 v1, September 3, 1999.

Liu, X.Y. and P.D. Sawant. 2002a. Mechanism of the formation of self-organized microstructures in soft functional materials. *Adv. Mater.*14: 421–426.

Liu, X.Y. and P.D. Sawant. 2002b. Determination of the fractal characteristic of nanofiber-network formation in supramolecular materials. *Chem. Phys. Chem.* 4: 374–377.

Magill, J.H. 2001. Spherulites: a personal perspective. *J. Mater. Sci.* 36: 3143–3164.

Marangoni, A.G. 2000. Elasticity of high volume fraction fractal aggregate networks: A thermodynamic approach. *Phys. Rev. B* 62: 13951–13955.

Marangoni, A.G., T. Aurand, S. Martini, and M. Ollivon. 2006. A probabilistic approach to model the non-isothermal nucleation behavior of triacylglycerol melts. *Cryst. Growth Des.* 6: 1199–1205.

Marangoni, A.G. and S.E. McGauley. 2003. The relationship between crystallization behavior and structure in cocoa butter. *Cryst. Growth Des.* 3: 95–108.

Marangoni, A.G. and M. Ollivon. 2007. Fractal character of triglyceride spherulites is a consequence of nucleation kinetics. *Chem. Phys. Lett.* 442: 360–364.

Marangoni, A.G. and M.A. Rogers. 2003. Structural basis for the yield stress in plastic disperse systems. *Appl. Phys. Lett.* 82: 3239–3241.

Marangoni, A.G., D. Tang, and A.P. Singh. 2005. Nonisothermal nucleation of triacylglycerol melts. *Chem. Phys. Lett.* 419: 259–264.

Martini, S., M.L. Herrera, and R.W. Hartel. 2002a. Effect of cooling rate on crystallization behavior of milk fat fraction/sunflower oil blends. *J. Am. Oil Chem. Soc.* 79: 1055–1062.

Martini, S., M.L. Herrera, and R.W. Hartel. 2002b. Effect of processing conditions on microstructure of milk fat fraction/sunflower oil blends. *J. Am. Oil Chem. Soc.* 79:1063–1068.

Mazzanti, G., S.E. Guthrie, A.G. Marangoni, and S.H.J. Idziak. 2007. A conceptual model for shear-induced phase behavior in crystallizing cocoa butter. *Cryst. Growth Des.* 7: 1230–1241.

Mazzanti, G., A.G. Marangoni, and S.H.J. Idziak. 2005. Modeling phase transitions during the crystallization of a multicomponent fat under shear. *Phys. Rev. E* 71: 041607.

Mazzanti, G., A.G. Marangoni, and S.H.J. Idziak. 2009. Synchrotron study on milkfat crystallization kinetics under shear flow. *Food Res. Int.* 42: 682–694.

Mazzanti, G., A.G. Marangoni, S.E. Welch, E.B. Sirota, and S.H.J. Idziak. 2003. Orientation and phase transitions of fat crystals under shear. *Cryst. Growth Des.* 3:721–725.

Mazzanti, G., A.G. Marangoni, S.E. Welch, E.B. Sirota, and S.H.J. Idziak. 2004. Effect of minor components and temperature profiles on polymorphism of milk fat. *Cryst. Growth Des.* 4: 1303–1309.

McGraw, R. and D.T. Hu. 2003. Kinetic extensions of the nucleation theorem. *J. Chem. Phys.* 118: 9337–9347.

Narine, S.S. and A.G. Marangoni. 1999a. Fractal nature of fat crystal networks. *Phys. Rev. E* 59: 1908–1920.

Narine, S.S. and A.G. Marangoni. 1999b. Mechanical and structural model of fractal networks of fat crystal at low deformations. *Phys. Rev. E* 60: 6991–7000.

Owen, A. and A. Bergmann. 2004. On the fractal character of polymer spherulites: An ultra-small-angle X-ray scattering study of poly[(R)-3-hydroxybutyrate]. *Polym. Int.* 53: 12–14.

Rothschild, W.G. 1998. *Fractals in Chemistry*. New York: Wiley-Interscience, pp. 19–65.

Rousset, P. 2002. Modeling crystallization kinetics of triacylglycerols. In: *Physical Properties of Lipids*, Marangoni, A.G., S.S. Narine, Eds. New York: Marcel Dekker, pp. 1–36.

Russ, J.C. 2002. *The Image Processing Handbook*, 4th edn., Boca Raton, FL: CRC Press.

Rybin, E.N. 2003. On the kinetics of non-isothermal nucleation. *Colloid J.* 65: 230–236.

Sato, K. 1999. Solidification and phase transformation behaviour of food fats—A review. *Fett/Lipid* 101: 467–474.

Sato, K. 2001. Crystallization behavior of fats and lipids—A review. *Chem. Eng. Sci.* 7: 2255–2265.

Sato, K. and S. Ueno. 2005. Polymorphism in fats and oils. In: *Bailey's Industrial Oil and Fat Products*, F. Shahidi, Ed., 6th edn., Hoboken, NJ: John Wiley & Sons, pp. 77–120.

Sharples, A. 1966. *Introduction to Polymer Crystallization*. London, U.K.: Edward Arnold, Ltd., pp. 44–59.

Small, D. 1986. Glycerides. In: *The Physical Chemistry of Lipids: From Alkanes to Phospholipids*, D. Small, Ed., 2nd edn., New York: Plenum Press, pp. 345–392.

Timms, R.E. 1984. Phase behavior of fats and their mixtures. *Prog. Lipid Res.* 23: 1–38.

Vicsek, T. 1992. *Fractal Growth Phenomena*. Singapore: World Scientific Publishing, pp. 9–132.

Wang, R., X.Y. Liu, J.Y. Xiong, and J.L. Li. 2006. Real-time observation of the fiber network formation in molecular organogel: Supersaturation-dependent microstructure and its related rheological property. *J. Phys. Chem. B.* 110: 7275–7280.

Weimann, P.A., D.A. Hajduk, C. Chu, K.A. Chaffin, J.C. Brodil, and F.S. Bates. 1999. Crystallization of tethered polyethylene in confined geometries. *J. Polym. Sci.* 37: 2053–2068.

Wright, A.J., R.W. Hartel, S.S. Narine, and A.G. Marangoni. 2000a. The effect of minor components on milk fat crystallization. *J. Am. Oil Chem. Soc.* 77: 463–475.

Wright, A.J., S.S. Narine, and A.G. Marangoni. 2000b. Comparison of experimental techniques used in lipid crystallization studies. *J. Am. Oil Chem. Soc.* 77: 1239–1242.

Wyslouzil, B.E. and J.H. Seinfeld. 1992. Nonisothermal homogeneous nucleation. *J. Chem. Phys.* 97: 2661–2670.

3 Intermolecular Forces in Triacylglycerol Particles and Oils

David A. Pink

3.1 INTRODUCTION

The intent of this chapter is to be a brief, focused outline of the dominant physical interaction in anhydrous fats and oils. Broadly speaking, physical interactions may be taken as comprising the ubiquitous (charge-fluctuation-induced) London van der Waals type interactions (Lifshitz, 1956; Landau and Lifshitz, 1968; Leckband and Israelachvili, 2001; Parsegian, 2005; Israelachvili, 2006), and "electrostatic" interactions—interactions between those moieties that carry electric charges on sufficiently large (long) spatial (time) scales and that can be considered as "permanent" (Cevc and Marsh, 1987; McLaughlin, 1989; Cevc, 1990). The last named encompass electric multipole (e.g., dipole–dipole) interactions. Although the distance dependence of electric dipole–dipole interactions can be similar to van der Waals interactions, the origins are different. Hydrogen bonding is a third, weaker, Coulombic effect, which is specific to certain proton-containing moieties and is short-range (<0.3 nm) and directional (Israelachvili, 2006). In nonquantum molecular dynamics, hydrogen bonding is treated as a combination of van der Waals forces together with attractive electrostatic interactions. Although it might be thought that fats and oils do not involve electrostatics, the presence of three carbonyl atomic dipoles per triglycerol together with the possibilities of hydrogen bonding in the case of all glycerols implies that electrostatic interactions should not be ignored.

Oils are composed of a variety of triglycerols. Accordingly, the possibility that separation of their components into many coexisting phases—possibly microphases, depending upon the environment—separated by interfaces of different characteristics and complexities, cannot be passed over. Indeed it is possible that it is exactly this phenomenon that makes different complex oils suitable for a range of different food preparation techniques (Sato, 2001; Marangoni, 2005; Himawan et al., 2006; Sato and Ueno, 2011).

This chapter deals largely with the physical interactions between crystals, or aggregates of them such as fractal structures, in a liquid oil environment. The interactions are all Coulombic of which the major components are van der Waals interactions though in cases where hydrogen bonding might play a role, electrostatic interactions become important. Other effective interactions might exist of which, perhaps, the most important are those brought about by depletion effects. This is an entropic interaction and requires at least two components comprising objects of different sizes. No description of physical forces is complete without an outline of how to model and calculate their effects in a given system. In an oil environment, the system is necessarily complex, a mix of assorted solids in a liquid which may exhibit complex (micro-)phase separation. The possibility that such a system can be realistically modeled by analytically solvable mathematical models is unlikely and, in recent years, computer simulation has provided a practicable avenue to obtain useful results. The philosophy of this chapter is practical, and can be summed up in the following sentences: (1) What are the fundamental interactions in an oil environment? (2) How does one go about modeling these interactions? This chapter will concentrate entirely on van der Waals interactions and some of its consequences. Examples of calculations will be given. Where necessary or desirable, models will be described: a reference to atomic scale molecular dynamics (AMD) will lead into a description of coarse-grained models. The level here is that of a graduate student or postdoctoral fellow in food science with a basic knowledge of physics, and the intent is that this chapter can be used as a general reference and as a basis for the theoretical section of a graduate course in modeling the physics of food systems.

In the following, we shall refer to crystalline nanoplatelets (CNPs) as appearing to be the fundamental components from which many systems of fats in oils are composed (Acevedo and Marangoni, 2010a,b; Acevedo et al., 2011) at room temperatures. These objects are typically highly anisotropic structures measuring ~500 nm on a side and being ~50 nm thick and their role in self-assembly is fundamental to understanding the detailed structures of oils. Much work has been done on the self-assembly of membranes, vesicles, and micelles from phospholipids (Antunes et al., 2009; Larsson, 2009; Weiss et al., 2009; McClements and Li, 2010; Sagalowicz and Leser, 2010; Fathi et al., 2011; Patel and Velikov, 2011) and work has been done on the related area of polar food lipids (Leser et al., 2006). However, the question of what structures arise from the interaction of CNPs with each other and with the oils in which they are embedded, and the pathways through which self-assembly comes about, is essentially completely unresolved.

3.2   VAN der WAALS INTERACTIONS

Van der Waals forces refer to interactions between materials that are brought about by interactions between fluctuating atomic electric dipoles (Hamaker, 1937) induced by fluctuations in the electromagnetic field (Dzyaloshinskii et al., 1961). The latter has been applied to problems in condensed matter by the work of Parsegian et al. (Parsegian, 2005 and references therein). A good, very readable introduction is that of Lee-Desautels (2005). What has come to be known as "Lifshitz theory" describes the induced dipole–dipole interaction in term of the electric permittivities of the two materials as well as those of any materials separating them. That permittivities arise reflects the fact that Lifshitz theory involves many-body correlations. This is in contrast to Lennard-Jones 6–12 potential theory which considers two infinitesimal volumes of material ("bodies") and writes the van der Waals interaction energy, $u(r)$, between them as a pairwise attractive potential in $1/r^6$ and a repulsive potential in

$1/r^{12}$ where r is the distance from one body to the other (Hamaker, 1937; Parsegian, 2005; Israelachvili, 2006)

$$u(r) = -\frac{C_6}{r^6} + \frac{C_{12}}{r^{12}} \tag{3.1}$$

The total interaction between the two materials would then be the sum of all these pairwise interactions:

$$U = \sum_j \sum_k u(r_{jk}) \tag{3.2}$$

where the sum is over all pairs of infinitesimal volumes labeled j and k, and one must not count interactions twice. Let us now restrict our attention only to the attractive part, $-C_6/r^6$, of the 6–12 potential since the short-range repulsion is merely a phenomenological interaction. The picture of the attractive interaction being described as a two-body effect is clearly inadequate as can be seen in Figure 3.1.

The creation of fluctuating dipoles anywhere inevitably takes part in creating fluctuating dipoles throughout the materials so that the total energy is not simply a sum of pairwise energies (Equation 3.2) but is a sum of energies due to all many-body correlations:

$$\tilde{U} = \sum_j \sum_k \tilde{u}_{jk} + \sum_j \sum_k \sum_l \tilde{u}_{jkl} + \sum_j \sum_k \sum_l \sum_m \tilde{u}_{jklm} + \cdots \tag{3.3}$$

Here we have indicated the summation up to four-body correlations (fourth "order"). This is the key difference between models which utilize the 6–12 potential approach and those which recognize that correlations to all orders are required. The two-body interactions of Equation 3.1 *clearly do not involve higher than two-body correlations in contrast to Lifshitz theory which involves correlations of all orders.*

The attractive term in the 6–12 potential is derivable from second-order perturbation theory which makes it clear why only two-body interactions enter. The repulsive term involving $1/r^{12}$ was chosen as a convenient phenomenological short-range

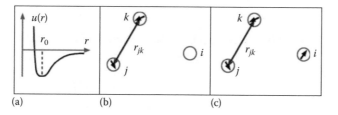

(a)   (b)   (c)

FIGURE 3.1 (a) The 6–12 potential (Equation 3.1). (b) Pairwise interaction between fluctuating dipoles (black arrows) at bodies j and k created by the fluctuating electromagnetic field. The sum of all these yields U (Equation 3.2). (c) Fluctuating dipoles anywhere create fluctuating dipoles everywhere: the fluctuating dipoles at j and k take part in creating a fluctuating dipole at i. This is a three-body interaction.

repulsion between electron "clouds." Lifshitz theory has provided many insights (French et al., 2010) and one might think that the pairwise 6–12 potential theory is now only of historical interest, except for one fact: 6–12 potential theory is used in atomic scale molecular dynamics (AMD) and it works: AMD is widely used to make predictions for gases and liquids which have been successfully compared to experiments.

One should note that, in the van der Waals energy, there are three terms describing the attractive interaction between two objects a distance r apart, all of which exhibit a dependence of $1/r^6$: the London dispersion term due to the interaction between induced fluctuating dipoles, the Debye interaction between such a fluctuation and a permanent electric dipole and the Keesom term which describes the interaction between a pair of electric dipoles. The Lifshitz approach includes all these terms in the expression for this electromagnetic field fluctuation induced interaction between two objects, but the 6–12 potential method includes only the London dispersion term and the other two have to be explicitly included.

Some recent work that could be relevant for food environments are Hanna et al. (2006) and Parsegian (2005) who calculated the distance dependence of the van der Waals interaction between (soft) polymeric or similar surfaces. Li et al. (2006) considered the van der Waals adhesion between rough surfaces. The equation for the van der Waals interaction between two spheres is especially useful because, under certain conditions, larger structures can be synthesized as a close-packed aggregate of spheres. For many years the Derjaguin approximation (Butt et al., 2003) has been used for this purpose but one should be aware of the Papadopoulos and Cheh (1984) approximation. However, Oversteegen and Lekkerkerker (2003) studied the depletion potential between two large hard spheres due to the presence of hard disks. They compared results of using the Derjaguin approximation with those obtained from the generalized Gibbs equation and concluded that, for hard disk aspect ratios smaller than 0.25, the error in the Derjaguin approximation was less than 1%.

3.3   MEAN FIELD MODELS

This section will outline mean field models: Lifshitz theory and the related Coupled Dipole Method (CDM), mean field methods using the 6–12 potential, and the Fractal and Semi-Classical models.

3.3.1   LIFSHITZ THEORY AND THE COUPLED DIPOLE METHOD

Lifshitz theory is an elegant approach for calculating the free energy of a mesoscale system. It assumes that one knows the electric permittivities as functions of frequency of all the components of a (possibly complex) system. Electric permittivity is defined as a parameter specifying the average electric dipole moment vector (polarization), $\vec{P}$, set up in some material by the application of an electric field, $\vec{E}$,

$$\vec{P} = \left[\varepsilon - \varepsilon_0 \right] \vec{E} = \varepsilon_0 \left[\varepsilon_r - 1 \right] \vec{E} \qquad (3.4)$$

where the permittivity is $\varepsilon = \varepsilon_0 \varepsilon_r$, with ε_r being the relative permittivity and ε_0 being the permittivity of "free space," that is, of the vacuum. Accordingly, the volume of space over which permittivity is a meaningful concept must be large enough that one samples sufficient atoms in order to define an average polarization, $\vec{P}$. "Large enough" will then depend upon the size and complexity of the molecules in that volume. Accordingly, using bulk values of permittivity when an oil is being studied in confined nanospaces should be undertaken cautiously.

There is another note of caution that is, perhaps, even more important in food systems. Lifshitz theory can be happily used in systems with fixed components such as fabricated solids (French et al., 2010), or at the mesoscale in systems with a one-component fluid. However, for a many-component fluid, in which phase separation can take place, the final system might involve fluid mixtures of compositions that have to be determined by the equality of chemical potentials. In such a case the use of Lifshitz theory could be problematical because the chemical potentials determine the various phases and depends upon the knowledge of the permittivities. But the permittivities depend upon the composition of the liquid phases so that although, in principle, it can be used to model phase separation, one would have to identify the composition of the various phases and deduce or measure the permittivity of each of the possible phases into which the system might separate. Furthermore, if some of the phases are "nanophases" involving nanoscale layers of fluid around solids, then one might have difficulties in specifying a permittivity of a nanoscale layer, as described in the last paragraph. With these reservations, what follows is a very simplified version of Lifshitz theory.

Lifshitz theory calculates the interaction energies of meso- or macro-phases and is based upon temperature-dependent fluctuations in the electromagnetic field which, in turn, give rise to fluctuations in atomic or molecular electric dipole moments. In it are both nonretarded and retarded interactions with the latter taking into account the fact that the speed of light, c, is finite and that, for distances sufficiently large, this must be taken into account. Here we shall consider only distances, s, where s/c is much less than all other timescales of the system. Let us consider a linear isotropic material characterized by an (angular) frequency-dependent permittivity, $\varepsilon(\omega)$. From Equation 3.4, we obtain the frequency-dependent polarization induced by an electromagnetic field of angular frequency, ω, $\vec{P}(\omega) = [\varepsilon(\omega) - \varepsilon_0]\vec{E}(\omega)$ showing that the frequency-dependent induced polarization, $\vec{P}(\omega)$, is proportional to the permittivity, $\varepsilon(\omega)$. Accordingly, Lifshitz theory relates the free energy to integrals over all frequencies of the frequency-dependent permittivities of a system.

An example. Let us consider two solids, 1 and 3, possessing relative permittivities $\varepsilon_{r1}(\omega)$ and $\varepsilon_{r3}(\omega)$ immersed in a liquid phase, 2, of relative permittivity $\varepsilon_{r2}(\omega)$.

Consider the solids to be two macroscopic homogeneous solid slabs with their faces smooth and parallel to each other, and filling entire half-spaces, $(-\infty < x, y < \infty)$ with one of the solids occupying $0 < z < \infty$ and the other occupying $-\infty < z < -d$, and the space between them, $(-\infty < x, y < \infty)$ and $-d \leq z \leq 0$ occupied by a material labeled 2, as shown in Figure 3.2, then the free energy per unit area of the solid faces is given by

$$G(d) = -\frac{A_{H123}}{12\pi d^2} \tag{3.5}$$

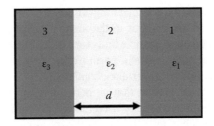

FIGURE 3.2   The system of two solid slabs, 1 and 3, modeled as filling two half-spaces and separated by a distance, d, with the intervening space, 2, occupied by an oil. The permittivities are frequency dependent with $\varepsilon_j = \varepsilon_0 \varepsilon_{rj}(\omega), j = 1, 2, 3$.

Here A_{H123} is the Hamaker coefficient given by (Hamaker, 1937; Dzyaloshinskii et al., 1961; Parsegian, 2005; French et al., 2010)

$$A_{H123} = -\frac{3}{2} k_B T \sum_{m=0}^{\infty} \int_0^{\infty} u\,du\,\ell n[1 - \Delta_{32}(i\xi_m)\Delta_{12}(i\xi_m)e^{-u}] \approx \frac{3}{2} k_B T \sum_{m=0}^{\infty} \Delta_{32}(i\xi_m)\Delta_{12}(i\xi_m) \tag{3.6}$$

where k_B is Boltzmann's constant, T is the absolute temperature, and

$$\Delta_{ab}(i\xi_m) = \frac{\varepsilon_{ra}(i\xi_m) - \varepsilon_{rb}(i\xi_m)}{\varepsilon_{ra}(i\xi_m) + \varepsilon_{rb}(i\xi_m)} \tag{3.7}$$

The sum is over imaginary (Matsubara) frequencies, $i\xi_m = 2\pi i k_B T m/\hbar$, where $\hbar = h/2\pi$ with h = Planck's constant, and the term for $m = 0$ has a prefactor of 1/2. In the case that the system responds predominantly only in the ultraviolet region of the electromagnetic spectrum, media 1 and 3 are identical, and the permittivities for media 1 and 2 are sufficiently similar, Equation 3.6 can be approximated (Israelachvili, 2006) by

$$A_{H123} = \frac{3}{4} k_B T \left(\frac{\varepsilon_{r1} - \varepsilon_{r2}}{\varepsilon_{r1} + \varepsilon_{r2}} \right)^2 + \frac{3h\nu_{UV}}{16\sqrt{2}} \frac{\left(n_1^2 - n_2^2\right)^2}{\left(n_1^2 + n_2^2\right)^{3/2}} \tag{3.8}$$

where
  n_k ($k = 1, 2$) are the refractive indices
  $\nu_{UV} = \omega_{UV}/2\pi$ is the electronic frequency in the UV region
  $\varepsilon_{rk} = \varepsilon_{rk}(0)$, the relative permittivity for frequency = 0

However, in general, a system will not respond only in the ultraviolet; systems also respond in both the microwave and infrared regions and these terms must be calculated and included in (3.8). French et al. (1995, 2010, and references therein) have developed techniques to calculate permittivities over the entire frequency range. It is necessary to know the constituents of the sample a priori. As earlier explained, if phase separation or nanoscale phases are involved then the Lifshitz approach can be problematical.

An approach was developed by Parsegian and Ninham (1970) who represented the permittivity for a system as a set of undamped harmonic oscillators (they have extended it to the damped case)

$$\varepsilon\left(i\xi_m\right) = 1 + \sum_j \frac{C_j}{1 + \left(\xi_m / \omega_j\right)^2} \qquad (3.9)$$

A plot of Equation 3.9 is shown in Figure 3.3.

By substituting Equation 3.9 into Equation 3.6 we can go beyond the approximations that lead to Equation 3.8 and calculate the effect of including infrared or microwave terms in the expression for the Hamaker coefficient. In order to do this we need to know about the coefficients of (3.9), the oscillator strengths $\{C_k\}$. The permittivities $\{\varepsilon_{rj}\}$ are related to the oscillator strengths $\{C_k\}$ by $C_j = \varepsilon_{rj} - \varepsilon_{rj+1}$ (Hough and White, 1980).

Although the original formulation by Dzyaloshinskii et al. (1961) was for spatially homogeneous systems, it has been extended to more complex geometries and spatial dependences (Parsegian, 2005). The complexity increases and it is not trivial to treat a system in which the permittivities might be continuously varying functions of spatial position. A relatively simple case is the Hamaker coefficient for the interaction between two spheres in the case that the permittivity of the spheres is a function of the radial distance from the centers of the spheres (Parsegian, 2005, p. 79).

Except for the work by Johansson and Bergenståhl (1992) who used Equation 3.8, Lifshitz theory has not been applied to fats and oils, even the simplest cases of a single type of fat in a known polymorphic form, in a single-component oil. We reiterate that the possibilities of phase separation on the nanoscale might render the use of Lifshitz theory inappropriate.

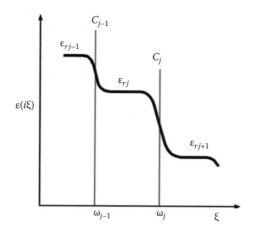

FIGURE 3.3 $\varepsilon(i\xi)$ as a function of ξ. The coefficient C_j is essentially the decrease in $\varepsilon(i\xi)$ as ξ increases from below ω_j to above ω_j.

The Coupled Dipole Method (CDM) (Kim et al., 2007; Verdult, 2010, for a very readable account) was developed in order to calculate van der Waals forces, in the spirit of Lifshitz theory, by including all correlations between dipoles but without invoking a mesoscale variable like the permittivity. The method begins by considering N vibrating atoms, $\{1, 2, 3,...\}$, which can be polarized by an electric field and which interact with each other via the Coulomb interaction. Atom j experiences an electric field set up at its position by all the other atomic dipoles; the dipole of atom j is created (induced) by the net electric field acting on it due to all the other dipoles. This picture is similar to that described by Figure 3.1c. One obtains a set of N coupled equations involving the unknown induced dipole moments $\{\vec{P}_j\}$ and the known atomic polarizabilities $\{\alpha_{j0}\}$ for $j = 1,..., N$ and any externally applied electric field, $\vec{E}_0$. One then solves for the induced dipole moments via matrix inversion.

To derive the energy of interaction between two solids, one writes down the quantum mechanical Hamiltonian operator, H, for the atomic harmonic oscillators of each solid in the usual way. We assume there are N_k atoms ($k = 1, 2$) in the two solids labeled 1 and 2, and we assume, for simplicity, that all atoms are identical so that we obtain

$$H = \sum_i^{N_1} \frac{p_i^2}{2m} + \sum_j^{N_2} \frac{p_j^2}{2m} + \frac{m\omega_0^2}{2}\left[\sum_i^{N_1} u_i^2 + \sum_j^{N_2} u_j^2\right] - \frac{q^2}{2\varepsilon_0}\sum_i^N\sum_j^N \vec{u}_i \cdot T_{ij} \cdot \vec{u}_j \quad (3.10)$$

where the first four terms comprise the usual Hamiltonian for $N_1 + N_2$ simple harmonic oscillators (e.g., Sakurai and Tuan, 1994; Messiah, 1999) and the last term describes the electromagnetic coupling between the atomic dipoles. Here the vector $\vec{u}_k$ describes the polarization vector of the kth atom, T_{ij} is a second-rank tensor that defines the electric field set up at atom j due to the atomic dipole i and $N = N_1 + N_2$. One then diagonalizes H, or uses perturbation theory, to obtain the eigenvalues (energies) of (3.10).

The advantage of the CDM approach is that it does not introduce a mesoscale variable like the permittivity as Lifshitz theory does, but uses atomic polarizabilities. One disadvantage is that it needs to obtain the eigenvalues, approximate or exact, for the Hamiltonian operator of the system. For the case of two solids in a vacuum, as done earlier, the problem can be solved. If, however, we have to introduce a fluid in which the two solids are immersed, the problem becomes more complicated and no such applications have yet been realized. Furthermore, for mixes of oils which can phase separate into mixtures of unknown a priori composition in a nanoscale environment, this approach suffers from the same drawback as Lifshitz theory. This is, however, an interesting approach that could find applications to fluids in confined nanospaces.

3.3.2 THE LENNARD JONES 6–12 POTENTIAL

The *Lennard-Jones (L-J) 6–12 potential*, $u_{AB}(r)$, describing the London van der Waals dispersion energy of interaction between two spherically symmetric atomic moieties, A and B are (Hamaker, 1937; Israelachvili, 2006)

$$u_{AB}(r) = -\frac{C_6^{AB}}{r^6} + \frac{C_{12}^{AB}}{r^{12}} \quad (3.11)$$

where we have denoted the energy between two atomic moieties, the smallest objects that go to make up larger systems, by a lowercase u. Here r is the center-to-center distance between A and B. While the attractive $1/r^6$ term can be justified from first principles, the form of the repulsive, $1/r^{12}$ term, representing the electron "overlap" repulsion between different atoms, has been chosen for computational convenience. Equation 3.11 has also been used to represent the interaction energy between two infinitesimal "chunks of matter" and has been utilized in mean field theories, such as the Derjaguin-Landau-Verveen-Overbeek (DLVO) theory of colloid interactions (Israelachvili, 2006). The attractive part of (3.11) for atom or molecules in their ground state can be derived from quantum mechanical second-order perturbation theory involving the dipole–dipole interaction as the perturbation. This yields (e.g., Wennerström, 2003)

$$u_{AB}(r) = -\frac{1}{24(\pi\varepsilon_0)^2}\frac{1}{r^6}\sum_l\sum_n\frac{\left|\langle A,l|\vec{P}|A,0\rangle\right|^2\left|\langle B,n|\vec{P}|B,0\rangle\right|^2}{(E_{A,l}-E_{A,0})+(E_{B,n}-E_{B,0})} = -\frac{C_6^{AB}}{r^6} \qquad (3.12)$$

Here, $|A, l\rangle$ is a statevector for the A molecule in state l, belonging to energy $E_{A,l}$, with $l = 0$ representing the ground state with energy $E_{A,0}$. A similar definition holds for the statevectors of the B molecule in state n. $\vec{P}$ is the electric dipole moment operator and r is the distance between the molecules. It is clear that Equation 3.12 involves only pairs of molecules so that the interaction between two objects composed of such molecules will be the sum of all the pairwise interactions. Nonadditive perturbation terms arise only in third-order perturbation theory (Axilrod and Teller, 1943).

If one knows the functional form of the number densities, $\Phi_i(\vec{r})$, of the solids ($i = 1, 3$) and the oil ($i = 2$), then one can use the Lennard-Jones 6–12 potential to compute the total energy of the system and, from it, deduce the Hamaker coefficient. The energy of the system of Figure 3.2 interacting via the 6–12 potential is (Israelachvili, 2006)

$$U_{123} = \sum_{j\neq i}\int_i d^3\vec{r}\int_j d^3\vec{r}'\left[\frac{-C_6^{ij}}{\left|\vec{r}-\vec{r}'\right|^6}+\frac{C_{12}^{ij}}{\left|\vec{r}-\vec{r}'\right|^{12}}\right]\Phi_i(\vec{r})\Phi_j(\vec{r}') \qquad (3.13)$$

where the sum is over $i, j = 1, 2, 3, j \neq i$ and the double-counting of energies is disallowed. The coefficients, C_n^{ij}, describe the interaction energy ($n = 6, 12$) between substances i and j. One has to make some assumptions about the number densities and the most common assumption is that they are constant, independent of the solid separation, d. However, if one requires that the chemical potential (free energy per molecule) of the slabs in Figure 3.2 is equal to the chemical potential of a pure-oil system (the "bulk") in contact with that slab system, then one need not assume that the number densities are constant. From this, one can obtain an average number density of oil between the two nanocrystals as a function of d.

If we consider the system of Figure 3.2 to be a pair of triacylglycerol crystals, 1 and 3, separated by a triacylglycerol oil with the same chain lengths, then it is reasonable to assume that the coefficients C_6 and C_{12} are equal for solid–solid, solid–liquid, and liquid–liquid interactions, and, assuming that the number densities are constant, then by a trivial integration we obtain the energy per unit cross section of the two solids:

$$U_{123}(d) = \left[-\frac{\pi C_6}{12}\frac{1}{d^2} + \frac{\pi C_{12}}{360}\frac{1}{d^8} \right](\Phi_1 - \Phi_2)^2 \qquad (3.14)$$

The attractive term then yields the Hamaker coefficient (Equation 3.5)

$$A_{H123} = \pi^2 C_6 (\Phi_1 - \Phi_2)^2 \qquad (3.15)$$

There are two other models of direct relevance to fats aggregation. One was developed to relate mechanical properties of fats aggregates to their structure, while the other deduces the value of the Hamaker coefficient from the disjoining energy of dividing a fats crystal into two parts.

3.3.3 Fractal Model and Semi-Classical Model

The Fractal Model. Vreeker et al. (1992) studied tristearin aggregates in olive oil using light scattering and deduced that they formed a fractal structure with dimension $D = 1.7$ which increased to $D = 2.0$ as the sample aged. They also studied the storage modulus, G' as a function of solid fraction (solid fat content), ϕ, and observed a power-law dependence in accord with a fractal dimension, $D = 2.0$. These experiments were carried out for low solid fat content (<10%), the "strong-link" regime, in which the interaction between the aggregating objects is stronger than those between the subunits of which the aggregates are composed. The value of $D = 1.7$ is characteristic of diffusion limited cluster-cluster aggregation (DLCA). They did not, however, identify the objects that were aggregating, nor did they report on how large the aggregating objects were. The first to attempt to make models of the process was van den Tempel (1961) but it was not until the work of Marangoni and Rousseau (1999) followed by Narine and Marangoni (1999a,b) and subsequently by Marangoni (2000, 2002) and Marangoni and Rogers (2003) that the fractal structure of aggregation was modeled.

From rheological experiments, as well as PLM observations, fractal models were developed (Rogers et al., 2008) and Figure 3.4 shows the interpretation of fat crystal networks with high solid fat content. It is composed of "flocs" which are themselves formed from fractal aggregates of smaller units.

While the low value of ϕ used by Vreeker et al. (1992) is the "strong-link" regime, the system studied by Marangoni's group is in the "weak-link" regime, corresponding to the large value of ϕ used; the interactions between the flocs (large gray spheres) are stronger (strong-link) or weaker (weak-link) than those within the fractal structures, composed of small spheres (black), that make up the flocs. The size of the flocs can be calculated from images (Rye et al., 2005, Figure 3.8). The size of the units making up flocs can also be deduced from the "Bragg peaks."

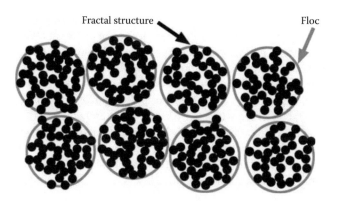

FIGURE 3.4 Schematic diagram of the weak-link fractal model. Small units (black) self-assemble into fractal structures, of fractal dimension and form flocs (gray). The weak-link describes the weak interaction between flocs compared to the stronger interaction between the small units of the fractal structures. (From Rye, G.G. et al.: *Bailey's Industrial Oil and Fat Products, Volume 4, Edible Oil and Fat Products: Products and Applications.* 2005. Copyright Wiley-VCH Verlag GmbH & Co. KGaA.)

The model relates the storage modulus, G', to the Hamaker coefficient, A_H,

$$G' = \frac{A_H}{6\pi a E d_0^2} \phi^{1/(3-D)} \tag{3.16}$$

$$A_H = 6\pi a E \lambda d_0^2 \tag{3.17}$$

where

λ is the slope of $ln[G']$ with respect to $ln[\phi]$ (Narine and Marangoni, 1999b; Marangoni, 2000) where ϕ is the solid fat content

D is the fractal dimension of the structure of which the floc is composed

a is the diameter of a fat floc

E is the extensional strain, $\Delta L/L$, at the limit of linearity (Tang and Marangoni, 2007)

d_0 is the equilibrium distance between the floc surfaces

For the cases of fully hydrogenated canola oil (FHCO, predominantly tristearin) and high oleic sunflower oil (HOSO, predominantly triolein), $E = 1.9 \cdot 10^{-4}$ (Hiemenz and Rajagopalan, 1997), $\lambda = 1.17 \cdot 10^7$ Pa (Ahmadi et al., 2008) and $a = 150$ nm (Acevedo and Marangoni, 2010b). The value of d_0 is unknown. This model, however, is not obviously represented by Figure 3.2.

Work has been done that could have relevance to modeling solids in oils. Of special note is that of Babick et al. (2011) who presented a scheme for calculating the van der Waals interaction between a pair of fractal aggregates and we shall return to this later. They make the point that, normally, it is assumed that the fractal aggregates are spherical, and they present results of simulations for which this is so. The point here is that, if the rotation time of an aggregate is very much smaller than its characteristic lateral movement time, then the spherical approximation could be valid. Otherwise, one cannot assume that it is even approximately valid: fractal structures can assume many different shapes. In the following we shall see that the CNPs identified by Acevedo and Marangoni (2010a) aggregate into long "multilayer sandwiches." It is possible that, on a microscale, such structures might appear as long rods or needles. Accordingly, other works of potential relevance are those of Dhont and Briels (2003) concerned with the hydrodynamics and ordering of long rods, and Li et al. (2011) concerned with the hydrodynamics of nanorods and nanoplatelets.

The *semi-classical model* relates the Hamaker coefficient A_H to the disjoining free energy of a fat crystal (Hiemenz and Rajagopalan, 1997) separated into two blocks at their equilibrium separation, d_e:

$$A_H = 24\pi\gamma d_e^2 \qquad (3.18)$$

where γ is the interfacial free energy per unit area. As in the case of the equilibrium distance between flocs, d_0, of the Fractal model, the value of d_e is unknown.

From Equations 3.17 and 3.18 we have two unknowns: the equilibrium distance between flocs, d_0 and the equilibrium distance between two blocks of fat crystals, d_e.

3.3.4 COARSE-GRAINED APPROACHES—1

(Müller et al., 2006; Marrink et al., 2007; Monticelli et al., 2008; Bennun et al., 2009; Laradji and Kumar, 2011) How we model a system depends upon what spatial scale interests us. As far as the Coulomb interaction in atomic systems is concerned, we could model a molecule by treating the nucleus as a point object and taking into account the electronic states most of which exhibit nonspherical distributions, as well as electronic polarizabilities and spin and orbital angular momentum states. However, many of these degrees of freedom are irrelevant to, for example, nano-platelet interactions. One intent of modeling must be to eliminate those system coor-dinates which undergo motions on timescales much less than the timescales in which we are interested. We then replace the detailed dynamics of the fat coordinates by average values. Such a procedure is called "coarse graining" and Example 3.3.4.1 illustrates this.

3.3.4.1 Example: Aggregation of Triacylglycerol CNPs

The smallest components of fats appear to be CNPs (Acevedo et al., 2011) and we begin with considering them as "fundamental components." These highly anisotro-pic objects have dimensions of hundreds of nanometers along two, approximately perpendicular axes and less than one hundred nanometers in the third dimension, and are composed of planes of triacylglycerol molecules (TAGs). In order to model CNPs we represented the flat nanocrystals by a close-packed lattice of three-dimensional solid unit structures composed of crystalline TAGs. It is convenient to represent each CNP as a close-packed structure of spheres, since the attractive part of the dispersion interaction between spheres has been established. Each sphere, repre-sented a continuum of TAG molecules with a density characteristic of the crystalline phase of interest, possesses radius R, and Figure 3.5 shows representations of a CNP.

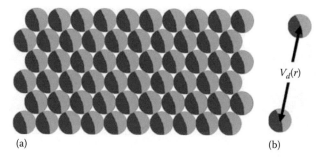

(a)   (b)

FIGURE 3.5 (a) Model of a CNP. They are defined by the integers $l \times m \times n$ specifying the number of spheres in each row (l) and column (m) and perpendicular to that plane (n). Here $l = 13$ and $m = 6$. (b) The interaction between two spheres, a center-to-center distance, r, apart, and belonging to different model CNPs. The total interaction between two CNPs is the sum of all pairwise interactions, $V_d(r)$.

Although we were unaware of it, such a model has been used elsewhere (Glotzer et al., 2005; Shim and McDaniel, 2010).

The interaction between two identical homogeneous spheres, each of radius R, a center-to-center distance, r, apart is the Hamaker hybrid form (Parsegian, 2005, Table S.3, p. 155)

$$V_d(r) = -\frac{A_H}{6}\left[2R^2\left(\frac{1}{s^2+4Rs}+\frac{1}{(s+2R)^2}\right)+\ell n\left(\frac{s^2+4Rs}{(s+2R)^2}\right)\right] \quad r \geq 2R \qquad (3.19)$$

where

$s = r - 2R$ is the surface-to-surface separation of the two spheres

A_H is the Hamaker coefficient

Aggregation of CNPs were modeled using Monte Carlo methods (Binder, 1997; Binder and Heermann, 2010) in an $L \times L \times L$ simulation box with periodic boundary conditions. One MC step involved translating and rotating CNPs and translating and rotating clusters of CNPs with respect to their centers of mass. We permitted movement of all clusters with the translational step size as well as the angle of rotation around a randomly chosen axis through the centre of mass, proportional to $M^{-1/2}$, where M is the mass of the cluster. This approach is known as diffusion limited cluster-cluster aggregation (DLCA). All CNPs were of size $m \times m \times 1$. We chose the radius of the spheres making up the CNPs to be $R = 0.5$ in arbitrary units. We carried out two kinds of simulation which would lead to a total of N CNPs taking part in aggregation, though not necessarily forming a single cluster. This case had all N CNPs present initially. The other case argued that CNPs were formed as the temperature dropped below the freezing point of those TAGs which went to form the nanoplatelets or spherical structures. This creation of CNPs is described by a characteristic time τ_{create}. As soon as they come into existence, however, such structures can begin to aggregate and relax with a characteristic time of τ_{relax}. This process was simulated by choosing an initial number of CNPs, N_I, and incrementing this by a number, ΔN, every ΔK MC

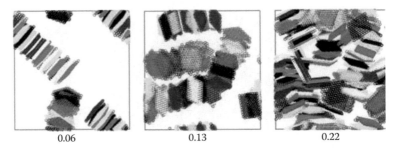

0.06   0.13   0.22

FIGURE 3.6 DLCA of CNPs via Monte Carlo simulations. Representative configurations of CNPs of size (10 × 10 × 1) with sphere radius of $R = 0.5$ for three mass concentrations, 0.06, 0.13, and 0.22.

steps. After a total of $(N - N_I)/\Delta N$ MC steps, all N CNPs would have been created. N is determined by the concentration of TAGs going to make up CNPs in the system.

We chose the total number of spheres to be $N = 8,000$, 16,000, or 24,000 yielding 80, 160, or 240 CNPs which corresponds to CNP concentrations of 0.06. 0.13, and 0.22. Figure 3.6 shows that the CNPs form long multilayer "sandwiches" which, taken together, form a structure filling all of space. We also considered the case in which the CNP sizes followed a Gaussian distribution, $\{m \times m \times 1\}$, with average, $<m> = 10$, and variance, $\sigma^2 = 5$. We found the same results as shown in Figure 3.6. We note that, for higher concentrations of CNPs, it is possible that the multilayer sandwiches could appear as rods or needles and that there is evidence for such structures in Figure 3.1 of Acevedo et al. (2011) and the following text (Van der Waals interactions and rheological characteristics, Janssen, 2004, Figure 3).

3.3.4.2 Application: Oils in Confined Nanospaces

Recently we carried out simulations in order to investigate the oil binding capacity of CNPs as well as the validity of the assumption that the liquid oil is homogeneous when the distance, d, between two CNPs, as shown in Figure 3.2, was a few nanometers. We modeled the radial interactions between nonbonded TAG atomic moieties using GROMACS (Berendsen et al., 1995; Kutzner et al., 2007) and used the force fields given by Berger et al. (1997). Although the NpT ensemble (constant number of molecules, pressure, and temperature) is the "natural" ensemble to use, the problem of simulating two sufficiently large slabs immersed in a liquid oil (bulk) is excessively compute-intensive. Instead, we represented the bulk and the fats particles separately, representing the fats particles as in Figure 3.2, and required that the chemical potential of the fats particles system be equal to that of the bulk. We represented the fats particles as tristearin nanocrystals and the oil as triolein. Because we know the density of a bulk triolein fluid, ~930 kg/m³, we used the NVT (constant number of molecules, volume, and temperature) ensemble for both the bulk and the system of Figure 3.2 with periodic boundary conditions in the planes perpendicular to the interfaces. The solids were represented as continua interacting with each other by replacing the individual molecules by average values of the 6–12 theory parameters. The coefficients, C_6^{ij} and C_{12}^{ij} ($i, j = 1, 2$), were replaced by their averages over the atomic moieties making up the solids. Figure 3.7 shows some of the results for the

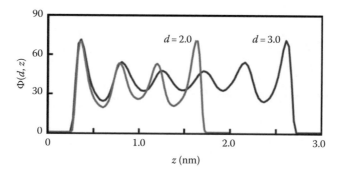

FIGURE 3.7   TAG number density, $\Phi(d, z)$, as a function of position, z (in nm) between the two solid slabs (Figure 3.2), for given slab separation, d.

TAG number density, $\Phi(d, z)$, as a function of position, z, between the two solid slabs of Figure 3.2, for given slab separation, d. The result from this is that the number density, and therefore the density, is oscillatory and not constant as assumed by the Lifshitz and the 6–12 potential theories, as well as any theories which make use of these, such as DLVO theory.

If we integrate the number density over z to obtain the area under the two curves of Figure 3.7, then we obtain the average number density between the two slabs, $\langle\Phi(d)\rangle$. We then discover that $\langle\Phi(2.0)\rangle < \langle\Phi(3.0)\rangle$. In fact, $\langle\Phi(d)\rangle$ exhibits a general decrease as $d \to 0$. What this means is that any oil between the two slabs will begin to leak away for finite values of d until the two slabs come into molecular contact. If we define the "oil binding capacity," $B(d)$, to be

$$B(d) = \frac{\langle\Phi(d)\rangle}{\langle\Phi_{bulk}\rangle} \tag{3.20}$$

where $\langle\Phi_{bulk}\rangle$ is the average number density of the bulk oil, then we would see that the oil binding capacity of tristearin CNPs in triolein is low.

Because of the number density oscillations (Figure 3.7), the permittivity and the refractive index will likely vary as functions of z. Huang and Levitt (1977) calculated the average permittivity of a phospholipid bilayer membrane perpendicular to bilayer plane and found that the value differed from bulk hydrocarbon chain values by a few percent. Such a membrane is ~5 nm thick which is larger than the number density oscillation wavelength of ~0.5 nm (Figure 3.2). The definition of permittivity and refractive index on the scale of tenths of nm, as required by Lifshitz theory, might be problematic.

Why does the 6–12 potential work for AMD? In the introduction to the section on van der Waals interactions, it was recognized that, even though the pairwise attractive potential is inadequate to describe the many-body interactions, it works in AMD computer simulations. The reason is likely to be that the C_6 coefficients used in the two-body attractive interactions of AMD arise from the "decoupling" of the many-body interactions of Equation 3.3, a technique that has long been used in, for example, solid- and liquid-state theories. This amounts to breaking up many-body interactions into sums of two-body interactions weighted by average values, indicated by brackets < >:

$$\tilde{U} = \sum_j \sum_k \tilde{u}_{jk} + \sum_j \sum_k \sum_l \tilde{u}_{jkl} + \cdots$$

$$\approx \sum_j \sum_k \tilde{u}_{jk} + \sum_j \sum_k \sum_l \left[<\tilde{u}_l> \tilde{u}_{jk} + <\tilde{u}_k> \tilde{u}_{jl} + <\tilde{u}_j> \tilde{u}_{kl} \right] + \cdots \quad (3.21)$$

3.3.5 COARSE-GRAINED APPROACHES—2

The principle of coarse graining is that, if one is interested in long timescales, then one attempts to replace those components possessing the fastest timescales by average values and average fluctuations. Naturally, correlations will be lost but in certain cases, for example, when one is not near a (generally continuous) phase transition, such a replacement might be justified. The most compute-intensive component of many food systems is water and any technique which can coarse grain water should be sought. In oil systems, the fastest timescales are possessed by individual TAG molecules so that, if one is interested in the aggregation of solid fats then one should attempt to replace individual TAG molecules by some "collective" entity that reflects the average values that characterize many TAG molecules. The new shortest timescale will then be those characterizing the motion of the fats particles. One technique is *Brownian Dynamics* (Ermak and McCammon, 1978) based on the Langevin equations. One replaces liquid-state molecules by a continuum which contributes a stochastic force to the molecular dynamics simulation. Still others simplify the motion by replacing the continuum space by a lattice on which all objects are constrained to move from site-to-site. Such models are the *Lattice Gas* or the *Lattice Boltzmann* approaches (e.g., Rajagopalan, 2001). The drawbacks of the latter approaches are (1) the stochastic forces do not conserve momentum and (2) the lattice models violate Galilean invariance. The latter states that, if Galilean invariance holds, then Newton's laws of motion hold in all inertial frames of reference. *Dissipative Particle Dynamics* (DPD) (Marsh, 1998; Flekkøy et al., 2000; Español and Revenga, 2003; Schiller, 2005 is readable; Pink et al., 2010 applies it to TAGs) was created to solve the Galilean invariance problem of these models; DPD is not lattice-based and it has been shown to be in accord with the Navier–Stokes equation, the fundamental equation of hydrodynamics. It represents all components by points possessing position, velocity, and mass on each of which is acting three forces derived from the surroundings of each point: a conserved force, $\vec{F}_C$, which represents elastic scattering; a dissipative force, $\vec{F}_D$, which represents a friction force (inelastic scattering) between the components, and a random force, $\vec{F}_R$, representing the effects of the temperature. $\vec{F}_D$ and $\vec{F}_R$ are related via the fluctuation-dissipation theorem (Marconi et al., 2008). Molecular Dynamics (Frenkel and Smit, 2002; Hess et al., 2008) is then used to obtain average quantities.

3.4 VAN der WAALS INTERACTIONS AND RHEOLOGICAL CHARACTERISTICS

The shear stress magnitude, σ, defines a force that a flowing liquid exerts on a surface, per unit area of surface, and its direction is that of the flow (Larson, 1999). If this force is sinusoidal with angular frequency, ω, and if we can represent the stress as

$$\sigma(t) = \gamma_0[G'(\omega)\sin(\omega t) + G''(\omega)\cos(\omega t)] \tag{3.22}$$

then the system is in its linear viscoelastic regime. Here $G'(\omega)$ and $G''(\omega)$ are the storage (elastic scattering) and loss (inelastic scattering) moduli, and are in phase and $\pi/2$ radians out of phase, respectively, with the shear strain, γ, the distortion produced by the shear stress. One can define the complex modulus, $G^*(\omega)$, and the loss tangent, $\tan\delta(\omega)$, as

$$G^*(\omega) = G'(\omega) + iG''(\omega)$$

$$\tan\delta(\omega) = \frac{G''(\omega)}{G'(\omega)} \tag{3.23}$$

with $\tan\delta(\omega) \gg 1$ for a liquid-like system and $\tan\delta(\omega) \ll 1$ for a solid-like system. From now on the functional dependence upon ω will be omitted for simplicity. Other important quantities are the shear rate, $\dot{\gamma} = V/h$ for shearing between two parallel plates, and the complex viscosity, $\eta^* = \eta' - i\eta'' = G^*/i\omega$ where V is the velocity of one plate with the other plate stationary and h is the distance between the plates. For a Newtonian fluid, a plot of $\dot{\gamma}$ versus σ is a straight line: $\sigma/\dot{\gamma} = \eta$, the shear viscosity. If a plot of G' and G'' as functions of ω exhibits a crossover at $\omega = \omega_c$ then the system possesses a characteristic relaxation time, $\tau_c \sim \omega_c^{-1} \sim \dot{\gamma}_c^{-1}$. A question that now arises is as follows: How do we relate the van der Waals interaction to measurable quantities such as G' and G''?

It should be stressed that this does not imply that a knowledge of molecular structure will necessarily point the way to the answer. This was summed up cogently a decade ago (Marangoni, 2002); a knowledge of the dominant structures which arise, on all length scales, is likely essential to be able to predict the rheological properties of materials. Furthermore, one cannot be certain until one has understood how different structures come about, whether the system being studied is even in thermodynamic equilibrium. When one realizes that oils used in foods are mixes of a number of components which can separate into micro- or nanoscale phases, one appreciates the complexity of the task. A taste of this complexity was given by Higaki et al. (2004). One thus returns to the beginning of this chapter; given that CNPs are the fundamental components of a TAG system, the structures which arise from their interactions with each other and with the oils in which they are embedded, and the pathways through which self-assembly comes about, is essentially unknown.

Van den Tempel (1961) introduced the Linear Chain model which gave rise to the Modified Linear Chain model (Van den Tempel, 1979). Although the latter fitted data better than the former, it was difficult to find independent determinations of some of the parameters of the models. The report of Vreeker et al. (1992) that tristearin aggregates in olive oil formed a fractal structure with dimension $D = 1.7$, in accord with DLCA, inspired the fractal model. The work of Shih et al. (1990), Rahman (1997), and Liang et al. (2006), together with that of Heertje (1993) and that of the Marangoni group (Marangoni and Rousseau, 1996; Narine and Marangoni, 1999a, 2000, 2002; Marangoni and Rogers, 2003; Tang and Marangoni, 2006), resulted in relating the storage modulus to the Hamaker coefficient and other characteristics of the sample. This led to Equations 3.16 and 3.17.

A key observation leading to Equation 3.16 is that fat crystal networks at high ϕ-values behave as if they are in the "weak-link" regime. A key assumption is that, if the flocs are not as shown in Figure 3.4 with unit-free (solid black circles) volumes between them but actually fill all of space, then the solid volume fraction within a floc is equal to that, ϕ, of the system as a whole (Tang and Marangoni, 2007). One then obtains the ϕ-dependence of Equation 3.16.

Work of possible relevance to the structure and rheology of oils was carried out by Janssen (2004) who used finite element analysis to study a model in which the solid fat was represented by interacting needles and concluded that this model best explained the scaling exponents for the elastic modulus. This conclusion is interesting in the light of the results obtained using DLCA that CNPs will aggregate into "multilayer sandwiches" that might appear as needles (Section 3.3.4.1, Figure 3.6). Interesting work on the elasticity of fractal media has been done by Carpinteri et al. (2004).

Finally, much work has been done in the field of powder technology that might be relevant to oils; the calculation of the van der Waals force between pairs of fractal objects. A recent paper by Babick et al. (2011) summarizes the latest advances in this field and describes an algorithm for calculating the van der Waals interaction energy between fractal aggregates when those fractal objects are aggregates formed from colloidal particles (Langbein, 1974; Naumann and Bunz, 1993). The algorithm of Babick et al. (2011) is specifically tailored for DLCA, the technique that was utilized in obtaining the results of Figure 3.6.

3.5 X-RAY SCATTERING AND FRACTAL DIMENSIONS

There is now convincing evidence from a number of sources that CNPs aggregate and ultimately give rise to the formation of fractal structures. In the past, fractal dimensions have been deduced from the analysis of two-dimensional structures (Marangoni, 2002, 2005). The structures observed were not in TAG oils and the question remained as to whether such fractal structures actually existed *in vivo*, so to speak. X-ray scattering offers a way to measure fractal dimensions (Beaucage, 2004) formed by CNPs in the presence of oils in *in vivo* three-dimensional samples. The x-ray scattering intensity, $I(q)$, for samples composed of structures with zero average orientation, is determined by, amongst other factors, the structure factor, $S(q)$, where q is the magnitude of the difference between the wavevector, $\vec{k}$, of the incident photons and the wavevector, $\vec{k}'$, of the elastically scattered ($|\vec{k}'| = |\vec{k}|$) photons: $q = |\vec{k} - \vec{k}'|$.

The key relation establishing the value of the fractal dimension, D, is that, for some range of q (Jullien, 1992),

$$S(q) \sim q^{-D} \quad \text{or} \quad \log[S(q)] \sim -D\log[q] \tag{3.24}$$

Experiments have recently been carried out at the Advanced Photon Source, Argonne. Details of these structures have not yet been elucidated but preliminary results (Peyronel et al., unpublished data) offer confirmation that, for values of solid fat concentration near 100%, such fractal structures do exist. Caution should be exercised in that computer simulations suggest that, under certain conditions, the value of D obtained from $S(q)$ might be overestimated (Lach-hab et al., 1998).

3.6 CONCLUSION

This chapter has attempted to describe aspects of the van der Waals interactions between solid fats in oils. The intent has been to raise questions in the readers' minds without always explicitly identifying those models, methods, and assumptions which might be worth a second skeptical look. This not only protects the writer from the wrath of his friends but also provides some stimulus to the reader.

ACKNOWLEDGMENTS

It is a pleasure to thank Fernanda Peyronel, Bonnie Quinn, and Erzsebet Papp-Szabo for their comments and assistance. Thanks also to AFMnet and NSERC for their support.

REFERENCES

Acevedo, N.C. and A.G. Marangoni. 2010a. Characterization of the nanoscale in triacylglycerol crystal networks. *Cryst. Growth Des.* 10: 3327–3333.

Acevedo, N.C. and A.G. Marangoni. 2010b. Towards nanoscale engineering of triacylglycerol crystal networks. *Cryst. Growth Des.* 10: 3334–3339.

Acevedo, N.C., F. Peyronel, and A.G. Marangoni. 2011. Nanoscale structure intercrystalline interactions in fat crystal networks. *Curr. Opin. Colloid Interface Sci.* 16: 374–383.

Ahmadi, L., A.J. Wright, and A.G. Marangoni. 2008. Chemical and enzymatic interesterification of tristearin/triolein-rich blends: Chemical composition, solid fat content and thermal properties. *Eur. J. Lipid Sci. Technol.* 110: 1014–1024.

Antunes, F.E., E.F. Marques, M.G. Miguel, and B. Lindman. 2009. Polymer-vesicle association. *Adv. Colloid Interface Sci.* 147–148: 18–35.

Axilrod, B.M. and E. Teller. 1943. Interaction of the van der Waals type between three atoms. *J. Chem. Phys.* 11: 299–300.

Babick, F., K. Schiesl, and M. Stintz. 2011. Van der Waals interaction between two fractal aggregates. *Adv. Powder Technol.* 22: 220–225.

Beaucage, G. 2004. Determination of branch fraction and minimum dimension of mass-fractal aggregates. *Phys. Rev. E.* 70: 031401–031410.

Bennun, S.V., M.I. Hoopes, C. Xing, and R. Faller. 2009. Coarse-grained modeling of lipids. *Chem. Phys. Lipids.* 159: 59–66.

Berendsen, H.J.C., D. van der Spoel, and R. Vandrunen. 1995. GROMACS: A message-passing parallel molecular dynamics implementation. *Comput. Phys. Commun.* 91: 43–53.

Berger, J., O. Edholm, and F. Jähnig. 1997. Molecular dynamics simulations of a fluid bilayer of dipalmitoylphosphatidylcholine at full hydration, constant pressure, and constant temperature. *Biophys. J.* 72: 2002–2013.

Binder, K. 1997. Applications of Monte Carlo methods to statistical physics. *Rep. Prog. Phys.* 60: 487–559.

Binder, K. and D. Heermann. 2010. *Monte Carlo Simulation in Statistical Physics: An Introduction.* Berlin, Germany: Springer.

Butt, H.-J., K. Graf, and M. Kappl. 2003. *Physics and Chemistry of Interfaces.* Weinheim, Germany: Wiley-VCH Verlag GmbH & Co. KGaA.

Carpinteri, A., B. Chiaia, and P. Cornetti. 2004. The elastic problem for fractal media: Basic theory and finite element formulation. *Comput. Struct.* 82: 499–508.

Cevc, G. 1990. Membrane electrostatics. *Biochim. Biophys. Acta.* 1031: 311–382.

Cevc, G. and D. Marsh (Eds). 1987. *Phospholipid Bilayers: Physical Principles and Models.* New York: John Wiley & Sons.

Dhont, J.K.G. and W.J. Briels. 2003. Viscoelasticity of suspensions of long rigid rods. *Colloids Surf. A Physicochem. Eng. Asp.* 213: 131–156.

Dzyaloshinskii, I.E., E.M. Lifshitz, and L.P. Pitaevskii. 1961. General theory of van der Waals' forces. *Sov. Phys. Uspekhi.* 73: 153–176.

Ermak, D.L. and J.A. McCammon. 1978. Brownian dynamics with hydrodynamic interactions. *J. Chem. Phys.* 69: 1352–1361.

Español, P. and M. Revenga. 2003. Smoothed dissipative particle dynamics. *Phys. Rev. E.* 67: 026705–026712, and references therein.

Fathi, M., M.R. Mozafari, and M. Mohebbi. 2011. Nanoencapsulation of food ingredients using lipid-based delivery systems. *Trends Food Sci. Technol.* 23: 13–27.

Flekkøy, E.G., P.V. Coveney, and G. De Fabritiis. 2000. Foundations of dissipative particle dynamics. *Phys. Rev. E.* 62: 2140–2157.

French, R.H., R.M. Cannon, L.K. DeNoyer, and Y.-M. Chiang. 1995. Full spectral calculation of non-retarded Hamaker constants for ceramic systems from interband transition strengths. *Solid State Ionics.* 75: 13–33.

French, R.H. et al. 2010. Long range interactions in nanoscale science. *Rev. Mod. Phys.* 82: 1887–1944.

Frenkel, D. and B. Smit. 2002. *Understanding Molecular Simulation: From Algorithms to Applications,* 2nd edn., Chapter 17. Amsterdam, the Netherlands: Elsevier, Inc., pp. 465–478.

Glotzer, S.C., M.A. Horsch, C.R. Iacovella, Z. Zhang, E.R. Chan, and X. Zhang. 2005. Self-assembly of anisotropic tethered nanoparticle shape amphiphiles. *Curr. Opin. Colloid Interface Sci.* 10: 287–295.

Hamaker, H.C. 1937. The London-van der Waals attraction between spherical particles. *Physica.* 4: 1058–1072.

Hanna, C.B., D.A. Pink, and B.E. Quinn. 2006. Van der Waals interactions with soft interfaces. *J. Phys. Condens. Matter.* 18: 8129–8137.

Heertje, I. 1993. Microstructure studies in fat research. *Food Microstruct.* 12: 77–94.

Hess, B., C. Kutzner, D. van der Spoel, and E. Lindahl. 2008. GROMACS 4: Algorithms for highly-efficient, load-balanced, and scalable molecular simulation. *J. Chem. Theory Comput.* 4: 435–447. See also www.gromacs.org, GROMACS users manual version 4.0, gromacs4_manual.pdf

Hiemenz, P.C. and R. Rajagopalan. 1997. *Principles of Colloid and Surface Chemistry,* 3rd edn. New York: Marcel Dekker.

Higaki, K., T. Koyano, I. Hachiya, K. Sato, and K. Suzuki. 2004. Rheological properties of β-fat gel made of binary mixtures of high-melting and low-melting fats. *Food Res. Int.* 37: 799–804.

Himawan, C., V.M. Starov, and A.G.F. Stapley. 2006. Thermodynamic and kinetic aspects of fat crystallization. *Adv. Colloid Interface Sci.* 122: 3–33.

Hough, D.B. and L.R. White. 1980. The calculation of Hamaker constants from Lifshitz theory with applications to wetting phenomena. *Adv. Colloid Interface Sci.* 14: 3–41.

Huang, W.-T. and D.G. Levitt. 1977. Theoretical calculation of the dielectric constant of a bilayer membrane. *Biophys. J.* 17: 111–128.

Israelachvili, J.N. 2006. *Intermolecular and Surface Forces*. New York: Academic Press.

Janssen, P.M.W. 2004. Modelling fat microstructure using finite element analysis. *J. Food Eng.* 61: 387–392.

Johansson, D. and B. Bergenståhl. 1992. The influence of food emulsifiers on fat and sugar dispersions in oils: 2. Rheology, colloidal forces. *J. Am. Oil Chem. Soc.* 69: 718–727.

Jullien, R. 1992. From Guinier to fractals. *J. Phys. I France.* 2: 759–770.

Kim, H.-Y., J.O. Sofo, D. Velegol, M.W. Cole, and A.A. Lucas. 2007. Van der Waals dispersion forces between dielectric nanoclusters. *Langmuir.* 23: 1735–1740.

Kutzner, C., D. van der Spoel, M. Fechner, E. Lindahl, U.W. Schmitt, B.L. de Groot, and H.J. Grubmuller. 2007. Speeding up parallel GROMACS on high-latency networks. *Comput. Chem.* 28: 2075–2084.

Lach-hab, M., A.E. Gonzalez, and E. Blaisten-Barojas. 1998. Structure function and fractal dimension of diffusion-limited colloidal aggregates. *Phys. Rev. E.* 57: 4520–4527.

Landau, L. and E. Lifshitz. 1968. *Electrodynamics*. Moscow, Russia: Nauka.

Langbein, D. 1974. *Theory of van der Waals Attraction*. Berlin, Germany: Springer-Verlag.

Laradji, M. and P.B.S. Kumar (Eds). 2011. Coarse-grained computer simulations of multicomponent lipid membranes. In: *Advances in Planar Lipid Bilayers and Liposomes*, Vol. 14. Oxford, U.K.: Academic Press, pp. 201–233. Although this reference is concerned with lipid bilayers, it does contain descriptions of various simulation techniques.

Larson, R.G. 1999. *The Structure and Rheology of Complex Fluids*. New York: Oxford University Press.

Larsson, K. 2009. Lyotropic liquid crystals and their dispersions relevant in foods. *Curr. Opin. Colloid Interface Sci.* 14: 16–20.

Leckband, D. and J.N. Israelachvili. 2001. Intermolecular forces in biology. *Q. Rev. Biophys.* 34: 105–267, and references therein.

Lee-Desautels, R. 2005. Theory of van der Waals forces as applied to particulate materials. *Educ. Reso. Part. Techn* 051Q-Lee 1–8. Available at http://www.erpt.org/051Q/leed-01.pdf

Leser, M.E., J. Sagalowicz, M. Michel, and H.J. Watzke. 2006. Self-assembly of polar food lipids. *Adv. Colloid Interface Sci.* 123–126: 125–136.

Li, J., M.G. Forest, Q. Wang, and R. Zhou. 2011. A kinetic theory and benchmark predictions for polymer-dispersed semi-flexible macromolecular rods or platelets. *Physica D.* 240: 114–130.

Li, Q., V. Rudolph, and W. Peukert. 2006. London-van der Waals adhesiveness of rough particles. *Powder Technol.* 161: 248–255.

Liang, B., J.L. Sebright, Y. Shi, R.W. Hartel, and J.H. Perepezko. 2006. Approaches to quantification of microstructure for model lipid system. *J. Am. Oil Chem. Soc.* 83: 389–399.

Lifshitz, E.M. 1956. The theory of molecular attractive forces between solids. *Sov. Phys. JETP.* 2: 73–83.

Marangoni, A.G. 2000. Elasticity of high-volume fraction fractal aggregate networks: A thermodynamic approach. *Phys. Rev. B.* 62: 13951–13955.

Marangoni, A.G. 2002. The nature of fractality in fat crystal networks. *Trends Food Sci. Technol.* 13: 37–47.

Marangoni, A.G. 2005. *Fat Crystal Networks*. New York: Marcel Dekker.

Marangoni, A.G. and M. Rogers. 2003. Structural basis for the yield stress in plastic disperse systems. *Appl. Phys. Lett.* 82: 3239–3241.

Marangoni, A.G. and D. Rousseau. 1996. Is plastic fat rheology governed by the fractal geometry of the fat crystal network? *J. Am. Oil Chem. Soc.* 73: 991–994.

Marangoni, A.G. and D. Rousseau. 1999. Plastic fat rheology is governed by the fractal nature of the fat crystal network and by crystal habit. In: *Physical Properties of Fats, Oils and Emulsifiers*, Wildak, N., Ed. Champaign, IL: AOCS Press, pp. 96–111.

Marconi, U.M.B., A. Puglisi, L. Rondoni, and A. Vulpiani. 2008. Fluctuation-dissipation: Response theory in statistical physics. *Phys. Rep.* 461: 111–195.

Marrink, S.J., H.J. Risselada, S. Yefimov, D.P. Tieleman, and A.H. de Vries. 2007. The MARTINI force field: Coarse-grained model for biomolecular simulations. *J. Phys. Chem. B.* 111: 7812–7824.

Marsh, C. 1998. Theoretical aspects of dissipative particle dynamics. PhD thesis, University of Oxford, Oxford, U.K., and references therein.

McClements, D.J. and Y. Li. 2010. Structured emulsion-based delivery systems: Controlling the digestion and release of lipophilic food components. *Adv. Colloid Interface Sci.* 159: 213–228.

McLaughlin, S. 1989. The electrostatic properties of membranes. *Ann. Rev. Biophys. Chem.* 18: 113–136.

Messiah, A. 1999. *Quantum Mechanics*. Mineola, NY: Courier Dover Publications.

Monticelli, L., S.K. Kandasamy, X. Periole, R.G. Larson, D.P. Tieleman, and S.-J. Marrink. 2008. The MARTINI coarse-grained force field: Extension to proteins. *J. Chem. Theory Comput.* 4: 819–834.

Müller, M., K. Katsov, and M. Schick. 2006. Biological and synthetic membranes: What can be learned from a coarse-grained description. *Phys. Reports.* 434: 113–176.

Narine, S.S. and A.G. Marangoni. 1999a. Fractal nature of fat crystal networks. *Phys. Rev. E.* 59: 1908–1920.

Narine, S.S. and A.G. Marangoni. 1999b. Mechanical and structural model of fractal networks of fat crystals at low deformations. *Phys. Rev. E.* 60: 6991–7000.

Naumann, K.-H. and H. Bunz. 1993. Van der Waals interactions between fractal particles. *J. Aerosol Sci.* 24: S181–S182.

Oversteegen, S.M. and H.N.W. Lekkerkerker. 2003. Testing the Derjaguin approximation for colloidal mixtures of spheres and disks. *Phys. Rev. E.* 68: 021404–021406.

Papadopoulos, K.D. and H.Y. Cheh. 1984. Theory on colloidal double-layer interactions. *AIChE J.* 30: 7–14.

Parsegian, V.A. 2005. *Van der Waals Forces: A Handbook for Biologists, Chemists, Engineers and Physicists*. New York: Cambridge University Press.

Parsegian, V.A. and B.W. Ninham. 1970. Temperature-dependent van der Waals forces. *Biophys. J.* 10: 664–674, and references 1–3 therein.

Patel, A.R. and K.P. Velikov. 2011. Colloidal delivery systems in foods: A general comparison with oral drug delivery. *LWT—Food Sci. Technol.* 44: 1958–1964.

Pink, D.A., C.B. Hanna, C. Sandt, A.J. MacDonald, R. MacEachern, R. Corkery, and D. Rousseau. 2010. Modelling the solid-liquid phase transition in saturated triglycerides. *J. Chem. Phys.* 132: 54502–54513.

Pink, D.A., M.S.G. Razul, C.J. MacDougall, F. Peyronel, C.B. Hanna, and A.G. Marangoni. Nanoscale characteristics of molecular fluids in confined spaces: Triacylglycerol oils. Submitted 2012.

Rahman, M.S. 1997. Physical meaning and interpretation of fractal dimensions of fine particles measured by different methods. *J. Food Eng.* 32: 447–456.

Rajagopalan, R. 2001. Simulations of self-assembling systems. *Curr. Opin. Colloid Interface Sci.* 6: 357–365.

Rogers, M.A., D. Tang, L. Ahmadi, and A.G. Marangoni. 2008. Fat crystal networks. In: *Food Materials Science. Principles and Practice*, Aguilera, J.M., P.J. Lillford, Eds. New York: Springer Science+Business Media LLC, pp. 396–401.

Rye, G.G., J.W. Litwinenko, and A. Marangoni. 2005. Fat crystal networks. In: *Bailey's Industrial Oil and Fat Products, Volume 4, Edible Oil and Fat Products: Products and Applications*, 6th edn., Shahidi, F., Ed., Chapter 4. Hoboken, NJ: John Wiley & Sons, Inc. pp. 121–135.

Sagalowicz, L. and M.E. Leser. 2010. Delivery systems for liquid food products. *Curr. Opin. Colloid Interface Sci.* 15: 61–72.

Sakurai, J.J. and S.F. Tuan. 1994. *Modern Quantum Mechanics*. Reading, MA: Addison-Wesley.

Sato, K. 2001. Crystallization behaviour of fats and lipids—A review. *Chem. Eng. Sci.* 56: 2255–2265.

Sato, K. and S. Ueno. 2011. Crystallization, transformation and microstructures of polymorphic fats in colloidal dispersion states. *Curr. Opin. Colloid Interface Sci.* 16: 384–390.

Schiller, U.D. 2005. Dissipative particle dynamics: A study of the methodological background. Diploma thesis, Condensed Matter Theory Group, Faculty of Physics, University of Bielefeld, Bielefeld, Germany, and references therein.

Shih, W.H., W.Y. Shih, S.I. Kim, J. Liu, and I.A. Aksay. 1990. Scaling behavior of the elastic properties of colloidal gels. *Phys. Rev. A.* 42: 4772–4779.

Shim, M. and H. McDaniel. 2010. Anisotropic nanocrystal heterostructures: Synthesis and lattice strain. *Curr. Opin. Solid State Mater. Sci.* 14: 83–94.

Tang, D. and A.G. Marangoni. 2006. Quantitative study on the microstructure of colloidal fat crystal networks and fractal dimensions. *Adv. Colloid Interface Sci.* 128–130: 257–265.

Tang, D. and A.G. Marangoni. 2007. Modeling the rheological properties and structure of colloidal fat crystal networks. *Trends Food Sci. Technol.* 18: 474–483.

Van den Tempel, M. 1961. Mechanical properties of plastic disperse systems at very small deformations. *J. Colloid Sci.* 16: 284–296.

Van den Tempel, M. 1979. Rheology of concentrated suspensions. *J. Colloid Interface Sci.* 71: 18–20.

Verdult, M.W.J. 2010. A microscopic approach to van-der-Waals interactions between nanoclusters: The coupled dipole method. Master thesis in Theoretical Physics, Utrecht University, Utrecht, the Netherlands.

Vreeker, R., L.L. Hoekstra, D.C. den Boer, and W.G.M. Agterof. 1992. The fractal nature of fat crystal networks. *Colloids Surf.* 65: 185–189.

Weiss, J., S. Gaysinsky, M. Davidson, and J. McClements. 2009. Nanostructured encapsulation systems: Food antimicrobials. In: *Global Issues in Food Science and Technology*, Barbosa-Canovas, G.V., A. Mortimer, D. Lineback, W. Spiess, K. Buckle, P. Colonna, Eds., Chapter 24. New York: Academic Press, pp. 425–479.

Wennerström, H. 2003. The van der Waals interaction between colloidal particles and its molecular interpretation. *Colloids Surf. A Physicochem. Eng. Asp.* 228: 189–195.

4 Rheology of Fats

Alejandro G. Marangoni and Suresh S. Narine

4.1 HOOKE'S LAW

The theoretical foundations for the characterization of the properties of solids were devised by Robert Hooke (1635–1703). Hooke observed that, for small elongations, the amount of elongation of a spring, ΔL, was directly proportional to the applied force. This observation is expressed mathematically as Hooke's law:

$$F = k\Delta L \qquad (4.1)$$

The range of elongation where Hooke's law is obeyed is called the *elastic region* of the spring (Figure 4.1). Within this elastic region, spring deformations are *reversible*, that is, once the force is removed, the spring returns to its original length (L_o). At elongations above the *elastic limit* of the spring, irreversible deformations take place, and Hooke's law is no longer applicable—no linear relationship between the applied force and the amount of spring elongation is observed. Beyond the *limit of elasticity*, the spring eventually reaches its *breaking point* (Figure 4.1).

Materials are usually analyzed rheologically assuming they behave as springs. Most materials display an elastic region at small deformations, usually in a range not exceeding 1%–3%. When these materials are taken above their *limit of elasticity*, they will be permanently deformed and will not recover to their original dimensions once the stress is removed. Substances that deform considerably beyond their *limit of elasticity* without breaking are termed plastic. For plastic deformations, a small increment in the applied force leads to a large deformation. These materials are termed ductile, for example, lead, copper. Materials that break shortly after the limit of elasticity is reached are termed brittle, for example, glass, steel. The *yield stress* corresponds to the value of the stress at the limit of elasticity.

Hooke's law can be used for the characterization of any solid material, as long as a linear relationship exists between the applied force and the amount of deformation of the material. Most solid materials obey Hooke's law for deformations up to 1%–3%. When referring to small deformations in this chapter, we will be referring to any deformation below a 1% change in dimension in the direction of the applied force.

4.2 STRESS–STRAIN RELATIONSHIPS AND ELASTIC

4.2.1 SHEAR AND BULK MODULI

The amount of deformation of a material will depend not only on the applied force and the nature of the material, but also on sample dimensions. Hooke's law is therefore of limited practical utility, since the constant k will be a function of the nature

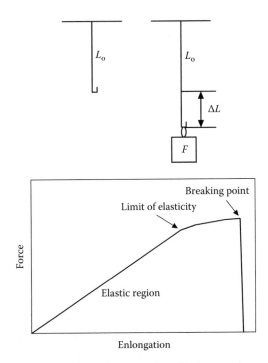

FIGURE 4.1   Idealized force-deformation curve for a Hookean spring.

of the material being studied and its dimensions. One way of solving this problem is to modify Equation 4.1 to take into account the amount of force per cross-sectional area, rather than force, being applied, and the resulting relative, rather than absolute, change in dimension of the material:

$$\frac{F}{A} = M \frac{\Delta L}{L_o} \tag{4.2}$$

We define the applied *stress* (σ) as

$$\sigma = \frac{F}{A} \tag{4.3}$$

where
   F is the applied force (N)
   A is the cross-sectional area (m^2)

We define the resulting *strain* (*s*) as

$$\gamma = \frac{\Delta L}{L_o} \tag{4.4}$$

where
   ΔL is the change in length
   L_o is the original length of the sample

Stress (σ) has units of (N/m²), or Pascals, *strain* (*s*), a dimensionless quantity, is usually expressed as a fraction or percentage, and the *modulus of elasticity* (*M*) has units of (N/m²), or Pascals. Equation 4.2 can therefore be rewritten as

$$\sigma = M \cdot s \tag{4.5}$$

The *strain* is directly proportional to the applied *stress*, and the *elastic modulus* is a constant that characterizes the elastic properties of the material. This constant is independent of sample dimensions, and dependent strictly on the nature of the material being studied.

4.3 TYPES OF STRESSES AND CORRESPONDING

4.3.1 DEFINITIONS OF MODULI

Tensile stress arises when a material is stretched in one dimension, compressive stress arises when a material is compressed in one dimension, and shear stress arises when opposite forces are applied on a material across its opposite faces (Figure 4.2), in one or two dimensions. A material can also be hydrostatically compressed in three dimensions. This last situation causes a change in volume.

For tensile and compressive stresses, the *Young's modulus of elasticity* is defined as

$$E = \frac{\sigma}{\varepsilon} \tag{4.6}$$

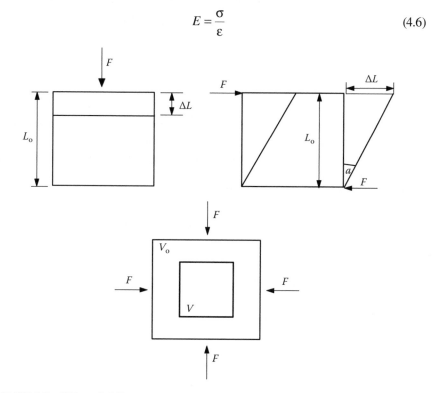

FIGURE 4.2 Effect of different types of stresses on a material's deformation.

where ε is the compressive strain, $\Delta L/L_o$, as shown in Figure 4.2. For shear stresses, the *shear modulus* is defined as

$$G = \frac{\sigma}{\gamma} \qquad (4.7)$$

where γ is the shear strain, $\Delta L/L_o$ or tan a, as shown in Figure 4.2. For small deformations, tan $a \approx a$ (radians).

For three-dimensional compression, the *bulk modulus* is defined as

$$K = \frac{\Delta P}{(\Delta V/V_o)} \qquad (4.8)$$

where

P is the change in pressure, that is, hydrostatic stress

$\Delta V/V_o$ is the relative change in volume of the solid, that is, volumetric strain, as shown in Figure 4.2

The reciprocal of the Young's, shear and bulk moduli describes the strain per unit stress, and is referred to as the *compliance* (J) of a material:

$$J = \frac{1}{\text{modulus}} \qquad (4.9)$$

Another fundamental quantity in the study of the rheology of solids is the Poisson's ratio, a dimensionless quantity defined as the change in diameter per unit original sample diameter ($\Delta D/D_o$) over the change in length per unit original sample length ($\Delta L/L_o$) under the influence of an external applied stress. In other words, the Poisson's ratio is the ratio of the lateral expansion or contraction, as a fraction of the diameter of the sample, over the longitudinal strain:

$$\mu = \frac{(\Delta D/D_o)}{(\Delta L/L_o)} \qquad (4.10)$$

When the volume of a material does not change upon stretching or contraction, $\mu = 0.5$. If the volume decreases, then $\mu < 0.5$. For materials which can be compressed with no change in diameter, $\mu = 0$. These fundamental moduli are inter-related, namely,

$$G = \frac{9EK}{(9K - E)} \qquad (4.11)$$

$$K = \frac{E}{3(1-2\mu)} = \frac{EG}{(9G-3E)} = \frac{2}{3}G\frac{(1+\mu)}{(1-2\mu)} \qquad (4.12)$$

$$E = \frac{9GK}{(3K+G)} = 2G(1+\mu) = 3K(1-2\mu) \qquad (4.13)$$

$$\mu = \frac{(E-2G)}{2G} = \frac{1}{2}\frac{(1-E)}{3K} \qquad (4.14)$$

These fundamental quantities which characterize the rheological behavior of a material have meaning only when derived using rheological tests within the elastic range of the material, when the material being tested is continuous, homogenous, and isotropic, and when the test piece is of uniform and regular shape.

4.4   ELASTIC BEHAVIOR

Elastic behavior can be explained in terms of intermolecular bonding mechanisms that provide coherence to the sample. An applied external force will cause spatial displacements, or compressions and tensions, of the structural elements within the solid. The nature of inter-molecular, internanostructural, or intermicrostructural interactions, and the geometry of the network can influence the mechanical properties of materials. The displacement or deformation of the microstructures results in an increase in the free energy of the system, as inter- or intra-microstructural bond tensions and compressions balance the applied external force. When the applied force is removed, the internal energy stored in compressed and extended inter- or intra-microstructural bonds is dissipated by restoring the microstructures to their original positions, or shape and dimension. For a reversible deformation process, the sample will return to its original state and dimensions upon removal of the applied stress. For an irreversible deformation process, the sample's original structure will have been changed, for example, compression could have occurred, and therefore it will not return to its original state and dimensions upon removal of the applied stress.

4.4.1   STRUCTURAL THEORY OF ELASTICITY

In most solids, atoms, and molecules are arranged in some order. How rigidly these atoms and molecules are held about their equilibrium positions depends on the relative strength of the short-range forces between them. In the classical picture of a regular material such as a metal, the extension of say a metal wire is due to a displacement of the molecules within the wire from their mean positions. If the displacement is small enough, the restoring force toward the mean positions of the molecules is proportional to the displacement. To understand this statement, one may consider the variation of potential energy, U, between two molecules at a distance r apart. Depending on the separation distance, the potential energy is either negative or positive. The general form of the potential energy function, therefore, is given by

$$U(r) = \frac{a}{r^p} - \frac{b}{r^q} \qquad (4.15)$$

where
p and q are powers of r
a and b are constants

Since the force F between the two molecules is given by

$$F = -\frac{dU}{dr} \qquad (4.16)$$

The positive term with the constant a in Equation 4.15 indicates a repulsive force and the negative term with the constant b represents an attractive force. It is difficult to formulate the exact mathematical form of the potential energy curve that describes the interaction of molecules, partly because there are different types of bonds or forces between atoms and molecules in solids, depending on the nature of the solid. A much-used general expression for the potential energy relationship for nonpolar molecules is given by the expression attributable to the English chemist J.E. Lennard-Jones:

$$U(r) = \left(\frac{\mu}{r}\right)^{12} - \left(\frac{\lambda}{r}\right)^{6} \qquad (4.17)$$

where λ and μ are adjustable parameters. In an ionic solid, for example sodium chloride, the potential energy relationship is given by

$$U(r) = \frac{a}{r^9} - \frac{b}{r} \qquad (4.18)$$

If one considers the Lennard-Jones form of the potential energy function, as is depicted in Figure 4.3, it can be seen that there is a value of r which corresponds to a potential energy minimum; this value of r, r_o, is the equilibrium spacing between the molecules. Figure 4.3 also shows the force as a function of r, given by Equation 4.16. At a distance r_o, the repulsive and attractive forces balance, that is, $F = 0$. If the separation r between the molecules is slightly increased from r_o, the attractive force between them will restore the molecules to their equilibrium position after the external force is removed. If the separation is decreased from r_o, the repulsive force will restore the molecules to their equilibrium position after the external force is removed. Therefore, the molecules of a solid oscillate about their equilibrium or mean position. If one examines Figure 4.3 closely, the graph of F versus r is

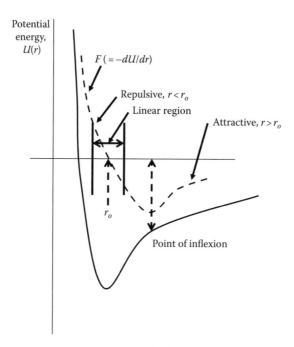

FIGURE 4.3   Schematic Lennard-Jones potential energy curve.

approximately linear at small distances from the equilibrium position r_o. This means that the extension of the bond is proportional to the applied force, when the displacement is small distances away from the equilibrium position of the molecules. The force constant, k, between the molecules is given by

$$F = -k(r - r_o) \tag{4.19}$$

where r is slightly greater than r_o. Therefore,

$$k = -\frac{dF}{dr} \tag{4.20}$$

or the force constant is given by the negative of the gradient of the tangent to the force-distance curve at $r = r_o$. In a macroscopic sense, this behavior of the molecules leads to Hooke's law, which states, "The extension is proportional to the force or tension in a material if the proportional limit is not reached." So the extension of the wire is due to the displacement of its molecules from their equilibrium positions. The molecules are therefore undergoing simple harmonic motion, vibrating about their equilibrium position, and up to the proportional limit, the energy gained or stored by the stretched wire is molecular potential energy, which is recovered when the external force is removed. Therefore, for systems such as a wire or a large sodium chloride crystal, where the arrangement of the molecules is such that the macroscopic structure is built up by the constant repetition of the molecular lattice, or unit cell, the elastic properties, character-ized by an elastic constant k, may be inferred from a molecular argument.

Hooke's law behavior is, however, also observed for materials where one cannot easily explain the elasticity from molecular considerations. Such materials include colloidal gels, fat crystal networks, composite materials such as concrete, and ceram-ics, etc. In such materials, there are distinct hierarchies of structure, and the stressing or elongation of the microstructure does not necessarily relate to a stressing of the molecular level. In such materials, the limit of elasticity is encountered at levels of structure that are supermolecular. Therefore, the structural entities that are stressed when the macrostructure is stressed may be aggregates of molecules, aggregates of crystals, single crystal entities, polymer strands, etc. However, the molecular picture does provide an important analogy in the pursuit of an understanding of the elas-ticity of such materials. It is not an unreasonable assumption that at some level of structure that is being stressed, the applied force varies with the change in distance between the structural entities of importance in a linear manner, which translates to the macroscale. It is never a trivial matter to attribute responsibility for the elastic-ity of a material to a particular structural level. Therefore, although it is implicitly understood that the material is composed of a large number of discrete structural entities that are responsible for its elasticity, it is advantageous to regard the system as a continuous distribution of matter.

To therefore develop an understanding of physical quantities of a material such as its elastic modulus, we consider the material as a deformable continuum. When exter-nal forces are applied to such a system, a distortion results because of the displacement of the "relevant structural entities" from their equilibrium positions, and the body is said to be in a state of stress. After the external force is removed, the body returns to the equilibrium position, providing the applied force was not too great. Therefore, the elasticity of a material is defined as the ability of a body of this material to return to its equilibrium shape after the application of an applied force. To reach a quantitative definition of elasticity, one must first understand the concepts of stress and strain.

If we consider that a body is acted upon by an external force, $\Delta \bar{F}$, which is neither tangential, nor normal to the surface upon which it acts, as is depicted in Figure 4.4, then the average stress on the surface with area ΔA is defined as the force per unit area:

$$\bar{S} = \frac{\Delta \bar{F}}{\Delta A} \qquad (4.21)$$

Therefore at some infinitesimal point p on the surface, the stress is given by

$$\bar{S} = \lim_{\Delta A \to 0} \frac{\Delta \bar{F}}{\Delta A} = \frac{d\bar{F}}{dA} \qquad (4.22)$$

If we consider that the external force $\Delta \bar{F}$ acting may be resolved in two components—one that is tangential to the surface, and one that is normal to the surface at the point p, then the stress $\bar{S}$ may also be resolved into tangential and normal stress. The tangential

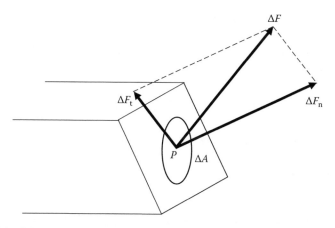

FIGURE 4.4 Schematic of a body being acted upon by an external force.

component of the stress is referred to as the shear stress, and the normal component is referred to simply as normal stress. The normal stress may either be in compression or tension, depending on whether the applied force is a push or pull, respectively. Referring to Figure 4.4, the normal stress, s, at point p is given by

$$s = \frac{dF_n}{dA} \tag{4.23}$$

and the shear stress, τ, at the point p is given by

$$\tau = \frac{dF_t}{dA} \tag{4.24}$$

Now, the effect of an applied stress on a body is to cause a change in size and shape of the body. Strain is the quantity that measures the relative change in size or shape of the body when under the applied stress.

If one considers a rectangular rod, with a cross-sectional area A, which is subject to a normal force, F, then the application of the force will induce a change in length of the rod. If the initial length of the rod is L_o and the increased length of the rod is L, then the change in length is given by $\Delta L = L - L_o$. Recall that the normal stress is given by

$$s = \frac{F_n}{A} \tag{4.25}$$

and the strain ε induced by this stress, called the longitudinal or tensile strain, is given by

$$\varepsilon = \frac{L - L_o}{L_o} = \frac{\Delta L}{L_o} \tag{4.26}$$

Within the elastic region of the material, the ratio of stress to strain is called the elastic modulus and is constant, and in the case of a normal stress and a tensile strain, the ratio of stress to strain is called the Young's modulus or tensile modulus. The Young's modulus, E, is given by

$$E = \frac{s}{\varepsilon} = \frac{F_n/A}{\Delta L/L_o} \tag{4.27}$$

The dimensions of the Young's modulus are the same as those of stress, since strain is a dimensionless quantity. These dimensions are N/m² or Pa. Now, within the region where stress/strain is constant, the material is obeying Hooke's law:

$$F_n = k\Delta L \tag{4.28}$$

where k is the stiffness of the material, sometimes referred to as the spring constant of the material. One may then relate the stiffness of the material to its Young's modulus by considering Equations 4.27 and 4.28:

$$k = \frac{EA}{L_o} \tag{4.29}$$

Therefore, for a given material, within the elastic region, the Young's modulus and stiffness constant are important parameters that yield information about the deformability of the material, and therefore about the sensory impact of the material when one deforms it in the mouth.

The proportionality relationship between s and ε holds only if stress is less than a certain maximum value. This point at which maximum value is reached is referred to as the proportional limit or yield point or yield value. Beyond the yield point, the material begins to undergo a plastic deformation and a small increase of stress causes the material not regain its original shape and length upon removal of the applied force, but rather, induces the material to begin flowing.

The shear elastic modulus, which is the ratio of the shear stress to the shear strain, may be derived in like manner to the Young's modulus. Figure 4.5 shows a schematic

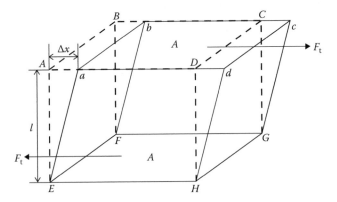

FIGURE 4.5 Schematic of a material deformed under the action of a pair of equal and opposite forces.

of a material deformed under the action of a pair of equal and opposite forces not acting along the same line of action. The resulting shear stress produces a change in the shape of the body (but no change in length) and therefore induces a strain, called the shear strain. The points *ABCD* under the action of the shear stress have moved to the points *abcd*, whilst the points *EFGH* are not displaced. The shear stress is therefore given by

$$\tau = \frac{F_t}{A} \tag{4.30}$$

and the shear strain, ν, is defined as the ratio of the displacement, Δx and the length *l* as shown in Figure 4.5. Therefore, the shear elastic modulus *G* (also called the modulus of rigidity or torsion modulus) is given by

$$G = \frac{\tau}{\nu} = \frac{F_t/A}{\Delta x/l} \tag{4.31}$$

The shear elastic modulus therefore also has units of N/m² or Pa. In general, the tensile modulus or Young's modulus, *E*, is related to the shear elastic modulus or rigidity modulus by

$$G = \frac{E}{3} \tag{4.32}$$

Fat crystal networks demonstrate a tensile modulus as well as a shear elastic modulus. Figure 4.6 shows an example of the shear elastic modulus of cocoa butter. For viscoelastic materials like fat crystal networks, both the loss modulus, G'' (related to the liquid portion of the network) and a storage elastic modulus, G' (related to the

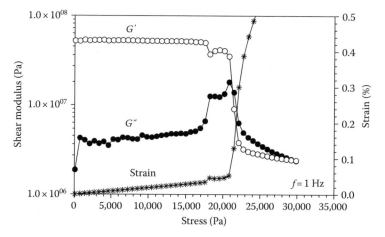

FIGURE 4.6 Effect of applied stress on the storage (open circles) and loss (closed circles) moduli, as well the strain (stars), of cocoa butter at 22°C.

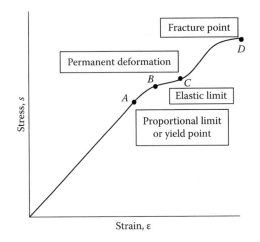

FIGURE 4.7   Stress–strain behavior of a typical elastic system.

solid network) can be measured. The range of stress/strain within which the elastic modulus is constant is called the linear viscoelastic range (LVR). As is demonstrated by Figure 4.6, this range for fat crystal networks is very small. Therefore, the system is elastic only for very small stress values. Figure 4.7 shows the behavior of a typical elastic system; up to point *A*, the system is perfectly elastic, and then between point *A* and point *B*, stress is not proportional to strain, but if the stress is removed, the body should return to its original shape. If the applied stress is beyond point *C*, then permanent deformation occurs in the body and leads to breakage of different levels of structure. Point *A* is usually taken as the yield point. Since the shear elastic modulus is the slope of the plot of stress versus strain, it is obvious that changes in elastic modulus will be produced by changes in the yield point. Thus, a relationship between these two parameters is self-evident.

The relationship of elastic modulus to yield value depends on the levels of structure that are affected in the network during elastic measurements and measurements of the yield value, and the relative strength of these levels of structure. In a classical material such as a metal wire, at stresses beyond the yield point, the elongation of the material is caused by the movement of crystal planes, called slip, and the origin of slip lies in crystal dislocations, which correspond to a lack of symmetry in the crystalline nature of the material. In a way, these dislocations may be thought of as fault lines or planes, along which the material flows under external forces large enough to stress the material beyond its elastic point. However, since the yield point is influenced by the ability of the material to flow beyond its elastic limit, it is not immediately evident that the magnitude of changes in the yield point of a material will be related in a straightforward manner to magnitudes of changes in elastic modulus for materials such as fat crystal networks. Even in classical materials such as a crystal of sodium chloride, the yield point would be affected by the amount of dislocations in the crystal, which is a function of the purity of the material and of the environmental conditions under which the crystal was formed. The relationship of elastic modulus to yield value would therefore depend on the levels of structure that are affected in

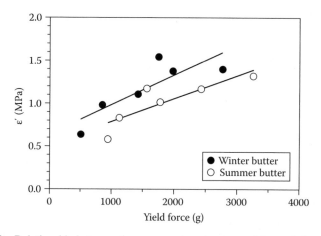

FIGURE 4.8 Relationship between the compression storage modulus and the yield force in two different types of milkfat at 5°C.

the network during elastic measurements and measurements of the yield value, and the relative strength of these levels of structure.

In most of our studies we have found that the shear storage modulus (G') and the yield force (F_y), or hardness index (HI) are directly proportional to each other (Figures 4.8 through 4.10). However, the proportionality constant between these

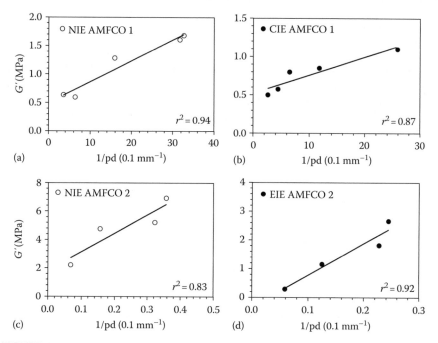

FIGURE 4.9 Relationship between the shear storage modulus (G') and the hardness index in anhydrous milkfat-canola oil (AMFCO) blends at 5°C. NIE, non-interesterified; CIE, chemically interesterified; EIE, enzymatically interesterified.

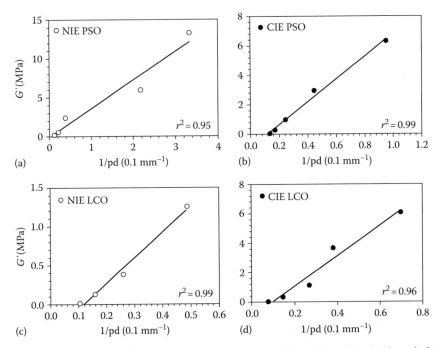

FIGURE 4.10 Relationship between the shear storage modulus (G') and the hardness index in palm-soybean oil (PSO) and lard-canola oil (LCO) blends at 5°C, NIE, non-interesterified; CIE, chemically interesterified.

two parameters is system specific (slopes are not the same). This becomes particularly evident when considering the relationship between G' and the F_y in milkfat crystallized at different cooling rates (Rye et al., 2005) as shown in Figure 4.11. In all cases, G' is directly proportional to F_y, however, the proportionality constant for the covariation of these two parameters is a function of the cooling rate of

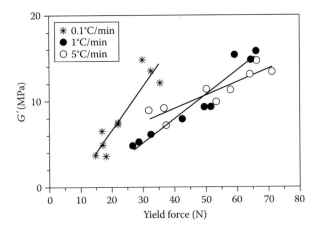

FIGURE 4.11 Relationship between the shear storage modulus (G') and the yield force in milkfat crystallized at three different cooling rates from the melt to 5°C.

the material. The microstructure of milkfat is a strong function of the cooling rate experienced upon crystallization (Haighton, 1959; Dixon and Parekh, 1979; Wright et al., 2001; Campos et al., 2002; Rye et al., 2005). Thus, different microstructures will influence the relative relationship between small and large deformation rheological parameters.

4.5 YIELD VALUE FROM CONSTANT FORCE CONE

4.5.1 PENETROMETRY MEASUREMENTS

The cone penetrometer is widely used in the fats and oils industry for the measurement of the consistency of plastic fats (American Oil Chemists Society, AOCS method Cc 16–60). This method involves the penetration of the fat network by a metal cone of known mass and geometry. By monitoring the depth reached by the cone and the time it takes to achieve that depth, a parameter referred to as the yield value of the network may be calculated. As explained previously, yield values are determined to be the point at which the stress on the network is such that the network is just at its elastic limit. If one considers a typical plastic material, at stresses below the yield point, the material will behave as a perfectly elastic solid, that is, the strains are proportional to the applied stress. However, from the yield point onward, this proportional relationship is lost, and the material begins to flow, that is, some level of structure has been broken. The relation of cone penetrometer measurements to the yield value of the network has been formulated by a series of investigators in the fats and oils field, but unfortunately, many of these formulations did not totally agree. However, it is important to provide a brief review of these developments, as is done in the following.

Rebinder and Semenko (1949) formulated the yield value as calculated from cone penetrometry measurements to be given by

$$Y = \frac{Mg\cos^2(\alpha)}{p^2\tan(\alpha)} \tag{4.33}$$

where
  M is the mass of the penetrating cone
  α is the half-angle of the cone
  p is the penetration depth of the cone in a prescribed amount of time (5 s)
  g is the acceleration due to gravity

Later, Haighton (1959) proposed the following formulation:

$$Y = \frac{KMg}{p^{1.6}} \tag{4.34}$$

where K is a constant depending on the cone angle, and the other parameters are as defined earlier. Following the work of Haighton, Mottram (1961) proposed the following formula:

$$Y = \frac{KMg^2}{p^n} \qquad (4.35)$$

where

    K is a constant

    n is an exponent which is close to two, but allowed to vary with the nature of the network

Following this work, Vasic and deMan (1968) proposed to define hardness as force divided by area of penetration (so, essentially defining hardness as a pressure, which has been done before). They related this hardness to penetration values by the following equation:

$$H = \frac{Mg \cdot 10^{-3}}{\left[p\pi(\tan(\alpha)/\cos(\alpha))(p + (2r/\tan(\alpha))) + r^2\pi \right] \cdot 10^{-4}} \qquad (4.36)$$

where

    H is the hardness

    r is the radius of the flat tip of the cone, and all other symbols are as defined earlier

Dixon and Parekh (1979) related hardness of butter from sensory investigations with a measure of hardness from cone penetrometer measurements, which they called the cone stress index (C_v), and found that C_v correlated very well with sensory impressions of hardness. They formulated the cone stress index as

$$C_v = \frac{MA^{-1.65}}{p^2} \qquad (4.37)$$

where A is the angle of the cone, and all other symbols are as defined earlier.

    As noted by Hayakawa and deMan (1982), most of the aforementioned equations suggest that the yield value used as a measure of hardness, or hardness expressed as some relationship to the yield value, follows a general form of dependence on the mass of the cone, the geometry of the cone, and the penetration depth, given by

$$H = C\frac{M}{p^n} \qquad (4.38)$$

where

    H is hardness or yield value

    C is a constant depending on the geometry of the cone, and other symbols are as defined earlier

Hayakawa and deMan (1982) also noted that if a cone of 20° angle is used as is specified in the AOCS method, then Equation 4.38 may be reduced to

$$HI = \frac{M}{p^n} \qquad (4.39)$$

where HI is the hardness index of the material.

Hayakawa and deMan (1982) presented experimental proof from their own experiments, and summarized the experiments of others to support that the hardness index of a plastic fat is best represented by Equation 4.39 when n is equal to 1. However, this relationship does not hold unless the penetration depth is between 15 and 150 units. Penetration depth is measured in units of 0.1 mm, therefore the actual range for the penetration depth for Equation 4.39 to be valid when $n = 1$, is 1.5–15 mm.

4.6 RHEOLOGY OF LIQUIDS

4.6.1 VISCOSITY

Fluid flow takes place as a fluid deforms when acted upon by an external force. This flow can be visualized as the movement of thin layers of fluid molecules. *Viscosity* is the frictional force between the different layers of the fluid as they move past one another. This frictional force arises due to cohesive forces between liquid molecules, or in gases, due to collisions between molecules in different layers.

The dynamic viscosity of a fluid is characterized quantitatively by the coefficient of viscosity (η), which is defined in the following way. Envision a thin layer of fluid contained between two flat plates (Figure 4.12). While the bottom plate is held stationary, a force is applied to the upper plate to make it move. Due to the frictional force between molecules in the different fluid layers, a fluid layer velocity gradient ($\partial v/\partial y$) is established between the stationary plate and the moving plate. This fluid drag arises due to attractive forces between fluid molecules. The fluid molecules in direct contact with upper plate move at the same speed as the plate, while the fluid molecules in direct contact with the bottom plate remain stationary. This stationary fluid layer retards the flow of an adjacent fluid layer. This slower moving fluid layer then itself retards the flow of an adjacent fluid layer, and so forth. This process gives rise to the fluid layer velocity gradient mentioned earlier. Hence, viscosity is usually defined as *the resistance to flow*, and is a direct consequence of the strength of interaction, or force of adhesion, between fluid molecules.

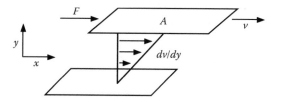

FIGURE 4.12 Idealized flow of two contiguous fluid layers depicting the gradient in fluid velocity as a function of distance from the walls.

For a particular fluid, the magnitude of the established velocity gradient $\partial v/\partial y$ (Figure 4.12) is directly proportional to the force applied on the upper plate, and inversely proportional to the area of the plate:

$$\frac{\partial v}{\partial y} \propto \frac{F}{A} \qquad (4.40)$$

The proportionality constant for this equation is defined as the coefficient of viscosity (η),

$$\frac{\partial v}{\partial y} = \eta \frac{F}{A} \qquad (4.41)$$

In a similar fashion as for the rheological analysis of solids, the quantity F/A is defined as the applied stress (σ), with units of N/m^2 or Pa. The shear strain (γ) for this system is defined as the change in position of a layer of fluid in the x axis (∂x) as a function of a change in position in the axis (∂y) in the region between the stationary and moving plates:

$$\gamma = \frac{\partial x}{\partial y} \qquad (4.42)$$

The rate at which the shear strain varies in time is defined as the shear strain rate ($\dot{\gamma}$), with units of (s^{-1}). Considering that velocity is the change in displacement as a function of time, the velocity gradient established between the two parallel plates is the shear strain rate, namely,

$$\dot{\gamma} = \frac{\partial \gamma}{\partial t} = \frac{\partial x}{\partial y \partial t} = \frac{\partial v}{\partial t} \qquad (4.43)$$

Equation 4.41 then becomes

$$\sigma = \eta \dot{\gamma} \qquad (4.44)$$

The viscosity η (Pa $\cdot$ s) characterizes the dynamic viscosity of a fluid. The still widely used CGS unit for dynamic viscosity is the Poise ($dyn/cm^2 \cdot$ s) For comparative purposes, 1 cP is equivalent to 1 mPa $\cdot$ s. Water has a dynamic viscosity of 1 mPa $\cdot$ s at 20.2°C.

4.7 TYPES OF FLUID FLOW

4.7.1 IDEAL, NEWTONIAN BEHAVIOR

Flow curves are linear and pass through the origin for ideal, Newtonian fluids (Figure 4.13a). The slope of the line is the coefficient of viscosity, which has a time-independent, constant value at all shear strain rates.

4.7.2 NONIDEAL, NON-NEWTONIAN BEHAVIOR

The different types of non-Newtonian fluids are summarized in Figure 4.14 and hypothetical flow curves shown in Figure 4.13. Each type will be discussed in turn.

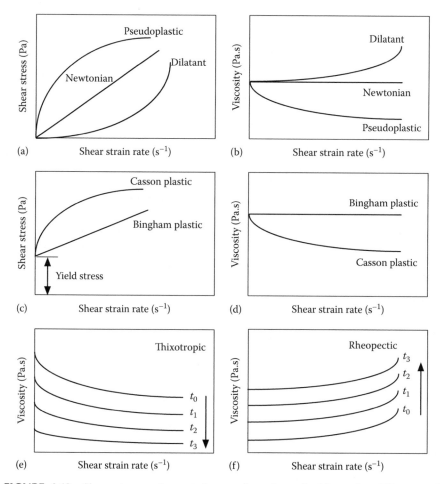

FIGURE 4.13   Shear stress – shear strain rate dependence for Newtonian, Dilatant and Pseudoplastic fluids (a) and their corresponding viscosity – shear strain rate dependence (b). Shear stress – shear strain rate dependence for Casson and Bingham plastic fluids (c) and their corresponding viscosity – shear strain rate dependence (d). Viscosity – shear strain rate dependence for thixotropic (e) and rheopectic (f) fluids.

4.7.2.1   Time-Independent Fluids

Flow curves for these fluids are nonlinear—the coefficient of viscosity is a function of the shear strain rate. The coefficient of viscosity, however, does not change with time. Two major types of flow behavior are observed within time-independent nonideal fluids, namely, shear-thinning or pseudoplastic, and shear-thickening or dilatant. For pseudo-plastic fluids, viscosity decreases as a function of increasing shear strain rate, while for dilatant fluids, viscosity increases as a function of increasing shear strain rates (Figure 4.13a and b). A special case of these time-independent fluids are those materials that display a yield stress before starting to flow (Figure 4.13c). The same Newtonian and shear-thinning behavior is observed once the material starts to flow (Figure 4.13d); however, the common names of these fluids is different.

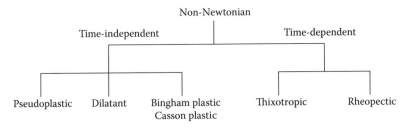

FIGURE 4.14 Types of non-Newtoninan flow behavior.

4.7.2.2 Time-Dependent Fluids

Flow curves for these fluids are nonlinear—the coefficient of viscosity is a function of the shear strain rate. In this case, the coefficient of viscosity does change with time. Two major types of flow behavior are observed within time-dependent nonideal fluids, namely, thixotropic and rheopectic. At a fixed shear strain rate, the coefficient of viscosity for thixotropic fluids *decreases* as a function of time (Figure 4.13e), suggesting fluid structural breakdown, while it *increases* for rheopectic fluids (Figure 4.13f), suggesting enhancement of fluid structure.

4.8 MODELING FLOW BEHAVIOR

The flow of most materials can be modeled using

$$(\sigma - \sigma_o)^m = \eta_{app}(\dot{\gamma})^n \tag{4.45}$$

or

$$m \log(\sigma - \sigma_o) = \log \eta_{app} + n \log \dot{\gamma} \tag{4.46}$$

where
  σ is the applied stress (Pa)
  σ_o is the yield stress (Pa)
  η_{app} is the apparent viscosity [$Pa^m \cdot s^n$]
  $\dot{\gamma}$ is the shear strain rate
  n is the flow index

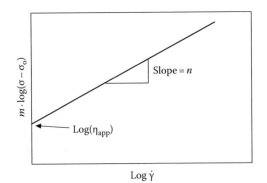

FIGURE 4.15 Linearized graphical representation of the viscosity model.

For a Newtonian fluid, $n = 1$, for a shear-thinning fluid, $n < 1$, while for a shear-thickening fluid, $n > 1$. An empirical parameter m can also be included in the model to improve fits, usually in the form of a fraction. For an ideal, Newtonian fluid, $n = 1$, $m = 1$, $\sigma_o = 0$, and thus Equation 4.45 reduces to Equation 4.44. An example of a plot of Equation 4.46 is shown in Figure 4.15.

REFERENCES

Campos, R., S.S. Narine, and A.G. Marangoni. 2002. Effect of cooling rate on the structure and mechanical properties of milk fat and lard. *Food Res. Int.* 35: 971–981.

Dixon, B.D. and J.V. Parekh. 1979. Use of the cone penetrometer for testing firmness of butter. *J. Am. Oil Chem. Soc.* 10: 421–434.

Haighton, A.J. 1959. The measurement of the hardness of margarine and fats with cone penetrometers. *J. Am. Oil Chem. Soc.* 36: 345–348.

Hayakawa, M. and J. deMan. 1982. Interpretation of cone penetrometer consistency measurements of fats. *J. Texture Stud.* 36: 345–348.

Mottram, F.J. 1961. Evaluation of pseudo-plastic materials by cone penetrometers. *Lab. Pract.* 10: 767–770.

Rebinder, P.A. and N.A. Semenko. 1949. Use of the penetrating cone method for the characterization of structural-mechanical properties of visco-plastic material. *Proc. Acad. Sci. (USSR).* 64: 835–838.

Rye, G.G., J.W. Litwinenko, and A.G. Marangoni. 2005. Fat crystal networks. In: *Bailey's Industrial Oils and Fats Products*, 6th edn., Shaihidi, F., Ed. Hoboken, NJ: John Wiley & Sons, Inc., pp. 121–160.

Vasic, I. and J.M. deMan. 1968. Effect of mechanical treatment on some rheological properties of butter. In: *Rheology and Texture of Foodstuffs*, Vol. 27, Society of Chemical Industry. New York: Academic Press, pp. 251–264.

Wright, A.J., M.G. Scanlon, R.W. Hartel, and A.G. Marangoni. 2001. Rheological properties of milkfat and butter. *J. Food Sci.* 66: 1056–1071.

5 Viscoelastic Properties of Fats

Elasticity and viscosity are just two ways in which materials can respond to externally applied stress. Materials can display solid or liquid behavior depending on their characteristic time and the time required for the process to occur, hereby referred to as process time. Whether a material exhibits elastic or viscous behavior depends on our frame of reference. We will clarify this point later.

The characteristic time of a material, t_c, is the time that it takes for stress, or strain, to be relaxed by $1/e$ of its original value, approximately a 37% drop. The Deborah number is the ratio between the characteristic time and the process time:

$$D = \frac{t_c}{t} \tag{5.1}$$

High Deborah numbers are associated with materials displaying solid-like behavior, while low Deborah numbers are associated with materials displaying liquid-like behavior. Take, for example, the case of glass. Glass has a long characteristic time, making its Deborah number relatively high for short process times. However, given a process time of hundreds of years, the Deborah number of glass becomes quite small, characteristic of liquids. Another example of the relativistic nature of the elasticity and viscosity of materials is that of water. Water has a very short characteristic time, making its Deborah number relatively low for a short process time. However, if the process time is shortened drastically, the Deborah number increases, making water feel as a solid. Having a bucket of water gently poured over one's head as opposed to being sprayed with water using a high-pressure nozzle should illustrate this point quite clearly.

Another interesting consideration is that of the rate of deformation. If a material is deformed very rapidly, it will respond predominantly as a solid, where most of the energy will be stored in the bonds between the structural elements responsible for the observed macroscopic rheological behavior. On the other hand, if the deformation takes place slowly, the structural elements will have sufficient time to rearrange, align themselves in the direction of the applied force, and flow.

Another peculiar property of viscoelastic materials is the existence of a normal force perpendicular to the plane of the applied shear deformation. The so-called Weissenberg effect is a manifestation of this normal force. The Weissenberg effect is evident when mixing flour and water in a home mixer. As the dough forms (a viscoelastic material), it will start climbing up the mixer's shaft. Mixing a Newtonian liquid, on the other hand, will merely result in the formation of a meniscus. The Weissenberg effect is consequence of a stress gradient in the viscoelastic fluid radially outward from axis of rotation (wall, tip of impeller, etc.). The fluid layers that are furthest way from the axis of rotation will experience a greater torque, namely, torque $= rF$, where F is tangential force experienced at the particular fluid layer, and r is the distance between the axis of rotation and the layer of rotating fluid. Thus, fluid layers furthest away from the axis of rotation will experience a greater force, or stress, and will therefore be deformed to a greater extent. In order to dissipate this stress gradient, the material pushes inward toward regions of lower stress, which results in fluid being pushed upward (there is nowhere for the material to go but up!).

Fat is a good example of a viscoelastic material. Fats have an underlying fat crystal network composed of polycrystalline particles arranged into larger clusters. A deformation will stretch intercluster, and/or interparticle bonds, raising the energy of the system. If the strain is within the elastic region, below the limit of elasticity, the material will respond as a solid. Upon removal of the applied stress, stretched bonds will return to their equilibrium, lower energy, state. On the other hand, if the stress is such that the strain at the limit of elasticity is exceeded, bonds will break, the material will undergo plastic deformation and flow. Therefore, at strains beyond the limit of elasticity, the material will behave as a fluid. Depending on the extent and rate of bond breakage, the material will display viscoelastic behavior, where the material's response will be both elastic as well as viscous. Bonds can also reform once the stress is removed, contributing to the elastic character of the material. Fats can therefore behave both as solids and liquids.

5.1 CREEP AND RECOVERY/STRESS RELAXATION

In a creep and recovery test for viscoelasticity, the response time of the strain dependence of both the elastic and viscous components of a material is determined. A creep and recovery test is carried out under constant stress conditions. On the other hand, a stress relaxation test is carried out under constant strain conditions. The response time of the stress dependence of both the elastic and viscous components of a material is determined. In this chapter we will restrict our discussion to the case of creep and recovery.

A solid will deform when subjected to an applied stress, and the energy of deformation elastically stored in the material's structure. This deformation will be fully recovered upon removal of the stress (Figure 5.1). For the case of an ideal Newtonian solid, both stress and strain are time independent ($\tau = G\gamma$). On the other hand, an ideal Newtonian liquid will deform at a particular rate (i.e., it will flow) when subjected to an external stress ($\tau = \eta \partial\gamma/\partial t$), but this deformation will be fully maintained once the stress is removed (Figure 5.1). The deformation of a liquid will never be recovered, since the energy that made the material flow is fully transformed into shear heat, and lost to the system surroundings. A viscoelastic material behaves rheologically as a combination of a solid and a liquid. Such material will initially deform elastically when subjected to an external stress, and given sufficient time, it will start flowing. Upon removal of the stress, the elastic component will recover, while the viscous component will not, and a permanent deformation will be observed.

What follows is a brief discussion of the different models developed in an attempt to understand and model the behavior of viscoelastic materials. Our discussion is strictly limited to materials within their linear viscoelastic region (LVR), where stress is linearly related to strain. This is usually the case at deformation of 1%

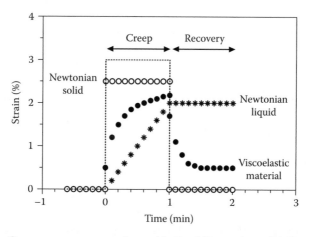

FIGURE 5.1 Creep–recovery curves for an idealized Newtonian solid, Newtonian liquid and a viscoelastic material.

or less. In the LVR, rheological parameters are a function the material's structure and not as much of instrumentation and test conditions.

5.1.1 KELVIN–VOIGT SOLID

A Kelvin–Voigt solid combines a spring and a dashpot in parallel (Figure 5.2). When stress is applied to such a system, the force is equally distributed between the spring and the dashpot. For this case, the total strain (γ) is equal to the strain in the spring (γ_e), which also equals the strain in the dashpot (γ_v), namely,

$$\gamma = \gamma_e = \gamma_v \tag{5.2}$$

The total stress applied on the system (τ), on the other hand, is equal to the sum of the stresses on the spring and the dashpot:

$$\tau = \tau_e + \tau_v \tag{5.3}$$

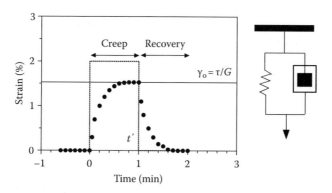

FIGURE 5.2 Creep–recovery pattern for a Kelvin–Voigt element.

Since $\tau_e = G/\gamma_e$ and $\tau_v = \eta \partial\gamma_v/\partial t$, in light of Equation 5.3, the total stress on the system can be expressed as

$$\tau = \frac{G}{\gamma} + \eta \frac{\partial\gamma}{\partial t} \tag{5.4}$$

This equation can be rearranged to

$$\frac{\partial\gamma}{\partial t} + \frac{1}{\lambda\gamma} = \frac{\tau}{G} \tag{5.5}$$

where λ corresponds to the relaxation time, and is defined as $\lambda = \eta/G$. The solution to the previous first-order ordinary differential equation is

$$\gamma = \frac{\tau}{G}(1 - e^{-t/\lambda}) \tag{5.6}$$

Notice that when the time tends to infinity, the strain will approach a limiting maximum value γ_o, where $\gamma_o = \tau/G$, the final maximum response of the spring. The viscous component thus merely retards the response of the spring. Knowledge of this limiting maximum strain value and the applied stress would allow for the determination of the modulus of the spring component of a Kelvin–Voigt solid ($G = \tau/\gamma_o$).

Upon removal of the stress, the Kelvin–Voigt solid will recover to its initial state (Figure 5.2). The strain will decay in a simple exponential fashion from the limiting maximum value (γ_o) to zero:

$$\gamma = \gamma_o e^{-((t-t')/\lambda)} \tag{5.7}$$

The relaxation times are the same for the creep and recovery phases when conditions of linear viscoelasticity prevail. Being an idealized viscoelastic solid, irreversible structural change does not take place, that is, the material does not flow.

5.1.2 MAXWELL FLUID

In this model, a spring and a dashpot are arranged in series rather than in parallel as for the case of the Kelvin–Voigt solid (Figure 5.3). For a Maxwell fluid, the total stress on the system (τ) is equal to the stress on the spring (τ_e), which also equals the stress on the dashpot (τ_v), namely,

$$\tau = \tau_e = \tau_v \tag{5.8}$$

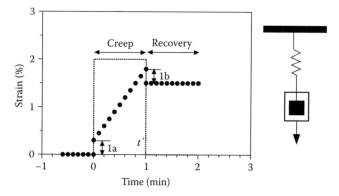

FIGURE 5.3 Creep–recovery pattern for a Maxwell element.

The resulting strain on the system (γ), on the other hand is equal to the sum of the strains on the spring and the dashpot:

$$\gamma = \gamma_e + \gamma_v \tag{5.9}$$

The change in strain as a function of time thus equals

$$\frac{\partial \gamma}{\partial t} = \frac{\partial \gamma_e}{\partial t} + \frac{\partial \gamma_v}{\partial t} \tag{5.10}$$

Since

$$\frac{\partial \gamma_v}{\partial t} = \frac{\tau}{\eta} \tag{5.11}$$

and

$$\frac{\partial \gamma_e}{\partial t} = \frac{1}{G} \frac{\partial \tau}{\partial t} \tag{5.12}$$

Equation 5.10 can be expressed as

$$\frac{\partial \gamma}{\partial t} = \frac{1}{G} \frac{\partial \tau}{\partial t} + \frac{\tau}{\eta} \tag{5.13}$$

Integration of the previous equation after variable separation for the boundary conditions $\gamma = 0$, $\tau = 0$ at $t = 0$,

$$\int_0^\gamma \partial \gamma = \frac{1}{G} \int_0^\tau \partial \tau + \frac{\tau}{\eta} \int_0^t \partial t \tag{5.14}$$

results in the expression

$$\gamma = \frac{\tau}{G} + \frac{\tau}{\eta} t \qquad (5.15)$$

When subjected to a stress, this viscoelastic liquid initially responds with an instantaneous strain (Figure 5.3, 1a), corresponding to the deformation of the elastic component of the material, that is, the spring ($\gamma_i = \tau/G$). This is immediately followed by a constant increase in strain, determined by the viscosity of the material, that is, the dashpot. Knowledge of this instantaneous strain and the applied stress would allow for the determination of the elastic modulus of the spring component of a Maxwell fluid ($G = \tau/\gamma_i$).

When the applied stress is removed, the strain drops immediately to a new constant level (Figure 5.3, 1b). This instantaneous drop is due to the recovery of the elastic component, that is, the spring, of the Maxwell fluid. Notice, however, how the strain levels never return to their original value. This is because the material has flowed, and obviously this irreversible deformation cannot be recovered.

5.1.3 BURGER MODEL

Neither the Kelvin–Voigt nor the Maxwell Fluid models can describe the rheological behavior of real viscoelastic materials. The Burger model comes a step closer to the description of a real viscoelastic material. The Burger model is mainly a Kelvin–Voigt element placed in series with a Maxwell element (Figure 5.4).

Since these elements are placed in series, the total stress on the system is the same for the Kelvin–Voigt and Maxwell elements, namely,

$$\tau = \tau_{KV} = \tau_M \qquad (5.16)$$

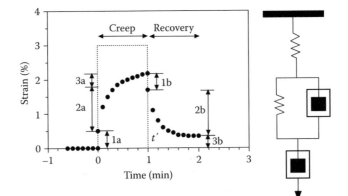

FIGURE 5.4 Creep–recovery pattern for a Burger element.

The total strain on the system, on the other hand, is an additive function of the strains on the Kelvin–Voigt and Maxwell elements, namely,

$$\gamma = \gamma_{KV} + \gamma_{M} \tag{5.17}$$

Substituting Equations 5.6 and 5.15 into the previous equation and rearrangement results in

$$e = \frac{\tau}{G_{M}} + \frac{\tau}{G_{KV}}(1 - e^{-(t/\lambda)}) + \frac{\tau}{\eta_{M}}t \tag{5.18}$$

This equation describes the time-dependent change in strain during the creep phase. The initial increase in strain observed in Figure 5.4 (1a) is due to the instantaneous deformation of the spring component of the Maxwell element. This is followed by an exponential increase in strain attributed to the Kelvin–Voigt element (2a). The purely viscous response that follows is related to the Maxwell dashpot (3a), after the Kelvin–Voigt element has reached its maximum strain value.

Thus, since the value of the stress is known, measurement of the instantaneous strain (γ_i) would allow for the determination of the spring modulus of the Maxwell element ($G_M = \tau/\gamma_i$). Equally straightforward is the determination of the viscosity attributed to the Maxwell dashpot. Knowledge of the slope of the linearly increasing strain region in the creep phase would yield an estimate of η_M ($\eta_M = \tau/\text{slope}$). Finally, an exponential curve-fit of Equation 5.6 to experimental γ versus time data would yield estimates of G_{KV} and the relaxation time λ, where $\lambda = \eta_{KV}/G_{KV}$. This exercise could be followed by a global curve-fit of Equation 5.18 to experimental data in order to improve the estimates of the rheological parameters.

Similar information can also be derived from the recovery phase of the experiment. Upon removal of the applied stress at time t', a certain proportion of the total strain will be immediately recovered (1b). This instantaneous recovery is due to the Maxwell spring. Thus, as before, the modulus of the Maxwell spring can be determined from knowledge of the instantaneous strain recovered (γ_i) and the applied stress ($G_M = \tau/\gamma_i$). The strain will then gradually decrease in an exponential fashion governed by the Kelvin–Voigt element (2b) to a limiting value dictated by the irreversible flow from the Maxwell dashpot (3b). Thus, the strain decay function for $t > t'$ will have the form

$$\gamma = \frac{\tau}{\eta_{M}} + \frac{\tau}{G_{KV}}e^{-((t-t')/\lambda)} \tag{5.19}$$

As time approaches infinity, the value of the strain will approach the τ/η_M limit. Thus, knowledge of this limiting minimum strain value (γ_L) would allow for the determination of η_M. ($\eta_M = \tau/\gamma_L$). A nonlinear exponential curve-fit of Equation 5.19 to experimental data would allow for the determination of G_{KV} and λ.

If during the rheological tests, conditions were restricted to the linear viscoelastic range of the material, then the elements that give rise to the elastic response will give similar results in the creep and recovery phases. The relaxation times in the creep and recovery phases are identical in this model.

5.1.4 REAL VISCOELASTIC MATERIALS

The rheological behavior of real viscoelastic materials is too complex to be modeled by the Kelvin–Voigt, Maxwell, and Burger models. Very popular is the construction of extended Maxwell and Kelvin–Voigt models where elements can be placed in an infinite number of arrangements. The problem with this approach is that, very quickly, the entire model building process degenerates into a curve-fitting exercise. This is not necessarily a bad thing, but one should exercise extreme care in describing the structure of a material as a combination of springs and dashpots since it is difficult to assign real physical meaning to these abstract structural elements.

Thus, for the creep phase of a creep–recovery experiment, a general function that describes the observed increase in strain upon loading has the following form:

$$\gamma = \frac{\tau}{G_{i,c}} + \frac{\tau}{G_{d,c}} \sum_{i=1}^{n} (1 - e^{-t/\lambda_i}) + \frac{\tau}{\eta} t \qquad (5.20)$$

where

$G_{i,c}$ corresponds to the apparent elastic modulus of structural elements responsible for the instantaneous elastic response of the material upon loading in the creep phase of the rheological test ($G_{i,c} = \tau/\gamma_{i,c}$)

$G_{d,c}$ corresponds to the apparent modulus of structural elements responsible for the time-dependent, delayed, elastic response of the material

η corresponds to the apparent viscosity of structural elements that flow during the creep phase of the rheological test

Notice how instead of a single relaxation time we now have a spectrum of relaxation times λ_i. All these elements need not be present in a particular material.

The situation is similar for the recovery stage of the rheological test. An equation that describes the decay in strain after removal of the externally applied stress at $t = t'$, that is, upon unloading is

$$\gamma = \frac{\tau}{\eta} + \frac{\tau}{G_{d,r}} \sum_{i=1}^{n} e^{-((t-t')/\lambda_i)} \qquad (5.21)$$

After an instantaneous strain recovery upon unloading ($\gamma_{i,r} = \tau/G_{i,r}$) at $t = t'$, the remaining strain will decay in a time-dependent, delayed, fashion to a limiting permanent value γ_p, where $\gamma_p = \tau/\eta$. As stated before, structural elements within the material have deformed irreversibly, that is, they have flowed, during the creep phase of the test. These elements have an apparent viscosity η. Notice how instead of a single relaxation time we now have a spectrum of relaxation times λ_i.

It is important to keep in mind that fats are extremely complex systems. As will be described in the following, the rheological response of fat crystal networks is quite time-dependent and is also a function of the applied stress. To make matters worse, fats also tend to flow during creep tests, and therefore $\gamma_{i,c} \neq \gamma_{i,r}$. It is for these reasons that I have explicitly defined parameters for the loading (creep) and unloading (recovery) phases of the rheological test.

5.1.5 CREEP–RECOVERY STUDIES OF FATS

Creep–recovery studies on fats are very rare. Of note are the study of DeMan et al. (1985) and Shellhammer et al. (1997). Fats are very difficult to study because the underlying fat crystal network is easily destroyed. DeMan et al. (1985) could measure creep–recovery curves in butter and margarine using a homemade rheometer (Figure 5.5). In these curves, he identified an instantaneous deformation at loading (A), an instantaneous deformation recovery upon unloading (B), a time-dependent deformation recovery after unloading (C), as well as a permanent deformation (D). All deformations (B, C, D) are expressed as strains, relative to the total deformation (B + C + D).

He also calculated viscoelastic parameters such as the instantaneously recovered elasticity (IRE),

$$IRE = \frac{\text{Loading stress (Pa)}}{B} \tag{5.22}$$

the time-dependent recovered elasticity (TDRE),

$$TDRE = \frac{\text{Loading stress}}{C} \tag{5.23}$$

and the viscous flow (VF),

$$VF = \frac{\text{Loading stress (Pa)}}{D} \times (\text{Time of flow, } s) \tag{5.24}$$

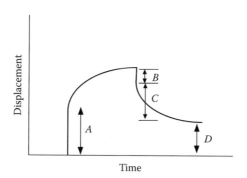

FIGURE 5.5   Idealized creep–recovery curve for milkfat.

In this study, DeMan and coworkers reported the viscoelastic response in terms of strains as well as the viscoelasticity parameters defined earlier.

This study demonstrated the experimental difficulties associated with these materials. For example, fats do not behave as typical viscoelastic materials since the instantaneous strain upon loading (*A*) is much greater than the strain instantaneously recovered after unloading (*B*). Thus, structural rearrangement takes place during the rheological test. Recovered strains, as well as permanent strains, were also a strong function of the loading time (Figure 5.6). As expected, as the loading time increased, so did the permanent strain (PS), while the instantaneously recovered strain (IRS) after unloading decreased. The time-dependent recovered strain (TDRS) after unloading, on the other hand, remained relatively constant. Thus, it is advisable that the loading time be as short as possible, while still retaining sensitivity. The effect of loading force on viscoelasticity parameters was quite pronounced as well (Figure 5.7). As the loading force increased (5°C, 10 min loading time), both the IRE and TDRE decreased. Surprisingly, the VF decreased as well.

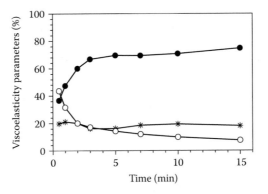

FIGURE 5.6 Effect of loading time at 5°C on viscoelasticity parameters. Open circles correspond to the IRS, closed circles correspond to the permanent strain, and asterisks correspond to the TDRS.

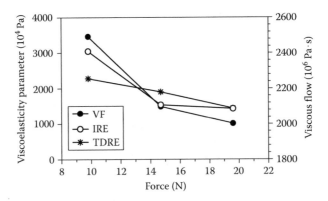

FIGURE 5.7 Effect of loading force at 5°C, 10 min loading time, on viscoelasticity parameters. Open circles correspond to the IRE, closed circles correspond to the VF, and asterisks correspond to the TDRE.

TABLE 5.1
Viscoelastic Parameters of Butter
and Margarine at 5°C (10 min Loading)

Sample	IRE 10^4 Pa	TDRE 10^4 Pa	VF 10^6 Pa·s
Butter A	1540	1903	2098
Butter B	928	837	1854
Margarine A	622	593	1774
Margarine B	994	767	2074

Even if plagued with uncertainties, the creep–recovery technique proved useful in detecting differences between butter and margarine samples. Surprisingly, the differences even between different butter types were quite dramatic (Table 5.1).

The viscoelastic properties of both butter and margarine were strongly dependent on temperature (Figure 5.8). The IRE, TDRE, and VF decreased as a function of increasing temperature—the materials became less elastic at higher temperatures. Another interesting effect is that of shear working on the viscoelasticity of butter and margarine (Figure 5.9). The elasticity of the material was decreased 4 h after shear

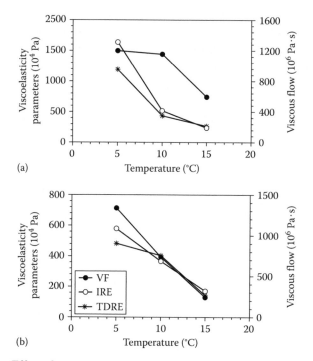

FIGURE 5.8 Effect of temperature, 10 min loading time, on viscoelasticity parameters for (a) butter and (b) margarine. Open circles correspond to the IRE, closed circles correspond to the VF, and starts correspond to the TDRE.

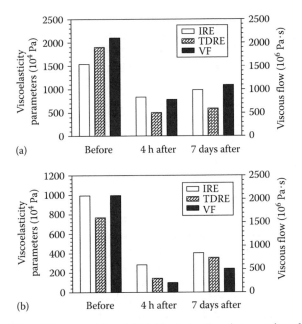

FIGURE 5.9 Effect of shear working at 5°C, 10 min loading time, on viscoelasticity parameters of (a) butter and (b) margarine.

working relative to the nonshear worked material, but was partially regained 7 days after shear working. Thus, this method was quite sensitive to changes in the structure of the underlying fat crystal network.

In all these experiments, one would have expected the VF to increase as the elasticity parameters decreased (and vice versa), however, this was not the case. This effect was probably due to the resetting of the material during the rheological test, which would affect the determination of the time of flow (Equation 5.24).

REFERENCES

DeMan, J.M., S. Gupta, M. Kloek, and G.E. Timbers. 1985. Viscoelastic properties of plastic food products. *J. Am. Oil Chem. Soc.* 62: 1672–1675.

Shellhammer, T.H., T.R. Rumsey, and J.M. Krochta. 1997. Viscoelastic properties of edible lipids. *J. Food Eng.* 33: 305–320.

6 Dynamic Rheological Studies of Fats

6.1 INTRODUCTION

Instead of applying a constant stress, leading to steady-state flow, it is possible to subject viscoelastic materials to oscillating stresses or strains. These dynamic tests offer an alternative to creep–recovery and stress-relaxation methods in the study of viscoelastic materials.

In a controlled stress rheometer, stress is applied as a sinusoidal time function, with maximum amplitude τ^* and angular velocity ω [rad/s],

$$\tau = \tau^* \sin(\omega t) \tag{6.1}$$

The rheometer will record the time-dependent strain function with maximum amplitude γ^*, and phase angle δ (relative to the input stress wave). The angular velocity ω is related to the frequency f [Hz] by

$$\omega = 2\pi f \tag{6.2}$$

Alternatively, in a controlled strain rheometer, strain is applied as a sinusoidal time function, with maximum amplitude γ^*, and angular velocity ω [rad/s],

$$\gamma = \gamma^* \sin(\omega t) \tag{6.3}$$

The rheometer will, in this case, record a time-dependent stress function with maximum amplitude τ^*, and phase angle δ (relative to the input strain wave).

An in-phase response ($\delta = 0°$ or $0\,\mathrm{rad}$) is considered an elastic response, while a 90° ($\delta = \pi/2\,\mathrm{rad}$) out-of-phase response is considered a viscous response, while a response with a phase angle between 0° and 90° ($0 < \delta < \pi/2\,\mathrm{rad}$) is characteristic of viscoelastic materials (Figure 6.1).

When working within the linear viscoelastic region (LVR) of a material, controlled-stress and controlled-strain experiments yield similar results.

It is possible to carry out dynamic tests in shear or compression mode. In controlled-stress/strain rotational rheometers, the bottom plate is stationary, while the upper plate or cone, is made to deflect alternatively for a small angle to the left and to the right in a sinusoidal, time-dependent fashion. In controlled-stress/strain dynamic mechanical analyzers, the bottom plate is stationary, while the upper plate is made to deflect alternatively for a small angle upward and downward in a sinusoidal, time-dependent fashion.

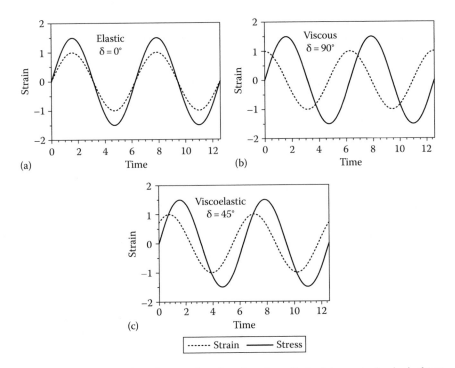

FIGURE 6.1 Stress- and strain-wave functions for a hypothetical dynamic rheological test. Shown here are typical patterns for an idealized elastic response (a), viscous response (b), and viscoelastic response (c).

A sample placed into that shearing or compression gap will be deformed in the same sinusoidal fashion causing stresses within the sample. These stresses will be sinusoidal, and their amplitude dependent on the nature of the material. In order to remain within the linear LVR of the material, deformation angles should not exceed 1°. At these small deformations, the structure of viscoelastic materials will not be irreversibly affected; hence, it is possible to probe their "at-rest" structure in this fashion.

6.1.1 THEORETICAL CONSIDERATIONS

The basic stress–strain relationships for a Hookean solid (elastic response) and a Newtonian fluid (viscous response) have been established before. For a Hookean solid (spring) within its elastic region, stress is directly proportional to strain:

$$\tau = G\gamma \tag{6.4}$$

where
  τ is the stress [Pa]
  G is the shear elastic modulus [Pa]
  γ is the shear strain [dimensionless]

For a Newtonian fluid, stress is directly proportional to shear strain rate:

$$\tau = \eta \frac{\partial \gamma}{\partial t} \tag{6.5}$$

where

τ is the stress [Pa]
η is the coefficient of viscosity [Pa · s]
$\partial \gamma / \partial t$ is the shear strain rate [s^{-1}]

The relaxation time [s^{-1}] of a material is defined as the ratio of the coefficient of viscosity over the shear elastic modulus:

$$\lambda = \frac{\eta}{G} \tag{6.6}$$

A high value for the relaxation time is characteristic of fluids, while a low value for the relaxation time is characteristic of solids.

In this section, we will establish a general stress–strain relationship for viscoelastic materials. However, before we do this, it is useful to establish such relationships for Hookean solids, Newtonian fluids, Kelvin–Voigt, and Maxwell viscoelastic materials.

6.1.1.1 Hookean Solids (Springs)

When a Hookean solid (spring) is subjected to a sinusoidal stress wave, τ, with maximum amplitude τ^*, and angular velocity ω [rad/s],

$$\tau = \tau^* \sin(\omega t) \tag{6.7}$$

it will give rise to an in-phase strain wave, γ, with maximum amplitude, or strain, γ^*, and angular velocity ω [rad/s]:

$$\gamma = \gamma^* \sin(\omega t) \tag{6.8}$$

Substituting Equation 6.8 into Equation 6.4 leads to the dynamic-stress function

$$\tau = G\gamma^* \sin(\omega t) \tag{6.9}$$

An important property of Hookean solids is that their elastic modulus is frequency independent, namely,

$$G = \frac{\tau}{\gamma} = \frac{\tau^* \sin(\omega t)}{\gamma^* \sin(\omega t)} \tag{6.10}$$

which simplifies to

$$G = \frac{\tau^*}{\gamma^*} \qquad (6.11)$$

The elastic modulus determined using a dynamic test has the same form as the elastic modulus determined using a static test, and its value is independent of frequency.

For a dynamic test, the shear elastic modulus, G, is referred to as the storage modulus, G'. The term storage modulus refers to the fact that the stress energy is temporarily stored during the test, but is fully recovered after the stress is removed.

6.1.1.2 Newtonian Fluids (Dashpots)

When a Newtonian fluid (dashpot) is subjected to a stress wave (τ), with maximum amplitude, or stress, τ^* [Pa], and angular velocity ω [rad/s],

$$\tau = \tau^* \sin(\omega t) \qquad (6.12)$$

it will deform, or strain, at a particular rate, that is, it will flow. As stated earlier, the shear stress wave will give rise to a shear strain wave:

$$\gamma = \gamma^* \sin(\omega t) \qquad (6.13)$$

The shear strain rate $\partial\gamma/\partial t$ is simply the time derivative of the shear strain function:

$$\frac{\partial\gamma}{\partial t} = \omega\gamma^* \cos(\omega t) \qquad (6.14)$$

For a Newtonian fluid, stress (τ) is linearly proportional to the shear strain rate $\partial\gamma/\partial t$, and thus, substituting Equation 6.14 into Equation 6.5 leads to the dynamic stress function

$$\tau = \eta\omega\gamma^* \cos(\omega t) \qquad (6.15)$$

Therefore, for Newtonian fluids (dashpots), the stress and strain waves are 90° out of phase. For a dynamic test, Equation 6.15 can be written as

$$\tau = G''\gamma^* \cos(\omega t) \qquad (6.16)$$

where G'' is referred to as the loss modulus [Pa], where

$$G'' = \eta\omega \qquad (6.17)$$

The term *loss modulus* refers to the fact that the stress energy which has been used to initiate the flow of the material has been irreversibly transformed into heat, and

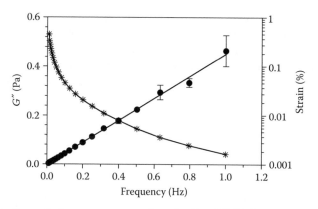

FIGURE 6.2 Dynamic frequency sweep of canola oil at 22°C using an oscillatory stress of 0.002 Pa. The geometry used was the DIN conical concentric cylinder.

therefore cannot be recovered after the strain is removed. An important property of Newtonian fluids (dashpots) is that the loss modulus (G'') is linearly proportional to frequency (see Equation 6.17 and Figure 6.2), in contrast to the case of Hookean solids (springs), where the storage modulus (G') is frequency independent (see Equation 6.11).

6.1.1.3 Kelvin–Voigt Viscoelastic Solid

The simplest model for a viscoelastic material is the Kelvin–Voigt model. In a Kelvin–Voigt element, a spring (elastic component) and a dashpot (viscous component) are arranged in parallel. For such an arrangement, the strain is equal for both the spring and the dashpot, while the stress is the sum of the individual stresses, the elastic stress (τ_e) and the viscous stress (τ_v),

$$\tau = \tau_e + \tau_v \qquad (6.18)$$

Substituting Equations 6.4 and 6.5 into Equation 6.18 yields

$$\tau = G\gamma + \eta\frac{\partial y}{\partial t} \qquad (6.19)$$

Substituting the respective expressions for the dynamic-strain (Equation 6.13) and strain-rate (Equation 6.14) functions into Equation 6.19, we obtain

$$\tau = G\gamma^* \sin(\omega t) + \eta\omega\gamma^* \cos(\omega t) \qquad (6.20)$$

or

$$\tau = G'\gamma^* \sin(\omega t) + G''\gamma^* \cos(\omega t) \qquad (6.21)$$

where G' and G'' are, respectively, the storage and loss shear moduli [Pa].

The dynamic-stress function (τ) in this model is given by a linear combination of an in-phase stress response and a 90° out-of-phase stress response. As for the case of Hookean solids (springs) and Newtonian fluids (dashpots), an important characteristic of a Kelvin–Voigt material is the frequency independence of the storage modulus (G'), and the linear dependence of the loss modulus (G'') on frequency $(G'' = \eta\omega)$.

Substituting the relaxation time (λ) for the coefficient of viscosity into Equation 6.17, we obtain

$$G'' = G'\omega\lambda \tag{6.22}$$

This equation implies that for a Kelvin–Voigt solid at low frequencies, the material's rheological properties are defined by its elastic component, while at high frequencies its rheological properties are defined by its viscous component. At intermediate frequency, values of the rheological properties of a Kelvin–Voigt material are defined equally by its viscous and elastic components.

6.1.1.4 Maxwell Viscoelastic Fluid

In the Maxwell model, a spring (elastic component) and a dashpot (viscous component) are arranged in series. For such an arrangement, the stress is equal for both the spring and the dashpot, while the strain is the sum of the individual strains, the elastic strain (γ_e), and the viscous strain (γ_v):

$$\gamma = \gamma_e + \gamma_v \tag{6.23}$$

By differentiation of the previous expression, we obtain

$$\frac{\partial \gamma}{\partial t} = \frac{\partial \gamma_e}{\partial t} + \frac{\partial \gamma_v}{\partial t} \tag{6.24}$$

Substituting the strain time derivatives from Equations 6.4 and 6.5 into Equation 6.24, results in

$$\frac{\partial \gamma}{\partial t} = \frac{1}{G}\left(\frac{\partial \tau}{\partial t}\right) + \frac{\tau}{\eta} \tag{6.25}$$

Finally, substituting Equation 6.14 into the left side of Equation 6.25 yields

$$\omega\gamma^* \cos(\omega t) = \frac{1}{G}\left(\frac{\partial \tau}{\partial t}\right) + \frac{\tau}{\eta} \tag{6.26}$$

The analytical solution of this familiar first-order ordinary linear differential equation has the following form:

$$\tau = \left[\frac{G\lambda^2\omega^2}{(1+\lambda^2\omega^2)} \right] \sin(\omega t) + \left[\frac{G\lambda\omega}{(1+\lambda^2\omega^2)} \right] \cos(\omega t) \tag{6.27}$$

where λ is the relaxation time, has been defined previously ($\lambda = \eta G$). This equation can be rewritten as

$$\tau = G' \sin(\omega t) + G'' \cos(\omega t) \tag{6.28}$$

where
  G' represents the storage modulus [Pa]
  G'' represents the loss modulus [Pa]

$$G' = \frac{G\lambda^2\omega^2}{(1+\lambda^2\omega^2)} \tag{6.29}$$

and

$$G'' = \frac{G\lambda\omega}{(1+\lambda^2\omega^2)} \tag{6.30}$$

As in the case for the Kelvin–Voigt model, the dynamic-stress function for a Maxwell viscoelastic fluid is a linear combination of an in-phase elastic stress function and a viscous, 90° out-of-phase stress function. For a Maxwell fluid, however, both storage and loss moduli are frequency dependent. We can, however, define two different frequency regimes.

At low frequencies, $\lambda^2\omega^2$ becomes very small and $(1 + \lambda^2\omega^2) \sim 1$. The expressions for the moduli then become

$$G' = G\lambda^2\omega^2 \tag{6.31}$$

and

$$G'' = G\omega\lambda \tag{6.32}$$

Therefore, in the low-frequency range, the storage modulus (G') increases as a function of frequency in a quadratic fashion, while the loss modulus (G'') increases linearly as a function of frequency. The slopes of the log–log plot of G' versus frequency and G'' versus frequency are therefore 2 and 1, respectively. These effects are

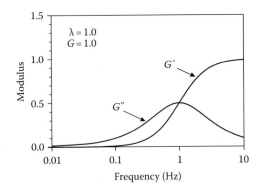

FIGURE 6.3 Simulations of the frequency dependence of the storage (G') and loss (G'') moduli of a Maxwell fluid.

clearly appreciated in a simulation of the behavior of the dynamic moduli versus the frequency (Figure 6.3).

At high frequencies, $\lambda^2\omega^2$ becomes very high and $(1 + \lambda^2\omega^2) \sim \lambda^2\omega^2$. The expressions for the moduli then become

$$G' = G \tag{6.33}$$

and

$$G'' = \frac{G}{\lambda\omega} \tag{6.34}$$

At high frequencies, the slopes of the log–log plot of G' versus frequency and G'' versus frequency are therefore 0 and –1, respectively. For a Maxwell material, therefore, the storage modulus (G') becomes independent of frequency, while the loss modulus (G'') decreases as a function of increasing frequency. These effects are clearly appreciated in Figure 6.3.

Closer examination of Equations 6.29 and 6.30 suggests that, at low frequencies, the loss modulus G'' is larger than the storage modulus G'. A Maxwell material will behave like a Newtonian fluid (viscous fluid) at low frequencies, since the viscous component has enough time to react to a given stress or strain. At high frequencies, however, this relationship is reversed—the material will behave like a Hookean solid (elastic spring) since the viscous component will not have enough time to react to the applied stress or strain.

6.1.1.5 Real Viscoelastic Materials—Generalization of the Model

Real viscoelastic materials are more complex than Maxwell and Kelvin–Voigt materials. It is possible to have different configurations of springs and dashpots, such as in the case of the Burger model.

6.1.2 COMPLEX MODULUS

The complex modulus (G^*) is the ratio of the maximum resulting stress amplitude to the maximum applied strain amplitude, as per our definition of a shear modulus (Equation 6.11)

$$G^* = \frac{\tau^*}{\gamma^*} \tag{6.35}$$

and represents the total resistance of the material to the applied strain. For real viscoelastic materials, the complex modulus and phase angle are both dependent on frequency. The phase angle increases as a function of increasing frequency while the complex modulus decreases.

The challenge then remains how to derive values for the storage (G') and loss (G'') moduli in the absence of a rheological model, that is, Kelvin–Voigt, Maxwell, and Burger.

We know from our treatment that a phase angle of 0° corresponds to a Hookean solid, a phase angle of 90° to a Newtonian fluid, while a phase angle between 0° and 90° corresponds to a viscoelastic material. It would stand to reason that if the complex modulus is a combination of the storage (G') and loss moduli (G''), then the following relationships would apply:

$$G' = G^* \cos \delta \tag{6.36}$$

and

$$G'' = G^* \sin \delta \tag{6.37}$$

If the phase angle (δ) is zero, as is the case for a Hookean solid, that is, the material is purely elastic, then $G' = G^*$ and $G'' = 0$. If the phase angle is 90°, as is the case for a Newtonian fluid, that is, the material is purely viscous, then $G' = 0$ and $G'' = G^*$. For phase angle values $0° < \delta^* < 90°$, storage and loss moduli correspond to proportions of G^* between 0% and 100%.

The most convenient way of establishing a quantitative relationship between complex modulus and the storage and loss moduli is by the use of complex numbers. A complex number is composed of a real and an imaginary part:

$$z = a + ib \tag{6.38}$$

where $i = (-1)^{1/2}$.

Therefore, as shown in Figure 6.4, the complex modulus G^* can be defined as a vector composed of a real part corresponding to the storage modulus (G'), and an imaginary part corresponding to the loss modulus (G''), namely,

$$G^* = G' + iG'' \tag{6.39}$$

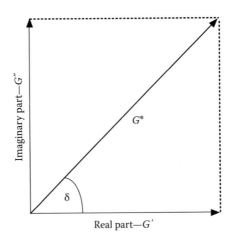

FIGURE 6.4 Argand diagram depicting the complex modulus as having a real part (G'), and an imaginary part (G''), weighted by the phase angle (δ).

The phase angle δ defines the relative weighting of G' and G'' in G^*. This definition of complex modulus is universal and a mechanical model is not required to derive values of G' and G''.

It is also possible to define a complex compliance, J^*,

$$J^* = \frac{1}{G^*} \tag{6.40}$$

and therefore,

$$J^* = J' + iJ'' \tag{6.41}$$

6.1.3 COMPLEX VISCOSITY

We can equally easily define a complex viscosity (η^*). The complex viscosity is defined as the ratio of the maximum resulting stress amplitude over the maximum applied strain amplitude times the angular velocity, as defined in Equation 6.15:

$$\eta^* = \frac{\tau^*}{\omega\gamma^*} \tag{6.42}$$

The complex viscosity η^* represents the total resistance of the material to flow. The complex viscosity and phase angle are both dependent on frequency. The phase angle decreases as a function of increasing frequency while the complex viscosity increases.

In the same fashion as for the complex modulus, the complex viscosity is a combination of a storage viscosity, η', the elastic component, and a dynamic viscosity, η'', the viscous component. The storage and dynamic viscosities are functions of the complex viscosity and the phase angle, and are related to the storage and loss moduli:

$$\eta'' = \eta^* \cos\delta = \frac{G'}{\omega} \tag{6.43}$$

and

$$\eta' = \eta^* \sin\delta = \frac{G''}{\omega} \tag{6.44}$$

The complex viscosity η^* can also be defined as a vector composed of a real part corresponding to the dynamic viscosity (η''), and an imaginary part corresponding to the storage viscosity (η'), namely,

$$\eta^* = \eta'' + i\eta' \tag{6.45}$$

The stress response in dynamic testing can now therefore be written in terms of moduli or of viscosities:

$$\tau = G'\gamma^* \sin(\omega t + \delta) + G''\gamma^* \cos(\omega t + \delta) \tag{6.46}$$

or

$$\tau = \eta'' \omega\gamma^* \sin(\omega t + \delta) + \eta' \omega\gamma^* \cos(\omega t + \delta) \tag{6.47}$$

6.1.4 SOME BASIC CONSIDERATIONS FOR RHEOLOGICAL STUDIES OF FATS UNDER DYNAMIC CONDITIONS

Under small deformation dynamic rheological testing (stresses below 5000 Pa and strains below 0.01%), fats behave very much like elastic solids. Storage (G') and loss (G'') moduli are frequency independent, and the tan $\delta(G''/G')$ is usually below 0.1 (Figure 6.5).

Having said this, G' does exhibit a slight frequency dependence as shown in Figure 6.6 (when using a linear, rather than logarithmic, scale for the y axis). Most practitioners of rheology would not consider this a significant effect. However, the slight dependence is evident. We usually fix the frequency at 1 Hz, where the material maintains its integrity during the lifetime of the rheological test.

In our experience, it is best to always carry out a stress sweep, at a fixed frequency for every sample measured (Figure 6.7). This ensures that the rheological parameters obtained are derived from within the LVR of the material, where stress and strain are linearly proportional to each other. We have found large variabilities in the range of stresses corresponding to the LVR and thus carry out a stress sweep every time. Some interesting features can be appreciated in Figure 6.7, such as the limit of linearity (LOL), where the stress–strain relationship departs from linearity. This

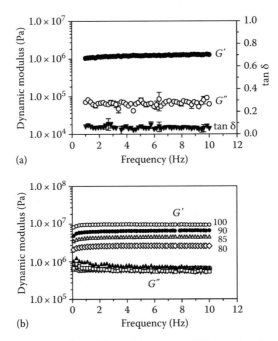

FIGURE 6.5 (a) Frequency dependence of the storage (G') modulus, loss (G'') modulus and the tangent of the phase angle (tan δ) for a commercial shortening at 5°C. (b) Frequency dependence of the G' and G'' for milkfat and milkfat–canola oil (100%–80% milkfat) blends at 5°C.

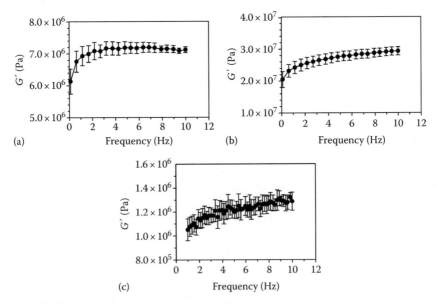

FIGURE 6.6 Frequency dependence of the storage modulus (G') for milkfat (a) at 5°C, cocoa butter (b) at 20°C, and a commercial shortening (c) at 5°C.

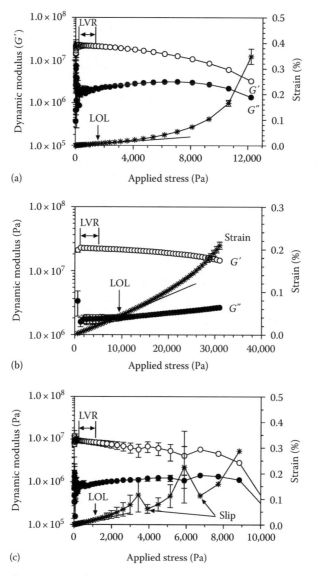

FIGURE 6.7 Stress sweeps for cocoa butter at 22°C (a), tallow at 5°C (b), and milkfat at 5°C. All stress sweeps were carried out at a frequency of 1 Hz. Evident in the graphs are the LVR as well as the LOL for ideal behavior. Evidence of slip and plastic yielding can be found in the strain data for milkfat (c).

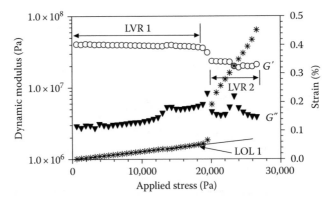

FIGURE 6.8   Stress sweep for cocoa butter at 22°C (f = Hz) demonstrating the existence of multiple LVR. Also depicted is the LOL 1 of the first LVR.

stress usually corresponds to the upper limit of the LVR, but not always. We can also find evidence of slippage of the plates used in this rheological test (Figure 6.7c). The sudden drop in strain corresponds to slippage. As well, drastic increases in the strain are indicative of plastic yielding of the material and the beginning of flow (Figure 6.7c).

An interesting aspect to keep in mind is that multiple LVRs can be present in the material at different stress levels (Figure 6.8). Different levels of structure within the material are being probed in this fashion.

7 Nanostructure and Microstructure of Fats

Alejandro G. Marangoni, Suresh S. Narine,
Nuria C. Acevedo, and Dongming Tang

7.1 INTRODUCTION

The macroscopic rheological properties of networks formed by lipids are of great importance in food products that contain significant amounts of fat. Such products include butter, margarine, chocolate, peanut butter, many spreads such as cream cheese, and ice cream. Many of the sensory attributes such as spreadability, mouthfeel, snap, and texture, and even flavor release are dependent on the rheological properties of the underlying fat crystal network. In addition to this obvious industrial importance, fat crystal networks form a particular class of soft materials, which demonstrate a yield stress and viscoelastic properties, rendering these materials elastoplastic. From a materials sciences point of view, the rheological behavior of these materials is also an important field of fundamental study.

This section provides a review of the development of techniques and models which attempt to relate the microstructural structural organization of fat crystal networks to their mechanical properties. In addition, the work provides a chronicle of our group's own attempts at the problem, focusing mainly on the effects of the microstructural level of structure on the macroscopic elastic moduli of fat crystal networks, and the quantification of this level of structure utilizing fractal geometrical analysis techniques.

Fat crystal networks, like many other materials, demonstrate distinct hierarchies of structural organization, the identification and quantification of which provides insight into the relationship of composition, processing, structure, and mechanical properties of the networks formed by these materials. As such, the macroscopic properties of the network are influenced by the different levels of structure as well as the processing conditions under which the network is formed. Figure 7.1 depicts the structural hierarchy defined during the crystallization of a typical fat crystal network, the factors that affect it as well as some indicators commonly used in the quantification of the relationship between structure and macroscopic properties.

Efforts to model the mechanical strength (Van den Tempel, 1961, 1979; Nederveen, 1963; Payne, 1964; Papenhuijzen, 1971, 1972; Kamphuis et al., 1984; Kamphuis and Jongschaap, 1985) of these networks have met with more failure than success over the past 50 years, mainly due to the lack of a comprehensive model to relate *all* structural network characteristics and solid/liquid ratios of lipid networks to their

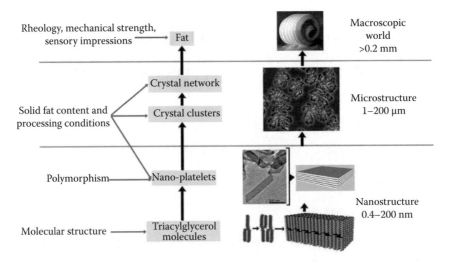

FIGURE 7.1 Schematic showing the structural hierarchy defined during the formation of a fat crystal network.

mechanical strength. This lack stemmed partly from the fact that many scientists in this area concentrated only on the lipid composition, polymorphism, and solid fat content (SFC) of the networks, in large part ignoring the *in situ* microstructure of the network. Lately, much work has been done in analysis of the microstructural and nanostructural levels of the network, leading to encouraging results, which suggests that consideration of this level of structure (together with the other levels previously studied) is absolutely essential in assessing the mechanical strength of the fat network.

7.2   MESOSCALE AND NANOSCALE IN FAT CRYSTAL NETWORKS

The microstructural level, or mesoscale, of a fat crystal network may be defined as those structures in the length scale between ~1 and 200 μm. This level of structure has an enormous influence on the macroscopic rheological properties of the network, noted as early as 1961 by van den Tempel, and has been studied in great detail using numerous microscopy techniques, in particular polarized light microscopy (PLM). Since fat crystals are birefringent, they appear bright between two crossed polarized filters, while the liquid oil remains dark. However, a limitation of this technique is its inherent low resolution, which is ~1 μm. Confocal laser scanning microscopy (CLSM) has been used to image the microstructure o fats (Heertje et al., 1987b, 1993; Marangoni and Hartel, 1998; Herrera and Hartel, 2000a–c), as well as multiple photon microscopy (Xu et al., 1996; Marangoni and Hartel, 1998). However, these techniques also suffer from inherently low resolution. Both polarized microscopy and confocal microscopy work for well for imaging crystal clusters (Figure 7.2), but do not work well to image true fat crystals. Moreover, CLSM and multiple photon microscopy also require the addition of a fluorescent dye, which is somewhat inconvenient.

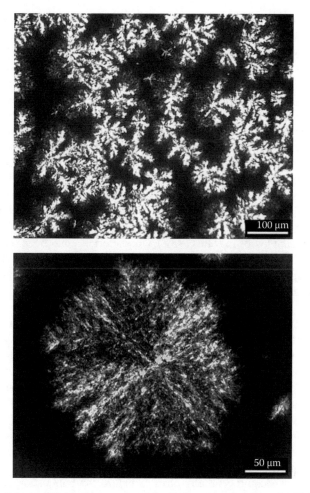

FIGURE 7.2 Polarized light micrographs of the high melting fraction of milk fat in triolein and a large milk fat spherulite.

Scanning electron microscopy (SEM) allows for the visualization of surface topography of materials and has been used in the imaging of fats. Since fats would melt under an electron beam, SEM of fats has to be carried out with good temperature control (Manning and Dimick, 1985) or under cryogenic conditions as for the case of freeze-fracture SEM (Heertje, 1993; Rousseau et al., 1996b). The problem with cryogenic SEM of freeze-fractured fat is that it has to be carried out under cryogenic conditions to avoid sample melting, and under these circumstances, the liquid oil present in the sample freezes. As a consequence, the contrast between the frozen oil and fat crystals is very poor, and therefore, this technique has an intrinsically low resolution as well. In order to overcome this limitation, oil removal from the crystalline matrix using either aqueous detergent solutions or cold organic solvent has been carried out. Jewell and Meara (1970) were able to de-oil lard and vegetable shortening surfaces using a 10%–35% aqueous solution of Teepol prior to making a Platinum-carbon replica of the exposed fat surface. They then successfully imaged micrometer and submicrometer microstructural elements in the fats by SEM. Organic solvents have also been used to separate the solid crystals from the liquid fat phase. Heertje et al. (1987a, 1988, 1993). Heertje (1987) reported the use of a 90:10 (v/v) mixture of 2-butanol and methanol at very specific temperatures to remove the liquid oil from a solid fat in a sample mounted on a special holder for visualization by SEM. In terms of finding an optimal organic solvent to carry out oil removal, while minimizing solid fat dissolution, Poot et al. (1975) reported successful extraction, separation, and isolation of fat crystals by dispersing the fat in a 5% aqueous solutions of sodium dioctylsulfosuccinate (AOT) using a special apparatus. John de Man's group (Chawla et al., 1990; Chawla and deMan, 1990) reported successful removal of oil and isolation of fat crystals using cold isobutanol. They reported that this technique was effective in separating the solid and liquid components of fats without significant losses in solids (~5%). Our experience has been similar and cold isobutanol (5°C–15°C) remains the best solvent we have tried to date. A recent paper from our group (Maleky et al., 2011) compares some of these oil-removal strategies for the visualization of the microstructure in sheared cocoa butter by freeze-fracture cryo-SEM.

In order to image *submicrometer* length scales, Jewel and Meara (1975) used a combination of oil removal and SEM imaging (discussed earlier), while Heertje and Leunis (1997) reported the observation of individual fat crystals using transmission electron microscopy (TEM). They smeared fat onto a copper surface, washed oil off with 2% AOT, froze the exposed fat crystals with liquid nitrogen, and then coated the surface with platinum/carbon. Any unfixed material was then washed off with 2% AOT, while the carbon films are floated and picked up with an uncoated grid, carbon side down. This technique allowed for proper imaging and quantification of nanocrystal size and shape.

The realization that the mesoscale arises from the aggregation of TAG nanoplatelets did not become generally obvious before the publication of the work of Acevedo and Marangoni (2010a). Nanoplatelets were shown to be the primary crystalline structural unit of a fat crystal network, in contrast to what was believed until then, that a primary "fat crystal" was a 1–3 μm sized object. As is the case for the other structural level in fats, the nanoscale has the potential to affect many of the functional properties of fats and fat-structured food products. Moreover, in that publication, we presented the first micrograph of TAG molecular lamellae stacking into crystalline domains which quantitatively correspond to the thickness of a TAG nanoplatelet (Acevedo and Marangoni, 2010a).

In order to be able to image and quantify this nanoscale, we had to develop a method to disrupt crystal aggregates and isolate individual nanoplatelets prior to TEM observation. Initially, several solvent, temperature, and homogenization treatment combinations were assessed. In the end, it was determined that cold (10°C) isobutanol in combination with mechanical matrix disruption dissolved the oil without significantly dissolving the crystalline matrix. Nanocrystals were successfully isolated and imaged. Figure 7.3 depicts some examples of cryo-TEM micrographs corresponding to two different fat samples.

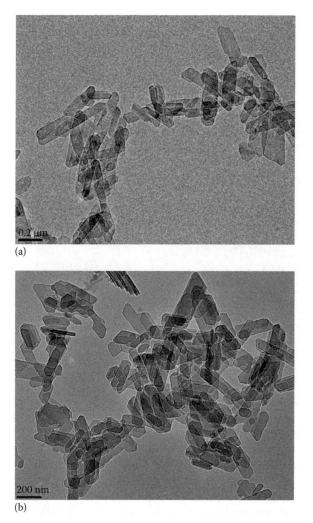

(a)

(b)

FIGURE 7.3 Examples of images obtained by cryo-TEM after using the novel cold-solvent sample treatment. (a) Extracted nanocrystal of a fat blend of 30:70 fully hydrogenated soybean oil (FHSO) and soybean oil (SO). (b) Nanoplatelets of a fully hydrogenated canola oil sample.

Interestingly, the extracted nanocrystal morphology was similar to that reported for pure TAG nanocrystals obtained by crystallization from dilute organic solvent solutions (Bunjes et al., 2007). A particularly visually dramatic result of this work was the observation of a well-defined layered internal structure within each individual nanoplatelet, which can be attributed to the stacking of 7–10 TAG molecular lamellae. Figure 7.4 shows the side view of a stack of nanoplatelets. Furthermore, using image analysis, the average thickness of the lamellae, as well as the platelets could be determined from the cryo-TEM micrographs. For example, Acevedo and Marangoni (2010a) obtained similar lamella thickness values by image analysis of cryo-TEM micrographs and small-angle powder x-ray diffraction (SAXRD) where

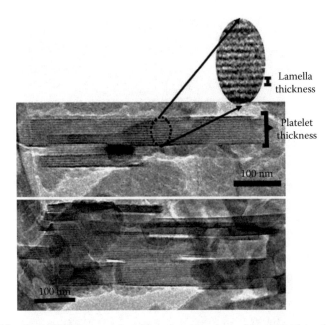

FIGURE 7.4 Cryo-TEM images depicting the side view of nanoplatelets showing their internal structure composed by stacks of TAG lamellae. The distance between consecutive internal lines corresponds to the thickness of an individual lamella. The thickness of a nanoplatelet is given by the stacking of several TAG lamellae.

lamella thickness is represented by their long spacing values. For a sample of fully hydrogenated canola oil, the average values were 4.23 ± 0.76 and 4.5 nm for cryo-TEM and SAXRD, respectively.

Powder x-ray diffraction (XRD) has been extensively used in the study of the crystal structure of fats. Figure 7.5 shows a characteristic XRD pattern of a fat sample in the wide and small-angle region. Typically, most of the information obtained from XRD are long and short spacings that define a polymorphic form, yet XRD patterns can also provide information on the characteristics of fat nanocrystals. For example, the width at the half maximum of a specific diffraction peak in the small angle region can be analyzed using the Scherrer analysis (West, 1984) to characterize the crystalline domain size (D) which corresponds to the thickness of the nanocrystals:

$$D = \frac{K\lambda}{\text{FWHM}\cos(\theta)} \tag{7.1}$$

where

K is a crystal shape factor (usually 0.9 for crystallites of unknown shape)
θ is the diffraction angle
FWHM is the full width at half the maximum intensity in radians (usually of the first small angle reflection corresponding to the [001] plane)
λ is the wavelength of the x-ray, 1.54 Å for a copper anode

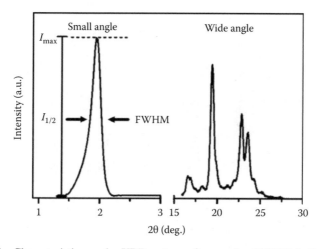

FIGURE 7.5 Characteristic powder XRD pattern of a sample of FHCO in the small- and wide-angle region. In the small-angle region it is possible to observe the (001) reflection peak where the full width at half maximum (FWHM) is indicated with arrows.

It is worth noting that the Scherrer equation is not applicable to sizes larger than ~100 nm. Thus, platelet lengths and widths, which are commonly larger than 100 nm cannot be estimated by using this analysis. Acevedo and Marangoni (2010a,b) found that when applying the Scherrer's analysis to the SAXRD patterns, nanoplatelet thickness values obtained were in close agreement to those obtained by image analysis of cryo-TEM micrographs.

Many research groups have studied the effects of composition and processing conditions on crystal size at the microstructural level. Their results have been explained by the effect exerted by the different processing conditions on the nucleation process (Martini et al., 2002; Singh et al., 2004). In addition, changes in matrix supersaturation induce alterations in the microstructure of TAG networks, for example, increases in cluster size and network density with decreases in supersaturation (Rodriguez et al., 2001; Ahmadi et al., 2008; Ribeiro, 2009). In the last 2 years we have also explored the possibility of engineering fat nanostructure in a predictable fashion in order to target specific functionalities. We reported that external fields strongly affect the nanostructure of TAG crystal networks (Acevedo and Marangoni, 2010b). In that work, we described that a high supersaturation in the melt as well as fast cooling rates and high shear rates translate to a more extensive nucleation process, which in turn yields a greater number of smaller nanoplatelets. Figure 7.6 shows changes in nanoplatelet dimensions in blends of fully hydrogenated canola oil and high oleic sunflower oil as a function of changes in supersaturation of the melt (Figure 7.6a) and shear and cooling rates (Figure 7.6b).

This new knowledge of the nanoscale in fats is important. It allow us to understand and place in proper context the tapping-mode atomic force micrographs of cocoa butter crystallized at room temperature for a year that we collected over a decade ago (Figure 7.7). We can clearly observe here the nanoplatelets on the left panel, 3–4 μm clusters of nanoplatelets in the middle panel, and aggregates of aggregates on the right panel. This suggests that the mechanism of aggregation could be one of diffusion-limited cluster aggregation.

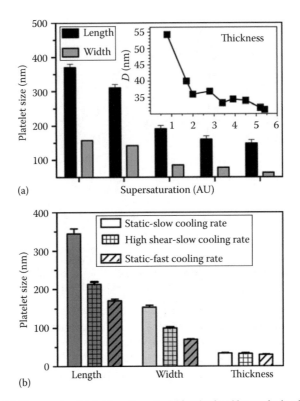

FIGURE 7.6 (a) Changes in platelet length and width (obtained by analysis of the cryo-TEM images); and thickness (obtained by the Scherrer analysis of the x-ray data) as a function of the supersaturation. (b) Example of the effect of crystallization under different cooling rates on the nanostructural level of fats. Platelet dimensions for a 1:1 (w/w) mixture of FHCO and HOSO crystallized statically, under slow cooling rate (1°C/min), fast cooling rate (10°C/min), or under a high laminar shear rate (300 s^{-1}).

However, this still remains conjecture and the mechanisms responsible for the creation of a mesoscale composed of clusters of nanoplatelets still remains ill-defined. Efforts such as those of David Pink (see Chapter 2) on modeling the aggregation of nanoplatelets into larger clusters/aggregates will hopefully shed some light on the process. Preliminary results from Pink's work and from our group by ultrasmall angle XRD (unpublished) suggest that diffusion-limited aggregation can explain many of the observed crystal cluster morphologies encountered in fat crystal networks, and that the aggregation structures are fractal in nature. In order to better understand what is meant by this statement, we must review the concept of fractality as it relates to colloidal aggregation.

7.2.1 Fractals

Classical or Euclidean geometry is based upon the use of regular shapes to describe objects. The reader will be familiar with the use of straight lines, circles, conic sections, polygons, spheres, quadratic surfaces, etc., and combinations of these elements to describe objects around us. However, many patterns in nature defy description by these

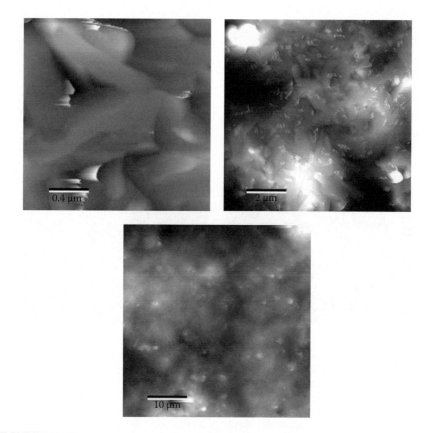

FIGURE 7.7 Tapping mode atomic force micrograph of cocoa butter crystallized statically at room temperature for 1 year as a thick slab.

regular shapes. The geometry of coastlines, mountains, trees, and vegetables (Figure 7.8), for instance, cannot always be defined adequately by spheres, cubes, or cones.

Fractal geometry was born out of this lack of geometrical tools. Benoit Mandelbrot is credited with having developed the field of fractal geometry to describe many of these natural shapes (Mandelbrot, 1982). According to Mandelbrot (1982), "…I conceived and developed a new geometry of nature and implemented its use in a number of diverse fields. It describes many of the irregular and fragmented patterns around us, and leads to full-fledged theories, by identifying a family of shapes I call fractals."

The unifying concept underlying fractals is the concept of self-similarity. Self-similarity essentially means invariance against changes in scale or size, and is demonstrated in many of the laws of nature. Self-similarity is one of a vast number of symmetries that exist in nature. Symmetry is usually taken to mean invariance against some change, that is, some aspect of an object stays the same regardless of changes in the state of the observer. For example, there is symmetry in regular bodies such that they may be operated upon by a number of operations, and yet after the operation is carried out, every point of the body in its original state is coincident

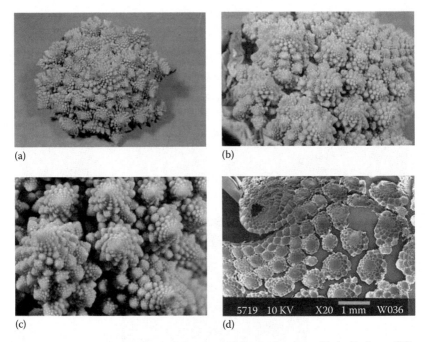

(a)

(b)

(c)

(d)

5719   10 KV   X20   1 mm   W036

FIGURE 7.8 (a–c) Photographs of a Minaret cauliflower showing self-similarity at different length scales. (d) Electron micrograph (SEM) of Alverda cauliflower on submillimeter scale. (Adapted from Grey and Kjems, *Physica D*, 38, 154, 1989.)

with an equivalent point in its altered state. We have been able to take advantage of underlying symmetries in nature—notably in the previous example in the application of group theory to the symmetry operations which render groups of molecules invariant against sets of operations consisting of reflections, rotations, translations, inversions, and combinations of these operations (Cotton, 1971). Equally useful and more popular symmetries in nature include the exploitation of invariance against uniform motion, which has spawned the theory of special relativity, and the equivalence of acceleration and gravity which is the basis of Einstein's general theory of relativity. Even classical physics is built along the lines of symmetry in nature—the electrostatic or gravitational attraction or repulsion between two bodies demonstrates mirror symmetry—there is no partiality shown to left or right. Suffice it to say, therefore, that symmetry in nature has aided us tremendously in understanding and quantifying our world. As will be demonstrated, the symmetry of self-similarity at different length scales is equally as useful as those mentioned earlier. To quote Schroeder (1991), "Yet, among all these symmetries flowering in the Garden of Invariance, there sprouts one that, until recently, has not been sufficiently cherished: the ubiquitous invariance against changes in size, called self-similarity."

The word fractal stems from the fact that fractal objects demonstrate fractional dimensions, rather than the integer dimensions for objects encountered in Euclidean geometry. The connection between fractional and self-similarity is that the easiest way to construct a set that has fractional dimension is through self-similarity (Crownover, 1995). In classical Euclidean geometry, objects have integer dimensions: the reader would be familiar with the reasoning that a line is a 1D object, a plane a 2D object, and a volume a 3D object. In this way, Euclidean geometry is suited for quantifying objects that are ideal, man-made, or regular.

One may imagine that if enough kinks are placed in a line or a plane, the result is to have an object that may be classified as being an intermediate between a line and a plane or a plane and a cube. The dimension of such an object is fractional (i.e., between 1 and 2 or between 2 and 3) and the object may be classified as a fractal object from the fact that instead of having a Euclidean dimension (integer) it has a fractional dimension. One of the most important features of fractal objects is that they are self-similar; that is, there is a repetition of patterns in the object at many different scales. For natural objects such as trees, clouds, coastlines, etc., Euclidean geometry fails to provide an adequate quantification, but many of these natural objects are self-similar at different scales. For example, a tree has branches, these branches have smaller branches, and so on, and if one changes the scale of observation of the tree, the same pattern is observed, at least in a statistical sense if not in a deterministic sense. Therefore, fractal geometry provides a good measure of such objects with nonintegrer dimensions.

The concept of fractional dimension was introduced by Hausdorf (1919). As early as 300 years ago, Leibniz (1721) used the scaling invariance of the infinitely long straight line for its definition. However, as self-similar entities with fractional dimensions started appearing in the mathematical literature, they were met with distaste. Charles Hermite, the famous mathematician, for example, labeled such entities monsters. However, largely due to the efforts of Benoit Mandelbrot (1982), fractal geometry is now an accepted and extremely useful method of describing and quantifying entities that demonstrate scale invariance and fractional dimensions.

The reviews by Jullien (1987), Meakin (1988), and Lin et al. (1989) as well as the books by Jullien and Botet (1987), Russ (1984), Falconer (1990), and Viczek (1992) on the subject of fractals and fractal aggregation are recommended.

For a disordered distribution of mass, such as in a clustering of stars in the Milky Way or the clustering of particles in a colloid, fractal geometry is also useful. A short example is useful. For a solid 1D line, a 2D plane, or disc (Figure 7.9), or a 3D cube (Figure 7.10), the relationship of mass (M) to the length (L) of the object is given by

$$M(r) \sim L^d \tag{7.2}$$

where d is the Euclidean dimension of the object. In this case, the dimension is an integer, and the object is therefore a Euclidean object.

However, for a disordered distribution of mass (Figure 7.11), if at different scales of observation the patterns are statistically self-similar, then the relationship of radius to mass may be given by (Mandelbrot, 1982; Jullien and Botet, 1987; Vreeker et al., 1992b; Uriev and Ladyzhinsky, 1996)

$$M(r) \sim L^D \tag{7.3}$$

where D is a fractional or *fractal dimension*, known as the mass fractal dimension. Here, the symbol $\sim$ is taken to mean "approximately proportional to."

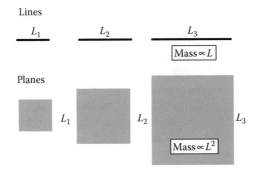

FIGURE 7.9   Mass-length scaling for lines and planes.

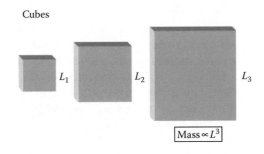

FIGURE 7.10   Mass-length scaling for cubes.

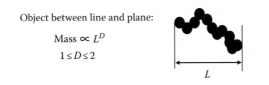

FIGURE 7.11   Mass-length scaling for non-Euclidean objects with noninteger dimensionality.

A classic example of a fractal structure is shown in Figure 7.12. This object is built by the aggregation of five particles, five aggregates, five clusters, five clusters of clusters, etc. In the end, just by following this simple algorithm, one obtains a "snowflake" looking object. As indicated in the figure, this object has a fractal dimension of 1.465 and it displays exact self-similarity.

The reader should be aware that questions have arisen as to whether natural systems should be considered fractal (Biham et al., 1998; Mandelbrot, 1998). Much of the

$$D = \log(N)/\log(L) = 1.465$$

FIGURE 7.12   Vicsek's snowflake, a famous fractal.

controversy stems from a lack of understanding of the meaning of a fractal dimension. For an object, or distribution of objects, to be fractal, certain conditions should be met:

1. The object or distribution of objects should display some degree of self-similarity (exact or statistical) within a specific range of dilations (or contractions). This means that the object looks similar at different magnifications.

2. A specific property, y (elastic modulus, number of particles, mass, etc.), should scale in a power-law fashion with length scale, x, within this specific range of magnifications ($y \sim x^{\mu}$). Thus a log-log plot of y versus x should be linear with a slope μ.

3. The value of this "scaling factor" or "dimension" (D) should be *fractional* and be less than the dimensionality of the corresponding Euclidean embedding space (d), thus $D < d$. If these conditions are satisfied, this scaling factor could be referred to as a fractal dimension.

First of all, exact self-similarity should not be considered a condition for fractality in natural systems. Randomness will always be introduced into a structure due to the inherent complexity of the processes responsible for structure formation. This will therefore negate exactness in symmetry properties, for example, self-similarity, at the nanostructural and microstructural levels. Secondly, in our view, there is no reason why an object should display statistical self-similarity over countless decades in order for it to be considered fractal. In colloidal gels, for example, fractality in the spatial distribution of network mass is usually encountered between the size of a primary colloidal particle and a floc of these primary particles (Dietler et al., 1986). These limits of fractality can be justified from knowledge of the material's structure. In principle, a narrow scaling range should not, in my opinion, negate fractality. As Mandelbrot taught us (Mandelbrot, 1982), a fractal dimension is not a panacea, but it is a powerful means of quantifying the structure of non-Euclidean objects. The beauty of a fractal dimension is that it can capture the complexity of a structure's geometry in a single number. The challenge, however, is to give physical meaning to the number obtained. This fact, however, does not diminish its usefulness.

7.2.2 SCALING THEORY AS APPLIED TO COLLOIDAL GELS

The irreversible aggregation of small particles to form clusters is a common natural phenomenon (Stanley, 1984); for example, this is seen in colloids (Medalia and Heckman, 1971), coagulated aerosols (Friedlander, 1977), chemical species precipitating from a supersaturated matrix, and crystals growing from a supercooled melt (Mullins and Sekerka, 1963). The final structure of such aggregates is important not only because it potentially can yield information about the mechanical strength of the resulting structures, but can also suggest methods to alter the structure kinetically. In many cases, the rate-limiting step of the formation of these aggregates is the diffusion of species toward a growing surface (mass-limited transfer), or the transfer of heat away from the growing surface, or a combination of these factors, depending on the stage of growth (Halsey, 2000).

As early as 1979, Forrest and Witten (1979) demonstrated a class of aggregates that were shown to have density correlations of a power-law form. These aggregates were formed when a metal vapor produced by heating a plated filament was quench condensed, causing metal particles of the order of 40 Å radius to drift down onto a microscope slide. The particles were found to arrange in aggregates of the order of 10^5 particles per aggregate. In 1981, Witten and Sander (1981) constructed a computer simulation model for random aggregation which is diffusion limited, and demonstrated that the density correlations within the model aggregates fall off with distance with a fractional power law, like those of the metal aggregates (Figure 7.13).

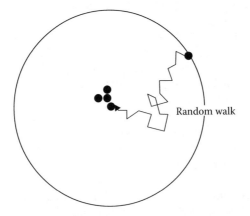

FIGURE 7.13 Witten and Sander's particle-cluster diffusion-limited aggregation. (From Witten and Sander, *Phys. Rev. Lett.*, 47, 1400, 1981.)

Following their earlier work, Witten and Sander (1983) showed that by construct-ing diffusion-limited aggregation models, the objects so formed are scale-invariant whose fractional dimensionality (Hausdorf dimension) is independent of short-range details. Additionally, they showed that diffusion-limited aggregation has no upper critical dimension. This study made the point that the properties associated with scale invariance are long range and universal, and that such long-range proper-ties do not arise from long range forces—rather, these long-range correlations are built up by short-range forces. In 1983, cluster–cluster aggregation models which are diffusion-limited were introduced by Meakin (1983) and Kolb et al. (1983), serv-ing to fuel the flurry of interest in analyzing the structural properties of aggre-gated colloids using fractal theories generated by the work of Witten and Sanders. Meakin's and Kolb et al.'s simulation studies suggested that the colloidal aggregates behave as stochastic mass fractals on a scale that is large compared to the primary particle size. Subsequent experimental studies by Weitz et al. (1984, 1985) on aque-ous gold colloids and Schaefer et al. (1984) on colloidal aggregates of small silica particles confirmed the behavior suggested by the simulations. Following this, in 1985, Brown and Ball produced a computer model, which simulated chemically limited aggregation, and suggested that the structures so formed should also behave as mass fractals. Much experimental work in this area ensued in the following years, with Aubert and Cannell (1986) performing further work on colloidal silica aggre-gates, Schaefer and Keefer (1986), Courtens et al. (1987), and Vacher et al. (1988) on silica aerogels, Rojanski et al. (1986) on mesoporous silica gels, Dimon et al. (1986) on gold colloids, Bolle et al. (1987) on polystyrene lattices, and in a Nature paper, Lin et al. (1989) investigated three different colloids—colloidal gold, col-loidal silica, and polystyrene latex. In all the experimental work detailed earlier, the fractal nature of the colloidal aggregates was well demonstrated. Additionally, colloidal-like gels such as casein gels (e.g., Bremer et al., 1989) have been shown to be composed of homogenous clusters of particles, with the structure within the clusters being fractal in nature.

For a particulate system that is composed of a number of aggregate clusters which are fractal, the number of particles, assumed identical, making up the fractal aggre-gate may then be given by (following Equation 7.3):

$$N_a(\xi) \sim \left(\frac{\xi}{a}\right)^D \tag{7.4}$$

where $N_a(\xi)$ is the number of particles in a fractal aggregate of size ξ, containing particles of size a, and the system is assumed to be fractal within the range bounded by the size of one primary particle and the size of the entire structure. The particle volume fraction of the object may be expressed in terms of the size of the aggregate and the fractal dimensionality, if one assumes a model such as is described later, originally developed by deGennes (1979) for polymer gels, independently by Brown (1987) for a network of fractal clusters, and shown experimentally to be applicable to colloidal gels by Dietler et al. (1986) and others. The following treatment is loosely adapted from Bremer et al. (1989), since these researchers provide an elegant devel-opment of the model originally developed by deGennes and Brown.

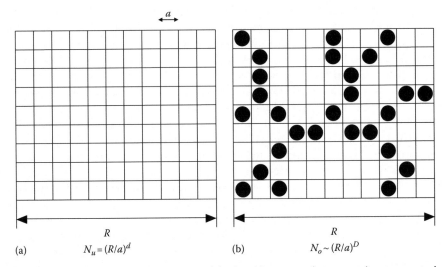

(a) $N_u = (R/a)^d$ (b) $N_o \sim (R/a)^D$

FIGURE 7.14 Square 2D microstructural lattice (a) representing a putative aggregate in which particles are embedded (b).

One may imagine that a regular square lattice is laid over the fractal object, where each lattice site is occupied by a primary particle, or by a volume element of solution (Figure 7.14).

If the particles are arranged in a fractal manner, the number of occupied lattice (or aggregate) sites is given by Equation 7.4. It is important to note that each particle does not completely fill each lattice site; we do not often encounter square particles in nature. The total number of available, unoccupied lattice sites in the 2D lattice shown in Figure 7.12 is given by

$$N_u = \left(\frac{\xi}{a} \right)^2 \tag{7.5}$$

The model assumes a square lattice, with square lattice elements, which is not usually the case, so that Equation 7.5 is more correctly written as

$$N_u \sim \left(\frac{\xi}{a} \right)^2 \tag{7.6}$$

Therefore, the volume fraction of particles within this 2D lattice (Φ) is given by the ratio of occupied lattice sites (N_o) to the total number of available, unoccupied lattice sites (N_u):

$$\Phi = \frac{N_o}{N_u} \sim \frac{(\xi/a)^D}{(\xi/a)^2} = \left(\frac{\xi}{a} \right)^{D-2} \tag{7.7}$$

According to the models for cluster–cluster diffusion-limited aggregation which were verified experimentally as discussed earlier, the aggregates grow until they become space filling, thus forming a gel. Therefore, the fractal dimensionality within each gel is maintained, while at length scales above the characteristic length of one aggregate, the colloid scales in a Euclidean manner. The sum of all available lattice sites in the gel (N_t), which is a space-filling collection of aggregates, occupied by both solution and particles is given by

$$N_t = \sum_{i=1}^{n} N_i \sim \sum_{i=1}^{n} \left(\frac{\xi}{a}\right)_i^2 \tag{7.8}$$

where n is the total number of aggregates in the gel. Therefore, the overall volume fraction of particles, Φ_t, is given by the total number of sites filled by particles, divided by the total number of sites in the gel:

$$\Phi_t = \frac{\Phi \sum_{i=1}^{n} N_i}{N_t} \sim \frac{\Phi N_t}{N_t} = \left(\frac{\xi}{a}\right)^{D-2} \tag{7.9}$$

This model assumes that the gel is a space-filling collection of aggregates, each having the same volume fraction of particles within the aggregates (Φ).

From Equation 7.9,

$$\xi \sim (\Phi_t)^{\frac{1}{D-2}} \tag{7.10}$$

Therefore, the characteristic diameter of an aggregate is related to the overall volume fraction of particles via the mass fractal dimension of the aggregates. As will be discussed in the next section, this relationship suggests that the two factors influencing this relationship are the degree of occupancy of the regular lattice by particles and the degree of order of the packing of the lattice sites by particles.

7.2.3 Elastic Properties of Colloidal Gels: Exploiting the Fractal Nature of the Aggregates

Colloidal gels respond to small amplitude deformations as elastic solids. In 1986, Sonntag and Russel (1986) showed that the storage modulus of a volume-filling network formed by Brownian flocculation of aqueous polystyrene lattices scaled with the volume fraction of particles in a power-law manner, given by

$$G' \sim \Phi^u \tag{7.11}$$

Sonntag and Russel were not the first to experimentally demonstrate this behavior for aggregated systems; other researchers such as Nederveen (1963) and Van den Tempel (1979) showed that storage modulus of dispersed systems of microcrystals formed from oils vary with the particle concentration in a power-law manner, and Payne (1964) showed that the storage modulus of systems formed from aggregates of carbon-black particles in mineral oil demonstrated a power-law relationship with particle concentration. Buscall (1988) demonstrated a similar relationship with systems of polystyrene lattices in water. However, Sonntag and Russel were among the first researchers to suggest that scaling arguments put forward by other researchers such as Kantor and Webman (1984) on the elastic properties of random percolating systems, and computer simulations of random lattices by Feng and Sen (1984) and Feng et al. (1984) may support the relationship that they observed experimentally. At this point, however, the scaling theory had not been developed sufficiently to totally explain the experimental behavior observed by Sonntag and Russel.

Brown (1987) soon after addressed the elasticity of a network of clusters. His method consisted of calculating the elasticity of an individual fractal cluster, and then supposing that in an overcrowded system, fractal behavior survived on a scale related to the overall volume fraction by

$$\xi \sim \Phi^{\frac{1}{D-3}} \tag{7.12}$$

which is the same relationship arrived at in the previous treatment (Equation 7.10). Brown predicted that the storage modulus should scale in the following manner:

$$G \sim \Phi^u \tag{7.13}$$

which is consistent with that seen by Sonntag and Russel and a number of other researchers, cited earlier. Additionally, Brown suggested that the exponent u is given by

$$u = \frac{3 + d_{\text{chem}}}{3 - D} \tag{7.14}$$

where d_{chem} is the so-called chemical length exponent, introduced earlier in simulation studies by Brown and Ball (1985). Soon after the model of Brown, Buscall et al. (1988) published experimental evidence supporting this model, for a system of silica particles. Later work by Ball (1989) on the elasticity of aggregates explains higher fractal dimensions (than was originally suggested by Kolb et al. (1983) and Meakin (1983) that had been reported by Courtens et al. (1987) and Vacher et al. (1988)), by considering that there may be consolidation beyond a critical size of the clusters. Soon after, Edwards and Oakeshott (1989) outlined broad guidelines for the treatment of the transmission of stress in an aggregate, without really achieving more than stating the complexity involved. However, this work is worth mentioning here, since it does point out that the model by Brown (1987) assumes that stress is transmitted in 1D paths, which are branched and are characterized by a fractal dimension, while in some cases the situation is more complicated, since the stress has to be spread among several neighbors in order to maintain stability of the aggregate.

Following Brown and Ball, Bremer et al. (1989, 1990, 1993) suggested elastic models for colloidal protein networks. These researchers envisioned the networks to be composed of strands of elastic material, where deformation causes a stretching of the strands of the network. The storage modulus is then a function of the number of stress-carrying strands per unit area, the geometry of the network, and the character of the bonds within the strands. Bremer et al. (1989) formulated two different scaling relationships for the storage modulus to the particle volume fraction, depending on the geometry of the strands. For straight stress-carrying strands, the relationship is

$$G' \sim \Phi^{\frac{2}{3-D}} \tag{7.15}$$

and for curved stress-carrying strands the relationship is

$$G' \sim \Phi^{\frac{3}{3-D}} \tag{7.16}$$

In 1990, Shih et al. (1990) developed a scaling theory to explain the elastic properties of colloidal gels well above the gelation threshold. The model of a colloidal gel visualized by Shih et al. corresponds well with the model of a colloidal gel described earlier: that is, they define a colloidal gel above the gelation threshold to be a collection of flocs or clusters, which are fractal in nature. These researchers defined two regimes based on the relative value of the elastic constant of the inter-floc links to that of the flocs themselves, and claim that the scaling in these types of gels is affected mainly by the structure of the individual flocs as opposed to percolation-type scaling (Figure 7.15).

An in-depth development of the model created by Shih et al. (1990) is provided later, since this model is used extensively in other parts of this section. In large part ignoring the progress made in describing the scaling relationship of the characteristic length of the cluster/floc size with particle volume fraction for colloidal gels described earlier, Shih et al. uses this identical relationship (given by Equations 7.10 and 7.12), but justifies its use because of the fact that it was found to be true for polymeric gels by deGennes (1979). They defend the use of this relationship for colloidal gels because colloidal gels are similar to polymeric gels by the fact that both are viscoelastic materials, and that both are formed by aggregation processes: polymeric gels by polymerization and/or cross-linking and colloidal gels by particle aggregation. However, Shih et al. does quote the experimental evidence of Dietler et al. (1986) for the validity of this relationship for colloidal gels. It is important to mention that there is indeed sound theoretical reason why this relationship may be used for colloidal gels as well, as was developed earlier (after Brown (1987) and Bremer et al. (1989)).

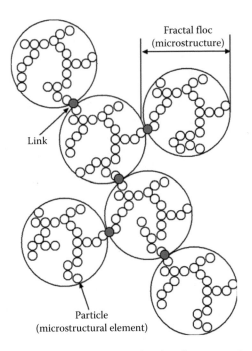

FIGURE 7.15 Model of a colloidal aggregate showing the aggregation of particles into a fractal cluster. Also shown are the links between the flocs.

According to Shih et al., the elastic properties of a floc/cluster are dominated by its effective backbone (connected path of particles responsible for transmission of stress), the size of which is ξ, same as the size of the floc/cluster. This backbone may be assumed to be a linear chain of springs, with each spring representing the bond between particles forming the backbone; a justifiable assumption, since the formation of aggregates via cluster–cluster aggregation which is either reaction limited or diffusion limited have very few loops. The point made by Edwards and Oakeshott (1989), as discussed earlier, concerning the limitations of an effective backbone argument must also be reiterated at this point—it is not entirely clear to the author whether this is a valid assumption, for the reasons given by Edwards and Oakeshott. An earlier publication by Kantor and Webman (1984) formulated the elastic constant of a linear chain of springs as

$$K_s = \frac{G}{N_{bb}S_{\perp}^2} \tag{7.17}$$

where
    G is the bending elastic energy
    N_{bb} is the number of springs in the chain
    $S_{\perp}$ is the radius of gyration of the projection of the nodes of the chain in the $\vec{F} \times \vec{Z}$
       direction, $\vec{F}$ being the applied force and $\vec{Z}$ being the normal of the plane within
       which the chain lies

Equation 7.17 ignores the stretching elastic energy of the chain, since it has been shown by Kantor and Webman (1984) to be negligible for long chains, and is only important for comparatively straight chains stretched along their long dimension. Therefore, when the chain in question is the effective backbone of a floc/cluster, the elastic constant of the floc is given by

$$K_\xi = \frac{G}{N_{bb}\xi^2} \qquad (7.18)$$

where the radius of gyration has now been replaced by the size of the effective backbone (or the size for the floc), since these two quantities would be the same. Now, the backbone of the flocs themselves are fractal objects, for which a fractal dimension between 1 and 2 can be defined, since instead of a straight chain, the backbone would have a certain tortuosity. In keeping with the fractal concept, where a line is kinked enough so that its dimension is raised to a fractional index between 1 and 2, the number of particles, and therefore the number of springs, in the backbone is given by

$$N_{bb} \sim \xi^x \qquad (7.19)$$

where x is the fractal dimension of the effective backbone, or the tortuosity of the effective backbone, and $x \geq 1$ in order to provide a connected path through the floc/cluster. Now, combining Equations 7.18 and 7.19, one may represent the elastic constant of the floc by

$$K_\xi \sim \frac{G}{\xi^x\xi^2} = \frac{G}{\xi^{2+x}} \qquad (7.20)$$

Now, if the elastic energy G is not a function of concentration, then the elastic constant of the flocs should decrease rapidly with increasing floc size. This point strikes the author as being suspect, since it is inconceivable that the elastic energy of a bent chain is not dependent on the density of the springs which comprises the chain. It is conceivable that the density of the particles (density of springs) never really changes with concentration, since the length of the effective backbone (length of the chain) also changes with concentration (via Equation 7.10). If the change in particle volume concentration of the entire system is compensated for in terms of an increased effective backbone length, resulting in no change of the density of particles in the effective backbone, then the bending elastic energy, G, may be considered independent of particle volume concentration. Therefore, if the flocs are allowed to grow larger, they behave as weaker springs. Now, the macroscopic elastic constant of the system of flocs may be expressed either as a function of the elastic constant of the flocs, K_ξ, or the elastic constant of the links between flocs, K_L, depending on the relative strength of K_ξ and K_L. Since from Equations 7.10 and 7.12 the characteristic length of the flocs/clusters, ξ, is related to the particle volume concentration, and since the elastic constant of the flocs will decrease with increasing ξ to a power of $2 + x$ (>3), the relative strengths of K_ξ and K_L

will be affected by the particle concentration of the system. To illustrate this, one may express the elastic constant of the flocs in terms of the particle volume concentration, by substituting for ξ in Equation 7.20:

$$K_\xi \sim \frac{G}{\xi^{2+x}} \sim \frac{G}{\left(\frac{1}{\Phi^{\frac{1}{D-3}}}\right)^{2+x}} = \frac{G}{\Phi^{\frac{2+x}{D-3}}} = G\Phi^{\frac{2+x}{3-D}} \tag{7.21}$$

Therefore, the elastic constant of the flocs will get larger with increasing particle volume concentration. Obviously, as the particle volume concentration increases, there will be a cross-over point at which the elastic constant of the flocs is greater than the elastic links between the flocs, resulting in the elastic constant of the entire system being determined by the nature of the links between clusters (this is called the weak-link regime). The converse, where the particle volume fraction decreases beyond the cross-over point, is also true, and in this case the elastic constant of the flocs grow weaker than the elastic constant of the links between flocs, leading to the elastic constant of the system being determined by the nature of the elastic constant of the flocs (this is called the strong-link regime).

Therefore, for the strong-link regime, the macroscopic elastic constant of the system is given by

$$K \sim \left[\frac{L}{\xi}\right] K_\xi \tag{7.22}$$

where L is the macroscopic size of the system. Therefore, for a constant macroscopic size of the system, combining Equation 7.22 with Equation 7.21, and again substituting for ξ from Equation 7.10:

$$K \sim \xi^{-1} G\Phi^{\frac{2+x}{3-D}} \sim \Phi^{\frac{1}{3-D}} \Phi^{\frac{2+x}{3-D}} = \Phi^{\frac{3+x}{3-D}} \tag{7.23}$$

assuming that the elastic energy of the effective backbone does not change with concentration. Therefore, the macroscopic elastic constant of a colloidal gel at comparatively low concentrations is given by (the strong-link relationship)

$$K \sim \Phi^{\frac{3+x}{3-D}} \tag{7.24}$$

As is evident from Equations 7.24 and 7.14, the strong-link formulation of Shih et al. is consistent with that developed by Brown (1987), if one assumes that the tortuosity, x, is the same as the chemical length exponent, a concept which does not deviate from the explanation offered by Brown. Shih et al. (1990) also studied two types of boehmite alumina gels, catapal and dispal powders, rheologically. The elastic behavior of both gels confirmed the strong-link relationship given by Equation 7.22.

Fractal dimensions calculated from the rheological measurements agreed well with those calculated from static light-scattering measurements.

For the weak-link regime, the macroscopic elastic constant of the system is given by

$$K \sim \left[\frac{L}{\xi} \right] K_L \tag{7.25}$$

Substituting for ξ from Equation 7.10,

$$K \sim \Phi^{\frac{1}{3-D}} \sim G' \tag{7.26}$$

where G' is the shear storage modulus of the system. Here, Equation 7.26 assumes that the links between flocs/clusters are of constant strength. It is important to note that the elastic constant of the links between flocs is not expressed in terms of the geometry of the network. However, Equation 7.26 does provide a scaling relationship of the elastic constant of colloidal gels at high concentrations, with the particle volume fraction.

Shih et al. (1990) also outlined a simple rheological test to determine whether a network is in the weak- or strong-link rheological regime based on the dependence of the strain at the limit of linearity (i.e., the elastic limit) on the particles' volume fraction. The strain at the limit of linearity is the point at which the weakest bond in a floc breaks upon applying a force beyond a critical value. Above the limit of linearity, the network is broken and the linear elastic behavior does not exist anymore. According to the model of Shih et al., the relationship between the strain at the limit of linearity (γ_o) and solids' volume fraction for the strong-link regime is

$$\gamma_o \sim \Phi^{[-(1+x)/(d-D)]} \tag{7.27}$$

From Equation 7.27, we can deduce that the strain at the limit of linearity γ_o should decrease with increasing the solids' volume fraction Φ. In contrast, γ_o increases with increasing Φ for the weak-link regime ($d = 3$) as

$$\gamma_o \sim \Phi^{1/(d-D)} \tag{7.28}$$

The stress at the limit of linearity (σ_o) (Pa) can be used interchangeably with the strain at the limit of linearity (γ_o) since $\sigma_o \sim \gamma_o$ in the linear viscoelastic region (LVR). Therefore, σ_o changes as a function of Φ could also be used to determine the fractal dimension of a network. Assuming the network is in the weak-link regime,

$$\sigma_o \sim \Phi^{1/(d-D)} \tag{7.29}$$

After the development by Shih et al. (1990), Chen and Russel (1991) studied the elastic behavior of a synthesized model system consisting of submicrometer silica spheres coated with octadecyl chains suspended in hexadecane. This study demonstrated that the storage modulus of the colloidal system increased with particle volume concentration in a power-law manner, and that the power-law exponent increased with increasing temperature, indicating a structural change in the network. It was obvious that the power-law exponent was sensitive to the packing of the particles (disturbed when the temperature is increased)—suggesting that the power was dependent on the fractal dimension. This work further served to lend credibility to the scaling models outlined by Brown (1987), Bremer et al. (1989, 1990), and Shih et al. (1990).

Vreeker et al. (1992a) showed that colloidal-like aggregates of whey proteins gels are fractal in nature, utilizing dynamic light scattering measurements. Additionally, they analyzed the protein gels rheologically, showing that the elastic moduli and yield stresses varied with protein concentration according to a power law. They related the power-law exponent to the equivalent models by Brown (1987) and (strong-link model) by Shih et al. (1990), and to the models by Bremer et al. (1989, 1990, 1993). By the value of the exponent measured experimentally, they were able to calculate fractal dimensions according to the various models. The fractal dimensions so calculated were all in reasonable agreement with those measured by the dynamic light scattering experiments. Therefore, these researchers concluded that there was no basis for the validity of one model versus another, based on their experiments. This partly stemmed from their inability to measure the chemical length exponent or tortuosity. In this work, they assumed x had a value which varied between 1.0 and 1.3 (suggested by Shih et al. (1990) based on conductivity measurements).

Hagiwara et al. (1997, 1998) and Marangoni et al. (2000) have analyzed a number of different types of protein gels using both rheological measurements to study the elasticity of the gels, and analysis of CLSM images of the gels, to study the structure of the gels. These researchers found that for the gels they studied, the weak-link regime of Shih et al. (1990) was valid. The gels demonstrated a power-law dependence on particle volume concentration, and the fractal dimensions calculated from the power-law exponent, using the weak-link formulation, agreed very well with fractal dimensions calculated from image-analysis of the CLSM of the gels.

Therefore, both the strong-link formulation and the weak-link formulation of Shih et al. (1990) have been demonstrated experimentally. The weak link has been demonstrated more convincingly, since with the strong-link regime, experimentalists have had to assume a value for the tortuosity of the backbone, and therefore the theory could not be tested well against those of Bremer et al. (1989, 1990, 1993).

More recently, Wu and Morbidelli (2001) developed an expression for the intermediate situation, where both the links and the flocs yield under an externally applied stress. These authors coined this regime the "transition" regime. The expression for the transition regime is

$$G_{TR} \sim \Phi^{\frac{(d-2)+(2+x)(1-\alpha)}{d-D}} \qquad (7.30)$$

where α is a constant in the range (0,1) which depends on the relative weighting of strong-link versus weak-link behavior. A value of $\alpha = 1$ corresponds to the weak-link regime, while a value of $\alpha = 0$ corresponds to the strong-link regime.

An experimental difficulty with this transition regime is the need to obtain an estimate of α. This can be done by determining the solids' volume fraction dependence of both the storage modulus (G') and the strain at the limit of linearity (γ_o).

More recently, Mellema et al. (2002) developed a complete categorization scheme for rheological scaling models of particulate gels. In this work, the authors assume protein particles form stress-carrying strands, in the spirit of the treatment by Bremer et al. (1990, 2003). This is followed by a categorization of the rheological behavior of gels into five classes, ranging from the weak-link regime to the strong-link regime. The general form of their model is

$$G' \sim a_{\text{eff}}^{-\alpha} \Phi^{\frac{\alpha}{(d-D)}} \tag{7.31}$$

where a_{eff} is the effective diameter of the particles which make up the strands, $\alpha = 2\varepsilon + \delta + 1$, and ε is a scaling exponent of the contour length of a cluster spanning strand with cluster size, with values in the range (0,1). The concept of a scaling exponent (δ) of the number of particles in a fractal cluster with the diameter of that cluster ($N \sim \xi$) was proposed originally in 1994 by deRooij and coworkers. The only practical problem with this scheme is the difficulty in obtaining unambiguous estimates of all parameters included in the volume fraction exponential term.

In all these treatments, the macroscopic elastic constant of the network is merely the product of the elastic constant of a basic mechanical unit (the flocs, the links between flocs, or a combination of both) and the number of these units present in the direction of an externally applied force (Shih et al., 1990). The fractal dimension defines the size of the cluster. Thus, for example, a higher fractal dimension will result in a larger cluster size. A larger cluster size translates to less cluster–cluster interactions per unit volume, resulting in a decrease in the elastic modulus of the material. At higher volume fractions, the average cluster size decreases, thus increasing the number of cluster–cluster interactions, which leads to an increase in the value of the elastic constant.

7.2.4 APPLICATION OF SCALING THEORY DEVELOPED FOR COLLOIDAL GELS TO FAT CRYSTAL NETWORKS

In 1992, Vreeker et al. (1992b) presented an interpretation of rheological data for aggregate fat networks in the framework of scaling theories developed for colloidal gels. These authors showed that the storage modulus of the network (G') varied with the volume fraction of solids (Φ_{SFC} = SFC/100), according to a power law, similar to that predicted by models for the elasticity of colloidal gels. It is important to remember that Φ_{SFC} is not equivalent to the particle volume fraction Φ. However, in the Vreeker et al. (1992b) paper and in many such publications following, these two parameters have been used as being equivalent. The fact remains that Φ_{SFC} and Φ are directly proportional to each other, and the difference between them is minimal. For the sake of simplicity in the treatment that follows, these two parameters will be considered equivalent. The article by Vreeker et al. provides an interpretation of rheological data for low Φ fats in terms of the strong-link model of Shih et al. (1990), developed, as is detailed earlier, for colloidal gels at low particle concentrations. From this rheological investigation of the fat network, a fractal dimension could be calculated, using the strong-link formulation given by Equation 7.24, and assuming that the tortuosity of the system, x, had a value between 1 and 1.3. This article also details the measurement of a fractal dimension of the network via light scattering methods, which agreed well with that calculated from the strong-link formulation. However, no attempt was made in this article to show that the structure of a fat crystal network at such low SFCs (low Φ) is organized in a similar manner to the way a colloidal gel is organized, in order to warrant the use of the strong-link formulation. It was unclear as well, in this article, what the primary particles of the gel constituted, and at what length scales the network was fractal (although the fact that light scattering methods were used to calculate the fractal dimension suggests that perhaps the length scales of importance lay in the microstructure region, that is, at a length scale greater than the crystalline level of structure). Therefore, this article served to demonstrate solely that the elastic constant of the network varied in a power-law manner, reminiscent of that demonstrated by colloidal gels. No justification for using the strong-link model was made from a structural perspective, mainly because these authors did not adequately define the structure of the network. However, this article remains one of the most important developments in the field, for it demonstrated conclusively that the storage modulus of low Φ fat crystal networks scale in a power-law fashion with the SFC. The physical and structural implications of the calculated fractal dimension using the strong-link formulation of Shih et al. (1990) was at this point unclear.

Work by Johansson (1995) provided strong experimental evidence for the fractal nature of fat crystal networks as well as showing for the first time that the relationship $\xi \sim \Phi^{1/(D-3)}$ held for fat crystal networks, which could then be considered as weak gels of crystals dispersed in oil.

The analyses of Vreeker et al. and Johansson were interesting enough for Marangoni and Rousseau (1996) to apply the model developed by Shih et al. (1990) for high concentration colloidal gels (weak-link theory described previously) to fat crystal networks of high solid fat concentration. As is explained earlier, the weak-link model offers the ability to relate small deformation rheological measurements (shear storage modulus) to the fractal dimension of the network. As was the case with the work by Vreeker et al. (1992b), Marangoni and Rousseau had no structural basis upon which to justify the application of the weak-link theory, developed for a particular geometry of colloidal aggregates, to fat crystal networks. The main reason for applying this type of analysis was the fact that the analysis by Vreeker et al. (1992b) produced a power-law relationship for low SFC fats. Additionally, as described before, the only variation of structure that accompanied the decrease in hardness index and elastic moduli of milk fat upon chemical interesterification was a change in the microscopically observed microstructure (Rousseau et al., 1996a–c). Marangoni and Rousseau (1996) attributed the weak-link model with a quantification of the microstructure of the network, but offered no explanation as to why the fractal dimension calculated from the application of the weak-link theory was a quantification of microstructure. However, the rheological application of the weak-link theory of Shih et al. (1990) to NIE and CIE milk fat blends demonstrated a power-law relationship between the shear elastic moduli of the networks and the SFC of the networks, given by $G' \sim \Phi^\mu$. Fractal dimensions calculated using the weak-link theory produced fractal dimensions of 2.46 and 2.15 for the NIE milk fat and CIE milk fat, respectively. Given that there was a qualitative change in microstructure upon chemical interesterification, and that there was a large change in the fractal dimension of these gels upon application of the weak-link theory, it did indeed seem possible that the fractal dimension was a measure of microstructure, in some undefined manner. At any rate, Marangoni and Rousseau (1996) attributed the change in structure indicated by a change in fractal dimension to the change in hardness of the milk fat upon chemical interesterification, a reasonable conclusion. In defense of the claim by Marangoni and Rousseau (1996) that the fractal dimension was in some manner a quantification of microstructure, it must be recalled that Vreeker et al. (1992b) utilized dynamic light scattering measurements to calculate a fractal dimension that agreed well with their rheological analysis, therefore suggesting that the structures that were indeed fractal are in the microstructural range.

Following their 1996 work, Rousseau and Marangoni (1998a,b) utilized the weak-link model to analyze the rheological behavior of non-interesterified (NIE) milk fat and enzymatically interesterified milk fat (EIE). The fractal dimension of milk fat changed from 2.59 to 2.50 upon interesterification. However, there was also a large decrease in SFC upon enzymatic interesterification, so that the drop in hardness index and shear elastic moduli observed by these authors was mainly due to the decrease in SFC. However, here again, the rheological scaling behavior was observed, and fractal dimensions could be calculated from the weak-link formulation. One is reminded by the results of this study that indicators of macroscopic mechanical properties such as SFC are indeed valid and necessary for a comprehensive description of the mechanical properties of the network, although as in the case of the CIE and NIE milk fat samples detailed earlier, such variables are not always direct indicators of changes in mechanical properties.

Marangoni and Rousseau (1998a,b) also studied the rheological scaling behavior of NIE and chemically interesterified (CIE) lard-canola oil blends and palm oil-soybean oil blends. The power-law relationship of storage modulus to SFC was again observed, and fractal dimensions could be calculated for all the systems, by using the weak-link formulation. No change was observed in hardness index or storage modulus of the palm oil-soybean oil blends upon chemical interesterification, and neither were there any changes in the fractal dimension upon chemical interesterification. SFC did not change upon chemical interesterification either. Therefore, it seems that indeed the fractal dimension was an indicator of macroscopic hardness: Whether this was true because it was an indicator of the microstructure could not reasonably be extrapolated from the evidence presented by Marangoni and Rousseau (1998b). The hardness index and elastic moduli of lard increased upon chemical interesterification, but the fractal dimension did not change, neither did the SFC. It therefore seems at first glance that not only is the SFC an insensitive indicator of macroscopic hardness, but so is the fractal dimension. Here again, one is reminded that there are many structural indicators of macroscopic mechanical properties, and one must consider them all in concert. However, are there other indicators of hardness that can be accessed through the rheological analysis? At this point, it is perhaps important to briefly examine the method of rheological analysis used by these authors. The weak-link theory of Shih et al. was tested by plotting ln G' as a function of ln Φ for various fat systems. According to the weak-link theory of Shih et al. (Equation 7.26) the slope of such a plot should yield $1/(3 - D)$. What about the intercept of the graph? Marangoni and Rousseau (1998b) wrote Equation 7.26 as

$$G' = \gamma \Phi^{\frac{1}{3-D}} \tag{7.32}$$

and defined γ as a constant which is related to the particles which make up the network. In this, Marangoni and Rousseau were influenced by Bremer et al. (1990), who in their publication, formulated a similar law, where the pre-exponential factor was a constant depending on the nature of the particles and the links between them. Equation 7.32 bears further scrutiny at this point—it should be evident that Equation 7.26 may be written as

$$G' \sim \Phi^{\frac{1}{3-D}} K_L \tag{7.33}$$

where, as the reader would recall, K_L is the elastic constant of the links between aggregates. Certainly, therefore, K_L would depend on the nature of the particles and the links between them, and would contribute to the value of γ. Although Marangoni and Rousseau (1998b) did not offer this justification for their description of the constant γ, it however seems to be appropriate. At any rate, the intercept of the $\ln(G')$ versus ln Φ graph should yield ln (γ) as the intercept, allowing a value of γ to be calculated. Therefore, if the fractal dimension offered a quantification of the spatial distribution of the microstructure (as yet undefined how), then the constant γ contained information on the influence of the particles and the links between them on the elastic constant of the network. Now, upon chemical interesterification, there was a fourfold increase in the value of the constant γ for the lard-canola oil system, which according to Equation 7.32, should have been accompanied by an increase in the storage modulus, as was observed experimentally by Marangoni and Rousseau. As for the palm oil-soyabean oil system, there was comparatively no change in the value of γ.

As a matter of fact, the decrease in G' and hardness index originally observed by Marangoni and Rousseau in 1996 was due to a decrease in γ rather than a decrease in the fractal dimension. A decrease in fractal dimension would have resulted in an increase in the elastic moduli and hardness. Thus, the parameter γ, at a constant SFC, was the true culprit in the observed decrease in mechanical strength upon chemical interesterification of milk fat.

Therefore, the scaling behavior of the fat crystal network, if the weak link is utilized, seems to suggest that there are three important indicators of macroscopic hardness: SFC, the fractal dimension, and the constant γ. It must be stated at this point that this does not mean that other indicators such as the crystalline nature of the network and the molecular ensemble of the network is being ignored by identifying D, Φ, and γ as the three important indicators. Since γ has been suggested to depend on the nature of the particles and the links between them, it almost certainly is dependent on the polymorphism of the network, which in turn is dependent on the molecular composition of the network.

7.2.5 NETWORK MODELS

As presented earlier, phenomenological investigations have been made on the rheology of fat crystal networks, the results of which have been interpreted by models developed for colloidal gels. Not only has this not been structurally justified, but this analysis excluded any structural and mechanical model of the fat network. In large part, the authors working in the field (mainly Marangoni and Rousseau et al. and Vreeker et al., detailed earlier) could not include a structural and mechanical model of the network, since there existed at the time no mechanical and structural model that predicted that the structure of the fat crystal network at any length-scale was fractal in nature. It must also be mentioned that the scaling behavior demonstrated experimentally by these authors was not adequately explained by the existing network models. In the following a brief introduction to the network models that existed in the literature before our group's contributions is presented. A useful review of these can be also found in deMan and Beers (1987).

Early work on a network model was performed by van den Tempel (1961). In his 1961 publication, he suggested modeling the network as a collection of particles held together by van der Waals-London forces. The structural model that was assumed by van den Tempel was that the network is made up of straight chains oriented in three mutually perpendicular directions, each chain consisting of a linear array of particles. The bonds between these large particles were formed from van der Waals-London forces (Hamaker, 1937; Vold, 1951). The relationship of the storage modulus to the SFC of the network arrived at by van den Tempel is

$$G' = \frac{5AD^{0.5}}{24\pi H_o^{3.5}}\Phi \qquad (7.34)$$

where
A is the Hamaker's constant
D is the diameter of the particles
H_o is the equilibrium distance between particles
Φ is the volume fraction of solids

However, the relationship of G' to Φ in most fats has been experimentally determined to be a nonlinear, power-law type of relationship, as shown for several different fats (Figure 7.16).

The work by van den Tempel failed to correctly predict this experimentally observed power-law relationship of G' to Φ because, according to Vreeker et al. (1992b), it did not take into consideration the fractal arrangement of the network at certain length scales. Furthermore, van den Tempel only considered the attractive forces in his treatment, choosing to ignore whatever repulsive forces were present.

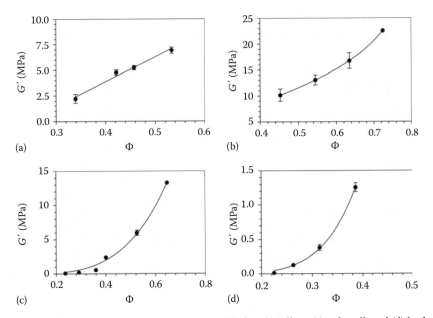

FIGURE 7.16   Plots of G' versus Φ_{SFC} for (a) milk fat, (b) tallow, (c) palm oil, and (d) lard.

In 1963 Nederveen used the same structural model as van den Tempel, but incorporated the repulsive forces acting between particles (he considered the Lennard–Jones potential between two particles). Nederveen found the following formula for the modulus of elasticity:

$$E = \left(\frac{A\Phi}{2\pi d_o} \right)\left(1 - \frac{11R\varepsilon}{d_o^2} \right) \tag{7.35}$$

where
   R is the particle radius
   ε is the strain of the deformation
   d_o is the equilibrium distance between two particles
   A and Φ are as defined earlier

Again, Nederveen's formulation failed to show the nonlinear dependence on Φ, due perhaps to an unrealistic structural model of the network which did not involve the fractal geometry detailed in the previous section.

In 1964, Payne (1964) tried to apply the linear chain concept of van den Tempel to carbon black networks in oil, but found that the shear modulus was inversely proportional to some power of the particle diameter, not in agreement with van den Tempel's or Nederveen's equation. This study may however be inconclusive about the validity of these equations, since the discrepancy may have been due to the non-homogenous distribution of the carbon black spheres.

Sherman (1968) proposed a different model for flocculated o/w emulsions involving chain-like configurations which form coils with cross-linkages, resulting in interlinked spherical structures (Shama and Sherman, 1970). In his model, localized regions of densely packed particles are joined to less densely packed regions; consequently the degree of interlinking is not the same throughout the entire flocculate. The formulation of G' for this model is

$$G' = \Phi\left(1+1.828v\right)\frac{A}{36\pi D^3 H_o^3} \tag{7.36}$$

where

v is the total volume of the continuous phase held in voids between the particles
D is the diameter of the particles
H_o is the minimum equilibrium distance between particle surfaces
A is called the interaction constant (Hamaker's constant)
Φ is as defined earlier

While Sherman's equation agreed with Payne's observations in terms of particle diameter, it also shows a linear relationship with Φ, contrary to what has been demonstrated experimentally.

In 1979, van den Tempel (1979) revised his structural model of fat crystal networks; he suggested that clusters (microstructures) of particles make up the chains, rather than particles. Additionally, he suggested that the forces holding the clusters together were due to common "chains" of particles between clusters. He used this argument to modify his original equation, but again did not take into consideration any fractal arrangement of the network. Here he suggested that the relationship between the shear elastic modulus and the volume fraction of solids was a nonlinear, power-law type of relationship ($G \sim \Phi^\mu$). Moreover, he observed that the scaling factor μ was proportional to the number of particles within the clusters and that several parameters that determine the state of aggregation of the system could be lumped into this scaling factor, μ. He also estimated that about 10^6 crystal particles made up the clusters and that increasing agitation lead to less aggregation. Various other researchers have tried to modify van den Tempel's work, in varying degrees of complexity (Papenhuijzen, 1971, 1972; Kamphuis et al., 1984; Kamphuis and Jongschaap, 1985). However, none of these models took into consideration a fractal arrangement of the network, and none demonstrated a power-law dependence of G' on Φ as is observed experimentally. Our model published in 1999 (Narine and Marangoni, 1999b) addressed this power-law dependence and introduced the concept of fractality in a structural-mechanical model of fats which explained well their rheological characteristics (Narine and Marangoni, 1999c).

7.3   WHERE LIES THE FRACTALITY IN FAT CRYSTAL NETWORKS?

In this section, the structure of fat crystal networks at the microstructural level will be analyzed and discussed so as to determine whether there is a structural organization present which warrants the application of the weak-link theory developed for colloidal gels to fat crystal networks. Furthermore, since the microstructure of fat networks has been rather loosely characterized before this work, effort will be made to utilize a number of different imaging techniques to attempt to provide a description of the network at the microstructural level that is supported by various types of microscopical evidence.

7.3.1 STRUCTURAL MODEL OF THE FAT CRYSTAL NETWORK

The microscopic evidence presented earlier forms a composite picture of the structure of fat crystal networks. No one method of microscopy on its own reflects the composite picture, but by considering the various images from the different types of microscopy, one can build a structural model of fat crystal networks which is supported by all the forms of microscopy. One of the needs in the fats and oils industry is the establishment of nomenclature that will unambiguously identify different levels of microstructure. Perhaps the best way to establish such nomenclature is to trace the development of the fat network as it forms from the melt. At the start of crystallization, there is a process of nucleation followed by growth of these nuclei into crystallites. Molecular thermodynamics and kinetics most probably control this process. As we have seen, these crystallites associate in dense packets, several intertwined crystallites tending to form "particles" of the order of 1–3 μm. It is proposed that these "particles" be called *microstructural elements*, since they form the smallest repeating structure at a length scale visible under a light microscope. These microstructural elements then continue to grow larger through further crystallization, but there is an aggregation process that takes place as well, leading to the formation of intermediately sized clusters, which further aggregate to form large clusters. This aggregation process is most probably controlled by mass and heat transfer limitations. The large clusters, which it is proposed be called *microstructures*, pack in an orthodox, space-filling manner to form the network, as has been shown by Heertje and coworkers. Additionally, evidence of the microstructures has been presented here, in CLSM, AFM, and PLM images, and to a less certain extent, in SEM images. Figure 7.17 is a schematic 2D representation of the structural hierarchy of the microstructure of fat crystal networks. The microstructural elements, the intermediately sized clusters, and the microstructures have all been represented by circles, indicating that they are spherical.

There is evidence from all of the methods of microscopy that these structures are spherical-like (which may be taken to mean stretched and distorted spheres as well). However, whether they are smooth spheres is something that cannot be determined well, and this factor will most probably differ with the type of fat. Certainly, SEM and AFM micrographs of the microstructural elements suggest that these are not smooth spheres, and that the sphericity is not perfect. However, representing these structural entities as spheres is probably fairly representative, providing one keeps in mind that this is an imperfect approximation. The packing of the microstructural elements within the microstructures is quite disordered, as has been seen in the images presented before. However, it is important to recall that at length scales ranging from the diameter of the microstructural elements to the diameter of the microstructures, these structures appear to be statistically self-similar, that

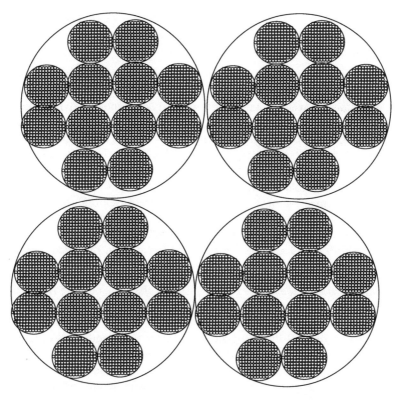

FIGURE 7.17 Idealized 2D schematic representation of the self-similar structural hierarchy of the microstructure of fat crystal networks.

is, the structures within this range of length scales, at different magnifications, appear to be similar. This of course is most strikingly obvious in the atomic force micrographs presented. Therefore, in an idealized attempt to demonstrate this self-similarity, Figure 7.17 shows that the packing of the microstructural elements within clusters are similar to the packing of the clusters themselves into the microstructures. In order to easily achieve this effect, the microstructural elements and the clusters are arranged in an ordered manner, but in reality, these arrangements are quite disordered, as is evidenced by the microscopy images shown before. The fact that there appears to be statistical self-similarity in the length ranges between the size of the microstructural elements and the microstructures is not surprising, since the method of growth/aggregation between these two length ranges is limited by the same physical constraints.

7.3.2 CHARACTERIZING MICROSTRUCTURE

A fractal dimension describes the self-similar or self-affine character of some objects. Fractal dimensions calculated by microscopy methods such as box-counting, particle-counting, and Fourier transforms of polarized light micrographs of fats can be used to quantify the microstructure of fat crystal networks. The fractal dimensions describe the combined effects of morphology and spatial distribution patterns of the crystal clusters in the fat crystal networks.

Fractal dimension by microscopy methods. In general, fractal dimensions by microscopy methods provide a description of how space is occupied by a particular curve or shape. A fractal dimension is an intensive property of an object (Avnir et al., 1985) and is used to quantify the variation in the length, area, volume, or other properties with changes in the scale of the measurement interval. To calculate the fractal dimensions of an object, a property of the fractal object, such as length L, area A, or volume V, is measured at different length scales r. For a self-similar fractal object, the length L (or area, volume) displays a power-law relationship to the length scale r, and the fractal dimension of the object can be derived from the exponential term. This principle is applied to calculate the box-counting, particle-counting, and the Fourier transform fractal dimensions. The only difference among these methods is the properties measured and the formula used to calculate the fractal dimensions from the exponential term.

Box-counting fractal dimension, D_b. To calculate the box-counting fractal dimension, grids with side length l_i are laid over the PLM of fat crystal networks. Any grid containing particles more than a threshold value m is considered to be an occupied grid. The number of the occupied grids, N_i, for side length, l_i, is counted. This process is repeated for grids with different side lengths. The box-counting fractal dimension, D_b, is calculated as the negative of the slope of linear regression curve of the log-log plot of the number of occupied boxes N_b with the side length l_b:

$$D_b = -\frac{\Delta \ln N_b}{\Delta \ln l_b} \tag{7.37}$$

To reduce errors and artifacts, the small and large box size should be exempted from the calculation.

Particle-counting fractal dimension, D_f (Narine and Marangoni, 1999a). The concept of particle-counting fractal dimension, D_f, is derived from the mass fractal dimension, D_m, where the D_m relates the number of particles N to the linear size of the fractal object R, and the linear size of one particle (microstructure element) σ as

$$N = \left(\frac{R}{\sigma}\right)^{D_m}, \quad N \gg 1 \tag{7.38}$$

If the average size of the microstructural element is assumed to remain constant (Narine and Marangoni, 1999a), then Equation 7.38 becomes

$$N \propto R^{D_f} \tag{7.39}$$

where D_f is the particle-counting fractal dimension. To calculate the value of D_f, 2D PLM of fat crystal networks is used. The logarithm of the number of microstructure elements $\log(N(R))$ is plotted against $\log R$ for varying values of R. The slope of the linear regression curve is the particle-counting fractal dimension D_f (Narine and Marangoni, 1999a). The particle-counting fractal dimension algorithm should be carried out within the range between 100% and 35% of the original image size (Marangoni, 2002).

Fractal dimension by Fourier transform method, D_{FT}. In image analysis, a 2D image is considered to be a discrete function, $f(x, y)$, where x and y are the coordinates of the object pixels in the horizontal and vertical direction. The 2D discrete Fourier transform is applied to transform a 2D image to its corresponding frequency domain image $F(u, v)$ as

$$F(u,v) = \frac{1}{MN} \sum_{x=0}^{M-1} \sum_{y=0}^{N-1} f(x,y) * e^{-j*2\pi(ux/M+vy/N)} \tag{7.40}$$

or expressing it in the form of sin and cos functions as

$$F(u,v) = \frac{1}{MN} \sum_{x=0}^{M-1} \sum_{y=0}^{N-1} f(x,y)\left[\cos(2\pi(ux/M + vy/N)) - j\sin(2\pi(ux/M + vy/N))\right] \tag{7.41}$$

where u and v are the coordinates of the pixels in the frequency domain image. The power spectrum of $F(u, v)$,

$$P(u,v) = |F(u,v)|^2 = R^2(u,v) + I^2(u,v) \tag{7.42}$$

where
$R(u, v)$ is the real part of the function $F(u, v)$
$I(u, v)$ is the imaginary part

The frequency domain image is the transformed plotting of $P(u, v)$.

Since a fractal profile by definition includes information at all frequencies, it might seem that it is difficult to calculate fractal dimensions from frequency domain images. In fact, the formula used to calculate fractal dimensions from the frequency domain image is simple. The logarithm of the magnitude, which is the square root of the power spectrum, in the frequency domain image shows a linear relationship with the logarithm of the frequencies. The slope of this linear relationship, β, is related to the fractal dimension D_{FT} as $D_{FT} = (4 + \beta)/2$ (Russ, 1994).

D_{FT} can be used to study both self-similar and self-affine fractal objects. The data at low frequencies (u and $v < 10$) is not to be included in the calculation of D_{FT} (Russ, 1994). Figure 7.18 illustrates how the D_b, D_f, and D_{FT} are calculated from polarized light micrographs of fat crystal networks.

Experimentally, to determine the fractal dimension by microscopy, one small drop of melted fat sample (about 10 μL) is placed on a glass slide and then covered using a cover slip. The sample is kept above its melting temperature for a certain period of time to remove any crystal memory effects, then cooled and the structure of the fat

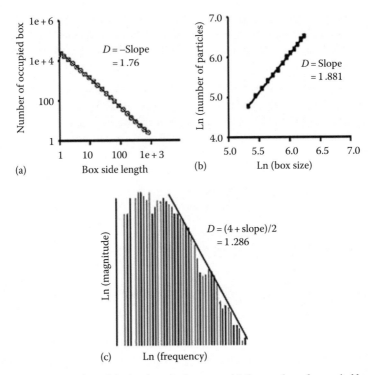

FIGURE 7.18 Illustration of the log-log plot between: (a) the number of occupied boxes with box size to calculate D_b, (b) the number of fat crystals inside each ROI (region of interest) with ROI size, and (c) magnitude and the frequency of the power spectrum image obtained by 2D Fourier transformation of the PLM of fat crystal networks.

samples is imaged using PLM. To determine the box-counting and particle-counting fractal dimensions, polarized light micrographs of fat samples need to be threshold to binary images. Commercially available software can be used to determine many kinds of fractal dimensions, except for the particle counting fractal dimension. We found Benoit 1.3 (TruSoft Int. Inc., St. Petersburg, FL) to be very powerful and effective, particularly for the box-counting fractal dimension. Care must be taken in the selection of the thresholding value. Because the global threshold method is applied by most of the commercial image analysis software, it is important to obtain a well-balanced illuminated image prior to thresholding. Other common practices to obtain high-quality PLM images include a background subtraction, where an image of the background liquid oil is taken and then subtracted from the images of the fat crystal networks, and frame averaging, where several images of the same sample area are taken and then averaged to remove the noise. In contrast to D_b and D_f, the fractal dimension by Fourier transform method, D_{FT}, can be determined directly from the grayscale images. To measure D_{FT}, the grayscale polarized light micrographs are transformed from a spatial domain to a frequency domain by applying a 2D discrete Fourier transformation. The D_{FT} can be calculated from the slope of the log-log plot of the magnitude against the frequency of the frequency domain image. All these manipulations can be done using Photoshop (Adobe, San Jose, CA) plug-ins.

All current microscopy fractal dimensions are derived from 2D polarized light micrographs, so they are 2D fractal dimensions. According to the commonly used method, fractal dimensions of the fat crystal networks in 3D space can be obtained by adding "1" to the 2D fractal dimensions (Russ, 1994). Tang and Marangoni (2006), however, demonstrated that this procedure may not be appropriate for many of these systems, since the distribution of network mass is also nonhomogenous in the z-axis. These authors accomplished this by succeeding to image volume elements of the network and determining a fractal dimension in 3D space.

Fractal dimensions have been used extensively to quantitatively describe the microstructure of the fat crystal networks (Wright and Marangoni, 2002; Marangoni and McGauley, 2003; Ahmadi et al., 2008b). Awad et al. (2004) measured the box-counting fractal dimension, D_b, of AMF (anhydrous milk fat), PO (palm oil), and CB (cocoa butter) (Awad et al., 2004).

The D_b of all the fat samples were found to increase with SFC up to a critical SFC value and then to a plateau. The trends in D_b of the samples correspond well with their polarized light micrographs: The lower D_b values correspond to larger microstructures at low SFC, while higher D_b values correspond to a large number of smaller microstructures at high SFC. When the SFC of the fat samples was above a critical point, fat crystals filled space homogeneously on the microscope slide and the D_b of the samples cannot increase anymore. For thick fat samples, the fractal dimensions of the fat crystal network D_b and D_f were found to have the highest value at an intermediate depth (Litwinenko et al., 2004).

The fractal dimensions calculated by different methods may have different values for the same fat crystal networks (Tang and Marangoni, 2005) and even display different trends when the microstructure of the fat crystal networks is changed (Litwinenko et al., 2004).

Through computer simulation in 2D (Tang and Marangoni, 2005) and 3D Tang and Marangoni (2008) found that, in general, D_f reflects the radial mass density spatial distribution in the fat crystal networks while D_{FT} is more sensitive to cluster size and morphology and D_b is affected by SFC, cluster size, and crystal morphology. For example, for 20% SFC and diamond-shaped crystals, the D_b of the fat crystal network increased from 1.44 to 1.58 when the radius of the crystals was increased from 3 to 42 pixels. In contrast, for 20% SFC, as the radius of the crystals was increased from 2 to 6 pixels, D_{FT} decreased from 1.748 to 1.45. By comparing the affecting factors of D_b, D_f, D_{FT}, the particle-counting fractal dimension, D_f, and the Fourier transform fractal dimension, D_{FT}, were found to be better parameters to represent the microstructure of the fat crystal networks than the box-counting fractal dimension, D_b.

7.3.3 FRACTALITY

Now that the structural organization of the microstructural level of a typical fat crystal network has been established, the next logical step is an attempt to quantify this level of structure. Apart from monitoring microstructural element size, cluster size, and microstructure size (not an easy task, because of the nature of the methods of microscopy), it seems important as well to ask the following question: What is the spatial arrangement of the mass of the network at the microstructural level? Certainly, the size of the various hierarchical levels of structure will determine the way in which forces holding the network together will be affected, but equally as certain, the spatial arrangement of the various levels of structural entities will also affect the forces that hold the network together. Of course, additional factors such as the nature of the crystalline material (polymorphism) and the nature of the constituent molecules themselves are also important factors, but from a microstructural perspective, the parameters that can be determined are concerned with the size, shape, and spatial arrangement of the structural entities. Certainly, the density of this level of structure will yield some information about the spatial arrangement of the mass of the network, but does not give information about the actual packing, that is, it is a rather macroscopic measurement, which does not yield structural information about the actual order of packing of the structural units of interest. Due to the extremely disordered nature of the packing of levels of structure below the size of the microstructures, it is difficult to assign lattice parameters to the positioning of the microstructural elements. Fractal analysis has in the past been used to characterize the spatial arrangement of such disordered structures that demonstrate self-similarity at different length scales. Additionally, based on the calculation of a fractal dimension from rheological measurements on fats which were fitted to a fractal model developed for colloidal gels (Narine and Marangoni, 1999a), it seems fitting that one tries to apply fractal principles to the study of the microstructural level of structure in fats. Certainly, the self-similarity of the structure suggested by the AFM, CLSM, SEM, and PLM images of the network encourages this endeavor. The challenge here is that each of the methods of microscopy that have been presented earlier demonstrate inherent artifacts, and none of them presents an unbiased, complete view of the network. What seems plausible is that the structure is self-similar at length ranges bounded by the size of the microstructural elements and the size of the microstructures.

Before leaving this section, we would like to leave the reader with the image in Figure 7.19. Self-similarity or self-affinity is an essential character of a fractal object. According to the definition of fractals by Mandelbrot (1982), a fractal is "… a rough or fragmented geometrical shape that can be subdivided in parts, each of which is (at least approximately) a reduced-size copy of the whole …." It means that the structure of a fractal object is similar at finer and finer length scales.

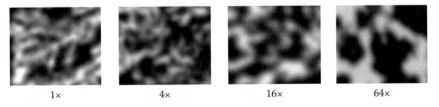

1x   4x   16x   64x

FIGURE 7.19 Polarized light micrograph of a fat crystal network (35% tristearin in sunflower oil) at different magnifications. 1x: the original image, 4x: 4 times enlarged image, 16x: 16 times enlarged image, 64x: 64 times enlarged image.

However, for a real fractal object, such as fat crystal networks, although the structure of the object is similar on average at different length scales, it is not exactly similar. Due to minor differences among fine structures, it is rare to find the exactly same structure at different length scales. A simple way to demonstrate the self-similar character of a fat crystal network (or any other fractal object) is to look at the object at different length scales after eliminating the fine structure of the object by using a blurring technique such as Gaussian blur. Figure 7.19 shows a polarized light micrograph of a fat crystal network at different magnifications. The original image was thresholded, enlarged to different magnifications, and blurred by applying the Gaussian blur filter in Photoshop (Adobe, San Jose, CA). By increasing the magnification of the image, the structure of the fat crystal network at finer length scales can be explored, and by applying the Gaussian blur filter, the overall structure of the fat crystal network at different length scales can be compared without focusing on fine structure details. From Figure 7.19, it is obvious that the overall structure of the fat crystal network at different magnifications and thus at different length scales is similar. It confirms the self-similar character of fat crystal networks.

7.3.4 WEAK LINK REVISITED

Now that the general structural arrangement of the microstructure of fat crystal networks has been established, it is relevant to examine how this structural organization relates to the weak-link theory. The model of the structure of fat crystal networks developed earlier identifies the microstructures as the largest structural building block of the network. It seems reasonable that any stress that is put upon the network will first be felt by the microstructures. The question is, how do the microstructures behave under this stress? Heertje and coworkers (Heertje et al., 1987; Juriaanse and Heertje, 1988; Heertje, 1993) have demonstrated that the microstructures are separated when the network is stressed, while remaining intact (maintaining their shape and size). It therefore seems reasonable to expect that when the network is stressed, the first level of structure that is stressed are the links between microstructures. Figure 7.20 shows a theoretical schematic of the network under stress. It seems reasonable to expect that the stressing of the network within the elastic limit results in a stressing of the links between microstructures, which are a repeating, regularly packed structural unit. Referring to Figure 7.20, if one were to express the force constant of the links between microstructures as K_L, then the macroscopic elastic constant, K, (in 1D) of the network could be written as

$$K = \left[\frac{L}{\xi}\right]^{d-2} K_L \tag{7.43}$$

where
ξ is the diameter of one microstructure
L is the macroscopic size of the system
d is the Euclidean dimension of the sample (=3)

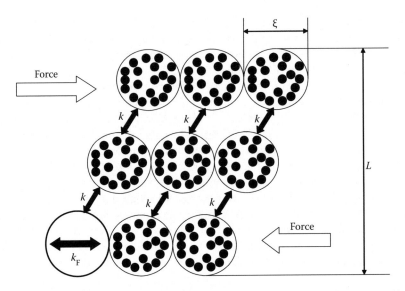

FIGURE 7.20   Theoretical schematic of the fat crystal network under a small stress which is insufficient to exceed the elastic limit of the network.

However, as have been shown earlier, the structure within the microstructures is fractal in nature. We can therefore relate the diameter of the microstructure (or aggregate) to the particle volume fraction of the entire network, according to the development made in Section 7.2. Therefore, using Equation 7.44:

$$\xi \sim (\Phi_t)^{1/(D-3)} \tag{7.44}$$

and substituting this expression in Equation 7.43:

$$K \sim \left[\frac{L}{(\Phi)^{1/(D-3)}} \right] k \tag{7.45}$$

Now, if the links between microstructures are constant, and the size of the system is a constant, we can write Equation 7.45 as

$$K \sim (\Phi)^{1/(3-D)} \tag{7.46}$$

Recognizing that the shear storage modulus of the network is related in a proportional manner to the tensile elastic constant, we can write Equation 7.46 as

$$G' \sim \Phi^{1/(3-D)} \tag{7.47}$$

The theory espoused by Equation 7.47 relates the shear storage modulus to the particle volume fraction via the fractal dimension of the network. The particle volume fraction of the network is not easily measured (in fact, the author knows of no experimental method that yields this value). It must be stated clearly here that particle volume fraction is used to mean microstructural element volume fraction—it is obvious from the development above that the "particles" refers to the microstructural elements. It will be shown later that the Φ_{SFC}, determined as SFC/100, is proportional to the particle volume fraction, Φ. Therefore, one may use Φ_{SFC} in the stead of Φ in Equation 7.47, bearing in mind that the nature of the proportionality constant is now changed. One can now replace the proportionality sign by the constant λ (γ in our previous nomenclature), bearing in mind that this constant would be dependent on the links between microstructures, and the relationship between ξ, Φ, and D, as well as the nature of the proportionality between Φ_{SFC} and Φ, given the origin of the proportionality constant, as developed earlier:

$$G' = \lambda\Phi^{\frac{1}{3-D}}$$

(7.48)

Equation 7.48 is of course equivalent to the expression arrived at by Shih et al. (1990) for the weak-link theory for colloidal gels. Therefore, from a structural perspective, it seems that the use of this formulation for fat crystal networks is warranted. Not surprisingly, therefore, the work by Rousseau and Marangoni showed that the weak-link theory when applied to fats yielded plausible results. Additionally, these authors were not incorrect in assuming that the fractal dimensions that they calculated from this rheological treatment were related to the microstructure; as have been shown earlier, such a fractal dimension is related to the way in which the microstructural elements are distributed in the microstructures of the fat network. The obvious challenge is therefore to compare the fractal dimensions calculated by rheological methods to those calculated by image analysis methods. The most important issue here of course is the need to emulate the same processing conditions for the rheologically prepared samples and the samples prepared for microscopy on a glass slide.

7.3.5   RELATING THE PARTICLE VOLUME FRACTION TO THE SOLID FAT CONTENT

The SFC, Φ_{SFC}, of a fat network may be defined as

$$\Phi_{SFC} = \frac{n_{ME}V_{ME}\rho_{ME}}{M_T}$$

(7.49)

where
   n_{ME} is the number of microstructural elements
   V_{ME} is the volume of one microstructural element
   ρ_{ME} is the solids density of a microstructural element
   M_T is the total mass of the network (solid mass + liquid mass)

Equation 7.49 can be rewritten as

$$n_{ME} = \frac{\Phi_{SFC} M_T}{V_{ME} \rho_{ME}} \tag{7.50}$$

Now, the total particle volume concentration, Φ_t of the network is given by

$$\Phi_t = \frac{n_{ME} V_{ME}}{V_T} \tag{7.51}$$

where V_T is the total volume of the network. Substituting for n_{ME} in Equation 7.50:

$$\Phi_t = \frac{\dfrac{\Phi_{SFC} M_T}{V_{ME} \rho_{ME}} V_{ME}}{V_T} \tag{7.52}$$

This yields

$$\Phi_t = \frac{\rho_t}{\rho_{ME}} \Phi_{SFC} \tag{7.53}$$

where $\rho_t = M_T/V_T$, and is the density of the network itself. The solids density of the microstructural elements, ρ_{ME}, is higher than the density of the network, ρ_t, since the microstructural elements are very densely packed, while the network itself contains a lot of liquid oil and dispersed microstructural elements. However, as established by Equation 7.53, the particle volume fraction of the network is proportional to the SFC of the network. Therefore, although it is difficult to measure the particle volume fraction of the network, the SFC may be used in its stead, bearing in mind that these two entities are not equal, but are proportional.

7.3.6 RHEOLOGY

Plots of G' versus Φ shown in Figure 7.16 support the power-law dependence of G' on Φ suggested by Equation 7.47. Figure 7.16a shows data for milk fat at 5°C. This is in the form of a straight line, which implies from Equation 7.47 that the fractal dimension must be equal to 2. Figure 7.16b shows data for tallow at 5°C, Figure 7.16c shows data for palm oil at 5°C, and Figure 7.16d shows data for lard at 5°C. All of Figure 7.21b through d shows a nonlinear, power-law dependence of G' on Φ. As has been explained, a plot of $\ln G'$ versus $\ln \Phi$ yields a straight line with slope equal to $1/(3 - D)$; such plots are shown in Figure 7.21 for milk fat (a), tallow (b), palm oil (c), and lard (d).

All the systems we have studied to date display the power-law scaling behavior characteristic of fractal networks. Figures 7.22 through 7.24 are a few examples of this power-law scaling behavior. Table 7.1 summarizes values of D and λ obtained

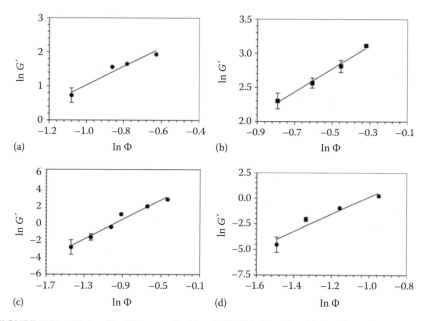

FIGURE 7.21   Plots of ln G' versus ln (Φ = SFC/100) for: (a) milk fat, (b) tallow, (c) palm oil, and (d) lard.

for a variety of systems crystallized isothermally at 5°C using the double logarithmic plots described earlier. Figure 7.25 and Table 7.2 shows an example of a fractal microstructural analysis of blends of fully hydrogenated canola oil with canola oil (Ahmadi and Marangoni, 2009).

7.3.7   PHYSICAL SIGNIFICANCE OF FRACTAL DIMENSION

The previous discussion of experimental data shows that fat crystal networks are fractal within certain length ranges, and that the structural arrangement of the microstructure of fat crystal networks makes it possible to apply the weak-link theory to analyze the mechanical properties of the networks. Central to this finding is the calculation of a fractal dimension. It is therefore relevant that we attach some physical significance to this quantity, if it is to be a useful parameter.

Note that for a higher fractal dimension, the density of the packing of the microstructural elements must be higher, since the number of microstructural elements is proportional to the characteristic length raised to a power equal to the fractal dimension. Density of packing of the microstructural elements does not refer to the traditional definition of density of the network (total mass/total volume), nor does it refer to the density *within* a typical microstructural element (which has been used in Equation 7.53 as the symbol ρ_{ME}). The density of packing of the microstructural elements refers to the density of the microstructural elements *within* one microstructure. Therefore, if one is observing a particular fat after it has been processed differently, then the increase or decrease of the

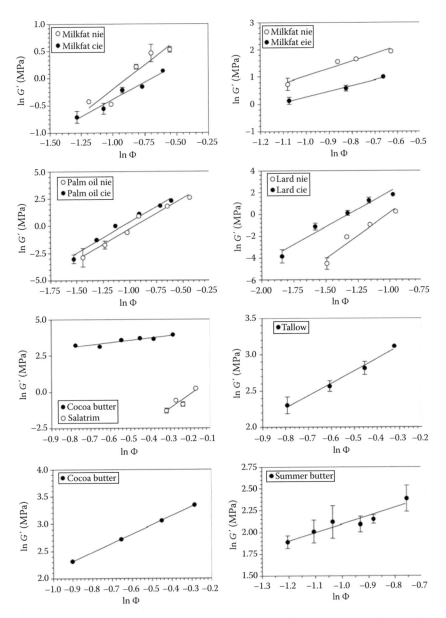

FIGURE 7.22 Plots of the ln G' versus ln Φ for a series of fats showing the universal adherence to a power-law behavior. NIE, non-interesterified; CIE, chemically interesterified; EIE, enzymatically interesterified.

amount of microstructural elements within a certain characteristic length will signal the increase or decrease of the fractal dimension. However, given the nature of the PLM images, this is not a parameter that is easily ascertained without image analysis. Additionally, as is discussed in the following, the fractal dimension is not only dependent on the density of the packing of the microstructural elements,

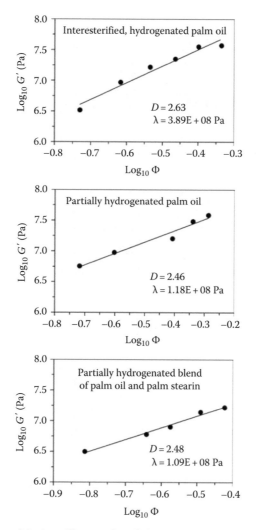

FIGURE 7.23 Plots of the $\log_{10}G'$ versus $\log_{10}\Phi$ for a series of palm oil-based fats showing adherence to a power-law behavior.

it is also dependent on the *order* in which they pack. Certainly, these two parameters are interrelated, since the order in which the elements pack also influences the density of the packing, and is itself influenced both by the nature of the inter-microstructural element forces and the mass and heat transfer limitations during the formation of the network.

If one considers a line, with microstructural elements placed at some equilibrium nearest-neighbor distance apart, then one has a picture represented by Figure 7.26a. The projection of the positions of the microstructural elements onto a line represents an ordered array. Now, the line containing the microstructural elements represents a 1D object. If one starts to put kinks in this line, one

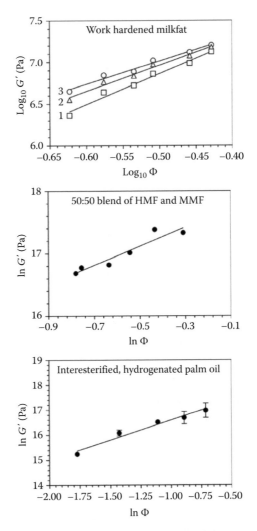

FIGURE 7.24 Log-log plots of G' versus Φ for a series of fat systems. Work hardening increases from 1 to 3 in the top panel. The acronyms HMF and MMF refer to, respectively, the high and medium melting fractions of milk fat.

starts to raise the dimension of the line to a value just above 1, and less than 2. This scenario is represented in Figure 7.26b. Now, the microstructural elements must still be placed at a nearest-neighbor equilibrium distance from each other on this new "space"; therefore, the projection of the positions of the microstructural elements onto a line begins to look disordered, as shown in the figure. It is useful to have a working quantitative definition of order in this analogy. When one looks at the projection of positions, one may define high order as the positions all being a common average distance apart from each other. Therefore, the standard deviation of the various distances apart of nearest-neighbor projection positions is a

TABLE 7.1

Fractal Dimensions (*D*) and Parameter λ Determined Rheologically for a Series of Fat Systems Crystallized Isothermally at 5°C

System	*D*	λ (MPa)
Milk fat-canola oil (MF-CO) 1_shear	2.63	43.1
Milk fat-canola oil (MF-CO) 2_shear	2.44	4.93
Milk fat-canola oil (MF-CO) 3_shear	2.73	37.0
Milk fat-canola oil (MF-CO) 4_shear	2.62	95.1
Milk fat-canola oil (MF-CO) 1_compression	1.96	21.2
Milk fat-canola oil (MF-CO) 2_compression	2.01	23.6
MF-CO EIE_shear	2.52	10.4
MF-CO CIE_shear	2.20	2.40
Milk fat TAGs-canola oil_shear	2.71	135
Cocoa butter-canola oil_shear	2.37	77.8
Cocoa butter-canola oil_compression	2.40	45.6
Salatrim-canola oil_shear	2.89	5.85
Tallow-canola oil_shear	2.41	37.4
Palm oil-soybean oil (PO-SBO)_shear	2.82	210
PO-SBO CIE_shear	2.82	398
Lard-canola oil (L-CO)_shear	2.88	4870
L-CO CIE_shear	2.84	4890
Interesterified, hydrogenated palm oil-canola oil_shear	2.63	389
Partially hydrogenated palm oil-canola oil_shear	2.46	118
Partially hydrogenated palm stearin/palm oil blend-canola oil_shear	2.48	109
Interesterified, hydrogenated palm oil-peanut oil_shear	2.36	77.8

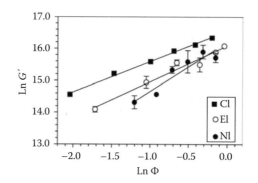

FIGURE 7.25   Log-log plot of the shear storage modulus (*G′*) versus the solids' volume fraction (Φ = SFC/100) at 30°C for the NIE, CIE, and EIE blends.

good representation of the deviation from order (lower standard deviation implies higher order). Figure 7.26c through e demonstrates situations in which more and more kinks are placed in the line containing the microstructural elements; therefore, the fractal dimension is increasing from (a) to (e). Table 7.3 shows the distance apart of nearest-neighbor projections of each situation (and lists an additional

TABLE 7.2
Fractal Analysis of the Relationship between the Storage Modulus (G') and the Volume Fraction of Solids (Φ), $\log G' = \log \lambda + \mu \log \Phi$, for the NIE, CIE, and EIE Blends of Fully Hydrogenated Canola Oil and Canola Oil at Different Temperatures

Temp. (°C)	μ			$\log \lambda$			r^2			D_f			λ (Pa)		
	NI	CI	EI	NI	CI	EI	NI	CI	EI	NI	CI	EI	NI	CI	EI
20	2.21[a]	1.31[b]	0.99[b]	7.31[c]	7.15[c]	6.86[d]	0.82	0.76	0.61	2.55	2.24	1.94	2.0×10^7	1.4×10^7	7.2×10^6
30	2.11[a]	0.96[b]	1.12[b]	7.23[c]	7.20[c]	6.99[d]	0.81	0.98	0.86	2.53	1.96	2.11	1.7×10^7	1.6×10^7	9.5×10^6
40	2.25[a]	0.97[b]	1.02[b]	7.52[c]	7.07[d]	6.77[e]	0.73	0.56	0.61	2.56	1.98	2.03	3.3×10^7	1.2×10^7	5.7×10^6
50	0.92[a]	1.12[a]	1.08[a]	6.98[c]	6.87[c]	6.82[c]	0.46	0.67	0.71	1.91	2.11	2.08	9.5×10^6	7.4×10^6	6.6×10^6

Different superscript letters (a, b) and (c–e) indicate significant differences ($p < 0.05$) for the slopes and y-intercepts, respectively, within each row.

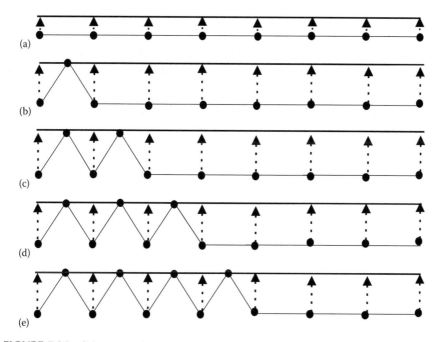

FIGURE 7.26 Schematic of microstructural elements placed on a line at equal distances apart. The dimensionality of the line increases from (a) to (e). Microstructural elements on the thin line are projected onto the thick line.

two situations following the same trend of kinks, not shown in Figure 7.26), and the standard deviation in these values for each situation. As can be seen, as the fractal dimension increases, the order initially (for a small amount of kinks in the line) decreases, and then order begins to increase as the fractal dimension gets larger and larger (i.e., closer and closer to 2). Of course, the fractal dimension referred to here is qualitatively getting larger (the actual dimension has not been computed), but it is generally accepted that more and more kinks in a line represents higher and higher fractal dimensions, since the line is closer approximating a plane. This situation is exactly what happens in the case where dimension is between 2 and 3. An interesting effect to note is also that as the fractal dimension and order increases, so does the density of packing of microstructural elements in the observed region. This is, of course, in agreement with the foregoing discussion on density and order.

7.3.8 OTHER METHODS FOR THE DETERMINATION OF THE FRACTAL DIMENSION

7.3.8.1 Fractal Dimension from Oil Permeability Measurements

In the oil migration method, the permeability coefficient B is measured for fat samples at different SFCs. The fractal dimensions of the fat crystal networks can be calculated from the slope of the log-log plot of the permeability coefficient of

TABLE 7.3

The Decrease in Standard Deviation (after an Initial Increase) from an Average Distance between Projected Position of Balls on a Line, with Increasing Amounts of Kinks in the Line[a]

Projected Position	Distance between Successive Projected Positions						
	1 Kink	2 Kinks	3 Kinks	4 Kinks	5 Kinks	6 Kinks	7 Kinks
1	20	10	10	10	10	10	10
2	20	10	10	10	10	10	10
3	20	20	10	10	10	10	10
4	20	20	10	10	10	10	10
5	20	20	20	10	10	10	10
6	20	20	20	10	10	10	10
7	20	20	20	20	10	10	10
8		20	20	20	10	10	10
9			20	20	20	10	10
10				20	20	10	10
11					20	20	10
12						20	10
13							20
Standard deviation	0	4.63	5.27	5.16	4.67	3.89	2.77

[a] Fractal dimension is proportional to the number of kinks—less kinks, fractal dimension is closer to 1, more kinks, fractal dimension is closer to 2.

the fat samples against their SFC. The permeability coefficient B is defined as "the rate of flow of water (liquid oil in our case) through a unit cross-sectional area under a unit hydraulic gradient at the prevailing temperature" (ASCE, 2005). The permeability coefficient B is calculated from the volumetric flow rate Q according to the Darcy's law:

$$Q = \frac{B^* A_c}{\eta} * \frac{\Delta P}{L} \tag{7.54}$$

where
A_c is the cross-sectional area through which flow takes place
η is the viscosity of the liquid oil in the fat sample
ΔP is the pressure applied to the permeating oil over the distance L

In addition, the flux through one fractal structure element also can be described by Poiseuille's law as $Q = \pi r^4 \Delta P/(8\eta L)$, where r is the inside radius of the element and the other symbols have the same meanings as that in Darcy's law. By combining

Darcy's law and Poiseuille's law and applying the fractal geometry principles, Bremer showed that the permeability coefficient B is related to network structure as (Bremer et al., 1989)

$$B = \left(\frac{\delta^2}{M}\right) * \Phi^{2/(D-3)} \tag{7.55}$$

where
    δ is the particle size
    M is a parameter similar to the tortuosity factor in the Koceny–Carman equation
        (Davis and Dollimore, 1980)
    Φ is the solid's volume fraction
    D is the fractal dimension of the network

Combining these two equations together, D is related to Q as

$$Q = \left(\frac{A_c}{\eta} * \frac{\Delta P}{L}\right) * \left(\frac{a^2}{K}\right) * \Phi^{2/(D-3)} \tag{7.56}$$

Thus, the fractal dimensions D can be calculated from the volumetric flow rate Q data at different solid's volume fraction, Φ. Dibildox-Alvarado et al. (2004) applied Bremer's equation to analyze the oil migration data for mixture of peanut oil and IHPO (chemically interesterified and hydrogenated palm oil) cooled at different cooling rates. Dibildox-Alvarado et al. (2004) measured the D_b and D_f of the fat samples and successfully predicted the increased permeability coefficient at a lower cooling rate according to Bremer's equation (7.55). Tang et al. (2006) studied the microstructure of the mixture of tristearin and sunflower oil using the oil migration method. Higher fractal dimensions were found for the samples cooled at higher cooling rates. A good fit ($r^2 = 0.99$) was found for the nonlinear regression of Bremer's equation to the data.

7.3.8.2 Fractal Dimensions by Light Scattering

In addition to the rheology method and the permeability coefficient method, the fractal dimensions of fat crystal networks can also be obtained using light scattering. When light travels through a dilute suspension of fat crystals in oil, it will be scattered by the crystallites. The intensity of the scattering is generally considered to be a function of two components, $S(q)$, which is the structure factor corresponding to the scattering due to the spatial correlation between particles, and $P(q)$, which is the form factor corresponding to the scattering arising from the properties of individual particles. The scattering intensity $I(q)$ is thus a product of the structure factor and the form factor, namely, $I(q) \sim S(q) * P(q)$. At small values of the scattering vector $\vec{q}$, $P(q)$ is considered to be a constant, so the variation of the scattering intensity mainly comes from the spatial distribution of the fat clusters in the fat crystal networks, which is expressed as $I(q) \sim q^{-D}$. The scattering vector has the following form: $q = (4\pi/\lambda)\sin(\theta/2)$, where λ is the wavelength of the light and θ is the scattering angle. The light scattering fractal dimension D is calculated as the negative of the slope of the log-log plot of the scattering intensity $I(q)$ with the scattering vector $\vec{q}$. This equation is only valid when q is within the range: $1/R_g \ll q \ll 1/r_0$, where R_g is the size of the aggregate and r_0 is the size of the primary particle. The light scattering technique has been employed to study the microstructure of fat crystal networks. Vreeker et al. (1992b) studied the micro-structure change of a dispersion of glycerol tristearate in olive oil during storage after rapid cooling from 90°C to 2°C. The fractal dimensions of the samples from light scattering method increased from 1.7 to 2.0 after 7 days storage. More compact fat crystal clusters were related to the increased fractal dimension. The fractal dimensions calculated using the rheology method was 2.0, corresponding well with the fractal dimensions from the light scattering method. To apply the light scattering method, the samples have to be optically transparent, so this method is useful only for fats with very low SFCs.

7.3.8.3 Thermomechanical Method for Determining Fractal Dimensions

The mass fractal dimension of a fat crystal network can be determined from the slope of the log-log plot of the storage modulus (G') versus the SFC, as described extensively earlier. Different SFCs were obtained by melting the fat, diluting it to different extents in liquid oil, and crystallizing the blends under specific conditions. One of the major drawbacks of this method is the inability of characterizing the native structure of an already crystallized fat. In the thermo-mechanical method developed by our group (Lam et al., 2009), SFC is changed via temperature variations instead. At each temperature, the G' and SFC were measured and the fractal dimension calculated as described earlier. The thermomechanical method proved to be a simpler and more reliable estimator of the fractal dimension of a fat crystal network than the dilution method. Figure 7.27 shows a comparison of log-log plots of G' versus SFC/100 carried out by standard dilution techniques or my melting, for the determination of the fractal dimension or milk fat (Figure 7.27a) and cocoa butter (Figure 7.27b). The agreement between the two methods was excellent, and possibly suggests that the fractal dimension of a fat crystal network can be determined from a single mechanical melting curve.

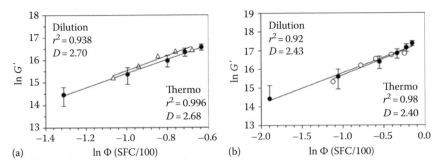

(a)   (b)

FIGURE 7.27 Comparison of the dilution and thermomechanical methods for the determination of the fractal dimension in milk fat (a) and cocoa butter (b).

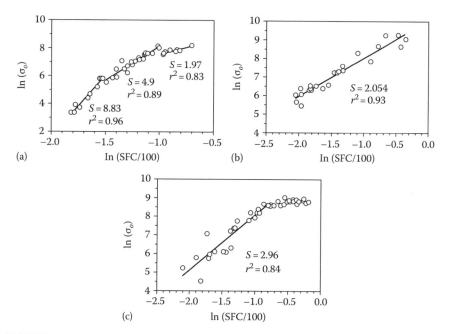

FIGURE 7.28 Relationship between the stress at the limit of linearity (σ_o) and the solids' volume fraction (Φ = SFC/100) for blends of (a) anhydrous milk fat, (b) palm oil, and (c) cocoa butter, clearly demonstrating that these fats are in the weak-link rheological regime.

7.3.8.4 Fractal Dimension from the Stress at the Limit of Linearity: Fats Are in the Weak-Link Rheological Regime

As mentioned previously, it is possible to determine the fractal dimension of a fat crystal network form log-log plots of the stress at the limit of linearity versus the volume fraction of solids (=SFC/100). This can be clearly shown in Figure 7.28, where linear scaling regions can be observed for (a) anhydrous milk fat, (b) palm oil, and (c) cocoa butter (Awad et al., 2004). More importantly, though, is the fact that there is a positive slope to this dependence, which, according to Shih et al. (1990), suggests that the network is in the weak-link rheological regime (see Equation 7.28).

7.3.9 Modified Fractal Model

The scaling relationship between the storage modulus (G') and the volume fraction of solids (Φ) in fat crystal networks has been explained by the fractal model developed by our group. However, many experimental results and simulation studies suggest that the stress distribution within a colloidal network is dramatically heterogeneous, which means that a small part of the network carries most of the stress, while the other part of the network does not contribute much to the elastic properties of the system. This concept was introduced into a modified fractal model (Tang and Marangoni, 2009). The volume fraction of solids term (Φ) in the original fractal model was replaced by Φ_e, the effective volume fraction of solids, in the modified fractal model, which represents the volume fraction of stress-carrying solids.

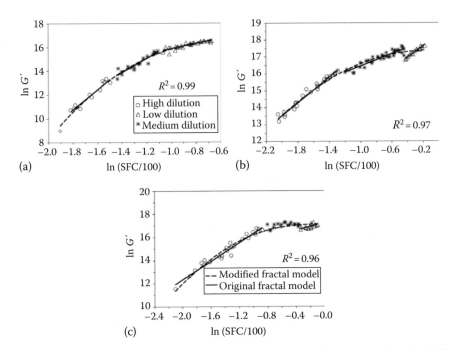

FIGURE 7.29 Fits of the original fractal model and the modified fractal model for G'—SFC data for CB, palm oil, and AMF samples: (a) mixture of AMF and canola oil, (b) mixture of palm oil and canola oil, and (c) mixture of cocoa butter and canola oil (R^2 corresponds to the goodness of fit of the modified fractal model to the data).

A proposed expression for Φ_e was given and a modified expression for the scaling relationship between G' and Φ was obtained, namely,

$$G' = c * (1 - e^{-k\Phi^b})^{1/(3-D)} \tag{7.57}$$

The modified fractal model fits the experiment data well and successfully explains the sometimes observed nonlinear log-log behavior between the storage modulus of colloidal networks and their volume fraction of solids (Figure 7.29).

7.4 CONCLUSIONS

It has been shown in this chapter, from a structural basis established by a multitude of microscopy techniques, that fat crystal networks crystallized statically are fractal within certain length ranges. Furthermore, it was shown that the weak-link theory is applicable to fat crystal networks. The next stage in this developmental process is to begin establishing links between heat and mass transfer conditions and the resulting microstructure of the fat from knowledge of the crystallization kinetics and phase behavior of the individual triacylglycerols present in the sample. This work is ongoing.

REFERENCES

Acevedo, N.C. and A.G. Marangoni. 2010a. Characterization of the nanoscale in triacylglycerol crystal networks. *Cryst. Growth Des.* 10(8): 3327–3333.

Acevedo, N.C. and A.G. Marangoni. 2010b. Towards nanoscale engineering of triacylglycerol crystal networks. *Cryst. Growth Des.* 10(8): 3334–3339.

Ahmadi, L., A.J. Wright, and A.G. Marangoni. 2008. Chemical and enzymatic interesterification of tristearin/triolein-rich blends: Chemical composition, solid fat content and thermal properties. *Eur. J. Lipid Sci. Technol.* 110: 1014–1024.

Ahmadi, L., A.J. Wright, and A.G. Marangoni. 2008. Chemical and enzymatic interesterification of tristearin/triolein-rich blends: Microstructure and polymoprhism. *Eur. J. Lipid Sci. Technol.* 110: 1025–1034.

Ahmadi, L., A.J. Wright, and A.G. Marangoni. 2009. Structural and mechanical behavior of tristearin/triolein-rich mixtures and the modification achieved by interesterification. *Food Biophys.* 4: 64–76.

ASCE (American Society of Civil Engineers). 1985. http://or.water.usgs.gov/projs_dir/willgw/glossary.html#p (June 06, 2005).

Aubert, C. and D.S. Cannell. 1986. Restructuring of colloidal silica aggregates. *Phys. Rev. Lett.* 56: 738.

Avnir D., D. Farin, and P. Pfeifer. 1985. Surface geometric irregularity of particulate materials: The fractal approach. *J. Colloid Interface Sci.* 103(1): 112–123.

Awad, T.S., M.A. Rogers, and A.G. Marangoni. 2004. Scaling behavior of the elastic modulus in colloidal networks of fat crystals. *J. Phys. Chem. B* 108: 171–179.

Ball, R.C. 1989. Fractal colloidal aggregates: Consolidation and elasticity. *Physica D.* 38: 13–15.

Biham, O., O. Malcai, D. Lidar, and D. Avnir. 1998. Is nature fractal? *Science.* 279: 784–786.

Bolle, C., C. Cametti, P. Codastefano, and P. Tartaglia. 1987. Kinetics of salt-induced aggregation in polystyrene lattices studied by quasielastic light scattering. *Phys. Rev. A* 35: 837.

Bremer, L.G.B., B.H. Bijsterbosch, R. Schrijvers, T. van Vliet, and P. Walstra. 1990. On the fractal nature of the structure of acid casien gels. *Colloid Surf.* 51: 159–170.

Bremer, L.G.B., B.H. Bijsterbosch, P. Walstra, and T. van Vliet. 1993. Formation, properties and fractal structure of particle gels. *Adv. Colloid Interface Sci.* 46: 117–128.

Bremer, L.G.B., T. van Vliet, and P. Walstra. 1989. Theoretical and experimental study of the fractal nature of the structure of casien gels. *J. Chem. Soc. Faraday Trans.* 85: 3359–3372.

Brown, W.D. 1987. *The Structure and Physical Properties of Flocculating Colloids.* Cambridge, U.K.: University of Cambridge.

Brown, W.D. and R.C. Ball. 1985. Computer simulation of chemically limited aggregation. *J. Phys. A* 18: L517.

Bunjes, H., F. Steiniger, and W. Richter. 2007. Visualizing the structure of triglyceride nanoparticles in different crystal modifications. *Langmuir.* 23: 4005–4011.

Buscall, R., P.D.A. Mills, J.W. Goodwin, and D.W. Lawson. 1988. Scaling behaviour of the rheology of aggregate networks formed from colloidal particles. *J. Chem. Soc. Faraday Trans.* 84: 4249–4260.

Chawla, P. and J.M. deMan. 1990. Measurement of the size distribution of fat crystals using a laser particle counter. *J. Am. Oil Chem. Soc.* 67: 329–332.

Chawla, P., J.M. deMan, and A.K. Smith. 1990. Crystal morphology of shortenings and margarines. *Food Struct.* 9: 329–336.

Chen, M. and B. Russel. 1991. Characteristics of flocculated silica dispersions. *J. Colloid Interface Sci.* 141(2): 565–577.

Cotton, F.A. 1971. *Chemical Applications of Group Theory.* New York: Wiley Interscience.

Courtens, E., J. Pelous, J. Phalippou, R. Vacher, and T. Woignier. 1987. Brillouin-scattering measurements of phonon-fracton crossover in silica aerogels. *Phys. Rev. Lett.* 58(2): 128–131.

Crownover, R.M. 1995. *Introduction to Fractals and Chaos.* Boston, MA: Jones & Bartlett.

Davies, L. and D. Dollimore. 1980. Theoretical and experimental values for the parameter k of the Kozeny-Carman equation, as applied to sedimenting suspensions. *J. Phys. D: Appl. Phys.* 13: 2013–2020.

Dibildox-Alvarado, E., J.N. Rodrigues, L.A. Gioielli, J.F. Toro-Vazquez, and A.G. Marangoni. 2004. Effects of crystalline microstructure on oil migration in a semisolid fat matrix. *Cryst. Growth Des.* 4(4): 731–736.

Dietler, G., C. Aubert, D.S. Cannell, and P. Wiltzius. 1986. Gelation of colloidal silica. *Phys. Rev. Lett.* 57: 3117.

Dimon, P., S.K. Sinha, D.A. Weitz, C.R. Safinya, G.S. Smith, W.A. Varady, and H.M. Lindsay. 1986. Structure of aggregated gold colloids. *Phys. Rev. Lett.* 57: 595.

Edwards, S.F. and R.B.S. Oakeshott. 1989. The transmission of stress in an aggregate. *Physica D* 38: 88–92.

Falconer, K. 1990. *Fractal Geometry.* Chichester, U.K.: Wiley.

Feng, S. and P. Sen. 1984. Percolation on elastic networks: New exponent and threshold. *Phys. Rev. Lett.* 52: 216.

Feng, S., P. Sen, B. Halperin, and C. Lobb. 1984. Percolation on two-dimensional elastic networks with rotationally invariant bond bending forces. *Phys. Rev. B* 30: 5386.

Forrest, S.R. and J.T.A. Witten. 1979. Long range correlations in smoke-particle aggregates. *J. Phys. A* 12: L109.

Friedlander, S.K. 1977. *Smoke, Dust and Haze.* New York: Wiley.

de Gennes, P.G. 1979. *Scaling Concepts in Polymer Physics.* Ithaca, NY: Cornell University Press.

Grey, F. and J.K. Kjems. 1989. Aggregates, broccoli and cauliflower. *Physica D* 38: 154–159.

Hagiwara, T., H. Kumagai, T. Matsunaga, and K. Nakamura. 1997. Analysis of aggregate structure in food protein gels with the concept of fractal. *Biosci. Biotechnol. Biochem.* 61: 1663–1667.

Hagiwara, T., H. Kumagai, and K. Nakamura. 1998. Fractal analysis of aggregates in heat-induced BSA gels. *Food Hydrocolloids.* 12: 29–36.

Halsey, T.C. 2000. Diffusion-limited aggregation: A model for pattern formation. *Phys. Today.* November, 53: 36–41.

Hamaker, H.C. 1937. The London-van der Waals attraction between spherical particles. *Physica [IV].* 10: 1058.

Hausdorf, F. 1919. Dimension und ausseres mass. *Math. Ann.* 79: 157–179.

Heertje, I. 1993. Microstructural studies in fat research. *Food Struct.* 12: 77–94.

Heertje, I., J. van Eendenburg, J.M. Cornelissen, and A.C. Juriaanse. 1988. The effect of processing on some microstructural characteristics of fat spreads. *Food Microstruct.* 7: 189–193.

Heertje, I. and M. Leunis. 1997. Measurement of shape and size of fat crystals by electron microscopy. *Lebensm. Wiss. Technol.* 30: 141–146.

Heertje, I., P. van der Vlist, J.C.G. Blonk, H.A.C.M. Hendrickx, and G. Brakenhof. 1987b. Confocal laser scanning microscopy in food research: Some observations. *Food Microstruct.* 6: 115–120.

Heertje, I., M. Leunis, W.J.M. van Zeyl, and E. Berends. 1987a. Product morphology of fatty products. *Food Microstruct.* 6: 1–8.

Herrera, M.L. and R.W. Hartel. 2000a. Effect of processing conditions on crystallization kinetics of a milk fat model system. *J. Am. Oil Chem. Soc.* 77: 1177–1188.

Herrera, M.L. and R.W. Hartel. 2000b. Effect of processing conditions on crystallization kinetics of a milk fat model system: Rheology. *J. Am. Oil Chem. Soc.* 77: 1189–1195.

Herrera, M.L. and R.W. Hartel. 2000c. Effect of processing conditions on crystallization kinetics of a milk fat model system: Microstructure. *J. Am. Oil Chem. Soc.* 77: 1197–1204.

Jewell, G.G. and M.L. Meara. 1970. A new and rapid method for the electron microscopic examination of fats. *J. Am. Oil. Chem. Soc.* 47: 535–538.

Johansson, D. 1995. Weak gels of fat crystals in oils at low temperatures and their fractal nature. *J. Am. Oil Chem. Soc.* 72: 1235–1240.

Jullien, R. 1987. Aggregation phenomena and fractal aggregates. *Contemp. Phys.* 28: 477–493.

Jullien, R. and R. Botet. 1987. *Aggregation and Fractal Aggregates.* Singapore: World Scientific Publishing Co. Pte Ltd.

Juriaanse, A.C. and I. Heertje. 1988. Microstructure of shortenings, margarine and butter—A review. *Food Microstruct.* 7: 181–188.

Kamphuis, H. and R.J.J. Jongschaap. 1985. The rheological behaviour of suspensions of fat particles in oil interpreted in terms of a transient-network model. *Colloid Polym. Sci.* 263: 1008–1024.

Kamphuis, H., R.J.J. Jongschaap, and P. Mijnlieff. 1984. A transient-network model describing the rheological behaviour of concentrated dispersions. *Rheol. Acta* 23: 329–344.

Kantor, Y. and I. Webman. 1984. Elastic properties of random percolating systems. *Phys. Rev. Lett.* 52(21): 1891–1894.

Kolb, M., R. Botet, and B. Jullien. 1983. Scaling of kinetically growing clusters. *Phys. Rev. Lett.* 51: 1123.

Lam, R., M.A. Rogers, and A.G. Marangoni. 2009. Thermo-mechanical method for the determination of the fractal dimension of fat crystal networks. *J. Therm. Anal. Calorim.* 98: 7–12.

Leibniz, G.W. 1721. *Principia Philosophiae, More Geometrico Demonstrata.*

Lin, M.Y., H.M. Lindsay, D.A. Weitz, R.C. Ball, R. Klein, and P. Meakin. 1989. Universality in colloid aggregation. *Nature.* 339: 360–362.

Litwinenko, J.W., A.P. Singh, and A.G. Marangoni. 2004. Effects of glycerol and tween 60 on the crystallization behavior, mechanical properties, and microstructure of a plastic fat. *Cryst. Growth Des.* 4(1): 161–168.

Maleky, F., A.K. Smith, and A.G. Marangoni. 2011. Laminar shear effects on crystalline alignments and nanostructure of a triacylglycerol crystal network. *Cryst. Growth Des.* 11(6): 2335–2345.

de Man, J.M. and A.M. Beers. 1987. Fat crystal networks: Structure and rheological properties. *J. Texture Stud.* 18: 303–318.

Mandelbrot, B.B. 1982. *The Fractal Geometry of Nature.* New York: Freeman.

Mandelbrot, B.B. 1998. Is nature fractal? *Nature.* 279: 783–784.

Manning, D.M. and P.S. Dimick. 1985. Crystal morphology of cocoa butter. *Food Microstruct.* 4: 249–265.

Marangoni, A.G. 2002. The nature of fractality in fat crystal networks. *Trends Food Sci. Technol.* 13: 37–47.

Marangoni, A.G., S. Barbut, S.E. McGauley, M. Marcone, and S.S. Narine. 2000. On the structure of particulate gels—The case of salt-induced cold gelation of heat-denatured whey protein isolate. *Food Hydrocolloids* 14: 61–74.

Marangoni, A.G. and R.W. Hartel. 1998. Visualization and structural analysis of fat crystal networks. *Food Technol.* 52(9): 46–52.

Marangoni, A.G. and S.E. McGauley. 2003. Relationship between crystallization behavior and structure in cocoa butter. *Cryst. Growth Des.* 3: 95–108.

Marangoni, A.G. and D. Rousseau. 1996. Is plastic fat rheology governed by the fractal nature of the fat crystal network? *J. Am. Oil Chem. Soc.* 73: 991–993.

Marangoni, A.G. and D. Rousseau. 1998a. The influence of chemical interesterification on physicochemical properties of complex fat systems 1. Melting and crystallization. *J. Am. Oil Chem. Soc.* 75: 1265–1271.

Marangoni, A.G. and D. Rousseau. 1998b. The influence of chemical interesterification on the physicochemical properties of complex fat systems. 3. Rheology and fractality of the crystal network. *J. Am. Oil Chem. Soc.* 75: 1633–1636.

Martini, S., M.L. Herrera, and R.W. Hartel. 2002. Effect of cooling rate on crystallization behavior of milk fat fraction/sunflower oil blends. *J. Am. Oil. Chem. Soc.* 79: 1055–1062.

Meakin, P. 1983. Formation of fractal clusters and networks by irreversible diffusion-limited aggregation. *Phys. Rev. Lett.* 1983(51): 1119.

Meakin, P. 1988. Fractal Aggregates. *Adv. Colloid Interface Sci.* 28: 249–331.

Medalia, A.I. and F.A. Heckman. 1971. Morphology of aggregates*VII. Comparison chart for electron microscopic determination of carbon black aggregates morphology. *J. Colloid Int. Sci.* 36(2), 173–190.

Mellema, M., J.H.J. van Opheusden, and T. van Vliet. 2002. Categorization of rheological scaling models for particle gels applied to casein gels. *J. Rheol.* 46: 1–9.

Mullins, W.W. and R.F. Sekerka. 1963. Morphological stability of a particle growing by diffusion or heat flow. *J. Appl. Phys.* 34: 323.

Narine, S.S. and A.G. Marangoni. 1999a. Fractal nature of fat crystal networks. *Phys. Rev. E* 59(2): 1908–1920.

Narine, S.S. and A.G. Marangoni. 1999b. Mechanical and structural model of fractal networks of fat crystals at low deformations. *Phys. Rev. E* 60(6): 6991–7000.

Narine, S.S. and A.G. Marangoni. 1999c. Relating structure of fat crystal networks to mechanical properties: A review. *Food Res. Int.* 32: 227–248.

Nederveen, C.J. 1963. Dynamic Mechanical behavior of suspensions of fat particles in oil. *J. Colloid Interface Sci.* 18: 276–291.

Papenhuijzen, J.M.P. 1971. Superimposed steady and oscillatory shear in dispersed systems. *Rheol. Acta* 10: 493–502.

Papenhuijzen, J.M.P. 1972. The role of particle interactions in the rheology of dispersed systems. *Rheol. Acta* 11: 73–88.

Payne, A.R. 1964. The elasticity of carbon black networks. *J. Colloid Sci.* 19: 744–754.

Poot, C., W. Dijkshoorn, A.J. Haighton, and C.C. Verburg. 1975. Laboratory separation of crystals from plastic fats using detergent solution. *J. Am. Oil Chem. Soc.* 70: 69–72.

Ribeiro, A.P.B., R. Grimaldi, L.A. Gioielli, A. Oliveira dos Santos, L.P. Cardoso, and L.A.G. Goncalves. 2009. Thermal behavior, microstructure, polymorphism, and crystallization properties of zero trans fats from soybean oil and fully hydrogenated soybean oil. *Food Biophys.* 4: 106–118.

Rodriguez, A., E. Castro, M.C. Salinas, R. Lopez, and M. Miranda. 2001. Interesterification of tallow and sunflower oil. *J. Am. Oil Chem. Soc.* 78: 431–436.

Rojanski, D., D. Huppert, H.D. Bale, X. Dacai, P.W. Schmidt, D. Farin, A. Seri-Levy, and D. Avnir. 1986. Integrated fractal analysis of silica: Adsorption, electronic energy transfer, and small-angle x-ray scattering. *Phys. Rev. Lett.* 56(23): 2505–2508.

de Rooij, R., M.H.G. van den Ende, and J. Mellema. 1994. Elasticity of weakly-aggregating latex dispersions. *Phys. Rev. E* 49: 3038–3049.

Rousseau, D., K. Forestiere, A.R. Hill, and A.G. Marangoni. 1996a. Restructuring butterfat through blending and chemical interesterification. 1. Melting behavior and triacylglycerol modifications. *J. Am. Oil. Chem. Soc.* 73: 963–972.

Rousseau, D., A.R. Hill, and A.G. Marangoni. 1996b. Restructuring butterfat through blending and chemical interesterification. 2. Microstructure and polymorphism. *J. Am. Oil Chem. Soc.* 73: 973–981.

Rousseau, D., A.R. Hill, and A.G. Marangoni. 1996c. Restructuring butterfat through blending and chemical interesterification. 3. Rheology. *J. Am. Oil. Chem. Soc.* 73: 983–989.

Rousseau, D., K.R. Jeffreys, and A.G. Marangoni. 1998. The influence of chemical interesterification on the physicochemical properties of complex fat systems. 2. Morphology and polymorphism. *J. Am. Oil Chem. Soc.* 75: 1833–1839.

Rousseau, D. and A.G. Marangoni. 1998a. Tailoring the textural attributes of butterfat/canola oil blends via *Rhizopus arrhizus* Lipase-catalyzed interesterification. 1. Compositional modifications. *J. Agric. Food Chem.* 46: 2368–2374.

Rousseau, D. and A.G. Marangoni. 1998b. Tailoring the textural attributes of butterfat/canola oil blends via *Rhizopus arrhizus* Lipase-catalyzed interesterification. 2. Modifications of physical properties. *J. Agric. Food Chem.* 46: 2375–2381.

Russ, J.C. 1994. *Fractal Surfaces.* New York: Plenum Press.

Schaefer, D.W. and K.D. Keefer. 1986. Structure of random porous materials: Silica aerogel. *Phys. Rev. Lett.* 56(20): 2199–2202.

Schaefer, D.A., J.E. Martin, P. Wiltzius, and D.S. Cannell. 1984. Fractal geometry of colloidal aggregates. *Phys. Rev. Lett.* 52: 2371.

Schroeder, M. 1991. *Fractals, Chaos, Power Laws.* New York: W.H. Freeman and Company.

Shama, F. and P. Sherman. 1970. The influence of worksoftening on the viscoelastic properties of butter and margarine. *J. Texture Stud.* 1: 196–205.

Sherman, P. 1968. The influence of particle size on the viscoelastic properties of flocculated emulsions. In: *Fifth International Congress on Rheology*, Kyoto, Japan.

Shih, W.H., W.Y. Shih, S.I. Kim, J. Liu, and I.A. Aksay. 1990. Scaling behavior of the elastic properties of colloidal gels. *Phys. Rev. A.* 42: 4772–4779.

Shukla, A. and S.S.H. Rizvi. 1996. Relationship among chemical composition, microstructure and rheological properties of butter. *Milchwissenschaft.* 51(3): 144–148.

Singh, A.P., C. Bertoli, D.R. Rousseau, and A.G. Marangoni. 2004. Matching Avrami indices achieves similar hardnesses in palm oil-based fats. *J. Agric. Food Chem.* 52: 1551–1557.

Sonntag, R.C. and W.B. Russel. 1987. Elastic properties of flocculated networks. *J. Colloid Interface Sci.* 116: 485–489.

Stanley, H.E. 1984. Fractal concepts in aggregation and gelation: An introduction. In: *Kinetics of Fractal Aggregation*, F. Family and D.P. Landau, Eds., Amsterdam, the Netherlands: Elsevier Science Publishers B.V., pp. 1–4.

Tang, D. and A.G. Marangoni. 2005. Computer simulations of fractal dimension of fat crystal networks. *J. Am. Oil Chem. Soc.* 83: 309–313.

Tang, D. and A.G. Marangoni. 2006a. Microstructure and fractal analysis of fat crystal networks. *J. Am. Oil. Chem. Soc.* 83(5): 377–388.

Tang, D. and A.G. Marangoni. 2006b. 3D fractal dimension of fat crystal networks. *Chem. Phys. Lett.* 433: 248–252.

Tang, D. and A.G. Marangoni. 2008. Fractal dimensions of simulated and real fat crystal networks in 3D space. *J. Am. Oil Chem. Soc.* 85: 495–499.

Tang, D. and A.G. Marangoni. 2009. Modified fractal model and rheological properties of colloidal networks. *J. Colloid Interface Sci.* 318: 202–209.

Uriev, N.B. and I.Y. Ladyzhinsky. 1996. Fractal models in the rheology of colloidal gels. *Colloids Surf. A* 108: 1–11.

Vacher, R., T. Woignier, J. Pelous, and E. Courtens. 1988. Structure and self-similarity of silica aerogels. *Phys. Rev. B* 37(11): 6500–6503.

Van den Tempel, M. 1961. Mechanical properties of plastic-disperse systems at very small deformations. *J. Colloid Interface Sci.* 16: 284–296.

Van den Tempel, M. 1979. Rheology of concentrated suspensions. *J. Colloid Interface Sci.* 71(1): 18–20.

Viczek T. 1992. *Fractal Growth Phenomena.* New York: World Scientific Publishing Company.

Vold, M.J. 1951. Van der Waals' attraction between anisotropic particles. *J. Colloid Sci.* 9: 451.

Vreeker, R., L.L. Hoekstra, D.C. den Boer, and W.G.M. Agterof. 1992a. Fractal aggregation of whey proteins. *Food Hydrocolloids.* 6: 423–435.

Vreeker, R., L.L. Hoekstra, D.C. den Boer, and W.G.M. Agterof. 1992b. The fractal nature of fat crystal networks. *Colloids Surf.* 65: 185–189.

Weitz, D.A., J.S. Huang, M.Y. Lin, and J. Sung. 1985. Limits of the fractal dimension for irreversible kinetic aggregation of gold colloids. *Phys. Rev. Lett.* 54: 1416.

Weitz, D.A. and M. Oliveria. 1984. Fractal structures formed by kinetic aggregation of aqueous gold colloids. *Phys. Rev. Lett.* 52: 1433.

West, A.R. 1984. *Solid State Chemistry and Its Applications.* Chichester, U.K.: John Wiley & Sons.

Witten, T.A. and L.M. Sander. 1981. Diffusion-limited aggregation, a kinetic critical phenomenon. *Phys. Rev. Lett.* 47(19): 1400–1403.

Witten, T.A. and L.M. Sander. 1983. Diffusion-limited aggregation. *Phys. Rev. B* 27(9): 5686–5697.

Wright, A.J. and A.G. Marangoni. 2002. The effect of minor components on milk fat crystallization, microstructure, and rheological properties. In: *Physical Properties of Lipids*, Marangoni, A.G., S.S. Narine, Eds., New York: Marcel Dekker, pp. 125–161.

Wu, H. and M. Morbidelli. 2001. A model relating structure of colloidal gels to their elastic properties. *Langmuir.* 17: 1030–1036.

Xu, C., W. Zipfel, J.B. Shear, R.M. Williams, and W.W. Webb. 1996. Multiphoton fluorescence excitation: New spectral windows for biological nonlinear microscopy. *Proc. Natl. Acad. Sci. USA.* 93: 10763–10768.

8 Yield Stress and Elastic Modulus of a Fat Crystal Network

The yield stress is one of the most important macroscopic properties of fats and fat-containing products since it is strongly correlated to sensory perception of hardness and spreadability, as well as to material stability. The apparent yield stress of a plastic solid is usually defined as the point at which, when the stress is increased, the deforming solid first begins to show liquid-like behavior (Barnes, 1999). Even though the concept of a yield stress has been incorporated into phenomenological rheological models, the relationship between a material's structure and its yield stress has not been established. In this chapter, we develop a general model relating the structure of a fat crystal network to its yield stress and elastic modulus.

8.1 MODEL

The change in free energy (∂G) of a system at constant pressure, volume and temperature equals the change in internal energy (∂U) minus the product of the change in entropy (∂S) times temperature (T):

$$\partial G = \partial U - T\partial S \tag{8.1}$$

The change in free energy of a flocculated colloidal network as a function of changes in extensional strain ($\partial \varepsilon$), at constant temperature, pressure (P), volume (V) and composition (μ), where deformation takes place without any rupture of bonds equals the product of stress (σ) times volume (Sonntag and Strenge, 1987):

$$\left(\frac{\partial G}{\partial \varepsilon}\right)_{T,P,V,\mu} = \left(\frac{\partial U}{\partial \varepsilon}\right)_{T,P,V,\mu} - T\left(\frac{\partial S}{\partial \varepsilon}\right)_{T,P,V,\mu} = \sigma \cdot V \tag{8.2}$$

The main assumption in this model is that the change in free energy of the network upon deformation arises due to changes in elastic energy ($\partial E_{\text{elastic}}$), namely,

$$\frac{\partial G}{\partial \varepsilon} = \frac{\partial E_{\text{elastic}}}{\partial \varepsilon} \tag{8.3}$$

233

The elastic energy of the network can be expressed in terms of deformation (ΔL), or strain ($\varepsilon = \Delta L/L$),

$$E_{\text{elastic}} = \frac{1}{2}k(\Delta L)^2 = \frac{1}{2}L^2 k\varepsilon^2 \tag{8.4}$$

where k is the elastic constant of the network. The elastic constant k can be substituted by $A \cdot E/L$, where E corresponds to the Young's modulus, A is the area over which the force is applied, and L is the size of the system. The product of the area times the length corresponds to the volume of the network, while the product of the Young's modulus times the strain corresponds to stress. Thus, the change in elastic energy as a function of strain can be expressed as

$$\frac{\partial E_{\text{elastic}}}{\partial \varepsilon} = \frac{\partial}{\partial \varepsilon}\left(\frac{1}{2}LAE\varepsilon^2\right) = VE\varepsilon = V\sigma \tag{8.5}$$

Sonntag and Strenge (1987) have shown that the entropic contribution to the change in free energy of a flocculated colloidal network upon a small elastic deformation is much smaller than the contribution from changes in the internal energy of that network. Thus, for the case where $\partial U \gg T\partial S$,

$$\frac{\partial G}{\partial \varepsilon} \approx \frac{\partial U}{\partial \varepsilon} = VE\varepsilon \tag{8.6}$$

Upon integration and rearrangement, an expression for the Young's modulus of the network can be obtained,

$$E = \frac{2\Delta U}{V\varepsilon^2} \tag{8.7}$$

where ΔU corresponds to the change in the internal energy of the network upon deformation, which equals the total interaction energy per floc.

The volume of the system can be expressed as a function of the particle volume fraction (Φ) and the total volume occupied by the flocs,

$$V = \frac{V_a N_a N_\xi}{\Phi} \tag{8.8}$$

where
    V_a is the volume of an individual particle
    N_a is the number of particles in a floc
    N_ξ is the total number of flocs in the system

Flocs are assumed to occupy space in a close-packed fashion and are thus completely space-filling. The number of particles in a floc is given by

$$N_a \sim \left(\frac{\xi}{a}\right)^D \tag{8.9}$$

where
  ξ is the diameter of the flocs
  a is the diameter of the particles within the floc
  D is the fractal dimension for the arrangement of particles within the floc

The volume fraction of particles within the floc (Φ_ξ) is therefore given by

$$\Phi_\xi \sim \frac{N_a V_a}{N_s V_s} \sim \left(\frac{\xi}{a}\right)^{D-d} \tag{8.10}$$

where
  N_s is the number of available embedding space elements within the floc ($N_s \sim (\xi/a)^d$)
  V_s is the volume of an element of embedding space
  d is the Euclidean dimension of the embedding space

At this point we will assume that a particle volume is equal to the volume of an element of embedding space, namely, $V_a = V_s$. Thus, the diameter of the flocs varies with the volume fraction of particles within the floc as

$$\xi \sim a\Phi_\xi^{1/(D-d)} \tag{8.11}$$

Flocs pack in a regular, close-packed, Euclidean fashion; hence, at the floc level of structure, the material can be considered as an orthodox amorphous substance. Within the flocs, however, particles pack in a non-Euclidean, fractal fashion. For such a structural arrangement, the volume fraction of particles in a floc (Φ_ξ) is equivalent to the volume fraction of particles in the entire system (Φ), namely, $\Phi_\xi = \Phi$. This well-known relation of polymer physics (De Gennes, 1979) has been experimentally shown to also apply to colloidal aggregates above their gelation threshold (Dietler et al., 1986).

  Thus, considering all of the above, for spherical particles packed in a fractal fashion within a floc, the Young's modulus of the network can be expressed as

$$E = \frac{12\Delta U}{\pi a^3 N_\xi \varepsilon^2} \Phi^{d/(d-D)} \tag{8.12}$$

In the weak-link rheological regime described by Shih et al. (1990), the links between flocs of colloidal particles yield under an external stress, that is, the flocs

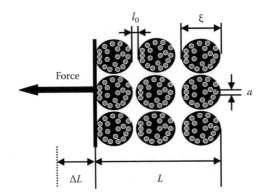

FIGURE 8.1 Idealized flocculated colloidal network under extension. Particles (a) are packed in a fractal fashion within flocs (ξ). A force (F) acting upon the network causes the links between flocs to yield, and the original length of the system in the direction of the applied force (L) to increase (ΔL). Thus, the inter-floc separation distance (l), also increases.

are mechanically stronger than the links between them. Thus, in this regime, the macroscopic deformation of the network (ΔL) can be related to the inter-floc deformation, $\Delta L = (n - 1)(l - l_o)$, where l_o is the equilibrium distance between flocs, l corresponds to the distance between flocs under an applied stress, and $(n - 1)$ corresponds to the number of links between flocs in the direction of the applied stress (Figure 8.1).

The number of flocs in the direction of the applied stress roughly equals the number of links between flocs for the case where $L \gg \xi$. The macroscopic strain terms can be expressed as $n(l - l_o)/L$, or $(l - l_o)/\xi$, since the term L/n corresponds to the size of the flocs, ξ, for the case where $\xi \gg l_o$. Substituting the strain terms with $(l - l_o)/\xi$, and ξ with a $\Phi^{-1/(d-D)}$, results in the expression

$$E = \frac{12(\Delta U_\xi / (l - l_o)^2)}{\pi a} \Phi^{1/(d-D)} \tag{8.13}$$

where $\Delta U_\xi (\Delta U_\xi = \Delta U / N_\xi)$ is the effective change in internal energy per floc.

Let us consider the case where a network is stretched to its elastic limit. Beyond this critical strain (ε^*), this material yields and begins to flow. The critical strain can be defined as

$$\varepsilon^* = \frac{l^* - l_o}{\xi} \tag{8.14}$$

where
   l^* is the critical deformation at the elastic limit
   l_o is the equilibrium separation distance between flocs
   ξ is the diameter of the flocs

The product of the Young's modulus (E) and ε^* equals the yield stress (σ^*), which can be obtained by multiplying both sides of Equation 8.13 by ε^*,

$$E\varepsilon^* = \frac{12(\Delta U_\xi/(l^*-l_o)^2)}{\pi a} \frac{(l^*-l_o)}{\xi} \Phi^{1/(d-D)} \tag{8.15}$$

yielding

$$\sigma^* = \frac{12(\Delta U_\xi/(l-l_o))}{\pi a\xi} \Phi^{1/(d-D)} \tag{8.16}$$

The term $\Delta U_\xi/(l - l_o)$ can be substituted with the expression of a force ($\Delta E/\Delta x = 1/2F$). The force required to pull two flocs apart corresponds to the force of adhesion (F_{ad}), which for two rigid, incompressible spheres of equal diameter can be estimated for the case where $\xi \gg l_o$ using the Derjaguin approximation (Israelachvili, 1992),

$$F_{ad} = \pi\delta\xi \tag{8.17}$$

where
  ξ corresponds to the diameter of the spheres
  δ to the surface free energy per unit area (J m^{-2})

This surface free energy is equivalent, in our case, to the crystal-melt interfacial tension (N m^{-1}).

Substitution of Equation 8.17 into Equation 8.16, results in an expression for the yield stress of a solid material structured as a space-filling collection of fractal flocs of particles

$$\sigma^* = \frac{6\delta}{a} \Phi^{1/(d-D)} \tag{8.18}$$

Moreover, the Young's modulus of the material can then be estimated from

$$E = \frac{\sigma^*}{\varepsilon^*} = \frac{6\delta}{a\varepsilon^*} \Phi^{1/(d-D)} \tag{8.19}$$

Equation 8.19 is basically equivalent to the model of Marangoni (2000) based on van der Waals arguments

$$E \sim \frac{A}{2\pi a\varepsilon^* d_o^2} \Phi^{1/(d-D)} \tag{8.20}$$

where
  A is Hamaker's constant
  d_o is the interfloc separation distance

The current model is a variation of this model using a semiclassical approach based on bulk properties. The purpose of developing an alternative form of the pre-exponential factor λ was to make it more accessible from computational as well as experimental points of view. Equation 8.20 can be easily obtained from Equation 8.19 by substituting δ with $F_{ad}/\pi\xi$, and F_{ad} with the attractive van der Waals' force between two spheres of equal diameter, $F_{vw} = A\xi/12d_o^2$. Keep in mind that $F = 2\partial E/\partial x$ when checking the derivation.

Values for the shear modulus (G) could be obtained from knowledge of the Poisson ratio (μ) of the material since $E = 2(1 + \mu)G$. For a material where no volume change takes place when it is stretched or compressed, the Poisson's ratio is 0.5 and thus $E = 3G$ and

$$G \sim \frac{A}{6\pi a\varepsilon^* d_o^2} \Phi^{1/(d-D)} \tag{8.21}$$

Thus, the yield stress of such a material is determined by the amount of network material present (volume fraction of solids), the structure of the network (primary particle size, fractal dimension) and intermolecular forces (solid-liquid interfacial tension). This universal expression can be applied to any flocculated particulate system including colloidal gels and crystals as well as pastes and other types of soft condensed matter.

Figure 8.2 shows simulations of the effects of δ, D and a on the yield stress. Of particular interest is the large effect of D on the calculated yield stress—a lower fractal dimension results in a higher yield stress. The δ for triacylglycerols has been determined experimentally to be about $0.01\,mJ\,m^{-2}$ (Phipps, 1964).

In this work we consider the stress at the limit of linearity as the yield stress. Small deformation mechanical testing is much more sensitive than large deformation techniques. In a large deformation mechanical test, the apparent yield stress is a strong function of the loading rate and loading history. Usually, high loading rates are used for the determination of the yield stress of soft, plastic solids due to sensitivity issues—the higher the loading rates used, the higher the apparent yield stress of the material. When the loading rate is higher than the rate of deformation of the material, a higher apparent yield force will be determined. For this reason, it is sometimes not possible to determine the yield stress of a soft, plastic material at low loading rates. This, however, does not mean that the material does not have a yield stress. The more sensitive, small deformation techniques should be used instead.

Figure 8.3 shows changes in the yield stress (σ^*) as a function of solids' volume fraction (Φ) for blends of milk fat (Figure 8.3a), cocoa butter (Figure 8.3b) and modified palm oil (Figure 8.3c) with canola oil crystallized for 24h at 5°C. The yield stress was determined using a small-deformation, controlled stress shear rheometer (TA Instruments AR2000, Mississauga, Ontario, Canada) as described in the methods section. The fractal dimensions of the three fat crystal networks were determined from the slope of the log-log plot of σ^* versus Φ (not shown), assuming a weak-link rheological

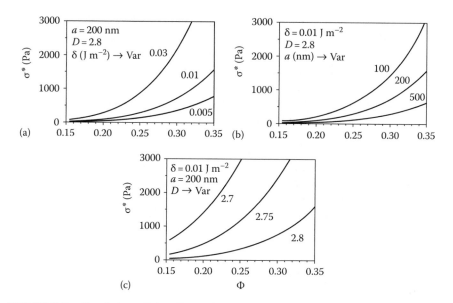

FIGURE 8.2 Simulations of the effects of (a) surface free energy, (b) primary crystal size, and (c) fractal dimension on the yield stress of a plastic disperse system at different solids' volume fractions (Φ).

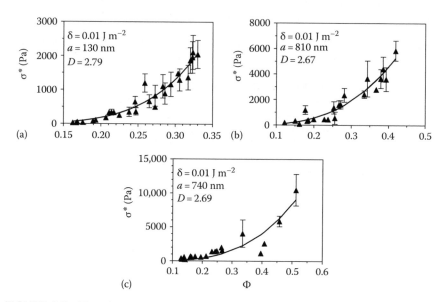

FIGURE 8.3 Experimentally determined changes in σ^* as a function of Φ for blends of (a) milk fat, (b) cocoa butter, and (c) modified palm oil with canola oil. Samples were crystallized statically from 40°C to 5°C at a cooling rate of 1°C/min, and annealed at 5°C for 24 h. Symbols represent the average and standard deviation of—two to six samples. The line through the data was generated by nonlinear least-squares minimization of the model to the experimental data. Indicated are the estimates of the model parameters. The surface free energy term (δ) was fixed as a constant.

regime $[D = 3 - (\text{slope})^{-1}]$. The $\sigma^* - \Phi$ scaling relationships for milk fat, cocoa butter and modified palm oil were, respectively, $\sigma^* \sim \Phi^{5.8}$, $\sigma^* \sim \Phi^{3.0}$, and $\sigma^* \sim \Phi^{2.1}$, yielding fractal dimensions of 2.82, 2.66, and 2.52. Our model was then fitted to $\sigma^* - \Phi$ data by nonlinear regression using Scientist 2.0 (Micromath Scientific Software, Salt Lake City, Utah), and parameter estimates obtained. As can be appreciated, the model fits the data quite well for the three different systems, and the values obtained for D and a were reasonable. In our experience it was necessary to either fix δ or a as a constant in order to obtain reliable parameter estimates—in our case, δ was fixed.

Thus, based on this work, we propose a structural definition for the yield stress of a particle network: The yield stress corresponds to the force required to move an ensemble of network structural elements from their equilibrium separation distance, l_o, characterized by an interaction energy, U_{eq}, to a critical separation distance, l^*, where the interaction energy equals zero (Marangoni and Rogers, 2003).

REFERENCES

Barnes, H.A. 1999. The yield stress—A review of "παντα"—everything flows? *J. Non-Newtonian Fluid Mech.* 81: 133–178.

De Gennes, P.G. 1979. *Scaling Concepts of Polymer Physics*. Ithaca, NY: Cornell University Press.

Dietler G., C. Aubert, D.S. Cannell, and P. Wiltzius. 1986. Gelation of colloidal silica. *Phys. Rev. Lett.* 57: 3117–3120.

Israelachvili, J. 1992. *Intermolecular and Surface Forces*, 2nd edn. New York: Academic Press, p. 163.

Marangoni, A.G. 2000. Elasticity of high-volume fraction fractal aggregate networks: A thermodynamic approach. *Phys. Rev. B* 62: 13951–13955.

Marangoni, A.G. and M.A. Rogers. 2003. Structural basis for the yield stress in plastic disperse systems. *Appl. Phys. Lett.* 82: 3239–3241.

Phipps, L.W. 1964. Heterogeneous and homogeneous nucleation in supercooled triglycerides and n-paraffins. *Trans. Faraday Soc.* 60: 1873–1883.

Shih, W.H., W.Y. Shih, S.I. Kim, J. Lin, and I.A. Aksay. 1990. Scaling behavior of the elastic properties of colloidal gels. *Phys. Rev. A* 42: 4772–4779.

Sonntag, H. and K. Strenge. 1987. *Coagulation Kinetics and Structure Formation*. New York: Plenum Press, pp. 345–392.

9 Liquid–Multiple Solid Phase Equilibria in Fats
Theory and Experiments

Leendert H. Wesdorp, J.A. van Meeteren,
S. de Jong, R. van der Giessen,
P. Overbosch, P.A.M. Grootscholten,
M. Struik, E. Royers, A. Don, Th. de Loos,
C. Peters, and I. Gandasasmita

9.1 INTRODUCTION AND PROBLEM DEFINITION

It is important to control the melting and solidification behavior of edible oils and fats for the production of fat-containing food products. The objective of this work is to develop a method to predict the melting range and solid phase composition of fats from their overall composition. Fats consist of triacylglycerols (TAGs), which show polymorphism in the solid phase. The polymorphic behavior is reviewed. The nomenclature for TAGs and groups of TAGs used in this work is explained. Existing methods for solid phase prediction are discussed.

9.1.1 SOLID–LIQUID PHASE EQUILIBRIA AND FATS

The fact that many languages have different words for the solid and liquid state of mixtures of triacylglycerols (triglycerides) indicates that the solid–liquid phase behavior of TAGs is something everyone encounters in daily life. Coconut "oil" can be conveniently used as table oil in many tropical countries, while in Northwest Europe it is considered a stone hard "fat."

Edible oils and fats usually consist of more than 95% of a complex mixture of TAGs. Typically an edible oil or fat can contain more than 500 different TAGs. Edible oils and fats therefore do not possess a distinct melting point, but exhibit a long melting range.

This melting range is one of the main factors determining the properties of fat-containing food products, like fat spreads, dressings, chocolate, cakes, ice cream, and cookies. A fat spread, for example, must contain enough liquid fat at refrigerator temperature in order to make it a spreadable, soft solid. At ambient temperature, it still must contain enough solid fat to prevent the spread from becoming pourable and oiling out. To give the spread a nice taste, the fat should be liquid at mouth temperature (Poot and Biernorth, 1986; Figures 9.1 and 9.2).

A fat with an optimal melting range for a particular application is obtained by carefully blending natural and modified oils and fats. Of course, the edible fats industry boasts of an ability to blend fats to a constant melting range regardless of the raw materials used. In other applications, like chocolate, cocoa butter is sometimes mixed with other fats, that require a very specific TAG composition to get a good chocolate. These are often obtained by fractional crystallization of natural fats like palm oil. Simulation of this fractionation process requires calculation of the dependence of the crystal composition on process conditions. Both melting range

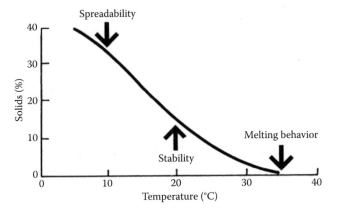

FIGURE 9.1 Requirements for the melting range of a fat spread.

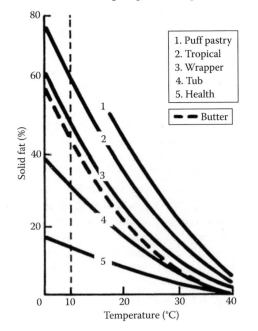

FIGURE 9.2 Melting ranges of some types of fat spreads.

and phase composition of a fat are primarily determined by the solid–liquid phase equilibrium in the fat. Describing the phase behavior of edible oils and fats is therefore a necessity for the edible fats industry.

The objective of this work is to develop a method that enables prediction of the melting range and the solid phase composition of edible oils and fats.

9.1.2 TRIACYLGLYCEROLS: NOMENCLATURE

Edible oils and fats consist mainly of triacylglycerols (TAGs). For convenience, TAGs will be identified by a three-letter code in this work. Each of the three characters in the code represents one of the fatty acids that is esterified with the glycerol. So glycerol-1-palmitate-2, 3-disterate will be represented by PSS. The middle character always indicates the fatty acid that is esterified on the 2-position of the glycerol. The characters used to represent the fatty acids are given in Table 9.1.

It is sometimes convenient to be able to refer to groups of TAGs. Table 9.2 defines a set of letter codes that represent a number of fatty acids.

Hence the TAG-group h_3 contains all TAGs that can be made from the four long chain fatty acids, like SSS, PPP, SPP, PPS, PSP, AAA, BBB, ASA, ASB, etc., while hOh stands for all TAGs having oleic acid on the 2-position and one of the four long chain fatty acids on the 1- and the 3-position of the glyceryl group like SOS, POP, POS, AOS, etc.

TABLE 9.1
Characters Used for Representing Individual Fatty Acids

Code	Fatty Acid	Code	Fatty Acid
2	Acetic acid	P	Palmitic acid (hexadecanoic acid)
4	Butyric acid	S	Stearic acid (octadecanoic acid)
6	Hexanoic acid	O	Oleic acid (*cis*-9-octadecenoic acid)
8	Octanoic acid	E	Elaidic acid (*trans*-9-octadecenoic acid)
C	Capric acid (Decanoic acid)	1	Linoleic acid (*cis–cis*-9, 12-octadecadienoic acid)
L	Lauric acid (Dodecanoic acid)	A	Arachidic acid (eicossanoic acid)
M	Myristic acid (Tetradecanoic acid)	B	Behenic acid (docosanoic acid)

TABLE 9.2
Letter Codes Used for Representing Groups of Fatty Acids

Code	Fatty Acids
m	Medium chain fatty acids (8 + C + L + M)
h	Long chain saturated fatty acids (P + S + A + B) ("hydrogenated")
u	*cis*-C18 unsaturated fatty acids (O + 1 + linolenic acid)

9.1.3 TRIACYLGLYCEROLS: POLYMORPHISM

The existence of a number of alternative crystal structures is a characteristic property of all lipids (alkanes, fatty acids, soaps, methyl esters of fatty acids, and TAGs) (Larsson, 1986). This is due to the fact that there are a number of different possibilities of packing the long hydrocarbon chains into a crystal lattice. This phenomenon is called polymorphism and each different crystal structure is called a polymorphic form or modification of the lipid. Two types of polymorphism occur in lipids. When each form is thermodynamically stable in a definite range of temperature and pressure, it is called enantiotropic polymorphism. Each enantiotropic polymorph transforms into another polymorph at the transition temperature. The opposite case, when only one polymorphic form is thermodynamically stable, is called monotropic polymorphism. TAGs show monotropic polymorphism, while long chain odd alkanes show enantiotropic polymorphism (see Section 9.9.4.1–9.9.4.3).

The polymorphism of TAGs was first observed by Duffy (1853), but only in the early 1960s, some agreement was reached about the number, structure, and nomenclature of the different polymorphic forms of TAGs (Chapman, 1962; Larsson, 1964). They occur in three different basic polymorphic forms, α, β′, and β, which are characterized by a particular carbon chain packing and stability. Recent accurate experimental techniques, combined with better purity of the samples, have brought about a new controversy about the existence, number, and nomenclature of submodifications of each polymorphic form (Hernqvist and Larsson, 1982; Simpson and Hagemann, 1982; Hagemann and Rothfuss, 1983; Gibon, 1984; Sato, 1987; Sato et al., 1989; Kellens et al., 1900).

It is not the objective of this work to enter into a discussion on submodifications. However the subject of this work, phase equilibria in TAGs, requires an opinion on TAG polymorphism.

9.1.3.1 Basic Polymorphic Forms of TAGs

This work will use the nomenclature proposed by Larsson (1964, 1986) in the form applied by de Jong (1980). Basically the fatty acid chains of TAGs can be packed into three main polymorphic forms, characterized by the short spacings in the x-ray diffraction pattern:

1. The α-modification, characterized by only one strong short-spacing line in the x-ray diffraction pattern near 0.415 nm. In the α-modification the chains are arranged in a hexagonal chain packing, without an order to the zig-zag chain planes. The chains do not have an angle of tilt (Figure 9.3a).
2. The β-modification, characterized by two strong short-spacing lines in the x-ray diffraction pattern near 0.38 and 0.42 nm. It also has a doublet in the $720 \, cm^{-1}$ region of the infrared absorption spectrum. The chain packing in the modification is orthorhombic, with a perpendicular arrangement of the zig-zag chain planes. The chains have an angle of tilt between 50° and 70° (Figure 9.3b).
3. The β-modification, characterized by a strong short-spacing line in the x-ray diffraction pattern near 0.455 nm and a number of other strong lines around 0.36–0.39 nm. The β-modification is the most densely packed polymorph. The chains are arranged in a triclinic chain packing, with a parallel arrangement of the zig-zag chain planes. The chains have an angle of tilt between 50°C and 70°C (Figure 9.3c).

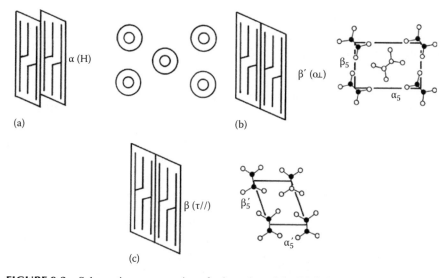

FIGURE 9.3 Schematic representation of orientation of the TAGs in their three basic polymorphic forms, together with the respective chain packing subcell. One zig-zag is seen in the direction of the hydrocarbon chains, open circles are hydrogen, filled circles carbon. (a) α: Unstable, lifetime <60 s present during process, (b) β': Metastable (>60 s → years) present in products, and (c) β: Stable. (From Hernquist, L., Thesis, University of Lund, Sweden, 1984.)

Based on the arrangement of the TAGs themselves in the crystal, two forms can be distinguished for each modification: a form with layers made up of two fatty acid chains and one with layers of three fatty acid chains (Figure 9.4). These forms are characterized by their long spacing in the x-ray diffraction pattern. To distinguish between the two forms, a suffix is added to the symbol that indicates the modification, for example, β-2 and β-3.

The -2 forms are the most stable and the -3 forms are therefore only found for TAGs of which the chain packing in the -2 form would be very unfavorable: in TAGs with *cis*-unsaturated fatty acids and in TAGs in which the fatty acid chains differ by 6 or more.

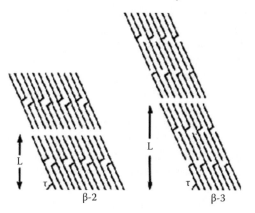

FIGURE 9.4 β-2 and β-3 arrangements of TAGs. (From de Jong, S., Thesis, Rijksuniversiteit Utrecht, the Netherlands, 1980.)

9.1.3.2   Submodifications

If submodifications exist, they will be indicated by a subscript in order of decreasing melting point, for example, β_1-2 and β_2-2.

9.1.3.2.1   Saturated TAGs

De Jong (1980) has done an excellent, extensive study of molecular packing possibilities in the β-modification. He shows that in theory a β-2-forming TAG can crystallize in at least two but often three different β-2 submodifications. For the β-3 modification, two submodifications are possible. He found that in practice, each TAG only occurs in one of these submodifications, although different TAGs crystallized in different submodifications.

Simpson and Hernqvist (Hernqvist and Larsson, 1982; Simpson and Hagemann, 1982) reported simultaneously the existence of two β′-submodifications for saturated TAGs, and somewhat later Hagemann and Rothfuss (1983) managed to distinguish even three β′ submodifications, two α-submodifications, and two β-submodifications in his differential scanning calorimetry (DSC) thermograms. The submodifications were never obtained as pure solid phases on their own, but they rapidly transformed to the most stable submodification. Consequently, reliable melting points or heats of fusion could not be determined. The x-ray diffraction patterns of the submodifications show only very subtle differences. The pattern of a less stable submodification has much broader peaks and shows less details than that of the most stable submodification.

Rapid polymorphic transitions used to be hard to study by x-ray. The polymorphic transition takes 5 min, while one x-ray scan takes 15 min. This makes the unambiguous identification of very unstable submodifications difficult. Very recent experimental techniques allow so-called time resolved x-ray diffraction and neutron scattering studies (Cebula et al., 1990; Kellens et al., 1900). These new techniques need only 5 s for a scan, which means that the actual order of events during a polymorphic transition can be followed.

Kellens et al. (1990) find that the polymorphic transitions take place in a very specific order. Reordering along the three different crystallographic axes is not achieved simultaneously but takes place sequentially. This is in agreement with computer simulation results of Hagemann and Rothfus (1988), who showed that, due to sterical hindrance, the minimal energy path for the rearrangement in the crystal packing follows a very specific order of events. Also Sato et al. (1989) has observed this phenomenon. Kellens states that melting the crystal in a transitional state, where it has not yet reached full crystallinity, can very well result in DSC thermograms similar to those from which Hagemann concluded the existence of unstable submodifications.

Kellens et al. also show that the melting entropies of Hagemann's and Hernqvist's submodifications are far too low to identify these submodifications as one of the differently packed β′ and β forms of de Jong (1980). Moreover, they found that the height of the shoulder or small prepeak in the DSC thermogram did not increase upon increasing scan rate, which is contradictory to the existence of an unstable submodification that rapidly transforms into a more stable one.

Hagemann could not show during all his experiments that the recrystallization from α to one of the β forms was actually complete. The observed less stable

β_2-form can therefore also be explained as the melting peak of the β-form in the presence of a little bit of liquid.

Kellens final conclusion is that these submodifications may very well not be separate polymorphic forms in a thermodynamic sense, but transitional variants with lower crystal perfection and crystallinity.

Due to the serious doubts that one may have on the actual occurrence of a large number of submodifications of saturated TAGs and the plausible alternative explanations available, this work assumes that only the three basic modifications occur.

9.1.3.2.2 Unsaturated TAGs

Only the polymorphism of hOh-type TAGs has been studied in more detail, because of their importance for explaining the phase behavior of cocoa butter. As before, there is no agreement at all on the existence, properties, and number of submodifications.

Perhaps the best results come from Sato et al. (1989). His x-ray diffraction results show very convincingly a fourth polymorph, the γ-modification, characterized by two strong short spacings, at 0.470 and 0.390 nm, and a weaker one at 0.45 nm, characteristic of orthorhombic chain packing with parallel orientation of the chain zig-zag planes. The thermodynamic properties are almost that of the perpendicular orthorhombic β'-modification. This γ-modification is of little practical importance; it is less stable than α-modification and converts readily into the β'-modification. It is not observed in mixtures. This γ-modification is also known under the names β'', sub-β-3, L_2, and form 4. This chapter will use Sato's nomenclature.

Sato, and other authors, also found $2\beta'$ and 2β-submodifications, which hardly differ in x-ray pattern and stability. The existence and properties of β'-submodifications, which only occurred in POP and not in SOS and BOB, strongly depend on the level of impurities.

Therefore similar doubts arise for the existence of these submodifications as for Hagemann's submodifications of saturated TAGs. This work will only consider one β'-modification, having properties of the most stable of the β'-submodifications that are reported.

The existence of two β-submodifications is quite certain (Sato et al., 1989), as they can actually be obtained in pure form. The transition to the most stable form is extremely slow and takes several weeks. X-ray diffraction shows that the differences in structure must be minor. De Jong (1980) suggests that the forms only differ in layer stacking. The most stable β_1-3 form has in this view a slightly higher symmetry and a heat of fusion that is only 1 kJ/mol more. This explains why the most stable form takes so long to form and why the 2 β-forms are only observed in pure components and mixtures of hOh-TAGs that are nearly isomorphous, like cocoa butter. In more complex mixtures that contain high amounts of hOh-type TAGs, like palm oil (Timms, 1984) and the mixtures that Hernqvist (1984) used, only one β-form is observed. Therefore, for most practical purposes, the possible existence of β-submodifications can be neglected. However, for systems with very high concentrations (>80%) of one hOh-type TAG, we need to be aware of the possible occurrence of β-submodifications.

9.1.3.2.3   Conclusion

The discussion on submodifications is still very confusing and clear evidence has not been put forward for their existence, nor is there any agreement about their properties. In most cases, alternative explanations, like those of Kellens et al. (1900), are possible for the observed changes in x-ray patterns and the shoulders and shifts in the DSC thermograms. This work it will therefore assume that each TAG crystallizes in three different polymorphic forms only.

9.1.3.3   Stability

The α-modification is the least stable and it converts within several minutes into the β'-modification. The α-form is found in edible fat products during their preparation.

The β'-modification is the stable modification for odd-chain TAGs and a number of even chain TAGs (like PSP). In other TAGs, it converts into the β-modification within several minutes to hours. In mixtures, this transition is often delayed to several months or years. It is therefore the modification that is encountered most frequently in normal edible fat products.

The β'-modification is the stable modification. It only occurs in edible fat products during their lifetime, if the fat is composed of TAGs that are nearly isomorphous, like hardened rape-seed oils (Nyvlt, 1967) and cocoa butter (Timms, 1984).

All three modifications can be obtained directly from the liquid, by varying the degree of supercooling. The transformation of α to β, however, always takes place via the β'-modification (Hernqvist, 1984; Hagemann and Rothfus, 1988; Kellens et al., 1990; Figure 9.5).

9.1.4   METHODS FOR PREDICTING SOLID PHASE COMPOSITION AND QUANTITY

The objective of this work is to develop a method to predict the melting range and composition of the solid phase of edible oils and fats from the overall composition of the fats. If the exact molecular composition of a fat is known, prediction of its properties is in principle possible. The vegetable oils and fats became only analytically accessible by the progress in high-performance liquid chromatography (HPLC) techniques made since the 1980s, while the analysis of animal oils and fats is still troublesome. Besides, the theoretical understanding of solid–liquid phase equilibria in multicomponent systems was not at all adequate to deal with even very simple oils. Nevertheless, some (semi) empirical methods for prediction of solid phase content were developed. They will be briefly discussed in this section.

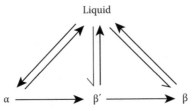

FIGURE 9.5   Simple picture of the possible polymorphic transitions in TAGs.

9.1.4.1 Linear Programming/Multiple Regression

This method is frequently used in the industry (Timms, 1984). The composition of the fat blend is defined in terms of the natural and modified oils of which it is made (e.g., 50% palm oil, 50% bean oil). The solid phase content of the mixture is expressed as a linear function of the concentration of these components:

$$\%S = k_1 x_{oil1} + k_2 x_{oil2} + \cdots \tag{9.1}$$

The coefficients k are obtained by multiple linear regression on the data of a number of mixtures. For limited ranges of compositions and solid phase contents, this method is very useful. It does have disadvantages:

1. The predictive equation has no theoretical basis at all. Extrapolation outside the range of experimental data is not possible and gives dangerously erroneous results.
2. The equations cannot be easily extended to incorporate new component fats. The complete regression analysis and experimental work must be repeated and all coefficients will change.
3. Understanding is not obtained about the actual phase behavior, which implies that a solution cannot be found for undesired recrystallization phenomena and other crystallization phenomena that may be observed in practice. Polymorphism is neglected.

9.1.4.2 Excess Contribution Method

Timms (1984) shows the application of an empirical method developed by Nyvlt (1967) to fats. The method is an extension of the multiple linear regression method of Section 9.1.4.1. The solid phase content is expressed as

$$\%S = \sum_{i=1}^{n} \%S_i \cdot x_i + \sum_{j=1}^{n} \sum_{i=1}^{j-1} I_{ij} \cdot \frac{x_i x_j}{x_i + x_j} \tag{9.2}$$

The concentrations x are weight fractions of the constituting oils, and the binary interactions must be determined from measurements of a number of mixtures of the two oils concerned. The advantages over the linear method are twofold:

1. The equation is in principle valid over larger concentration ranges.
2. If a new component fat is to be added, only the binary mixtures with all other component fats have to be measured in order to determine the interaction parameters. All other interactions remain unaffected.

Timms (1984) shows that the predictions are considerably worse than with the linear method, with deviations of 30% around the measured values. Again the model is totally empirical, the meaning of the parameters is unclear and it offers no real insight in the underlying crystallization phenomena.

9.1.4.3 TAGs Inductors de Crystallization Method

The first method that attempts to predict solid phase content from TAG composition is the TAGs Inductors de Crystallization Method (TGIC method) of Perron (1986). He assumes the following:

- A crystallized fat consists of a number of solid phases in equilibrium with each other and with the liquid phase present.
- Each solid phase consists mainly of one TAG, called the TGIC. All solid phases contain impurities, which modify its melting point and heat of fusion.
- Upon heating, each solid phase melts independently. The fraction of the solid phase that has disappeared at a certain temperature is given by a point on a Kessis curve. A Kessis curve is a general mathematical representation of a DSC or differential thermal analysis (DTA) melting peak of a single component in a liquid solvent. By adjustment of the parameters, it can be used to describe a measured DSC curve, including the thermal lag.
- The solid phase content at a certain temperature can be obtained from an addition of all contributions from the Kessis curves of each TGIC at that temperature.

Although Perron's notion of a number of coexisting solid phases is correct, the method he has developed is not sound at all:

1. His selection of the number and composition of the coexisting solid phases has no physical ground.
2. There is no reason why the solid phases should melt according to a Kessis curve. Moreover, the width of the Kessis curve is normally determined by the DSC apparatus and the scan rate selected. It is therefore not a property of the fat and any assumption about the width of such a curve is arbitrary.

The criticism Timms (1984) applied to an earlier version of Perron's TGIC method is still valid. Contrary to the two empirical methods mentioned, the method is not even useful for practical purposes.

9.1.4.4 Classification of TAGs Method

Wieske (1970) proposed a method in which he groups the TAGs into all possible TAG-groups that can be formed from elaidic acid (E) and the three fatty acid groups of Table 9.2: h, m, and u. These TAG classes are supposed to form separate solid phases in a crystallized fat. The fraction of each class that has not crystallized is given by the Hildebrand equation, assuming an average melting point and heat of fusion for each TAG-class. The Hildebrand equation reads as follows:

$$\ln x_i^L = \frac{\Delta H_{f,i}}{R}\left(\frac{1}{T_{f,i}} - \frac{1}{T}\right) \tag{9.3}$$

This method has a physical background. It assumes that oil can be considered as a mixture of a limited number of pseudocomponents that do not form mixed crystals. However, the predictions from this method are of a similar poor quality as those of the excess contribution method. Wieske's choice of the pseudocomponents, as well as the assumption of solid immiscibility, and the selection of melting points and heats of fusion can be criticized. This method does not offer any understanding of the actual, underlying solid–liquid phase behavior of TAGs.

9.1.4.5 Other TAG-Based Methods

Several attempts have been made to find an empirical correlation between the TAG composition of a fat and its solid phase content (Wieske, 1970; Wieske and Brown, 1986). None of these attempts was very successful. Moreover, the methods have the same disadvantages as the linear method of Section 9.1.4.1.

9.1.5 CONCLUSION

The objective of this work is to develop a method that predicts the melting range and solid phase composition in an edible oil from the overall fat composition.

Existing predictive methods have no sound physical basis, are not generally applicable, and offer no understanding at all of the actual solid–liquid phase behavior of TAGs.

TAGs show polymorphism in the solid phase. This work will assume that each TAG crystallizes in three different polymorphs only: the unstable α-modification, the β′-modification, and the stable β-modification.

Due to their poor analytical accessibility, animal fats will not be considered in this work.

9.2 APPROACH TO THE PROBLEM

The ultimate amount and composition of the solid phase in a fat are determined by the position of the thermodynamic equilibrium solely, but the crystallization process may lead to significant deviations from the equilibrium composition in practical situations. Yet it appears that the starting point for any general predictive method of solid phase content in fats is a description of the solid–liquid phase equilibrium for all three polymorphic forms in which fats can crystallize. The thermodynamic equations that describe these equilibria are worked out. A number of steps are identified that need to be taken before a complete thermodynamic description can be obtained.

9.2.1 SOLID–LIQUID EQUILIBRIUM THERMODYNAMICS

In the end, the amount and the composition of the solid phase in a crystallized fat will be determined by the position of the solid–liquid phase equilibrium in that fat. Therefore, a general method for prediction of the solid fat content and the fat crystal composition from the fat composition must be based on a description of the solid–liquid phase equilibrium thermodynamics in TAGs.

Fats are mixtures of many different TAGs. It is known that many pairs of TAGs only show limited miscibility in the solid phase (Timms, 1984). Therefore it is very likely that "solid fat" often will consist of a number of different coexisting solid phases. These solid phases do not necessarily have the same polymorphic form. A thermodynamic description of this complicated liquid–multiple solid equilibrium can be developed as follows.

Suppose a crystallized fat in equilibrium consists of N components and contains P phases (a liquid phase and $P-1$ solid phases). The phase equilibrium must satisfy the following equations (Prausnitz, 1986):

1. The condition of thermodynamic equilibrium: the chemical potential of each component i in each phase must be equal to that in any other phase:

$$\mu_i^{\text{solid}} = \mu_i^{\text{liquid}} \tag{9.4}$$

for each solid phase. This equation can be worked out for solid–liquid equilibria as follows:

$$\mu_1^{0,S} + RT \ln \gamma_i^S x_i^S = \mu_i^{0,L} + RT \ln \gamma_i^L x_i^L \tag{9.5}$$

γ is the activity coefficient, and x the mol fraction.

$$\ln\left(\frac{\gamma_i^S x_i^S}{\gamma_i^L x_i^L}\right) = \frac{1}{RT}\left(\mu_i^{0,L} - \mu_i^{0,S}\right) \tag{9.6}$$

$$\ln\left(\frac{\gamma_i^S x_i^S}{\gamma_i^L x_i^L}\right) = \frac{\Delta H_{f,i}}{R}\left(\frac{1}{T} - \frac{1}{T_f}\right) - \frac{\Delta c_{p,i}}{R}\left(\frac{T_{f,i}-T}{T}\right) + \frac{\Delta c_{p,i}}{R}\ln\frac{T_{f,i}}{T} \tag{9.7}$$

With $\Delta c_p = 0.2\,\text{kJ/mol}$ and $T_f - T$ never greater than 70 and usually between 0 and 20 the terms with Δc_p are comparatively small and tend to cancel due to their opposite sign. As an approximation, they can therefore be neglected and there remains

$$\ln\left(\frac{\gamma_i^S x_i^S}{\gamma_i^L x_i^L}\right) = \frac{\Delta H_{f,i}}{R}\left(\frac{1}{T} - \frac{1}{T_f}\right) \tag{9.8}$$

2. The mole balance: the sum of the amount of each compound i in each phase f, present in fraction Φ^f must be equal to the overall amount of i, z_i (P is the total number of phases):

$$\sum_{f=1}^{P} x_i^f \Phi^f = z_i \tag{9.9}$$

3. The stoichiometric condition: the sum of the concentrations of the components in each phase must be equal to 100%:

$$\sum_{i=1}^{n} x_i^f = 1 \qquad (9.10)$$

For P phases, this results in $PN + P$ equations with $PN + P$ unknowns. ($P*N$ mol
fractions $\times$ and the quantity of P phases). This set of equations can in principle be
solved to obtain the number of phases, the phase quantities and the composition of
each phase from the overall composition and the temperature.

However, in order to solve these equations, four things are needed:

1. Values for the pure component properties: the heat of fusion and the melting
point
2. Knowledge of the activity coefficients in the liquid phase
3. Knowledge of the activity coefficients in the solid phase
4. A method to solve this complex set of nonlinear equations.

Each of these points will be handled in separate sections of this chapter, so that in
the end a full description of the liquid–multiple solid equilibrium in fats will evolve.

9.2.2 KINETICS OF CRYSTALLIZATION

Although the ultimate amount of solid phase and the solid phase composition are
determined by thermodynamics solely, it is well known that due to the extremely
slow diffusion rate in solids the equilibrium state is not always reached in practical situations (Zief and Wilcox, 1967). Hence both crystal composition and amount
of solid phase that are observed may deviate considerably from what is predicted
by thermodynamics. To avoid possible pitfalls, the effect of kinetic factors on the
amount and composition of the solid phase of fats should be considered.

9.2.2.1 Polymorphism and Kinetics of Crystallization

When fats crystallize, they usually first crystallize in the most unstable polymorph,
the α-modification, followed by slow recrystallization to more stable polymorphs.
Direct crystallization into the β'- or β-modifications only takes place under conditions where little or no supercooling of the less stable modification is present.
Palm oil crystallizes into the α-modification when supercooled to 10°C, into the
β'-modification when cooled to 25°C, and into the β-modification when crystallized
at 32°C (van Putte and Bakker, 1987). In all three cases, the thermodynamically
most stable state is the β-modification.

Obviously solid fat content and crystal composition calculated for the
β-modification using thermodynamics are poor predictions if the fat has crystallized in another polymorphic form. For a lot of practical situations, the most stable
thermodynamic state is irrelevant: the residence time in the process line after the

onset of fat crystallization is only a few minutes for many edible fat products, so the unstable α-modification is the phase that should be considered. During the life of most edible fat products, the recrystallization to the β-modification does not take place: the product is already consumed while it is still in the β′-modification.

In spite of their limited lifetime, the β′-modification and even the α-modification may coexist very well during their existence in thermodynamic equilibrium with the liquid oil. Hence thermodynamics can be applied to predict the amount and composition of these intermediate solid phases. Application of equilibrium thermodynamics to unstable states is quite common: a mixture of benzene and air is not thermodynamically stable but should disintegrate into carbon dioxide and water, yet a vast amount of the literature exists about vapor–liquid phase equilibria with benzene.

9.2.2.2 Shell Formation

If a fat is slowly cooled while it is crystallizing, shell formation may occur. At each instant during the crystallization process, the surface of the growing crystals has the equilibrium composition. As temperature decreases, the equilibrium composition changes, but the composition of the inner part of the crystal does not change due to the low solid state diffusion rate. An inhomogeneous solid phase results, having a concentration gradient from the center of the crystals outward. The solid phase composition deviates from the equilibrium composition. It is reported that the solid fat content can decrease to only 80% of the equilibrium value (Timms, 1984) due to shell formation.

If diffusion limitations do not occur during crystallization, shell formation can be prevented by crystallizing isothermally. The rate-determining step in normal fat crystallization is the surface incorporation (Knoester et al., 1968; van Putte and Bakker, 1987). Indeed, when the crystal composition of an isothermally crystallizing fat is plotted against the reaction coordinate, shell formation seems absent (Figure 9.6).

Phase equilibrium thermodynamics will only give a reliable prediction of the solid phase composition and content if shell formation is absent. However, the effect of shell formation on solid phase content can be calculated from the equilibrium composition as a function of temperature, as will be shown in Section 9.9.3.1.

9.2.2.3 Poor Crystallinity

If fat is crystallizing rapidly, it may result in poorly packed crystals. The clear point of such poorly packed solid phases may be considerably lower than the clear point predicted by thermodynamics. Through recrystallization, these badly packed crystals can rearrange into well-packed crystals. The results of Gibon (1984) show that without the presence of a liquid phase, badly packed crystal forms may persist for years. However, recrystallization via the liquid phase can occur relatively easily (Zief and Wilcox, 1967). Norton (Norton et al., 1985b) shows that partially hydrogenated palm oil initially crystallizes in a very badly packed β′-form, but that in the presence of about 50% of liquid oil, it only takes a few hours to rearrange into well-packed crystals. Also the rate of polymorphic transitions was observed to be much larger in the presence of liquid (Norton et al., 1985a).

Most fats used in edible fat products only contain 0%–40% of solid fat that is purposely crystallized into very small (1 μm and less) crystals. Deviations from thermodynamic predictions due to bad crystal packing will therefore be of minor influence.

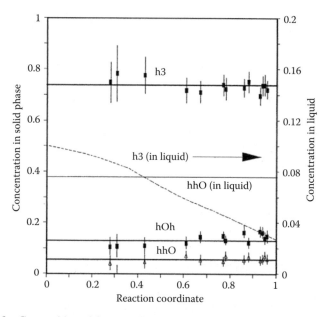

FIGURE 9.6 Composition of fat crystals from palm oil against the reaction coordinate of crystallization at 30°C. (Overall palm oil composition: h3, 0.1; hOh, 0.295; hhO, 0.075; hlh, 0.104; hOO, 0.246; liquid, 0.176.) The palm oil was heated to 80°C, rapidly cooled to 30°C, and stored in a stirred tank. At regular time intervals over totally 21 h samples were taken. The solid fat content of the sample was measured and a small amount of liquid phase was rapidly filtered off and analyzed by $AgNO_3$–HPLC. Crystal composition is calculated from overall composition, solid fat content, and the TAG analyses of the liquid phase of the samples. The error margin is indicated.

9.2.3 CONCLUSION AND APPROACH TO THE PROBLEM

The ultimate amount and composition of the solid phase in a fat are determined by the position of the thermodynamic equilibrium solely. The set of equations that describe a solid–liquid phase equilibrium can be solved, if a number of requirements are fulfilled.

For prediction of the amount and the composition of the solid phase that can crystallize from an oil in practical situations, calculation of the most stable state by solid–liquid phase equilibrium thermodynamics is insufficient. The solid–liquid phase equilibria of unstable polymorphic forms must be considered as well.

The crystallization route that is followed may lead to inhomogeneous solid phases, of which amount and composition can deviate significantly from the equilibrium composition. Yet in order to calculate the effect of the crystallization process on the solid phase composition, the equilibrium solid phase composition as a function of temperature must be known.

From this, it is clear that for a prediction of the amount and composition of the solid phase in fats as a function of temperature, the first requirement is a good description of the solid–liquid phase equilibrium for all three modifications in which fats crystallize. This work will therefore concentrate upon the development of this description for liquid–multiple solid phase equilibria in fats. The end of this chapter will briefly come back to the influence of the crystallization process.

In Section 9.2.1, it was shown that for the description of the solid–liquid phase equilibrium four steps must be taken:

1. Find a method to solve the complex set of nonlinear equations that describe the liquid–multiple solid phase equilibrium.
2. Find values for the pure component properties: the heat of fusion and the melting point for all of the hundreds of TAGs involved.
3. Predict the activity coefficients of TAGs in the liquid phase.
4. Predict the activity coefficients in all possible solid phases.

These steps, in the order in which they are listed, are the subjects of the next chapters of this book.

9.3   FLASH CALCULATIONS

Solving the set of nonlinear equations that describe both the phase equilibrium between a liquid phase and a number of solid solutions in order to obtain the number and amount of coexisting phases and the composition of each phase present from a given overall composition and temperature is called a "solid flash" calculation. There is not any literature on such solid flash calculations. This section will try to alter the best existing algorithms for vapor–liquid flash calculations so that they can deal with polymorphism and a number of solid solutions.

9.3.1   INTRODUCTION

A flash calculation requires the simultaneous solution of the set of nonlinear equations (Equations 9.8 through 9.10). As this set cannot be solved analytically, an iterating procedure has to be used, that involves the following steps:

1. First make an estimate of the number of phases that will be present, the amount of phases and their composition.
2. Calculate the activity coefficients.
3. Make a new estimate, applying Equations 9.4 through 9.9, using the activity coefficients of 2.
4. Repeat steps 2 and 3 until a convergence criterion is met.
5. Perform a stability test to check whether the initial estimate of the number of phases is correct. If not, an extra phase is added with an estimate of its composition.
6. Repeat steps 2 through 4 until convergence is obtained and a stability criterion is met.

Whether quick convergence is obtained in steps 2 and 3 depends on the quality of the initial estimate and on the reliability of the stability test. If the initial estimate is poor, convergence may be slow, to a local, rather than a global minimum or to a trivial solution with two phases having the same composition. A poor stability test will lead to an incorrect number of phases. A robust convergence procedure leads to the solution, even with a poor initial estimate. It follows that for composing a flash algorithm for fats, one needs the following:

1. A procedure to test the phase stability and provide an initial estimate for the phase compositions
2. An iterating procedure to solve the phase equilibrium and mass balance equations

9.3.2 INITIAL ESTIMATES AND STABILITY TESTS

There are no algorithms that directly give an estimate of the total number of phases that coexist. All methods start with assuming a single phase, either liquid or vapor. A simple calculation of the overall Gibbs energy learns which of the two is the most stable. For fats, the equivalent procedure is a comparison between the molar Gibbs energy of the fat in the liquid state and those in the α, β', and β polymorphic forms:

$$g^f = \sum_{i=1}^{n} z_i \left(\mu_i^{0,f} + RT \, \ln\gamma_i^f z_i \right) \qquad (9.11)$$

For convenience the authors set in the remainder of this chapter the chemical potential in the pure liquid reference state arbitrary to 0, so that Equation 9.11 reduces for the liquid state to

$$\frac{g^L}{RT} = \sum_{i=1}^{n} z_i \left(\ln z_i \right) \qquad (9.12)$$

and for the 3 solid states to

$$\frac{g^m}{RT} = \sum_{i=1}^{n} z_i \left(\frac{\Delta H_f^m}{R} \left(\frac{1}{T} - \frac{1}{T_f^m} \right) + \ln \gamma_i^m z_i \right) \qquad (9.13)$$

(with $m = \alpha$, β', or β and neglecting all terms with Δc_p).

The phase that has the lowest molar Gibbs energy is the starting point for the stability test. The stability test checks whether addition of a new phase giving a decrease in the overall Gibbs energy is possible. Two types of stability tests for multicomponent, multiphase systems can be found in the literature: methods using a "splitting component" and methods based upon the tangent plane criterion of Gibbs.

9.3.2.1 Splitting Component Method

An example of this method is that of Gautam and Seider (1979). Shah (1980) uses a nearly similar procedure. Asselineau and Jacq (1989) proposes a simpler procedure where the "splitting component" has to be known beforehand. That is not the case for fats. Gautam and Seider first search all phases to locate the component with the highest activity. That component is named the "splitting component." Secondly, the phase in which the splitting component is found is searched for the component that has the highest activity in a binary mixture with the splitting component, taking concentrations proportional to those in the splitting phase. Next, these two components are distributed over two trial phases by solving Equations 9.8 through 9.10 for this binary system using the well-known two phase flash Equation (Prausnitz, 1986) (Box 9.1), starting with two pure phases.

BOX 9.1   METHOD TO SOLVE A TWO-PHASE FLASH EQUATION

If an initial amount S is assumed for the first trial phase, then Equations 9.8 through 9.10 (the material balance combined with the phase equilibrium equation) give

$$x_i^B = \frac{z_i}{1 + S\left(K_i^{AB} - 1\right)} \tag{9.14}$$

and

$$x_i^A = \frac{z_i K_i^{AB}}{1 + S\left(K_i^{AB} - 1\right)} \tag{9.15}$$

where K is the distribution constant x^A/x^B that follows directly from Equation 9.8.

Subsequently a function $f(S)$ is defined:

$$f(S) = x_1^B + x_2^B - x_1^A - x_2^A = \sum_{i=1}^{2} \frac{z_i\left(1 - K_i^{AB}\right)}{1 + S\left(K_i^{AB} - 1\right)} \tag{9.16}$$

The equilibrium value for the phase split S is determined by calculating the zero of this function by a Newton–Raphson iteration:

$$S_n = S_{n-1} - \frac{f(S)_{n-1}}{\left(df(S)/dS\right)_{n-1}} \tag{9.17}$$

When S is found, the mole fractions are calculated and next the values of the distribution constants K are recalculated, using the new values for x. If the new K-values differ too much from the old ones, the function $f(S)$ is solved again, using the new values of K. If the value of S is not between 0 and 1, the split is considered unsuccessful.

Source: From Prausnitz, J.M., *Molecular Thermodynamics of Fluid Phase Equilibria*, Prentice Hall, New York, 1986.

If this calculation results in a split, then the remaining components are distributed over the two trial phases in order of decreasing binary activity with the splitting component. The distribution coefficients that are needed are obtained using the composition of the trial phases calculated so far for getting the activity coefficients.

$$K_i^{A,B} = \frac{x_i^A}{x_i^B} = \frac{\gamma_i^B}{\gamma_i^A} \tag{9.18}$$

This procedure is repeated, taking the component with the second highest activity as splitting component, etc. The Gibbs energy of all these trail splits is compared and that one with the lowest Gibbs energy is taken as initial estimate for a subsequent iterating procedure. If none of the binary flashes is successful, no trial phases can be formed and the solution is considered stable.

This procedure can be easily extended to deal with polymorphism in the solid phase. The number of trial splits needs to be extended: instead of one binary flash at each trial split, four flashes need to be considered; for example, if a component in a β phase is selected as "splitting" not only a β–β split but also β–β' split, a β–α split, and a β–liquid split must be considered. For a P-phase, N-component system maximally $4P(N-1)$ trial splits must be evaluated.

9.3.2.2 Michelsen's Tangent Plane Criterion Method

The second approach is based on an extension of the well-known tangent plane criterion of Gibbs (Prausnitz, 1986) for phase stability to the multicomponent, multiphase situation. The tangent plane criterion says that if the tangent to the Gibbs free energy curve at the solution at no point lies above the Gibbs free energy curve then the Gibbs free energy is at a global minimum. The mixture is stable and will not show further demixing. In Figure 9.7, the tangent to the Gibbs free energy curve at F lies partially above the curve: F is an unstable mixture and will demix in A and B. The tangent to the curve at G lies at no point above the curve and G represents a stable mixture.

The general formulation of the tangent plane criterion is presented by Michelsen (1982) and the parts of interest for this work will be given next.

The total Gibbs energy of an original P-phase system ($P \geq 1$) apparently in equilibrium ($\mu_i^B = \mu_i^A$) and to be tested for stability is

$$G^{(I)} = \sum_{j=1}^{P} \sum_{i=1}^{n} n_i^j \mu_i^j = \sum_{i=1}^{n} n_i \mu_i^{(I)} \tag{9.19}$$

A different phase split into $L = P + 1$ phases with mole fractions y_i^j and totally N_1 mole in phase has a Gibbs energy of

$$G^{(II)} = \sum_{j=1}^{L} \sum_{i=1}^{n} N_j y_i^j \mu_i^j \tag{9.20}$$

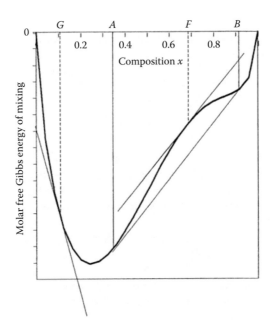

FIGURE 9.7   Gibbs free energy of mixing as a function of composition.

The energy difference between the two situations is

$$G^{(II)} - G^{(I)} = \sum_{j=1}^{L}\sum_{i=1}^{n} N_j y_i^j \mu_i^j - \sum_{i=1}^{n} n_i \mu_i^{(I)}$$

$$= \sum_{j=1}^{L}\left(\sum_{i=1}^{n} y_i^j \left(\mu_i^j - \mu_i^{(I)}\right)\right) = \sum_{j=1}^{L} N_j F_j \qquad (9.21)$$

A phase split into L phases will occur if the Gibbs energy change is negative. That can only be so if at least one of the F_j is negative. Hence the stability criterion becomes: a system is stable if for any extra trial phase L with trial composition y:

$$F_L = \sum_{i=1}^{n} y_i^L \left(\mu_i^L - \mu_i^{(I)}\right) \geq 0 \qquad (9.22)$$

This is the general form of the tangent plane criterion.

F_L is positive for any composition y if the minimum of F_L is positive. The composition of the extra phase L at that minimum of F_L is found by simple differentiation of F_L to the $n-1$ independent mole fractions y_i and setting all derivatives to O:

$$\frac{\partial F_l}{\partial y_i} = \frac{\partial\left(\left(\mu_n^L - \mu_n^{(I)}\right) + \sum_{i=1}^{n-1} y_i\left[\left(\mu_i^L - \mu_i^{(I)}\right) - \left(\mu_n^L - \mu_n^{(I)}\right)\right]\right)}{\partial y_i}$$

$$= \left(\mu_i^L - \mu_i^{(I)}\right) - \left(\mu_n^L - \mu_n^{(I)}\right) = 0 \tag{9.23}$$

which gives $\mu_i - \mu_i^{(I)} = \mu_j - \mu_j^{(I)} = K$, a constant. The minimum of F should be greater than zero for stability:

$$F(\min) = \sum y_i K = K \geq 0 \tag{9.24}$$

It is often simpler to deal with activity coefficients rather than chemical potentials. The activity coefficients of the trial phase L are introduced with the help of Equation 9.5:

$$k = \frac{K}{RT} = \frac{\mu_i^{0,L} - \mu_i^{(I)}}{RT} + \ln \gamma_i^L + \ln y_i^L \tag{9.25}$$

If addition to a trial phase L gives a negative value of k, then the addition of an infinitesimal amount of phase L (infinitesimal, in order not to change composition of the other phases and so $\mu^{(l)}$ will lower the Gibbs energy. The original situation was unstable. The composition of the trial phase L with minimum value for k can be found from an initial estimate of L by iterating

$$\ln\left(y_i^L e^{-k}\right) = \frac{\mu_i^{(I)} - \mu_i^{0,L}}{RT} - \ln \gamma_i^L \tag{9.26}$$

From the calculated values of $y_i^L e^{-k}$, the composition of the new trial phase is obtained from

$$y_i = \frac{y_i^L e^{-k}}{\sum_i y_i^L e^{-k}}$$

If the denominator in this equation is smaller than 1, it corresponds to $k > 0$ and so to stability.

If there is no minimum of K located on the plane between the initial estimate of the trial phase and one of the existing phases, the iteration procedure converges to a trivial solution, in which the trial phase has the same composition as one of the existing phases. Therefore a number of initial estimates are necessary, to ensure that the minimum of K is found if it exists.

Michelsen recommends taking a pure phase of each component as initial estimate plus a trial phase with a composition that is the average of the composition of all phases present. He has shown that the iterations need not be continued to convergence: for each initial estimate four iterations suffice. The trial phase with the most negative value of K is used as the initial estimate for the new phase, which will form after the split. If no negative values of K are found for all initial estimates, then only the one with the least positive value of K is converged. If this trial phase still yields positive values of K, then the original situation is considered stable.

Sometimes the authors found in highly nonideal systems after four iterations a phase with a negative K, indicating instability. But upon continuing iterations to convergence, it occurred that K turned greater than O, indicating stability. Stopping after four iterations would have led to a false conclusion. Therefore always converge the trial phase that has the smallest K after four iterations.

Michelsen's stability test can be applied to solid–liquid equilibria in TAGs simply by increasing the number of initial estimates for the trial phases: pure β, β', and α phases of each component plus a β, a β', an α, and a liquid phase (if not already present) with a composition being the average of all phases already present. In an N-component, P-phase system maximally $3N + 4$ trial phases must be considered, about a factor P less than with the splitting component approach.

Michelsen's stability test has the advantage of being derived from the proven thermodynamic tangent plane criterion and requiring only a small number of trial phases, while the method of Gautam and Seider and like methods have no fundamental guarantee that phase instability is always detected. However, in Michelsen's procedure the true minimum of K may be overlooked because of the selection of starting values for the trial phases. The algorithm of Michelsen and a number of algorithms similar to that of Gautam and Seider have been tested on several different vapor liquid and vapor liquid–liquid–liquid equilibria (Swank and Mullins, 1986). Generally Michelsen's algorithm proved to be the most reliable. In this chapter, the performance of both methods with solid–liquid equilibria in fats will be tested.

9.3.3   ITERATING PROCEDURES

Once a good initial estimate of the number of phases, their polymorphic form, and their composition is obtained, the equilibrium composition and phase quantities can be calculated by solving Equations 9.8 through 9.10. The iterating procedures to solve these phase equilibria and mass balance equations fall into two categories:

1. The direct substitution methods
2. Methods involving the minimization of the Gibbs free energy

9.3.3.1   Direct Substitution

This is the multiphase multicomponent analogue of the two- phase two-component flash equation, described in Box 9.1 (Equations 9.15 and 9.16).

The function $f(S)$ is redefined for the P-phase analogue as a set of $P - 1$ equations:

$$\sum_{i=1}^{n} \frac{z_i \left(K_i^{mP} - 1 \right)}{1 + \sum_{f=1}^{P-1} \Phi_f \left(K_i^{fP} - 1 \right)} = 0 \quad (m = 1, 2, \ldots, P - 1) \tag{9.27}$$

where

m ranges from 1 to $P - 1$

K^{mP} is the distribution coefficient of a component over the phases m and P

The solution of these sets of equations to obtain the $P - 1$ phase fractions is carried out by Newton–Raphson iteration. Then using these phase fractions, new compositions are obtained. Next, new values for the distribution constants are calculated with the appropriate thermodynamic models. If these new values do not agree with the previous ones within a certain tolerance, the set equations is solved again for the $P - 1$ phase fractions, using the new values of K.

Direct substitution is a very fast and reliable method for phase equilibria where the values of the activity coefficients do not depend strongly on the phase composition. It therefore seems most suited for calculating a phase equilibria, where all activity coefficients are 1 (see Section 15.5) and the values of K only depend on temperature, so that only one iteration will be sufficient.

When the activity coefficients strongly depend on the phase composition and/or when phase envelopes are very narrow, like it is often the case in the β- and β′-modifications, the direct substitution method converges very slowly. Examples of the need for several hundreds of iterations are known (Crowe and Nishio, 1975; Michelsen, 1982). Various acceleration procedures exist, the general dominant eigenvalue method (GDEM) of Crowe and Nishio (1975) being recommended by Michelsen (1982) and by Swank and Mullins (1986). Even with these methods, convergence is not always obtained in the case of vapor–liquid equilibria, and other methods had to be used (Michelsen, 1982).

The authors found for the systems in this work that three GDEM acceleration steps are sufficient, each after five iterations and using more than two eigenvalues, even when more than three phases were present, offered no significant improvement in acceleration. This is in agreement with the findings of Michelsen (1982) and Swank and Mullins (1986).

9.3.3.2 Gibbs Free Energy Minimization

The other method to solve the phase equilibrium and material balance equations is to minimize the Gibbs free energy. When the Gibbs energy is at its minimum the requirement for equilibrium, Equation 9.4 is satisfied. The problem can hence be reformulated for a N-component, P-phase system as

Find the minimum of

$$G_{\vec{n}} = \sum_{j=1}^{P} \sum_{i=1}^{N} n_i^j \mu_i^j = \vec{n} \cdot \vec{\mu} \tag{9.28}$$

subject to the constraints of the mass balance and the requirement of all $n_i^j \geq 0$.

The two vectors are defined as

$$
\vec{n} = \begin{pmatrix} n_1^1 \\ n_2^1 \\ \cdot \\ \cdot \\ \cdot \\ n_N^P \end{pmatrix}, \quad \vec{\mu} = \begin{pmatrix} \mu_1^1 \\ \mu_2^1 \\ \cdot \\ \cdot \\ \cdot \\ \mu_N^P \end{pmatrix}
\tag{9.29}
$$

The Gibbs free energy can be minimized by several variations on Newton's method (Box 9.2).

BOX 9.2   NEWTON'S METHOD FOR FINDING A MINIMUM IN THE GIBBS ENERGY FUNCTION

The Gibbs free energy close to an initial estimate of the solution is given by a Taylor expansion:

$$
G_{\vec{n}+\overrightarrow{\Delta n}} = G_{\vec{n}} + \overrightarrow{\nabla G_{\vec{n}}} \cdot \overrightarrow{\Delta n} + \frac{1}{2} \overrightarrow{\Delta n}^T \cdot \nabla^2 G_{\vec{n}} \cdot \overrightarrow{\Delta n}
$$

$$
= G_{\vec{n}} + \sum_{j=1}^{P} \sum_{i=1}^{N} \frac{\partial G_{\vec{n}}}{\partial n_i^j} \Delta n_i^j + \frac{1}{2} \sum_{j=1}^{P} \sum_{i=1}^{N} \sum_{k=1}^{P} \sum_{h=1}^{N} \frac{\partial^2 G_{\vec{n}}}{\partial n_i^j \partial n_h^k} \Delta n_i^j \Delta n_h^k
\tag{9.31}
$$

Since the first term of this equation is constant, the minimum Gibbs energy is obtained when the two right-hand side terms

$$
\Phi_{\overrightarrow{\Delta n}} = \overrightarrow{\nabla G_{\vec{n}}} \cdot \overrightarrow{\Delta n} + \frac{1}{2} \overrightarrow{\Delta n}^T \cdot \nabla^2 G_{\vec{n}} \cdot \overrightarrow{\Delta n} \text{ have a minimum.}
$$

The minimum of this quadratic function Φ is easily calculated by setting its derivative to zero:

$$
\nabla^2 G_{\vec{n}} \cdot \overrightarrow{\Delta n} = -\overrightarrow{\nabla G_{\vec{n}}}
\tag{9.32}
$$

This set of linear equations in Δn_i^j is readily solved and the resulting vector $\overrightarrow{\Delta n}$ is called the Newton direction. $\vec{n} + \overrightarrow{\Delta n}$ is an improved estimate of the composition at the minimum.

The next step would be to calculate new values for $\overrightarrow{\nabla G}$ and $\nabla^2 G$ at this improved estimate, determine the Newton direction from this point, and repeat this until convergence is obtained.

Two criteria are added to the standard Newton method:

1. The solution should be within the constraints imposed by the mass balance and not result in negative concentrations.
2. It should be verified that the extreme obtained is actually a minimum and not a maximum or a saddle point.

The mass balance Equation 9.9 can be equivalently formulated for each component i as follows:

The elements i of the vector $\overrightarrow{\Delta n}$ should obey

$$\sum_{j=1}^{P} \Delta n_i^j = 0 \tag{9.30}$$

This set of N linear constraints can be used to eliminate N variables from Equation 9.32, so that indeed the mass balance is satisfied. Box 9.3 describes a general way of doing this.

BOX 9.3   MODIFICATION OF NEWTON'S METHOD OF BOX 9.2 TO SATISFY MASS BALANCE CONSTRAINTS

The set of linear mass balance constraints can be alternatively formulated as follows:

$$A \cdot \overrightarrow{\Delta n} = \vec{0}, \quad A = (I_1, I_2, \ldots, I_P) \tag{9.33}$$

The $(P*N) \times N$ matrix A consists of P identity matrices with dimension N.

If a second matrix Z of which its columns form a basis for the set of all vectors orthogonal to the rows of A is defined, then any vector $\overrightarrow{\Delta n} = Z \cdot \overrightarrow{\Delta n}^z$ will satisfy the mass balance constraints: $A \cdot \overrightarrow{\Delta n} = A \cdot Z \cdot \overrightarrow{\Delta n}^z = 0 \cdot \Delta n^z = 0$.

Such a matrix Z of which the columns are orthogonal to the rows of A is

$$Z = \begin{pmatrix} I_1, I_2, \ldots, I_{p-1} \\ -I\mu \end{pmatrix} \tag{9.34}$$

in which I is an identity matrix of dimension N and I_μ is an identity matrix of dimension N and I_μ is an identity matrix of dimension $(N-1)*P$.

If Δn in Equation 9.31 is substituted by Δn^z, then the solution to the constrained problem of Equation 9.32 is given by

$$\nabla^2 G_{\bar{n}}^Z \cdot \overrightarrow{\Delta n}^Z = -\overrightarrow{\nabla G_{\bar{n}}}^Z, \quad \nabla^2 G^Z = Z^T \cdot \nabla^2 G \cdot Z$$

$$\overrightarrow{\nabla G}^Z = Z^T \cdot \overrightarrow{\nabla G} \tag{9.35}$$

The solution $\overrightarrow{\Delta n}^z$ is transformed to the original $\overrightarrow{\Delta n}$ by multiplication with Z.

Non-negative values for the mole numbers can be assured by introduction of a so-called step size λ. The new estimate for the composition is set to: $\vec{n} + \lambda \vec{\Delta n}$. The step size λ is taken as the largest possible value not exceeding 1 that

- Still results in a new estimate of the solution in which all mole numbers n_i^j are greater than a small positive number δ
- Gives a decrease in Gibbs free energy

The value that is chosen for δ depends on the precision of the computer used. In this way, Δn^z is guaranteed that no negative mole numbers will occur and that the calculations remain numerically stable. λ should differ from unity only in the first few Newton iterations.

A Newton direction that points to a minimum in the Gibbs energy is only obtained when the matrix of the second derivative of the Gibbs energy to all mole numbers, $\nabla^2 G_{ji}$, the so-called Hessian, is positive definite, that is, has only positive eigenvalues. In the case of activity coefficients only slightly different from 1 this is so. However, in case of solid β and β' phases, it is not necessarily true. If the Hessian is not positive definite, the iterations may converge to a maximum in the Gibbs energy or even not converge at all.

Gautam and Seider (1979) solve this problem by ignoring the second derivative of the excess Gibbs energy (ignoring the compositional derivatives of the activity coefficients). This is called the "Rand method." The Hessian is in that case always positive definite. Because of the inaccurate Hessian used, the Rand method will converge slower than when the full-second derivative is used.

Michelsen (1982) recommends the method of Murray et al. (1981). In cases where the Hessian is not definite, an approximate Hessian is calculated that is positive definite and looks as much as possible like the original Hessian (Box 9.4).

BOX 9.4 MURRAY'S METHOD TO FORCE A POSITIVE DEFINITE SECOND DERIVATIVE OF THE GIBBS ENERGY TO ALL MOLE NUMBERS

The Hessian is decomposed by a so-called Cholesky decomposition into a lower triangular matrix L and a diagonal matrix D:

$$\nabla^2 G = L \cdot D \cdot L^T \tag{9.36}$$

The jth column of the L matrix is defined from the previous columns by the following equations:

$$d_j = \nabla^2 G_{jj} - \sum_{s=1}^{j-1} d_s l_{js}^2 \tag{9.37}$$

BOX 9.4 (continued)

$$l_{ij} = \frac{1}{d_j}\left(\nabla^2 G_{ij} - \sum_{s=1}^{j-1} d_s l_{js} l_{is}\right) \quad (9.38)$$

If the Hessian is positive definite, all the elements of the diagonal of D are positive. In the other case, the decomposition results in a matrix D with some negative elements and a matrix L with sometimes extremely large values. Murray's method modifies the elements of the L and D matrices during decomposition, and if during decomposition an element d_j becomes smaller than a small positive number Δ, d_j is replaced by Δ. The value of delta is determined by computer precision. If, by using this value of d_j for the calculation of the next column of L, one of the values of the elements of L next exceeds a certain maximum, the value of d_j is increased such that the elements of L will be below that maximum. The maximum of the elements of the column of L to be calculated is given by

$$l_{ij}^2 \le \frac{1}{d_j} \cdot \max\left(\gamma, \frac{\xi}{\sqrt{N^2-1}}, \in_M\right) \quad (9.39)$$

where
γ is the largest diagonal element of the Hessian
$\in$ is the largest off-diagonal element of the Hessian
$\in_M$ is the precision of the computer used

Using the modified forms of the L and D matrices, a modified Hessian is calculated. This modified Hessian is positive definite. The decomposition and recombination of the original Hessian is numerically stable. The resulting modified Hessian is a very close positive definite approximation of the original Hessian.

Instead of using the original Hessian, the modified Hessian is subsequently used when solving Equation 9.32; then the resulting Newton direction always points to smaller values of the Gibbs energy.

9.3.3.3 Removal of Phases

Occasionally, it is necessary to remove one of the phases while solving the flash equations. This can occur in cases where initially a liquid phase is most stable, next a phase split occurs into liquid and β, thereafter a second split into liquid, β and β', and during subsequent iterations, the amount of liquid starts to approach zero. Calculations become increasingly inaccurate in that case. Therefore, if the amount of any of the phases present drops below 5% during iterations, it is examined whether two phases can "coalesce" such that a reduction in Gibbs energy is obtained. In that case, iterations are continued with a reduced number of phases.

9.3.4 COMPARING METHODS

9.3.4.1 Criteria

The procedures that were outlined and modified in the previous section have to be compared on their performance with multicomponent multiphase solid–liquid flashes. Unfortunately, experimental data do not exist in which the number of coexisting solid phases and their composition are known. It is probably impossible to obtain such data for TAG systems. As an alternative the following procedure is adopted.

Both stability tests are applied and the initial estimates are converged using all three convergence methods. Criteria for performance of the stability tests are as follows:

1. Indication of instability where this is not the case: the test indicates instability and results in a phase split while during subsequent convergence this phase is removed again.
2. Failure to predict instability: one test indicates instability, and the other does not, while the resulting converged phase split indeed has a reduced Gibbs free energy.
3. Number of iterations needed to converge the initial estimate.

Performance criteria for the convergence methods are as follows:

1. Convergence
2. The total computing time to reach convergence

As test cases the following systems were considered:

1. Fully hardened palm oil (PO58), a six-component system containing all TAGS that can be formed from palmitic and stearic acids, at temperatures between 40°C and 70°C.
2. A ternary system of SSS, PPP, and SES, where two components that are completely immiscible in the solid phase are combined with a component that is partially miscible with both. Temperature is varied between 50°C and 75°C.

As description of the excess Gibbs energy both the 2-suffix and the 3-suffix Margules equation were used (see Section 9.7.7.1). The necessary binary interaction coefficients were obtained by fitting the data of de Bruijne (Knoester et al., 1972; Section 9.7.2.4). Calculations were performed three times, first only allowing the formation of the α-modification, secondly only allowing the β'-modification, and finally allowing all modifications.

Swank and Mullins (1986) performed a similar exercise with a number of vapor–liquid phase equilibria. They judged Michelsen's stability test as the most reliable and direct substitution as the quickest convergence method. However, not all problems could be solved by direct substitution.

A combination of Michelsen's stability test plus a Murray minimization always lead to the correct solution.

9.3.4.2   Test Results

A computer program was written in Turbo Pascal 5.0 for MS-DOS PCs that implements the two initial estimate and stability test procedures and the three methods to solve the flash equations.

The stability test of Michelsen clearly performs better than the algorithm of Gautam and Seider. It is faster and more reliable; especially in the ternary test system the splitting component approach indicates instability, where this is not the case.

Both tests were perfectly able to deal with polymorphism in the solid phase. The coexistence of a number of stable β and β' phases in PO58 was obtained without problems.

The three convergence algorithms are compared in Table 9.3.

"No conv." in the table means that convergence to a gradient norm of 10^{-12} was not obtained within 50 iterations. Except for the α phase and cases where no convergence was obtained, direct substitution converged within 6–25 iterations, while the Murray method needed 2–6 iterations.

For α phase calculations (in Section 15.5 is shown that miscibility in the α-modification is ideal, so only one solid phase plus a liquid phase are present), direct substitution is especially for larger systems ($N = 6$–50) the quickest method that will converge safely.

The Rand method has, as expected, problems with converging in highly nonideal systems and in systems where the initial estimate contains concentrations close to zero. Performance is inferior to both other methods.

TABLE 9.3

Average Time per Flash (of 60 Flashes) for the Test Systems (on a Compaq 386/25 PC under MS DOS)

System	Number of Phases	Direct Substitution	Rand	Murray (s)
PO58 (α form)	2	0.1 s	0.3 s	0.3
PO58 (β', 2 suffix)	3	0.5 s	no conv.	0.8
PO58 (β', 3 suffix)	3	no conv.	no conv.	1.4
PO58 ($\beta' + \beta$, 2 suffix)	4	no conv.	no conv.	2.0
PO58 ($\beta' + \beta$, 3 suffix)	5	no conv.	no conv.	2.7
(SSS/PPP/SES) α	2	0.08 s	0.12 s	0.12
(SSS/PPP/SES) β'	2	0.16 s	0.3 s	0.12
(SSS/PPP/SES) β	3	0.16 s	0.2 s	0.24

The Murray method always resulted in a reasonably fast and safe convergence and is therefore the best method to be used for the β'- and β-modifications, where demixing in the solid phase can occur. It is slower than the direct substitution method but is more reliable.

These conclusions are in line with those of Michelsen: for vapor–liquid flashes with one liquid phase, he suggests the use of direct substitution method, while for VLL and VLLL flashes a Murray minimization works out better.

For flash calculations in fats, it is therefore recommended to use Michelsen's stability test for initial estimates and Murray minimization of the Gibbs free energy for obtaining the final solution. For the α-modification, the direct substitution method should be used.

9.3.5 Calculation of Differential Scanning Calorimetry Curves

DSC is one of the most frequently used techniques to study the solid phase behavior of TAGs.

With DSC, the apparent heat capacity of a sample is measured as a function of the temperature. It is often used as a tool to characterize a crystallized fat (Juriaanse, 1985; Christophersen, 1986; Busfield et al., 1990; Wesdorp and Struik, 1990) and slowly replaces the more traditional methods for quality control of cocoa butter equivalents and replacers (Smith, 1988). Comparison of experimental and calculated DSC curves enables a better interpretation of those curves.

Make the simplifying approximation that the heat capacity of solid and liquid fat is equal, then the equilibrium DSC curve at infinitive slow scanning rate of any fat mixture can be calculated from

$$c_p^{\text{apparent}} = c_p + \left(\frac{\partial H}{\partial T} \right)_n \tag{9.40}$$

$$H = H^E + \sum_{j=1}^{P} \sum_{i=1}^{N} n_i^j H_i^{0,j}$$

For convenience, set H (liquid) to zero for each component, which implies that H^0 becomes equal to the heat of fusion of the pure component in that modification. Another assumption is to neglect the excess entropy, which will be most likely comparatively small. The DSC curve can then be calculated from

$$c_p^{\text{apparent}} = c_p + \frac{\partial G^E}{\partial T} + \sum_{j=1}^{P} \sum_{i=1}^{N} H_i^j \frac{\partial n_j^i}{\partial T} \tag{9.41}$$

The two partial derivatives in this formula are easily obtained by numerical differentiation, which requires two flash calculations for each point on a DSC curve. Calculations are speeded up considerably when the outcome of the flash calculation for one point is used as initial estimate for calculation of the next point.

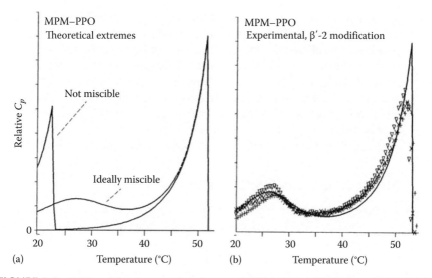

FIGURE 9.8 DSC melting curves of a mixture of 25% MPM, 25% PPO, and 50% OOO: (a) theoretical extremes: ideal and no solid state miscibility; (b) experimental (points) and calculated (line) curves.

Norton et al. (1985) constructs a DSC curve by using the Hildebrand equation for calculating the partial derivatives. The applicability of this approach is rather limited, because the Hildebrand equation describes only the phase behavior of pure (= completely demixed) solid phases. Solid phase miscibility has a large influence on the shape of a DSC curve as is shown in Figure 9.8 for the ternary system 25 MPM/25 PPO/50 OOO in the β′ form. This system cannot be described by the Hildebrand equation.

As will be shown in Section 9.7.5.4, the ternary of Figure 9.8 can be described by the 2-suffix Margules equation in which the binary interaction coefficient $A_{MPM-PPO}$ is 1.8 + 0.3. Although no demixing of the solid phase takes place, still two peaks occur in the DSC curve. In the literature, the appearance of two peaks in DSC thermograms of a fat is often used as indication for the presence of two solid phases (Murray et al., 1981; Timms, 1984; Yap et al., 1989; Wesdorp and Struik, 1990). In fact, it only indicates the presence of two groups of TAGs in the fat that have a clear difference in melting point, but may have cocrystallized.

Figure 9.8 also shows that DSC, together with the calculation procedures of this report, forms an elegant method for determination of binary interaction parameters and verification of excess Gibbs energy models.

9.3.6 CONCLUSION

A flash calculation consists of two parts: a stability test that gives an initial estimate of the phase compositions when it detects that a phase split can occur, and a convergence method that determines the phase compositions and phase quantities at equilibrium starting from the initial estimate. As no procedures for solid–liquid flashes were available, several existing methods for vapor–liquid flashes were adapted and tested for their performance with solid–liquid flashes in fats.

1. The best stability test is the so-called stability test of Michelsen, that, with some small changes, can perfectly handle solid fats and polymorphism.
2. Only in the case of the α-modification (ideal solid miscibility) the normal direct substitution method for solving the flash equation gives a quick, reliable result.
3. For the β'- and β-modifications (highly nonideal solid phases), direct substitution is unreliable. The flash problem can better be solved with a Gibbs free energy minimization using Murray's method (a modified Newton method). This method is somewhat slower but very reliable.
4. Flash calculations can be applied for simulation of DSC curves of fat blends.

9.4 PURE COMPONENT PROPERTIES

For a thermodynamic description of the solid–liquid phase equilibrium in a fat, the enthalpy of fusion and melting point of each modification of each TAG in the mixture are needed. These pure component properties can impossibly be measured for all TAGs. Therefore correlations between structural characteristics and the properties must be developed. Existing correlations for heat of fusion and melting point are only reliable for mono-acid TAGs. Therefore, after having appended literature data with a set of experimental data, correlations for the heats of fusion and the melting points of TAGs in the α-, β'-, and β-modifications are developed.

9.4.1 Literature Data and Correlations

9.4.1.1 Correlating Enthalpy of Fusion and Melting Points of Lipids

It is often assumed (Bailey, 1950; Broadhurst, 1962; Flory and Vrij, 1963; Wurflinger, 1972; Billmeyer, 1975; Zacharis, 1977; Timms, 1978; Dollhopf et al., 1981; de Bruijne and Eedenburg, 1983; Perron, 1984; Bommel, 1986; Larsson, 1986) that the enthalpy and entropy of fusion of lipids (alkanes, fatty acids, methyl esters, TAGs) can be seen as the sum of a contribution of the hydrocarbon chains that depends linearly on the chain length and a contribution of the end and head groups that is independent of chain length (Equations 9.42 and 9.43, and Figure 9.9).

$$\Delta H_f = hn + h_0 \qquad (9.42)$$

$$\Delta S_f = sn + s_0 \qquad (9.43)$$

where n is the carbon number of the component.

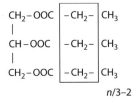

$$n/3-2$$

FIGURE 9.9 Schematic representation of the head group, the hydrocarbon chain and the end group of a TAG.

TABLE 9.4

Values for the Incremental Hydrocarbon Chain Contribution to the Enthalpy of Fusion (h) and the Entropy of Fusion (s)

Compound	Polymorph	h (kJ/mol CH$_2$)	s (J/K, mol CH$_2$)
n-Alkanes	α	2.64[43], 2.5[44]	6.4[43], 6.0[44]
TAGs	α	2.5[45], 3.4[52], 3.6[53]	6.1[45]
n-Alkanes	β'	3.2[43], 3.8[44], 3.98[46], 3.86[47]	7.7[43], 9.7[44], 9.52[46]
Methylesters	β'	3.78[48]	9.2[48]
TAGs	β'	3.87[45], 3.25[49]	9.8[45]
n-Alkanes	β	4.11[43], 4.12[50]	9.9[43], 9.89[50]
Methylesters	β	4.28[48]	10.3[48]
Fatty acids	β	4.2[2], 4.3[49]	11.1[49]
Diglycerides	β	4.28[49]	10.5[49]
Monoglycerides	β	4.28[49]	10.5[49]
TAGs	β	4.28[49], 4.20[45]	10.5[45], 10.6[49]

Note: Numbers in superscript are references.

In this view, the incremental hydrocarbon chain contributions h and s do not depend on the nature of the compound but only on the way the hydrocarbon chains are packed: h and s are universal lipid constants that only depend on the polymorphic form in which the lipid has crystallized. This is experimentally confirmed (Table 9.4). Only the values for h of Bailey (1950) and de Bruijne and Eedenburg (1983) for the α-modification of TAGs deviate. Perron (1984) showed that these numbers were derived from very unreliable data.

The end group contributions h_0 and s_0 are specific to each class of lipids. Correlating the properties of alkanes and methylesters results in different values for h_0 and s_0 for these two groups of components.

The melting point T is simply given by the ratio of the enthalpy and entropy of fusion:

$$T_f = \frac{\Delta H_f}{\Delta S_f} = \frac{hn + h_0}{sn + s_0} \tag{9.44}$$

Expanding the denominator in this equation into a power series of $1/n$ gives

$$T_f = \frac{h}{s}\left(1 + \left(\frac{h_0}{h} - \frac{s_0}{s}\right)\frac{1}{n} - \frac{s_0}{s}\left(\frac{h_0}{h} - \frac{s_0}{s}\right)\frac{1}{n^2} + \cdots\right) \tag{9.45}$$

This equation can be cut off after the second or the third term:

$$T_f = T_\infty\left(1 + \frac{A}{n} - \frac{AB}{n^2}\right) \tag{9.46}$$

$$T_f = T_\infty \left(1 + \frac{A}{n}\right) \tag{9.47}$$

The constants T_∞, A, and B are given by

$$T_\infty = \frac{h}{s}, \quad A = \frac{h_0}{h} - \frac{s_0}{s}, \quad B = \frac{s_0}{s} \tag{9.48}$$

This implies that if the melting points of a class of lipids have been correlated, only one data point for the enthalpy of fusion is in principle sufficient to obtain a correlation for the enthalpy of fusion of the complete class of lipids.

Equation 9.44 has successfully been used by Zacharis (1977) to correlate the melting points of a large number of lipids: n-alkanes, methylesters, ethylesters, fatty acids, mono-acids, mono-, di-, and tria-cylglycerols, phosphoglycerides, and dicarboxylic acids. He has found values for T_∞ that vary between 390 and 410 K.

9.4.1.2 Data and Correlations for TAGs

The main difficulty in the development of correlations for the thermal properties of TAGs is that TAGs do not belong to one class of lipids in the sense of Section 15.4.1.1. Not all TAGs share the same end group. This becomes clear when a saturated TAG is formally denoted by the lengths p, q, and r of its three fatty acid chains, that is, p, q, and r. P is always the shortest of the fatty acid chains on the 1- and 3-position of the glyceryl group. For example, myristoyl-stearoyl-palmitoyl-glycerol (MSP) is denoted as 14.18.16. The chain length differences x and y are defined as follows:

$$x = q - p$$

$$y = r - p$$

Only TAGs having the same value of x and y belong to a family of TAGs that share the same end group. For TAGs that differ in x and y, the value of the head and end group contributions h_0 and s_0 from Equations 9.39 and 9.40 are different.

The dependency of the "head and end group" contributions h_0, s_0, A, and B from x and y and from the presence of unsaturation in the hydrocarbon chains must be accounted for in the development of general correlations for all TAGs. In this work, this will be accomplished by the addition of two extra terms to Equations 9.39 and 9.40:

$$\Delta H_f = hn + h_0 + h_{xy} f_{xy} + h_{unsat} f_{unsat} \tag{9.49}$$

$$\Delta S_f = sn + s_0 + s_{xy} f_{xy} + s_{unsat} f_{unsat} \tag{9.50}$$

The functions f_{xy} and f_{unsat} should account for the effects on the thermodynamic properties of differences in chain length and degree of unsaturation. Their functional form cannot a priori be established.

9.4.1.2.1   Literature

The literature on the enthalpy of fusion of TAGs was reviewed up to 1975 by Timms (1978). Timms provided the enthalpy of fusion of 42 triglycerides, 19 saturated (16 beta and 3 beta'), and 23 unsaturated TAGs. The data were correlated using Equation 9.39:

$$\Delta H_f^\beta = 4.28n - 32.6 \text{ kJ/mol} \tag{9.51}$$

$$\Delta H_f^{\beta'} = 0.76\Delta H_f^\beta \tag{9.52}$$

For mixed acid saturated TAGs the enthalpy of fusion should be reduced by 18.3 kJ/mol. The enthalpy of fusion of unsaturated TAGs was described by setting an effective carbon number for each unsaturated fatty acid (O [oleic acid] = 10.4, E [elaidic acid] = 13.9, l [linoleic acid] = 8.9). The root mean square error between experimental and predicted data is 4.3 kJ/mol. The values for *h* agree with those reported for other compounds (Table 9.4).

Since 1975, various data on the enthalpy of fusion of the stable beta modification of saturated fatty acids have been reported (Hagemann, 1975, 1988; Zacharis, 1975; Lovegren, 1976; Gray and Lovegren, 1978; Ollivon and Perron, 1979, 1982; Hagemann and Rothfuss, 1983; Norton, 1984; Perron, 1984; Gibon, 1986; Garti et al., 1988; Sato et al., 1989). The majority of these data concern mono-acid TAGs. There is only one observation for the α-enthalpy of fusion of a mixed acid saturated TAG (PSP, Gray and Lovegren, 1978). The data for the β'-modification of saturated TAGs mainly concern PSP and LML. Some data on the mono-acid saturated TAGs are available, but they show large deviations (Ollivon and Perron, 1982; Hagemann and Rothfuss, 1983; Perron, 1984), which are ascribed to the existence of a second, less stable β'-form. Only the data for the most stable β'-form were taken. The number of data for the lesser stable forms was too small and the data showed no consistency.

The data of Hagemann (1975) for AAA to 30.30.30 for the β' form and for BBB to 26.26.26 for the β'-form look very unreliable: the enthalpy of fusion levels off and starts to decrease with increasing carbon number, which is in contradiction to the generally observed trend in lipids. The same holds for the data of Perron (Ollivon and Perron, 1982) for the β'-form of AAA and BBB, which is also admitted by Perron in a later article (Perron, 1984).

The data reported by Garti, Schlichter, and Sarig (Gray and Lovegren, 1978) for all modifications differ completely from all other values reported. Therefore they are disregarded in this report.

Perron (Bommel, 1986) has carried out the most extensive work on development of correlations for the enthalpy of fusion and melting points of TAGs since Timms. His correlations for the enthalpy of fusion of mono-acid TAGs are as follows:

$$\Delta H_f^\alpha = 2.5n - 27.5 \text{ kJ/mol} \tag{9.53}$$

$$\Delta H_f^{\beta'} = 3.87n - 19.2 \text{ kJ/mol} \tag{9.54}$$

$$\Delta H_f^\beta = 4.20n - 29.9 \text{ kJ/mol} \tag{9.55}$$

TABLE 9.5

Number of TAGs for Which Melting Points and Enthalpies of Fusion Are Available in the Literature

	ΔH_f			T_f		
	α	β'	β	α	β'	β
Saturated	10	10	30	65	54	87
Unsaturated	8	11	16	50	47	49

Perron modeled the enthalpy of fusion of an unsaturated TAG as that of the corresponding saturated TAG minus a contribution of the double bonds:

$$\Delta H_f(\text{unsat}) = \Delta H_f(\text{sat}) - 115(1 - e^{-0.706\Delta}) \qquad (9.56)$$

where Δ = number of double bonds in the TAG.

Perron's values for h agree very well with those for other compounds given in Table 9.4.

Perron did not give relations for the enthalpy of fusion of mixed acid TAGs. However, he correlated the α- and β-melting points of 19 different TAG families using Equation 9.44. It resulted, as expected, in large fluctuations of the constant A.

The total collection of data from the literature and from the database of Unilever Research Vlaardingen resulted in 152 values for the enthalpy of fusion and 944 melting points. Many of these data are multiple measurements for the same TAG. The polymorphic form of the solid phase is not always given.

The number of different TAGs for which data are available is much less (Table 9.5).

9.4.1.2.2 Conclusion

There is a disappointing lack of data for the heat of fusion of TAGs. The heat of fusion data that are available for the unstable modifications are mainly for mono-acid TAGs. Consequently, the correlations that have been developed are only valid for mono-acid TAGs. An experimental program is required before these correlations can be extended to mixed acid TAGs.

A considerable amount of melting points that is reasonably well spread over the modifications and TAG families is available. Correlations have been developed for a number of individual TAG families, but a general correlation is not available. This correlation will be developed in Section 9.4.3.

9.4.2 EXPERIMENTAL WORK

The enthalpy of fusion and melting points of 42 saturated mixed acid TAGs and 9 unsaturated TAGs were determined for all three polymorphic forms. The pure TAGs were taken from the stock of reference materials of Unilever Research Vlaardingen. Their purity exceeds 95%.

The experimental program was carried out on a Perkin-Elmer differential scanning calorimeter (DSC-7). This DSC-7 is equipped with a TAG 7/3 datalogger and a liquid nitrogen cooling accessory. Control of the apparatus is achieved by means of a PE computer. A more extensive description of DSC is given in Section 15.7.

The thermal data for each modification were obtained by the following procedures:

The data for the β-modification were obtained from the melting curve of the samples as they were delivered after several months to years of storage at room temperature.

The α-modification was obtained by rapidly quenching the TAGs from 10°C above the β-melting point to at least 20°C below the α-melting point. The enthalpy of fusion was obtained from the cooling curves. The melting curves were normally not suited for determination of α-thermal data, as the α-melting peak interfered with the β′- and β-crystallization peaks.

The β′-modification was most difficult to prepare as a pure substance; interference of the β′-melting peak with the α–β′ or the β′–β recrystallization peaks was often obtained. Although the authors did not always succeed in preparing pure β′, in general one of the two following methods gave good results:

1. The TAG was crystallized into the α-modification by quenching. Next it was heated till α-melting set in and stabilized at that temperature for, depending on the TAG used, 1 min to 1 h. The melting curve was taken.
2. If a β′–β transition interfered in procedure A, the TAG was melted to 10°C above the β-melting point, rapidly cooled to 1°C or 2°C above the α-melting point and stabilized for 30 min to 1 h. The melting curve was taken.

The results are reported in Appendix 9.A, marked by an asterisk (*).

9.4.3 Development of the Correlation

In the development of the correlations, data for TAGs containing acetic acid, butyric acid, and hexanoic acid are not used. The chain length of these fatty acids is so short, that they cannot be looked upon as long chain hydrocarbons.

9.4.3.1 Saturated TAGs

9.4.3.1.1 Melting Enthalpy

If the data for mono-acid TAGs are regressed against the carbon number n (Equation 9.39), leaving out data that have a residual more than two times the root mean square error (RMSE) of the regression through all mono-acid data, we obtain

$$\Delta H_f^\alpha = 2.4n - 17.6 \text{ kJ/mol}, \quad RMSE = 4.6, \quad r^2 = 0.98 \tag{9.57}$$

$$\Delta H_f^{\beta'} = 3.95n - 59.2 \text{ kJ/mol}, \quad RMSE = 4.7, \quad r^2 = 0.97 \tag{9.58}$$

$$\Delta H_f^\beta = 4.13n - 27.6 \text{ kJ/mol}, \quad RMSE = 6.1, \quad r^2 = 0.97 \tag{9.59}$$

The RMSE is in the order of magnitude of the experimental error (5–8 kJ/mol) and the values of the constants h agree with those of Perron (1984) and those for other lipids, given in Table 9.4.

If all data are used for a regression of ΔH_f against the carbon number, the values of the incremental chain contribution h do not change significantly, but the RMSE goes up to 20 and the correlation coefficient decreases to 0.7. A RMSE of 20 is too large for a reliable correlation and a correction term for chain length differences must be introduced.

It cannot be assumed that the end group contribution and the variables x and y that define a TAG family are neatly correlated.

Neither is there any function f_{xy} that is self-evident for use. When the residual error of data predicted by the mono-acid relation with the experimental data is plotted against x and y, it becomes clear that the end group contribution and (x, y) are correlated. The enthalpy of fusion drops more or less quadratically when the difference in chain lengths, represented by x and y, increases. x and y behave similarly and their effect seems additive. From the data for the β-modification, it can be seen that the effect of x and y levels off at values of x or y of 6 and more. This is probably associated with the transition from the β-2 form to the β-3 form. At large x or y, the β-3 forms become more stable than the corresponding β-2 forms.

A function that can describe the influence of chain length differences on the melting enthalpy that is observed is a general quadratic function:

$$\Delta H_f = hn + h_0 + h_x x + h_{x2} x^2 + h_{xy} xy + h_y y + h_{y2} y^2 \qquad (9.60)$$

In order to introduce the leveling-off at high x or y, a cut-off value of 6 is used in the calculations instead of the real value of x or y when x or y exceeded 6. The resulting fit to the experimental data was very good: a RMSE nearly equal to that of a fit to the mono-acid data only was obtained. Melting enthalpies deviating more than 20 kJ/mol from the model predictions were rejected. The resulting values for h are in agreement with those of other lipids given in Table 9.4.

9.4.3.1.2 Melting Points

If Equation 9.43 is used to fit the mono-acid data, it results in a RMSE of 1.5°C for the α- and β-modifications and 3°C for the β'-modification. This compares very well with the experimental error in the melting points of 1°C–2°C. The RMSE increases to 7°C–10°C when Equation 9.43 is fitted to all data of saturated TAGs. A closer look to the differences between experimental melting points and those predicted by the fit to the mono-acid data suggests a quadratic relationship between x and y and the model parameters A and B:

$$A = A_0 + A_x x + A_{x2} x^2 + A_{xy} xy + A_y y + A_{y2} y^2$$

$$B = B_0 + B_x x + B_{x2} x^2 + B_{xy} xy + B_y y + B_{y2} y^2$$

$$(9.61)$$

If the model of Equation 9.57 is fitted to the data, the RMSEs that result are most satisfying. A subdivision of the modifications into their -2 and -3 forms (β-2 and β-3, etc.) does not give an improved fit.

The values for T_∞ are in agreement with the values that were found by Zacharis (1977).

9.4.3.1.3 Simultaneous Fit of Melting Points and Melting Enthalpies

Although the empirical relations that were derived are perfectly useful as such, a few unsatisfying aspects must be mentioned:

1. According to Equation 9.45, the values for the parameters of relationship (Timms, 1978) for the melting enthalpy can be derived from the parameters values of relation 43 plus 58 for the melting point. If this is actually done so, the resulting parameters for the melting enthalpy are completely different from those given in Table 9.6. The melting enthalpies that are calculated in this way do not at all agree with the experimental data. Apparently, the parameters of Table 9.7 are not consistent with those in Table 9.6.
2. The melting enthalpy data for the β-modification showed clearly that the influence of x and y levels off when the values of x or y are greater than 6. The melting points did not seem to show such effect. However, if the correlation for the melting points extrapolated to extreme values of x or y, absurd results are obtained, indicating that leveling off of the influence of x and y should be introduced.

Therefore, this section will attempt to derive a unified model for both the enthalpies of melting, the entropies of melting and the melting points of saturated TAGs, by fitting the ΔH_f and T_f data simultaneously.

TABLE 9.6
Parameters That Result When Equation 9.57 Is Fitted to the Experimental

Parameter	α-Modification Estimate	SE	β′-Modification Estimate	SE	β-Modification Estimate	SE
h	2.39	0.1	4.17	0.2	4.03	0.1
h_0	−16.3	5	−68.4	8	−24.4	4
h_x	1.98	0.3	17.3	1.6	2.16	0.4
h_{x_2}	−0.54	0.07	−3.25	0.3	−0.63	0.07
h_{xy}	—		−1.07	0.3	—	
h_y	—		−9.03	1.2	−7.28	0.5
h_{y_2}	−0.64	0.08	—		—	
RMSE	5.2		6.0		7.1	

TABLE 9.7
Parameters That Result When Equations 9.44
and 9.58 Are Fitted to the Experimental
Melting Points of Even Saturated TAGs

Parameter	α	β'	β
A_0	−9.0581	−8.4543	−8.0481
A_x	0.00290	−0.10360	0.074130
A_{x_2}	−0.0619116	−0.018881	−0.0348596
A_{xy}	0.115128	0.0739411	0.00771420
A_y	−0.453461	−0.49721	−0.404136
A_{y_2}	−0.005827	0.0115995	0.0111938
B_0	−4.4841	−0.26501	2.66923
B_x	−0.00111	0.54997	−0.31675
B_{x_2}	0.148938	0.074136	0.085967
B_{xy}	−0.365917	−0.340928	0.040642
B_y	1.41154	2.34238	0.55040
B_{y_2}	−0.001766	−0.135735	−0.000945
T_∞	401.15	401.15	401.15
RMSE	2.3°C	2.9°C	3.0°C

The function f_{xy}, as defined in Equations 9.46 and 9.47, should account for the observed decrease in melting enthalpies and melting points as differences between the three chain lengths increase. It should be chosen in such a way that this effect does not grow indefinitely with increasing x and/or y but levels off. The general quadratic relations with cut-off value, which were previously used, did not perform well. The function finally arrived at is

$$f_{xy} = 2 - \exp\left\{-\left(\frac{x - x_o}{k_x}\right)^2\right\} - \exp\left\{\left(\frac{y}{k_y}\right)^2\right\} \tag{9.62}$$

which increases from 0 (for small values of both x and y) via 1 (for a large absolute value of either x or y) to 2 (for large values of both x and y). We have investigated a more complex variation of this function:

$$f'_{xy} = 2 - \left(1 + \frac{1}{2}\delta\right)\exp\left\{-\left(\frac{x' - x_0}{k_x}\right)^2\right\} - \left(1 - \frac{1}{2}\delta\right)\exp\left\{-\left(\frac{y' - y_o}{k_y}\right)^2\right\} \tag{9.63}$$

with
$x' = \cos\theta x + \sin\theta y$
$y' = \cos\theta y - \sin\theta x$

The more general Equation 9.60 allows for a rotation over an angle of θ of the (x, y) axes to (x', y')-axes a difference in scaling along these directions (k_x and k_y) instead of a common k), a non-zero offset y_0, and a relative difference in the maximal effects along the two directions expressed by the parameter δ. It appeared, however, that Equation 9.59, which is a special case of the more general Equation 9.60 (viz. $\theta = 0$, $k_x = k_y = k$, $y_0 = 0$, $\delta = 0$) gave a satisfactory fit to the data. Therefore, the results reported further on refer to this simpler model.

One must also take into account the possible effect of asymmetry on melting points. This is done by including an additional $R \ln 2$ term in the expression for the melting entropy to distinguish symmetric ($y = 0$) from asymmetric ($y \neq 0$) TAGs. The work of de Jong and van Soest has shown that for the β-phase the inclusion of such a term is appropriate (de Jong and van Soest, 1978; Birker et al., 1991). In the β-modification, random mixing of the two mirror images is not possible. For the loosely packed α phase, we do not expect that this term is needed. The β' phase takes an intermediate position and it is not clear beforehand whether the inclusion of the $R \ln 2$ term will improve or worsen the fit.

Finally, consider the phenomenon of melting point alternation: odd mono-acid TAGs tend to melt systematically lower than expected on the basis of interpolation from even mono-acid TAGs. Again, anticipate that allowance for this effect will improve the fit for melting points of β-phase TAGs. For the α-phase TAGs, do not expect any benefit and it is not known for the β'-phase TAGs.

The full expressions for the enthalpy and entropy of melting then read as follows:

$$\Delta H_f = hn + h_0 + h_{xy}f_{xy} + h_{odd} \cdot odd \tag{9.64}$$

$$\Delta S_f = sn + s_0 + s_{xy}f_{xy} + s_{odd} \cdot odd + R \ln 2 \cdot asym \tag{9.65}$$

Here "asym" and "odd" are indicator variables taking the value 1 ("true") when an asymmetric TAG or an odd TAG is involved and the value 0 ("false") otherwise. A TAG is considered "odd"' when at least one of the fatty acid chains has an odd number of carbon atoms chains in the TAG. ΔH_f^{sat} is the melting enthalpy of the corresponding saturated TAG that can be obtained from the data in Tables 9.6 or 9.7.

The eight parameters h_0, h_n, s_0, s_n, h_{xy}, s_{xy}, k, and x_0 were estimated by simultaneously fitting the ΔH_f model, Equations 9.59 and 9.61, to observed melting enthalpy data and the T_f model based on Equations 9.41, 9.59, 9.61, and 9.62 to observed melting points. The models were fit using weighted nonlinear least-squares regression taking a factor 6 for the ratio of the variances of melting enthalpies and melting points.

Enthalpies of fusion deviating more than 20 kJ/mol from the model fit have been rejected. Exclude all melting points which could not be fit well, that is, that deviated more than 10°. In all, the enthalpies of some 40 TAGs and the melting points of about 70 TAGs were fitted for the α, β', and β phases separately.

TABLE 9.8

Estimates and Standard Errors of the Parameters from Equations 9.61 and 9.62 When Simultaneously Fitted to Melting Points and Melting Enthalpies of All Saturated TAGs

Parameter	α-Modification		β′-Modification		β-Modification	
	Estimate	SE	Estimate	SE	Estimate	SE
h_0	−31.95	3.00	−35.86	5.88	−17.16	4.83
h	2.70	0.07	3.86	0.13	3.89	0.10
s_0	−19.09	10.56	−39.59	19.28	31.04	15.81
s	6.79	0.25	10.13	0.42	9.83	0.34
h_{xy}	−13.28	2.42	−19.35	2.81	−22.29	2.08
s_{xy}	−36.70	7.79	−52.51	8.76	−64.58	6.45
k	4.39	0.50	1.99	0.24	2.88	0.35
x_0	1.25	0.27	2.46	0.19	0.77	0.27
t_∞	397	4.1	381	3.6	395	3.3
h_{odd}	—	—	—	—	2.29	0.44
RMSE (ΔH)	8.6		9.2		10.3	
RMSE (T)	2.5		3.7		3.7	

The main results are collected in Appendix 9.A. Measured values are averaged over several observations with the number of independent observations given in the column labeled FREQ. Zero FREQ values refer to cases excluded from the estimation procedure. The heats of fusion that were measured in this work are separately listed. The estimated parameter values along with their standard errors are given in Table 9.8.

It turned out that the inclusion of the symmetry/asymmetry term ($R \ln 2$) improved the fit for the β phase in contrast to the α phase and the β′ phase where the fit became worse. A similar result was obtained for the even/odd term: no significant improvement for the α and β′ phase and a much better fit for the β-phase odd TAGs.

The resulting RMSEs for the melting points are only slightly more than those obtained by fitting the melting points solely. The RMSEs for the melting enthalpy have increased by 50%, but are still acceptable when compared to the experimental error of 5–7 kJ/mol. The values for h and s agree very well with those of other lipids, given in Table 9.4. The values for h_{xy}, which represent the maximum decrease in ΔH_f that is caused by chain length differences and which are likely to be related to the enthalpy difference between the β-2 and β-3 form of a mono-acid TAG, compare well with the stability difference between the two forms that was calculated by de Jong (1980) (10–20 kJ/mol). The values of h_{xy} and s_{xy} increase from α to β, in accordance with the expectation that the influence of chain length differences is more pronounced in a more densely packed polymorph. In the β-modification, the influence of x and y levels off at x or y = 6 (2 k), as was found previously.

A much more consistent correlation has been obtained at the expense of only a relatively small increase of the RMSEs.

9.4.3.2   Unsaturated TAGs

9.4.3.2.1   Melting Enthalpy

The authors have chosen to model the effect of unsaturation on the melting enthalpy of a TAG as a correction to the melting enthalpy of the corresponding saturated TAG (Equation 9.46). The model proposed by Perron (Bommel, 1986) did not perform very well; it resulted in a RMSE of more than 24 kJ/mol. After having investigated several other functions, the following model evolved:

$$\Delta H_f^{\text{unsat}} = \Delta H_f^{\text{sat}} + h_0 n_0 + h_E n_E + h_l n_l \qquad (9.66)$$

Here n_0 stands for the number of oleic chains, n_E stands for the number of elaidic chains and n_l for the number of linoleic chains in the TAG.

The authors have fitted this model for the α-, β-, and β'-modifications. The complete data set contained over 80 melting enthalpies of only 16 different TAGs: 20 observations for 12 TAGs in the α-modification, 21 observations for 13 TAGs in the β'-modification, and 44 observations for 16 TAGs in the β-modification. The authors rejected data that deviated more than 30 kJ/mol from the predicted values. The data and the predictions are given in Appendix 9.A (see also Figure 9.10). The resulting parameters are given in Table 9.9.

The values of the parameters that were obtained for the β-modification agree very well with the effective carbon numbers that Timms derived for the unsaturated fatty acids. The large RMSE for the β'-modification is solely due to the β'-melting enthalpy of POP (104 kJ/mol vs. a predicted value of 128 kJ/ mol). If POP is left out, the RMSE decreases to 9.7, while the parameter values do not change. POP is the only *cis*-unsaturated TAG in the data set that crystallizes in the β'-2 form (Sato et al., 1989), which may explain its deviating behavior.

There is a lack of data for the unstable modifications of TAGs with linoleic acid, so that the model parameter h_l for these modifications could not be calculated. The parameter that describes the effect of the fatty acid that is most similar to linoleic acid (h_0) is nearly independent from the modification. Assume that this also holds for h_l.

A good fit to the available data was altogether obtained; the RMSEs are nearly equal to those of the saturated TAGs. A better general model can only be developed if more experimental data are made available, over a wider range of carbon numbers, including linoleic acid and including mixed unsaturated fatty acid TAGs.

9.4.3.2.2   Melting Point

An approach to model the melting points of unsaturated TAGs is start from a model for the saturated TAGs and add a correction term that accounts for the presence of O, E, l, and linolenic acid (le) (Equations 9.15 and 9.46). After exploring various simpler models, it was learned that it was necessary to model interactions between

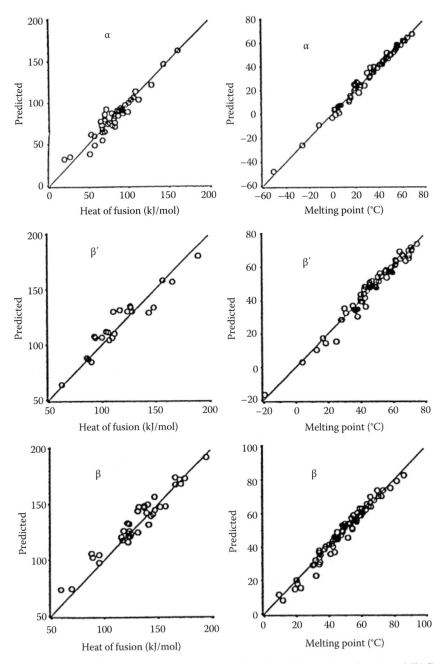

FIGURE 9.10 Values of the melting enthalpy and the melting points of saturated TAGs, calculated using the relations (9.62) and (9.63) plotted against the experimental data.

TABLE 9.9

Estimates and Standard Errors of the Parameters of Equation 9.63 Fitted to the Melting Enthalpies of Unsaturated TAGs

Parameter	α-Modification		β′-Modification		β-Modification	
	Estimate	SE	Estimate	SE	Estimate	SE
h_0	−31.7	1.8	−28.3	1.8	−30.2	1.4
h_E	−11.7	1.3	(−15.9)		−15.9	0.9
h_1	(−37.7)		(−37.7)		−37.7	2.5
RMSE	8.3		22		11.1	

Note: SE standard error. Values in brackets were guessed, due to lack of data.

unsaturated chains. The final model giving the best results is Equation 9.44 for which A is given by

$$A = A_{sat} + A_O n_O + A_E n_E + A_l n_l + A_{le} n_{le}$$

$$+ A_{OO} n_{OO} + A_{EE} n_{EE} + A_{ll} n_{ll} + A_{lele} n_{lele}$$

$$+ A_{Ol} n_{Ol} + A_{Ole} n_{Ole} + A_{lle} n_{lle} \qquad (9.67)$$

and for which B is given by the much simpler expression:

$$B = B_{sat} + B_O n_O + B_l n_l + B_{le} n_{le} \qquad (9.68)$$

Here, for example, n_O stands for the number of oleic chains in the TAG and n_{ol} for the number of O–l pairs. A_{sat} and B_{sat} can be obtained from Table 9.7 and Equation 9.58 or from Table 9.8, using Equations 9.45 and 9.60 through 9.62.

The authors have fit this model for the α-, β′-, and β-modifications. The complete data set contained over 120 melting points of *cis*-unsaturated TAGs, of which 18 were left out, because they involved unsaturated fatty acids other than o, l, and le. Of the remaining 102 melting point data, 13 were associated with the α-modification, 16 with the β′-modification, 13 with the β-modification, and for 60 data points the modification was not specified. The modification was not known for 13 of the 47 melting points of the *trans*-containing TAGs.

In these cases, the authors chose to assign the melting point to that modification, which gave the best fit. This assignment had to be done in an iterative manner using the results of the previous regression to determine the best modification input for the next regression analysis and carrying this process through till internal consistency was achieved. This laborious procedure can be viewed as a maximum likelihood estimation of both the melting point–structure relationship

TABLE 9.10

Estimates and Standard Errors of the Parameters in Equations 9.64 and 9.65 Fitted to the Melting Points of Unsaturated TAGs

Parameter	α-Modification		β′-Modification		β-Modification	
	Estimate	SE	Estimate	SE	Estimate	SE
A_0	3.46	0.39	2.20	0.36	2.93	0.24
A_E	1.38	0.16	1.34	0.22	1.68	0.11
A_1	3.35	0.66	2.5	1.1	4.69	0.50
A_{le}	4.2	2.1	2.2	2.1	5.2	1.9
A_{OO}	0.11	0.16	−0.27	0.14	−0.89	0.10
A_{EE}	0.01	0.19	−0.04	0.34	−0.40	0.13
A_{ll}	3.68	0.87	−0.55	0.26	−1.21	0.17
A_{lele}	1	1.9	−1.51	0.92	−1.38	0.60
A_{Ol}	−0.53	0.24	1.0	0.32	−0.71	0.15
A_{Ole}	−0.83	0.34	−0.76	0.22	−0.69	0.18
A_{lle}	3.0	1.3	−1.12	0.36	−0.73	0.52
B_O	0	1.6	−4.3	1.5	−3.7	0.8
B_1	5.4	2.3	−7.8	5.3	−1.5	1.6
B_{le}	2.6	8	−13.7	11.8	−1.8	7.0
RMSE	3.2°C		4.1°C		2.6°C	

and the unknown modifications. The data and the resulting assignments are given in Appendix 9.A. Data deviating more than 10°C from the predicted value were disregarded. The parameters are given in Table 9.10.

The standard deviations are only slightly larger than those for the saturated TAGs. We do not expect that these values can be substantially improved by another model. It should be realized that the reliability of the input data probably varies considerably. The data have been collected from the literature spanning nearly a century. The authors have screened the data to some extent and discarded a number of a priori very unlikely cases. In general, however, it is very difficult to assess the quality of the data reported. Thus, some of the input data may have a substantial error and this will cause some lack of fit.

In future work, we will extend the number of available melting enthalpy data and attempt to correlate the melting enthalpy and melting points of unsaturated TAGs simultaneously.

9.4.4 CONCLUSION

A compilation of literature data of melting points and melting enthalpies of TAGs was made. This data set was extended by measurements of melting enthalpies of 51 mixed acid saturated and unsaturated TAGs in all three modifications.

Reliable relations were developed that give the melting point and melting enthalpy as a function of the carbon number, chain length differences, and degree of unsaturation for the α-, β′-, and β-modifications.

9.5   MIXING BEHAVIOR IN LIQUID STATE

Before a study on solid TAG phases that are in equilibrium with a liquid TAG phase can be started, knowledge of the mixing behavior of TAGs in the liquid state is required. Although it is generally assumed that TAGs mix ideally in the liquid state, experimental evidence is lacking. Therefore the activity coefficients in mixtures of TAGs in the liquid state are measured using gas-liquid chromatography with a liquid stationary TAG phase. The results are compared with activity coefficients calculated with the UNIFAC group contribution method.

9.5.1   LITERATURE

It is generally stated in the literature on the phase behavior of TAGs that the miscibility of TAGs in the liquid state is ideal (Hannewijk et al., 1964; de Bruijne et al., 1972; Timms, 1984). Yet the experimental evidence on which this conclusion is based is only very minor:

1. In dilatation experiments on pure TAGs and binary mixtures of TAGs no volume effect was observed (Hannewijk et al., 1964; Hendriske). However, although ideal miscibility implies a zero excess volume of mixing, the opposite is not necessarily true.
2. For liquid alkanes, it was found that the heat of mixing is less than 0.1 kJ/mol (Wurflinger, 1972; de Bruijne et al., 1972), even for hydrocarbons considerably differing in chain length. Compared to the heat of fusion this heat of mixing is negligible. It was reasoned that if alkanes have almost no heat of mixing, then TAGs, which are chemically very similar to alkanes, will also have no heat of mixing. Again, ideal miscibility implies the absence of a heat of mixing, while the reverse is not necessarily true.
3. Mixtures of high-melting saturated TAGs and liquid unsaturated TAGs were often found to obey the Hildebrand solubility equation (Equation 9.69):

$$\ln x_i^{\mathrm{L}} = \frac{\Delta H_{\mathrm{f}}}{R}\left(\frac{1}{T_{\mathrm{f}}} - \frac{1}{T}\right) \tag{9.69}$$

It is stated that therefore liquid mixtures of TAGs are ideal and if the Hildebrand equation is not obeyed, it is often ascribed to liquid phase nonideality (de Bruijne et al., 1972; Norton et al., 1985). The Hildebrand equation assumes no solid phase miscibility and ideal liquid phase miscibility. In Sections 9.6 and 9.7, it will be shown that TAGs do show solid phase miscibility. Whether the Hildebrand equation is obeyed or not is therefore no indication for ideality of the liquid state, as the solid state behavior may not be disregarded.

Summarizing, it is clear that the evidence for ideal miscibility of liquid TAGs is not very convincing. Yet, before any description of solid–liquid phase behavior and nonideality in the solid phase can be given, the mixing behavior of TAGs in the liquid state needs to be known. In the next sections, measurements of activity coefficients of TAGs in the liquid state are reported and compared with activity coefficients that are estimated using the UNIFAC group contribution method.

9.5.2 Model Calculations

In order to enable the estimation of activity coefficients in mixtures for which no experimental data are available, Fredenslund, Rasmussen, and coworkers developed the UNIFAC method (Fredenslund et al., 1977). The UNIFAC method was shown to perform quite well in phase equilibrium calculations for many hydrocarbon systems (Fifth International Conference on Phase Equilibria for Chemical Process Design, 1989).

The basic idea of the UNIFAC method is that although there are thousands of different chemical components of interest for the chemical technology, the number of functional groups that constitute these compounds is much smaller. A natural oil contains thousands of different TAGs, but it only consists of CH_3-, $-CH_2-$, $-CH-$, $-CH=CH-$, $-OH$, and CH_2COO- functional groups. UNIFAC assumes that the activity coefficient of each component in the mixture is a function of the individual contributions of the components' functional groups. Therefore, activity coefficients can be calculated from a limited number of parameters for the functional groups. An excellent and clear description of the method is given by Fredenslund et al. (1977).

The activity coefficients of TAGs in a set of TAG mixtures were calculated: in three binary mixtures of SSS and MMM (Table 9.11), three binary mixtures of MMM and 8C8 (Table 9.12), three binary mixtures of SSS and 8C8, and in a mixture of the six main TAGs and the two main partial glycerides occurring in palm oil (Table 9.13). The authors used a computer program and a database that were obtained from the department of Prof. J.M. Prausnitz (University of California, Berkeley) for the calculations.

The results confirm the assumption made in the literature and show that the activity coefficients of all TAGs are unity, so ideal mixing of TAGs is predicted. In view of the chemical similarity between all components, this is not too surprising. The two partial glycerides are predicted to have activity coefficients considerably larger than unity, as may be expected for polar compounds in an apolar solvent (Table 9.13).

TABLE 9.11

Activity Coefficients in the Liquid Phase for Binary Mixture of SSS with MMM and 8C8, Calculated with UNIFAC

SSS [1] and MMM [2]			SSS [1] and 8C8 [2]		
T (°C)	x_1	y_1	T (°C)	x_1	y_1
100	0.25	1.007	100	0.25	1.091
100	0.50	1.003	100	0.50	1.020
100	0.75	1.000	100	0.75	1.004
120	0.25	1.006	120	0.25	1.079
120	0.50	1.002	120	0.50	1.022
120	0.75	1.000	120	0.75	1.003

TABLE 9.12

Activity Coefficients in the Liquid Phase for Binary Mixtures of MMM with 8C8, Calculated with UNIFAC

MMM [1] and 8C8 [2]		
T (°C)	x_1	y_1
100	0.25	1.046
100	0.50	1.015
100	0.75	1.003
120	0.25	1.041
120	0.50	1.014
120	0.75	1.002

TABLE 9.13

Activity Coefficients in the Liquid Phase for the Six Main TAGs and the Two Main Partial Glycerides of Palm Oil at 50°C, Calculated with UNIFAC

Palm Oil without Partial Glycerides			Palm Oil with Partial Glycerides		
Component	x	y	Component	x	y
PPP	0.08	0.9974	PPP	0.08	1.0005
POP	0.26	0.9997	POP	0.26	1.0018
PPO	0.14	0.9997	PPO	0.12	1.0018
PLP	0.10	1.0004	PLP	0.08	1.0008
POO	0.25	0.9998	POO	0.23	1.0009
OOO	0.17	0.9982	OOO	0.17	0.9982
Mono-P	—	—	Mono-P	0.01	4.2000
Di-P	—	—	Di-P	0.05	1.3518

9.5.3 EXPERIMENTS

9.5.3.1 Method for Determination of Activity Coefficients of Mixtures of Nonvolatile Liquids

Activity coefficients in the liquid phase are often determined by measuring to what extent the vapor pressure of a liquid mixture deviates from Raoult's law.

$$p_i = p_i^* x_i^{\mathrm{L}} \tag{9.70}$$

However, the vapor pressure of TAGs is less than 1 Pa at temperatures between 0°C and 80°C, the temperature region in which solid fat crystallizes. This is unmeasurably small and therefore another method for determination of activity coefficients must be developed.

A well-known method for determination of activity coefficients at infinite dilution of volatile compounds in nonvolatile liquids is the use of gas-liquid chromatography (GLC). The nonvolatile liquid is used as stationary phase and the volatile component is injected into the carrier gas stream. The activity coefficient is calculated from the retention time. The GLC method is discussed in many textbooks on gas chromatography (Box 9.5; Desty et al., 1962; Grob, 1985).

BOX 9.5 GLC METHOD TO DETERMINE ACTIVITY COEFFICIENTS OF NONVOLATILE LIQUIDS

The net retention volume of an injected sample is defined as the product of the net retention time and the flow rate of carrier gas, corrected for the pressure drop over the chromatographic column:

$$V_n = t_n \Phi_v \cdot \frac{3(p_{in}/p_{out})^2}{2(p_{in}/p_{out})^3} \tag{9.71}$$

The specific retention volume is defined as the net retention volume per unit mass of liquid stationary phase at 273.15°C. It is given by

$$V_g = \frac{273.15 V_n}{m_L T} \tag{9.72}$$

It is assumed that

- The sample concentration in mobile and stationary phase always have the equilibrium values
- That the sample volume is very small in comparison with mobile and stationary phase volume (infinite dilution)

In that case the mole fraction of the sample that is in the mobile phase must be equal to the ratio of the volume of the mobile phase and the retention volume (=net retention volume + mobile phase volume). The equilibrium distribution coefficient of the sample over mobile (L) and stationary phase (G) is therefore given by

$$K_{LG} = \frac{x^L}{x^G} = \frac{V_n}{n_L} \cdot \frac{n_G}{V_G} \tag{9.73}$$

The condition for equilibrium between stationary and mobile phase for infinite sample dilution is (if the mixture of carrier gas and sample behaves ideally)

$$px^G = p^* \gamma^\infty x^L \tag{9.74}$$

BOX 9.5 (continued)

By rearranging the latter two equations, one obtains an expression for the activity coefficient at infinite dilution of the sample in the stationary phase as a function of net retention volume:

$$\gamma^\infty = \frac{n_L}{p*V_n} \cdot \frac{pV_G}{n_G} \tag{9.75}$$

The last term in this equation reduces to RT if the ideal gas law may be applied to a mixture of a carrier gas and a sample.

When the influence of the pressure drop over the column and the nonideality of the mobile gas phase are taken into account, a correction term must be added to Equation 9.72. Replacing the net retention volume by the specific retention volume and the vapor pressure by the corrected vapor pressure according to Prausnitz (1986), Desty et al. (1962) obtains the following corrected expression for the activity coefficient at infinite dilution:

$$\ln\gamma^\infty = \ln\left(\frac{273.15R}{p*V_gM_L}\right) - p*\left(\frac{B_{11} - v}{RT}\right) \tag{9.76}$$

The second virial coefficient B_{11} and the molar volume v of the liquid probe at the measuring temperature can be obtained from one of the correlations based on the principle of corresponding states given in Prausnitz (1986) and Smith and ness (1987).

Although the correction term is significant for the activity coefficients at infinite dilution of the samples that are used in this work, it is of no influence on the calculated interaction between the nonvolatile components in the stationary phase, as will be clear from Equations 9.77 and 9.82 in the next section. In that case, the correction term can therefore be neglected.

Desphande et al. (1974) have extended the use of this GLC method, using so-called "probes" to measure the interaction between two nonvolatile liquids:

First, two GLC columns are prepared, each containing a pure nonvolatile liquid as stationary phase and the activity coefficient of a volatile "probe" in these two pure stationary phases is determined by the method outlined earlier.

Next a number of GLC columns are prepared, containing mixtures of the two nonvolatile components as stationary phase. The activity coefficients of the probe in these mixed stationary phases are determined.

The interaction between the two nonvolatile components will affect their affinity for the probe, which will be expressed in the retention time of the probe. The interaction between the two liquid components can therefore be obtained by comparing the activity coefficients of the probe in the pure liquids with those in mixtures of the liquid components.

For simple regular solutions, this can be worked out as follows: A system of two non-volatile components [1] and [2] and a probe [pr] are considered. The activity coefficient of the probe in a mixture of the two other components is given by (Prausnitz, 1986)

$$RT \ln \gamma_{pr,12} = A_{pr,1} x_1^2 + A_{pr,2} x_2^2 + (A_{pr,1} + A_{pr,2} - A_{1,2}) x_1 x_2 \qquad (9.77)$$

At infinite dilution, we obtain for a binary system of the probe and one of the non-volatile components

$$RT \ln \gamma_{pr,1}^\infty = A_{pr,1} \qquad (9.78)$$

and for the complete ternary mixture

$$RT \ln \gamma_{pr,12}^\infty = A_{pr,1} x_1 + A_{pr,2} x_2 - A_{1,2} x_1 x_2 \qquad (9.79)$$

The interaction between the two nonvolatile components can now be determined from the experimental activity coefficients at infinite dilution of the probe by

$$\frac{A_{1,2}}{RT} = \frac{x_1 \ln \gamma_{pr,1}^\infty + x_2 \ln \gamma_{pr,2}^\infty - \ln \gamma_{pr,12}^\infty}{x_1 x_2} \qquad (9.80)$$

It is clear that this elegant method can be applied straightforwardly to TAGs. The result should be independent of the probe used. If indeed TAGs mix ideally, the interaction coefficient $A_{1,2}$ must turn out to be zero.

9.5.3.2 Experimental Work

The mixing behavior of the binary SSS–MMM and of the binary SSS–8C8 was studied. The TAGs were obtained from Dr. A. Fröhling of Unilever Research Vlaardingen and were GC-pure.

9.5.3.2.1 Experimental Procedure

The stationary phases were prepared by mixing a predetermined amount of the carrier material Chromosorb W (80–100 mesh, acid washed, DCMS treated) that was suspended in chloroform, with a predetermined amount of the TAG or TAG mixture, dissolved in chloroform. The chloroform was slowly evaporated at 60°C under occasional stirring. The last remnants of chloroform were removed by heating 3 h at 80°C. It was checked by weighing whether no loss of carrier material or TAGs had occurred. The stationary phase was brought into a glass column of 1 m length and 2 mm diameter and the column was conditioned for 10 h at 150°C. Measurements were carried out at the Technical University of Delft (Gandasasmita, 1987) in a modified VARIAN 3700 gas chromatograph with a thermal conductivity detector. The retention times were determined with a VARIAN CDS-11 integrator. Helium was used as carrier gas.

In order to confirm that that the probe that is used has no influence indeed on the calculated TAG–TAG interaction, measurements were repeated using eight different probes:

1. *n*-pentane
2. *n*-hexane
3. *n*-heptane
4. 2-methyl-pentane
5. 3-methyl-pentane
6. benzene
7. toluene
8. cyclohexane

All measurements were repeated at two carrier gas flow rates (17 and 38 mL/min) and with two columns, containing 10% and 20% of the stationary phase on carrier. About 0.1 μL of the probe was injected.

9.5.3.2.2 Measurements

The specific retention volumes of the eight probes were determined at 82°C, 92°C, 102°C, 112°C, 123°C, and 133°C for the following stationary phases (concentrations in mole fractions):

SSS	0.4487 SSS, 0.5513 MMM	0.1557 SSS, 0.8443 8C8
MMM	0.7116 SSS, 0.2884 MMM	0.3593 SSS, 0.6407 8C8
8C8		0.6023 SSS, 0.3977 8C8

The results are given in Appendix 9.B. The average experimental error in the retention volumes is 3%.

9.5.3.3 Results and Discussion

The activity coefficients at infinite dilution of the probes can be calculated from the specific retention volumes with Equation 9.73. It turns out that they do not depend on temperature, as is illustrated for some of the probes and some stationary phases in Figure 9.11. The variation as a function of temperature is less than the experimental error of 3% in the activity coefficients. This indicates that the probe–TAG mixtures behave athermal, that is, that the excess Gibbs energy is nearly equal to the excess entropy of mixing.

Because a dependency on temperature was not found, the averages over all temperatures of the activity coefficients are used in further discussions. They are given in Table 9.14.

The interaction coefficients between the TAGs for the regular solution theory can now be calculated using Equation 9.77. The results are given in Table 9.15.

The results for the binary SSS–MMM clearly indicate that within experimental error these two TAGs mix ideally. The results are very consistent for both binary mixtures and all probes used.

The results for the binary SSS–8C8 clearly indicate that this binary does not mix ideally in the liquid phase. The interaction parameters significantly differ from zero.

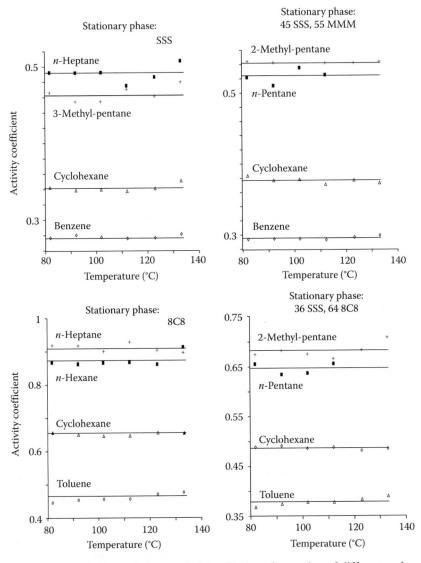

FIGURE 9.11 Activity coefficients at infinite dilution of a number of different probes in stationary TAG phases as a function of temperature.

And even worse, they also depend on the concentration of the TAGs. The simple regular solution model is therefore not a correct description for the nonideal behavior of this binary pair.

9.5.3.3.1 Interpretation with the Flory–Huggins Theory

The main difference between MMM and 8C8 is their difference in molecular size. This must be the cause of the large differences found in mixing behavior with SSS. An excess Gibbs energy model that accounts for such size differences

TABLE 9.14

Average Activity Coefficients at Infinite Dilution of Several Probes in a Number of Liquid TAG Mixtures Determined by GLC

x_{SSS}	1			0.449	0.712	0.156	0.359	0.602
x_{MMM}		1		0.551	0.288			
x_{8C8}			1			0.844	0.641	0.398
Probe								
n-Pentane	0.45	0.55	0.85	0.52	0.49	0.75	0.65	0.56
n-Hexane	0.47	0.57	0.87	0.52	0.50	0.75	0.66	0.57
n-Heptane	0.49	0.59	0.91	0.54	0.52	0.78	0.69	0.60
2-Methyl-pentane	0.49	0.59	0.90	0.54	0.52	0.77	0.68	0.59
3-Methyl-pentane	0.46	0.56	0.87	0.52	0.49	0.73	0.66	0.56
Benzene	0.28	0.32	0.43	0.30	0.29	0.39	0.36	0.32
Toluene	0.29	0.33	0.46	0.31	0.31	0.41	0.38	0.34
Cyclohexane	0.34	0.41	0.65	0.38	0.37	0.55	0.49	0.42

Note: The experimental error in the activity coefficients is 3%. Concentrations in mole fractions.

TABLE 9.15

Regular Solution Interaction Parameters A_{12}/RT for the Interaction between the TAGs Determined from GLC Results, Using Different Probes

x_{SSS}	0.449	0.712	0.156	0.359	0.602
x_{MMM}	0.551	0.288			
x_{8C8}			0.844	0.641	0.398
Probe					
n-Pentane	−0.15	−0.11	−2.61	−0.99	−0.58
n-Hexane	0.03	−0.03	−2.23	−0.93	−0.47
n-Heptane	0.01	−0.04	−2.29	−0.98	−0.50
2-Methyl-pentane	−0.01	−0.02	−2.25	−0.94	−0.48
3-Methyl-pentane	−0.02	−0.03	−2.29	−1.01	−0.47
Benzene	0.05	−0.02	−1.69	−0.73	−0.36
Toluene	0.02	−0.06	−1.77	−0.79	−0.41
Cyclohexane	0.03	−0.05	−2.41	−0.99	−0.50
Average	−0.01	−0.05	−2.19	−0.92	−0.47
Stand. dev.	0.06	0.03	0.29	0.10	0.06

Note: Experimental error in the parameters is 0.1. Concentrations in mole fractions.

is the Flory–Huggins theory for polymer solutions 89. According to this theory, the activity coefficient in the ternary system of probe [pr] and the two TAGs [1] and [2] is

$$\ln \gamma_{pr,12} = \ln \left(\frac{v_{pr}}{x_{pr}v_{pr} + x_1 v_1 + x_2 v_2} \right) + \left(1 - \frac{v_{pr}}{v_1} \right) \phi_1 + \left(1 - \frac{v_{pr}}{v_2} \right) \phi_2$$

$$+ x_{pr,1}\phi_1^2 + x_{pr,2}\phi_2^2 + (x_{pr,1} + x_{pr,2} - x_{1,2})\phi_1\phi_2 \tag{9.81}$$

The volume fraction ϕ_1 is defined as

$$\phi_i = \frac{x_i v_i}{\sum\limits_{j=1}^{N} x_j v_j} \tag{9.82}$$

in which v_i represents the molar volume.

At infinite dilution, we obtain for a binary system of the probe and one of the TAGs

$$\ln \gamma_{pr,1}^{\infty} = \ln \left(\frac{v_{pr}}{v_1} \right) + \left(1 - \frac{v_{pr}}{v_1} \right) + x_{pr,1} \tag{9.83}$$

and for the complete ternary mixture

$$\ln \gamma_{pr,12}^{\infty} = \ln \left(\frac{v_{pr}}{x_1 v_1 + x_2 v_2} \right) + \left(1 - \frac{v_{pr}}{v_1} \right) \phi_1 + \left(1 - \frac{v_{pr}}{v_2} \right) \phi_2 + x_{pr,1}\phi_1 + x_{pr,2}\phi_2 - x_{1,2}\phi_1\phi_2 \tag{9.84}$$

The Flory–Huggins interaction parameter x_{12} for the two TAGs can be determined from the experimental activity coefficients at infinite dilution by

$$x_{12} = \frac{\phi_1 \left(\ln \gamma_{pr,1}^{\infty} - \ln \left(\frac{v_1}{x_1 v_1 + x_2 v_2} \right) \right) + \phi_2 \left(\ln \gamma_{pr,2}^{\infty} - \ln \left(\frac{v_2}{x_1 v_1 + x_2 v_2} \right) \right) - \ln \gamma_{pr,12}^{\infty}}{\phi_1 \phi_2} \tag{9.85}$$

When the molecular volumes of the two TAGs are equal, the Flory–Huggins interaction parameter x becomes equal to the regular solution interaction parameter A_{12}/RT and Equation 9.82 reduces to Equation 9.77 for the regular solution theory. The Flory– Huggins interaction parameters are given in Table 9.16.

The interaction parameters of both binaries do not depend on the probe used and not on the concentration of the two TAGs in the binary mixture. The values of interaction parameters, -0.04 ± 0.1 for SSS–MMM and 0.07 ± 0.1 for SSS–8C8,

TABLE 9.16

Flory–Huggins Interaction Parameters X_{12} for the Interaction between the TAGs Determined from GLC Results, Using Different Probes

x_{SSS}	0.449	0.712	0.156	0.359	0.602
x_{MMM}	0.551	0.288			
x_{8C8}			0.844	0.641	0.398
Probe					
n-Pentane	−0.17	−0.15	−0.07	−0.02	−0.08
n-Hexane	0.01	−0.05	0.17	0.04	0.05
n-Heptane	−0.01	−0.06	0.17	0.02	0.04
2-Methyl-pentane	−0.02	−0.04	0.21	0.06	0.08
3-Methyl-pentane	−0.04	−0.05	0.24	0.02	0.11
Benzene	0.05	−0.03	0.14	0.04	0.09
Toluene	0.01	−0.07	0.14	0.01	0.04
Cyclohexane	0.01	−0.07	0.17	0.04	0.07
Average	−0.02	−0.07	0.15	0.03	0.05
Stand. dev.	0.06	0.03	0.09	0.02	0.06
Total av.		−0.04		0.07	
Stand. dev.		0.05		0.08	

Note: Experimental error in the parameters is 0.1. Concentrations in mole fractions.

do not differ significantly from zero, which implies that the TAGs have no specific interaction in the mixtures. The deviation from ideality that was found for SSS–8C8 is entirely explained by the extra entropy of mixing that arises from the large difference in molecular size of the two TAGs. The activity coefficients at infinite dilution of the three possible binary TAG mixtures, calculated with Equation 9.80, and in agreement with measurements, are given in Table 9.17.

TABLE 9.17

Activity Coefficients at Infinite Dilution of the Three TAGs Studied, Calculated Using the Flory–Huggins Theory for the Excess Entropy of Mixing (Equation 9.80)

	$1ny_{I,SSS}^{\infty}$	$1ny_{I,MMM}^{\infty}$	$1ny_{I,SCS}^{\infty}$	$y_{I,SSS}^{\infty}$	$y_{I,MMM}^{\infty}$	$y_{I,SCS}^{\infty}$
SSS	—	−0.02	−0.16	1	0.98	0.85
MMM	−0.03	—	−0.07	0.97	1	0.93
8C8	−0.24	−0.09	—	0.79	0.91	1

9.5.3.3.2   Implications for Natural Edible Oils

The results imply that deviations from ideal miscibility only become noticeable when an oil contains reasonable amounts of TAGs that have a carbon number that differs more than about 10 with the average carbon number of the mixture.

In normal vegetable oils and fats, the differences in molecular size of the TAGs are of the order of the differences between MMM and SSS. For practical purposes, liquid vegetable oils can therefore be treated as ideal mixtures of TAGs.

In animal oils and fats, like fish oil, butter fat, and edible tallow, the spread in molecular size of the TAGs is much larger, although the concentrations of very small and very large TAGs are limited. Treating these oils as ideal mixtures may, depending on the situation, lead to errors of about 10%–15% in calculation results.

This chapter considers the TAGs that occur in normal vegetable oils and fats. The liquid TAG phase may therefore safely be treated as an ideal mixture.

9.5.4   CONCLUSION

TAGs that differ not too much in molecular size mix ideally in the liquid state. The deviation from ideality that becomes noticeable at differences in carbon number greater than about 15–20 can entirely be ascribed to the extra entropy of mixing that occurs in mixtures of molecules that differ considerably in size. No specific TAG–TAG interactions were found.

In the liquid state, the vegetable oils that are normally used in the edible fats industry may be treated as ideal mixtures of TAGs.

9.6   MIXING BEHAVIOR IN THE α-MODIFICATION

Contrary to the β- and β′-modifications, in the α-modification the fatty acid chains still appear to oscillate and rotate with considerable molecular freedom. The formation of mixed crystals will therefore hardly disturb the α-crystal packing. As a consequence, mixing in the α phase may be nearly ideal. With this assumption α-melting ranges of a number of common fat blends are calculated and compared with experimental data.

9.6.1   EVIDENCE FOR PARTIAL RETAINED CHAIN MOBILITY IN THE α-MODIFICATION

Not only TAGs, but many lipids, like alkanes, *n*-alcohols, and simple esters solidify upon quick cooling from the melt in a crystal form with hexagonal chain packing, which for TAGs is called the α-modification. In 1932, it was suggested by Müller (1932) that the hexagonal polymorph of *n*-alkanes has complete rotational disorder of the chains in the crystal. Therefore the hexagonal polymorph of *n*-alkanes has been called the "rotator phase." At present, the rotational disorder in the rotator phase of *n*-alkanes has been extensively studied and is well established (Broadhurst, 1962; Flory and Vrij, 1963; Wurflinger, 1972; Larsson, 1986). A curious property of lipid crystals in the rotator phase is their plasticity. Due to this property, esters occurring in this form have been called waxes.

Contrary to *n*-alkanes, the chains in the α-modification of TAGs cannot have complete freedom of rotation. For sterical reasons, chain mobility near the glyceryl group must be restricted. Yet the heat of fusion and the entropy of fusion of the

TABLE 9.18

Enthalpy and Entropy of Fusion of Some TAGs and Some *n*-Alkanes

TAG Name	ΔH_f (kJ/mol)			ΔS_f (J/mol K)		
	α	β'	β	α	β'	β
MMM	84	107	145	275	334	440
PPP	98	132	169	309	399	501
SSS	113	156	193	343	464	561
n-C19	45	60	—	147	197	—
n-C21	48	63	—	153	203	—

α-modification of TAGs correspond closely to that of the rotator phase of alkanes. Moreover, the entropy of fusion is only 60% of that of the very crystalline β and 75% of that of the β-modification, indicating that in the remaining part of the molecule still a considerable disorder must exist (Table 9.18).

The melting dilatation (the volume increase upon melting) of the α-modification is only 60%–70% of that of the β-modification (Small et al., 1984). This clearly demonstrates a less dense packing of the TAG molecules in the α-modification.

In pulse NMR for the determination of solid fat content (Human et al., 1989) the so-called *f*-factor, which is inversely proportional to the relaxation time, is only 1.2–1.3 for the α-modification, compared to 1.4–1.5 for the β'-modification and about 1.6 for the β-modification. This indicates a more disordered and liquid-like packing of the fatty acid chains in the α-modification.

Hernqvist and Larsson (1982) observed that the long spacing of the α-modification of SSS depends on temperature. A temperature dependence of the same order of magnitude was one of the main arguments for Luzatti et al. in their classical work on liquid crystallinity (Luzatti et al., 1960) to suggest a liquid state of hydrocarbon chains in liquid crystals. Hernqvist and Larsson also measured Raman spectra of TAGs. The β'- and β-modifications showed sharp peaks at 1065 and 1130 cm^{-1} (C–C stretching vibrations), while liquid TAGs have a broad band near 1090 cm^{-1}. The Raman spectra of the α-modification had a significant liquid-like character.

The most convincing evidence for chain mobility comes from Norton et al. (1985a). He concludes from high-resolution CP/MAS^{13}C-NMR spectra of the liquid state and the three polymorphs of SSS and PPP that in the α-modification, like in the liquid phase and contrary to the β'- and β-modifications, the 1- and 3-position of the glycerol are equivalent. He was able to study molecular motion by application of interrupted decoupling (switching of the ^{13}C–^{1}H decoupling for a short time, which suppresses the signal of immobile protonated carbons). The results clearly show that the mobility of the fatty acid chains near the glyceryl group is very limited in all three polymorphs. In the α-modification, the main hydrocarbon chain still possesses some mobility, while the carbons near the methyl end plane are very mobile. Main chain mobility is still present to some extent in the β'-modification, while the β-modification shows no main chain mobility. Even in the β-modification, the methyl end group has retained some mobility.

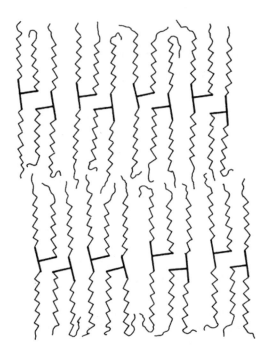

FIGURE 9.12   Chain disorder in the α-modification.

The good correspondence with the rotator phase of alkanes and all facts that are mentioned earlier clearly support the view of Hernqvist (1984) on the α-modification. In the α-modification, close to the methyl end plane, the fatty acid chains are disordered like a lamellar liquid crystalline phase. Close to the glyceryl group, chain mobility is absent (Figure 9.12).

9.6.1.1   Supercooling of the α-Modification

One of the other arguments that is often mentioned in favor of chain mobility in the α-modification is the observation of van den Tempel (1979) that the α-modification, like lamellar liquid crystalline phase, cannot be supercooled (Muller, 1932).

This is not entirely correct. The authors have observed that in oil-in-water emulsions with a relatively small droplet size, supercooling of the α-modification could be obtained. Moreover the author has found a whole class of very slowly crystallizing fat blends, where the main crystallizing TAGs are of the unsymmetric hhu type, in which even in a bulk phase under shear supercooling of the α-modification could be obtained (Wesdorp and van Meeteren, 1989). Crystallization to equilibrium was extremely quick and took place in about 1 min. A noticeable transition to the β'-modification only occurred after about 1 h.

In all other fat blends investigated, the authors could not obtain supercooling of the α-modification in sheared bulk phases.

This effect in the experiments is described in Section 9.6.2.

9.6.1.2 Excess Gibbs Energy in the α-Modification

In normal crystalline phases, ideal mixing is only found when the components that are mixed and are nearly isomorphous (Zief and Wilcox, 1967). In all other cases, incorporation of a second component causes disturbances in the extremely regular crystal lattice, leading to nonideal mixing. However, the α-modification may be an exception. The large degree of liquid-like disorder that exists, especially in the methyl end-plane region, may very well enable the incorporation of TAGs with considerably longer or shorter fatty acid chains into the crystal "lattice," without causing much extra disorder or misfittings. In other words, the excess Gibbs energy of mixing is likely to be very small, or even zero in the α-modification.

There exists some evidence in the literature that in the α-modification of TAGs, considerable solid solubility occurs (Murray et al., 1981; Norton et al., 1985a), but the data are too inaccurate and too scarce to allow any quantification. However, the phase diagrams of the rotational phase–liquid equilibrium of n-alkanes (Wurflinger, 1972) indicate clearly that mixing in the rotational phase must be nearly ideal at atmospheric pressure. The topic will come back to the solid–liquid phase behavior of n-alkanes in Section 15.9.

Assume that TAGs mix ideally in the α-modification and with this assumption the α-melting range of a fat can be predicted. In the next section, it will be investigated whether such predicted α-melting ranges agree with the experimental melting ranges.

9.6.2 COMPARISON OF EXPERIMENTAL AND CALCULATED α-MELTING RANGES

9.6.2.1 Experimental Procedure

The α-modification is extremely unstable. In normal fats it transforms to the β′-modification within 3–20 min. This instability poses some problems in the determination of α-phase equilibria. The normal thermal techniques for studying solid–liquid equilibria, like DSC and DTA, are too slow. A faster method must be defined.

The fact that it is very hard to supercool the α-modification in a bulk fat phase that is subjected to shear, while it is very easy to supercool the β′- and β-modifications under those conditions, can be used to determine the α-melting range of a fat blend. The following procedure takes advantage of this phenomenon:

1. About 2 kg of the fat blend is heated in a stirred tank to a temperature at least 10°C above the slip melting point of the fat. Next the fat is circulated through a gear pump, two lab-scale scraped surface heat exchangers (SSHE) for cooling and a tubular heat exchanger for remelting, in the setup depicted in Figure 9.13. The volume of the SSHEs is 18 mL, the total cooling surface 7×10^{-3} m². As coolant ethanol is used at temperatures between 5°C and −20°C, samples can be taken directly after the SSHEs. The solid fat content of these samples can be measured in a Bruker Minispec p120i pulse NMR device. The NMR is operated with a 90_x-T-90_y pulse sequence, which not only gives the solid fat content, but also the f-factor, a number that is characteristic for the modification in which the fat has crystallized (van den Tempel, 1979).
2. The fat is cooled in the SSHEs to 1°C below its α-cloud point. Next the throughput is increased, while keeping the SSHE exit temperature constant,

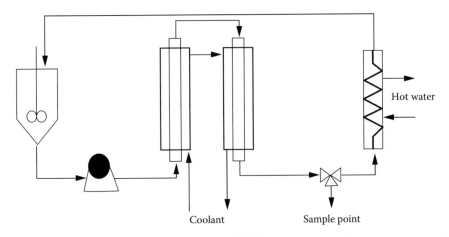

Coolant   Sample point

FIGURE 9.13 Experimental setup with two SSHEs and a remelter for measurements of α-melting ranges.

until no decrease in solid content is observed upon further increase of throughput. Usually this is obtained at about 16 kg/h (4.4 g/s). Thus it is made sure that the transition to the β′-modification has not yet started in the SSHEs and that the solid phase is still completely in the α-modification. Sample temperature and solid phase content are recorded.

3. Next the fat temperature is decreased in steps of about 2°C. When the solid phase content and temperature have stabilized, a sample is taken and the solid phase content and temperature are recorded. Eight data points are usually taken in duplicate for each fat blend.

4. The plot of solids content against temperature is called an α-line and represents the α-melting range of the fat blend.

During experiments, the pulse NMR always returned f-factors of about 1.25. An f-factor of 1.25 is indicative for the α-modification.

It may be argued that the α-line obtained in this way is not a line representing the α-liquid-phase equilibrium, but a line representing the maximum possible supercooling of the fat under the conditions used. Such lines of maximal supercooling usually depend strongly on the specific conditions, like impeller speed, that are applied (Pamplin, 1980; Grootscholten, 1986) (Figure 9.14a). However the α-lines, which were measured, were found to be totally independent of the rotational shaft speed of the SSHEs, coolant temperature, and throughout, once the throughput was high enough to prevent the formation a β′ phase in the SSHE (Figure 9.14b).

Nine fat blends were selected that cover the area of compositions and melting ranges normally occurring in household and industrial fat spreads. The TAG compositions of the fat blends were obtained by calculation from average TAG compositions of the fat blend components. The average TAG compositions of the fat blend components, like sunflower oil or hardened palm oils, were taken from the Edible Fats Database of Unilever Research Vlaardingen (Table 9.19).

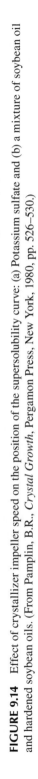

FIGURE 9.14 Effect of crystallizer impeller speed on the position of the supersolubility curve: (a) Potassium sulfate and (b) a mixture of soybean oil and hardened soybean oils. (From Pamplin, B.R., *Crystal Growth*, Pergamon Press, New York, 1980, pp. 526–530.)

TABLE 9.19

Summarized TAG Compositions of Nine Fat Blends of Commercial Fat Spreads

TAG	1	2	3	4	5	6	7	8	9
SSS				0.005	0.002			0.002	0.010
S_2P				0.009	0.005	0.001		0.004	0.020
SP_2	0.006	0.002	0.006	0.006	0.005	0.011	0.002	0.003	0.014
PPP	0.011	0.004	0.019	0.001	0.002	0.032	0.004		0.003
SES		0.001			0.007		0.002	0.004	0.025
PES	0.006	0.009			0.012	0.002	0.006	0.006	0.034
PEP	0.018	0.020			0.007	0.001	0.009	0.002	0.011
SSE		0.002			0.007		0.003	0.008	0.025
PSE	0.002	0.006			0.011	0.002	0.009	0.012	0.034
PPE	0.006	0.009			0.006	0.002	0.010	0.005	0.011
P_2m	0.004		0.001	0.006	0.002	0.004			0.004
PSm	0.002			0.020	0.004				0.013
S_2m				0.010	0.002				0.009
SOS	0.001	0.001	0.003	0.001	0.003	0.001	0.001	0.002	0.007
POS	0.020	0.015	0.052	0.002	0.007	0.021	0.004	0.004	0.010
POP	0.078	0.057	0.230	0.001	0.004	0.090	0.005	0.002	0.004
SIS			0.001	0.002	0.002			0.001	
PIS	0.008	0.005	0.009	0.004	0.004	0.008	0.003	0.003	0.002
PIP	0.027	0.014	0.035	0.003	0.003	0.029	0.004	0.002	0.002
SSO		0.001		0.002	0.003	0.001	0.002	0.005	0.007
PSO	0.010	0.007	0.008	0.004	0.008	0.013	0.006	0.009	0.010
PPO	0.033	0.016	0.031	0.003	0.005	0.052	0.007	0.004	0.004
SSI	0.001		0.001	0.003	0.002			0.002	0.001
PSI	0.011	0.005	0.005	0.008	0.005	0.007	0.005	0.006	0.003
PPI	0.030	0.009	0.009	0.006	0.004	0.020	0.007	0.005	0.004
SE_2	0.002	0.018			0.041	0.010	0.031	0.027	0.084
PE_2	0.009	0.034			0.036	0.014	0.050	0.021	0.056
Pm_2	0.013			0.011	0.001				0.002
Sm_2	0.003			0.018	0.001				0.004
EEE	0.003	0.037			0.059	0.020	0.063	0.021	0.047
mmm	0.014			0.007					
SmE					0.010				0.034
PmE					0.012				0.022
E_2m					0.017			0.001	0.034
SEO	0.002	0.018			0.029	0.010	0.031	0.031	0.054
PEO	0.010	0.034			0.035	0.014	0.049	0.025	0.036
SEI		0.002			0.023	0.001	0.004	0.003	0.005
PEI	0.001	0.004			0.021	0.002	0.006	0.003	0.004
SmO	0.002			0.005					
PmO	0.012			0.003					
mOm	0.006								

TABLE 9.19 (continued)
Summarized TAG Compositions of Nine Fat Blends of Commercial Fat Spreads

TAG	1	2	3	4	5	6	7	8	9
mlm	0.001								
mmO	0.013			0.002					
mml	0.003			0.001					
E_2O	0.004	0.055			0.049	0.030	0.094	0.037	0.053
E_2l		0.007			0.058	0.004	0.012	0.004	0.005
SO_2	0.20	0.007	0.009	0.008	0.021	0.013	0.004	0.012	0.009
PO_2	0.120	0.039	0.049	0.012	0.028	0.101	0.011	0.014	0.010
Liquid	0.485	0.560	0.530	0.836	0.437	0.481	0.557	0.706	0.272

9.6.2.2 Calculations

The α-melting ranges were calculated using the standard multicomponent two-phase flash algorithm and Michelsen's stability test for the initial estimate, both described in Section 15.3. Ideal mixing in the solid and liquid phases was assumed; therefore, all activity coefficients were unity. The TAG compositions from Table 9.8 were used, while all pure component properties were estimated using the correlations developed in Section 15.4.

9.6.2.3 Results

The results of the calculations and the measurements are given in Figures 9.15 through 9.17.

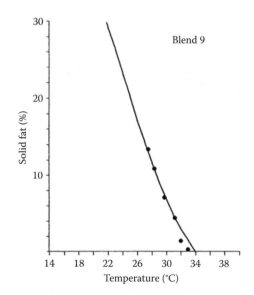

FIGURE 9.15 Measured (points) and calculated (lines) α-melting ranges of the commercial fat-blends of Table 9.19. The blend number indicated above the graphs corresponds to the compositions of Table 9.19.

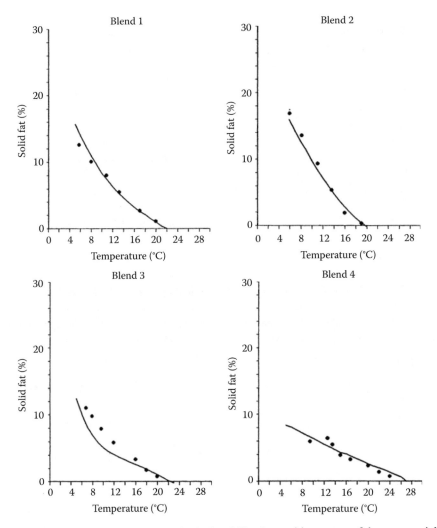

FIGURE 9.16 Measured (points) and calculated (lines) α-melting ranges of the commercial fat-blends of Table 9.19. The blend number indicated above the graphs corresponds to the compositions of Table 9.19.

The outcome is very surprising: it appears possible to predict the behavior of the extremely unstable α-modification by phase equilibrium thermodynamics. The agreement between predictions and measurements is striking.

9.6.3 CONCLUSION

A good description of the phase behavior of TAGs in the very unstable α-modification is obtained when it is assumed that TAGs form ideal solid solutions in the α-modification.

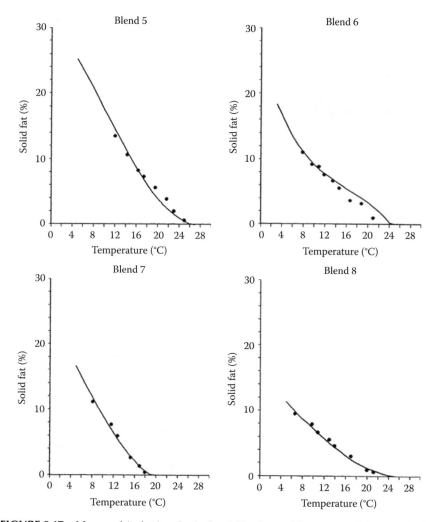

FIGURE 9.17 Measured (points) and calculated (lines) α-melting ranges of the commercial fat-blends of Table 9.19. The blend number indicated above the graphs corresponds to the compositions of Table 9.19.

9.7 MIXING BEHAVIOR IN THE β′- AND β-MODIFICATIONS

The miscibility of TAGs in the β′- and β-modifications is highly nonideal. It is attempted to describe this nonideal mixing with the simplest possible excess Gibbs energy models. A compilation of existing binary T, x, y solid–liquid phase diagrams is used to determine the binary interaction parameters that occur in these models. A new method is developed to determine binary interaction parameters from a single, complete DSC melting curve of ternary mixtures. Some ternary phase diagrams are considered.

9.7.1 Excess Gibbs Energy

It is well known (Murray et al., 1981; Timms, 1984) that the mixing behavior of TAGs in the β'- and β-modifications is highly nonideal. Therefore, in order to be able to solve the "solid-flash" problem of Chapter 3, we need to know the activity coefficient of each TAG in the β'- or β-solid phase as a function of the phase composition. Usually this is obtained from an excess Gibbs energy model:

The Gibbs energy of a phase is given by (Prausnitz, 1986)

$$G = \sum_{i=1}^{N} n_i \mu_i = G^{\text{ideal}} + RT \sum_{i=1}^{N} n_i \ln \gamma_i \tag{9.86}$$

when Equation 9.5 is used for the chemical potential. The excess Gibbs energy of this phase is subsequently defined as

$$G^{\text{E}} = G - G^{\text{ideal}} = RT \sum_{i=1}^{N} n_i \ln \gamma_i \tag{9.87}$$

As the chemical potential is by definition the partial molar Gibbs energy, it follows that an activity coefficient is a partial molar excess Gibbs energy:

$$\mu_i = \left(\frac{\partial G}{\partial n_i} \right)_{P,T,n_j} \Rightarrow RT \ln \gamma_i = \left(\frac{\partial G^E}{\partial n_i} \right)_{P,T,n_j} \tag{9.88}$$

This implies that once a model for the excess Gibbs energy has been formulated, the activity coefficients, needed for solving the set of equilibrium Equations 9.4 through 9.8 to obtain the number of phases and the amount and composition of each phase present, are readily obtained.

9.7.1.1 Excess Gibbs Energy Models

Hardly any literature on excess Gibbs energy models for solid mixtures is available. In thermodynamic calculations, it is nearly always assumed that solid phases are pure phases. Deviations between experiments and calculations are explained with "mixed crystal formation", but without any quantification.

Kitaigorodskii (1973) and Haget et al. (1985) propose a parameter, the "degré d'isomorphisme crystallin" that expresses how well a molecule will fit into the crystal of another compound. The parameter was successfully used for qualitatively predicting complete, partial, or no miscibility in the solid phase of mixtures of several substituted naphthalenes, but could not be used for quantitative predictions.

In a few cases, existing liquid phase excess Gibbs energy models were applied to solid mixtures: a regular solution model for petroleum waxes (Won 1986, 1989; Hansen et al., 1987) and van Laar and Wilson equations (Prausnitz, 1986) for sodium carbonate and sodium sulfate (Null, 1980). General guidelines cannot be distilled from the literature.

The excess Gibbs energy of a pure phase is zero: $G^E = 0$ (if $x_i \rightarrow 1$). The simplest relation meeting this requirement is the regular solution or two suffix Margules model. For a binary system, it is given by (Prausnitz, 1986)

$$g^E = A_{12}x_1x_2 \rightarrow RT \ln \gamma_1 = A_{12}x_2^2 \tag{9.89}$$

and for multicomponent system by

$$g^E = \sum_{i=1}^{N} \sum_{j=i+1}^{N} A_{ij}x_ix_j$$

$$RT \ln \gamma_i = -g^E + \sum_{j=1, j\neq i}^{N} A_{ij}x_j \tag{9.90}$$

It contains one interaction parameter per binary.

Usually the very regular crystal lattice of a pure component will be disturbed when a molecule of another size is incorporated. This results in a positive excess Gibbs energy. Therefore, normally the interaction parameters A_{12} in the solid phase will be positive and the activity coefficients will be greater than unity.

The two suffix Margules equation is symmetric: a mixture of 10% PPP and 90% SSS would have the same excess Gibbs energy as a mixture of 10% SSS with 90% PPP. However, symmetric behavior in the solid phase is unlikely; the effect of incorporating a large molecule in a crystal lattice of smaller molecules probably differs from the effect of the reverse case. The two suffix Margules equation is therefore not the most obvious excess Gibbs energy model to be used for solid fats.

The simplest models, able to describe demixing, that can account for the expected asymmetric behavior are the van Laar and the three suffix Margules equations. Both contain two interaction parameters per binary pair. More complex models are not justified (Prausnitz, 1986): they require very accurate and extensive data, which are not available for TAGs and in view of the experimental difficulties probably impossible to obtain. Because of its somewhat wider versatility (Prausnitz, 1986) the 3-suffix Margules equation is used in this work. For a binary system it is given by

$$g^E = (A_{21}x_1 + A_{12}x_2)x_1x_2$$

$$RT \ln \gamma_1 = x_2^2 \left[A_{12} + 2(A_{21} - A_{12})x_1 \right] \tag{9.91}$$

$$\ln \gamma_1^\infty = \frac{A_{12}}{RT}, \quad \ln \gamma_2^\infty = \frac{A_{21}}{RT}$$

The main disadvantage of 3-suffix Margules equation (and also of the van Laar equation) is its lack of a rational base for extension to multicomponent systems. An extra assumption has to be made. Assuming that an i, j pair in the multicomponent system gives the same contribution to g^E as in the binary mixture at the same relative concentrations (Fischer and Moeller, 1985), then the multicomponent 3-suffix Margules equation becomes

$$g^E = \sum_{i=1}^{N} \sum_{j=i+1}^{N} \left(A_{ij} \frac{x_j}{x_i + x_j} + A_{ji} \frac{x_i}{x_i + x_j} \right) x_i x_j$$

$$RT \ln \gamma_i = -g^E + \sum_{j=1, j \neq i}^{N} x_j \left(\frac{A_{ji}(x_i^2 + 2x_i x_j) + A_{ij}x_j^2}{(x_i + x_j)^2} \right)$$

(9.92)

When both interaction parameters are equal, the equations reduce to the 2-suffix Margules equations.

9.7.1.2   Regular or Athermal?

The parameters in the Margules equations are usually taken to be independent of temperature. This is equivalent to assuming that the excess entropy equals zero. Mixtures with this behavior are called regular. The opposite case is assuming that the excess enthalpy is zero. Mixtures with this property are called athermal. The interaction parameters are inversely proportional to temperature. In reality, the situation usually lies in between: both excess enthalpy and entropy deviate from zero.

The data of Haget and Chanh (Haget et al., 1985) for substituted naphthalenes in the solid state as well as the data of Maroncelli et al. (1985) and Snyder for n-alkanes in the solid state suggest that these mixed solid phases of chemically very similar compounds show nearly regular behavior. The temperature differences that are considered in this chapter are probably too small to be able to discriminate between both extreme situations. The authors could not obtain any improvement of fit to binary phase diagrams of TAGs by replacing the assumption of regular behavior with athermal mixing. Therefore, it is arbitrarily assumed in this work that TAGs form regular mixtures.

9.7.1.3   Phase Diagram

The solid–liquid phase behavior of two components is usually represented by a T, x, y phase diagram. Figure 9.18a and b illustrates the types of binary T, x, y phase diagrams that can be obtained assuming an ideal liquid phase (Section 15.4) and a nonideal solid phase that is described by the 2- or the 3-suffix Margules equation.

The following aspects feature:

1. The eutectic composition is determined almost completely by the difference in melting point of the two components when the 2-suffix Margules equation is used. Using the 3-suffix Margules equation, the eutectic composition can still be shifted about 0.05 mol fraction in both directions.
2. Peritectic diagrams cannot be obtained using the 2-suffix Margules equation.
3. The position of the liquidus is hardly influenced by the magnitude of the interaction coefficients:
   - If the value of the interaction coefficients A/RT exceeds 4
   - If the difference between the melting points of the components is more than about 15°C

In the first case, that of a eutectic diagram, the magnitude of the interaction coefficients only influences the position of the solidus at mole fractions less than 0.1 or greater than 0.9. In the second case, the position of the solidus hardly depends on the value of the interaction parameters at mole fractions of the lowest melting component less than or equal to 0.5. Only at very low concentrations of the highest melting component there is influence of the interaction parameters on the liquidus. However, in this region, the experimental errors are normally most pronounced.

According to Timms (1984), four types of phase diagrams are observed for TAGs: monotectic with continuous solid solubility, eutectic, monotectic with partial solid solubility, and peritectic diagrams. Only the 3-suffix Margules equation is able to describe all these diagrams.

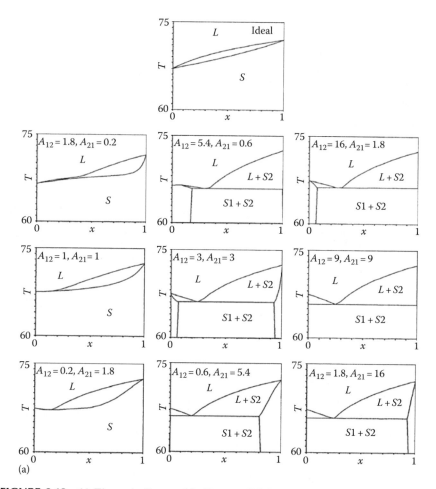

FIGURE 9.18 (a) Theoretically possible T, x, y solid–liquid phase diagrams of PSS–SSS calculated using the 3-suffix Margules equation. The values of the binary interaction parameters are indicated.

(continued)

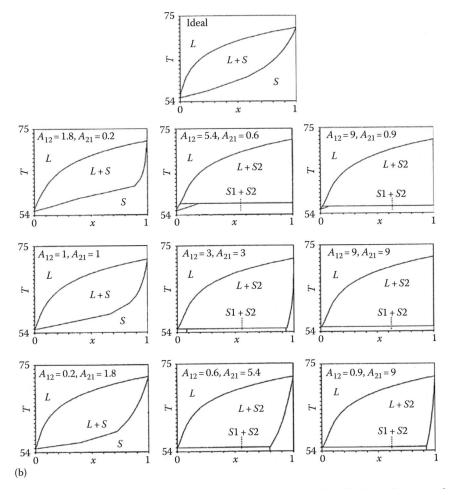

FIGURE 9.18 (continued) (b) Theoretically possible T,x,y solid-liquid phase diagrams of MMM-SSS calculated using the 3-suffix Margules Equation 9.91. The values of the binary interaction parameters are indicated in the figure.

9.7.1.3.1 Nomenclature

Henceforth the liquidus and solidus that are obtained when ideal mixing in the solid phase is assumed will be called ideal liquidus and ideal solidus. The interaction parameters of the excess Gibbs energy models are zero. The liquidus and the solidus that are obtained in the opposite case, when no mixing in the solid phase occurs, will be called eutectic liquidus and eutectic solidus. Then the interaction parameters of the G^E models are infinite, which is in practice equivalent to values of A/RT greater than or equal to 9.

The eutectic liquidus is also obtained from the well-known Hildebrand equation (Equation 9.39), which assumes the absence of mixed solid phases. In the literature, this liquidus is often referred to as the ideal solubility line, in spite of the highly nonideal solid phase behavior that is implicitly assumed.

A point on the liquidus is called the clear point and a point on the solidus will be referred to as the softening point of a mixture.

9.7.1.3.2 Use of Phase Diagrams to Determine Interaction Parameters

Experimental binary phase diagrams of TAGs can be used to determine the interaction parameters that occur in an excess Gibbs energy model. Using a fitting procedure, the interaction parameters are adjusted until the calculated and measured phase behavior agree.

There is one important limitation: in the aforesaid cases where the position of the liquidus hardly depends on the magnitude of the interaction parameters, it will be very hard to obtain reliable values of those interaction parameters from a phase diagram. That would require a number of solidus points at mole fractions less than 0.1 and greater than 0.9. Such data points are seldom available.

Therefore, the use of phase diagrams to determine interaction parameters is in practice limited to binaries of components that still show considerable solid miscibility and that differ less than about 15°C in melting point. It is possible to fit the phase diagram of other binaries; however, without learning much of the performance of the G^E model that is applied.

9.7.2 EXPERIMENTAL PHASE DIAGRAMS OF TAGs

9.7.2.1 Measuring Phase Diagrams

Rossell (1967) reviews all phase diagrams of TAGs that were published before 1967. Although he gives no quantitative interpretation of the phase diagrams, he recognizes the main problems that occur when determining a binary solid–liquid T, x, y phase diagram:

1. Impurities
2. Incomplete and incorrect stabilization
3. Other experimental errors

9.7.2.1.1 Impurities

Impurities generally lead to an increase of the observed melting range and may cause an erroneous picture of mutual solid solubility. The position of both liquidus and solidus are affected by the presence of impurities. Good determination of TAG purity by GLC, TLC, or HPLC is only possible since about 1965. Therefore older data may be suspect.

9.7.2.1.2 Stabilization

Due to the extremely low diffusion rates in the solid phase, lengthy stabilization procedures are required to ensure that the solid phase composition has its equilibrium value. The polymorphic behavior of TAGs complicates the stabilization even more. Part of the solid phase may persist in an unstable polymorphic form with a deviating equilibrium composition. Except for the combination of very short stabilization with very quick measurements, the position of the liquidus usually does not depend very strongly on the stabilization procedure. The small amount of solid phase at temperatures just below the clear point can relatively easily recrystallize via the liquid phase to the equilibrium solid phase composition. Stabilization strongly influences the position of the solidus.

Proper stabilization requires lengthy schemes of temperature cycling between clear point and softening point. These can take several months to a year.

Often the mixtures are stabilized by long storage at a temperature several degrees below the softening point. This procedure is certainly not sufficient; in this way unstable polymorphic forms are "frozen." It was shown that they could persist several years (de Bruijne and Reckweg).

9.7.2.1.3   Other Experimental Problems

In older reports, the technique usually applied for measuring phase diagrams is the thaw-melt technique with visual observation of softening and clear points. Visual observation of clear points and especially of softening points is very inaccurate and large errors both in liquidus and solidus may occur.

The method most applied at present is some form of thermal analysis, DTA or DSC. Problems that can occur when using DSC are as follows:

1. Thermal lag because of a large sample size or scan rate. The softening point remains in place, but the observed clear point is shifted to a higher temperature. It becomes more difficult to detect small heat effects.
2. Failure to detect melting peaks. Combination of a large (>15°C) difference in melting point, low concentration of one of the components, and some solid solubility leads to broad, diffuse humps rather than sharp melting peaks in the DSC thermogram. The determination of the exact starting and end temperature of these humps is very difficult. The hump itself is easily overlooked. A set of false liquidus and solidus points results.

An alternative method, which is sometimes applied, is measurement of the solid phase content of a set of binary mixtures at different temperatures by wide-range pulse NMR (or before 1970 by dilatometry) (Rossell, 1967). The phase diagram can be constructed from the resulting melting lines or solubility curves (Grootscholten, 1987; Smith, 1988). This method is much more time consuming than the thermal methods. Due to the inaccuracy of the measurements (0.7% solids), the solidus and liquidus are obtained with a relatively large experimental error.

9.7.2.1.4   Illustration

The influence of the experimental problems on the quality of the phase diagram is illustrated in Figure 9.19. Two sets of simulated (Section 9.3.5) DSC curves of the binary POP-PEP are given. The first set was generated using the 2-suffix Margules equation with A/RT of 3, resulting in a mutual solid solubility of about 15%. For the second set the 3-suffix Margules equation was used with $A_{pop\text{-}pep}/RT$ of 0 and $A_{pep\text{-}pop}/RT$ of 6, resulting in zero solubility of PEP in solid POP and 50% solubility of POP in solid PEP. Moreover 3% of impurities, a thermal lag of 0.2°C, and 3% random noise were included. These levels of impurities, thermal lag, and noise are quite normal for literature data on TAGs.

Perfect (i.e., in absence of thermal lag, impurities, etc.) measurements on the first system would result in DSC curves with a sharp spike at the eutectic temperature followed by a broader hump that ends abruptly at the clear point. Due to the presence of impurities and thermal lag the eutectic spike is broadened to 2°–4°.

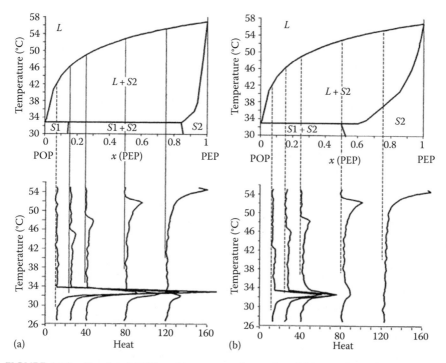

FIGURE 9.19 Simulated DSC melting curves (bottom) and binary phase diagrams (top) for a number of different mixtures of POP and PEP. The trajectory through the binary phase diagram that is followed is indicated for each mixture. (a) simulation with $A_{12} = A_{21} = 3$. (b) simulation with $A_{12} = 0$ and $A_{21} = 6$.

This causes an uncertainty in the position of the solidus of the same size, while the shape of the solidus remains correct.

In the second system, the sharp spike at the eutectic temperature disappears and is replaced by a broad hump at PEP concentrations over 50%. The onset of this hump is hard to determine exactly. At 75% PEP, the hump becomes very broad. In the simulated curve, it is no longer detectable. The softening point of this mixture is 36°C. However, the softening point that one would read from the simulated DSC curve is the onset of the second melting peak, at about 46°C, a difference of 10°C.

The influence that the experimental errors can have the measured phase diagram becomes clear in Figure 9.20. In this figure, the simulated phase diagram is plotted together with pseudoexperimental solidus and liquidus points that are determined from the simulated experimental. DSC melting curves. The original simulated liquidus and the experimental liquidus points correspond very well, but the solidus points deviate considerably from the original solidus. At the POP side of the diagram, the experimental points lie far below the original solidus, while they are positioned above the ideal solidus at the PEP side of the phase diagram.

The POP-PEP binary was measured by Lovergren (de Bruijne et al., 1972). As in the simulated DSC curve, it is impossible to detect a start melting point between

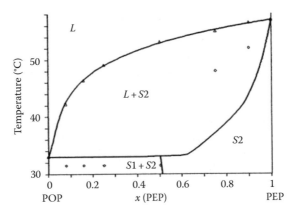

FIGURE 9.20   Solidus and liquidus points read from the simulated DSC curves of Figure 9.19 with the phase diagram that formed the basis for the simulation of the DSC curves.

30°C and 40°C in Lovegren's curve for 75% PEP. Whatever the solubility of POP in PEP would be, the start melting point would have to lie in this temperature range, which implies that the softening point that is read from this curve is wrong.

Solidus points that lie below the eutectic solidus at the side of lowest melting component and suddenly shift to values far above the ideal solidus on the other side of the phase diagram are frequently reported for TAGs with a large difference in melting point (Rossell, 1967). Often, in about half of the cases, contradicting reports exist, showing no solidus points that lie above the ideal solidus. An example from the data of Kerridge for SSS-LLL (Kerridge, 1952), which are in contradiction to the data of Lutton (1955).

A thermodynamic consistency test on the data is not possible. Heats of mixing, which are needed to check if the isobaric, nonisothermal Gibbs–Duhem equation is obeyed, are not available.

Compound formation, which would explain a solidus that lies above the ideal solidus, is very unlikely: compound formation is only known between pairs of TAGs made up of the same fatty acids, but attached to different positions of the glycerol group (like POP and PPO). The binaries for which deviating solidi are reported normally have a large difference in melting point, which stems from large differences in size or shape. It is therefore much more likely that the deviations are caused by impurities, improper stabilization, and failures in detecting the true softening point.

9.7.2.2   Literature Overview

It is clear that the determination of solid–liquid T, x, y phase diagrams is a tedious and unrewarding job, which is probably the reason why since Rossell's review (Rossell, 1967) so few phase diagrams were reported.

De Bruijne et al. (1972) have published the phase diagrams of all binary combinations of TAGs with palmitic (P) and stearic (S) acid. They put much effort in stabilization and purification. Their data seem therefore to be of good quality. They always expected the sharp double-peaked DSC curve that is characteristic for eutectic behavior. Therefore, they missed the solidus in those cases where due to solid solubility the DSC melting curve had another shape.

De Bruijne also determined the phase diagrams of tribehenate (BBB) with tristearate (SSS) and tripalmitate (PPP) (de Bruijne and Reckweg), and of POP with PPO in the β'-modification (de Bruijne and Reckweg). Unfortunately, these diagrams are only of little value as hardly any stabilization was used.

Perron (Perron et al., 1971) and coworkers have also measured the systems of de Bruijne. They do not report any purity of their TAGs and did not stabilize the mixtures at all. Their DSC curves are full of polymorphic transitions, partially overlapping with melting peaks. They report β'-stable PSS and PPS while these TAGs are known to be beta stable (de Bruijne et al., 1972; Gibon, 1984). Consequently, the resulting diagrams can better be disregarded.

In later work, Perron (Ollivon and Perron, 1979) again reports some of these phase diagrams (SSS–PSP, PSP–SPP, and SSS–SPP), but now with very pure components and a longer stabilization procedure.

Gibon (1984) also measured the same systems and those of some of the TAGs occurring in palm oil (POP, PPO, OPO, and POO). Although she worked with very pure samples, she put no effort at all in stabilization. Most of her phase diagrams are therefore, like the early Perron work, useless for our purpose. Gibons work is continued by De Smedt and Gibon (1990) with TAGs that contain elaidic acid (SES, SSE, PEP, PPE, EPE), but unfortunately in the same manner.

Krautwurst (1972) gives the binary diagram of PPP–MMM and some data of the ternary PPP–MMM–LLL. He does not give purities and followed a reasonable stabilization procedure.

The binary SSS-trioctanoate (888) was determined by Barbano and Sherbon (1977) with a stabilization of 23 months and very pure TAGs.

Grootscholten (1987) determined the liquidi of SSS–SES, SSS–SSE, SSE–SES, SSS–SEE, and of PPP and SSS with OOO using a reasonable stabilization procedure.

Smith (1988) reports the only known good quality ternary diagram: a well-stabilized ternary diagram of POP/SOS/POS in their most stable polymorphic form determined by NMR.

Lovegren (de Bruijne et al., 1972) gives the DSC melting curves of the binary POP-PEP after several stabilization procedures, from which a diagram can be constructed.

Norton et al. (1985) gives some liquidus points of PPP and SSS with OOO determined by DSC. His end melting peaks show a strange tail for which the reason is not clear.

Only Grootscholten interpreted his phase diagrams applying an activity coefficient approach (Grootscholten). All other authors only state, if something, whether their data deviate from the Hildebrand equation (eutectic liquidus and solidus).

Also outside the area of edible oils and fats phase diagrams of organic compounds that show similar phase behavior as TAGs, like substituted naphthalenes (Haget et al., 1985) or sulfolane with compounds as dimethysulfoxide and dioxane (Janelli, 1985) are only interpreted with the Hildebrand equation for pure solids. Solid solutions have not been described quantitatively. Related phase diagrams, between a liquid crystalline phase and a gel phase of phospholipids (e.g., stearyl-palmityl-glycero-phosphocholine and dimyristyl-glycero-phosphocholine, 2 lecithins) have been described quantitatively with the 2-suffix Margules equation. In these systems, stabilization is consider ably less important. Values of A/RT between 0 and 1.5 are found (Hagemann, 1975).

9.7.2.3 Fitting Experimental Phase Diagrams

The ability of the 2-suffix and 3-suffix Margules equations to describe non ideal mixing in solid fats was tested by fitting these G^E models to the experimental binary phase diagrams. The Simplex method (Brumbaugh et al., 1990) was used for parameter adjustment. In Section 9.7.2.1, it was concluded that solidus data may be very inaccurate. Therefore, only the liquidus was fitted.

Clear points were calculated using Michelsens' stability test (Section 9.3.2.2). This guaranteed that the solid phase was in the right polymorphic form. Next the solidus data were used to fine-tune the interaction parameters within the error margin that resulted from the fit to the liquidus.

As pointed out in Section 9.7.1.3, accurate values of binary interaction parameters can only be obtained from binaries of components that have a small difference in melting point and a reasonable mutual solid solubility. Typically the requirements are an interaction parameter A/RT less than 3 and a difference in melting point less than 15°C. For all other cases, only rough indications of the values of the interaction parameters can be given, using the solidus points that are available.

The phase diagram was rejected as a whole if

- The reported melting points differ more than 4°C from the accepted values
- The reported liquidus lies on average more than 2°C above the ideal liquidus or more than 2°C below the eutectic liquidus
- The eutectic composition differs 0.08 mol fraction or more from the composition that follows from the melting points alone

Solidus points that lie more than 3°C above the ideal solidus were neglected: as is outlined in Section 15.7.3.1, it is likely that the actual softening point was not detected in the experiments. When published or when a lab journal was still available, the original data were used, but usually the data had to be read from the graphs in the publications (Appendix 9.D).

9.7.2.4 Saturated TAGs

9.7.2.4.1 SSS, SSP, PSP, SPS, PPS, and PPP

These are the TAGs that occur in fully hydrogenated vegetable oils as palm oil, soybean oil, sunflower oil, rape seed oil, and safflower oil (Figures 9.21 through 9.25). The interaction parameters and residual errors in predicted liquidus lines (the square root of the quotient of the sum of squared errors and the number of degrees of freedom) are given in Table 9.20, and the phase diagrams in Figures 9.21 through 9.25.

Taking into account that the experimental error that is claimed is 0.3°C–0.4°C, the fit to the data is usually most satisfying. In spite of this good fit, the experimental error causes a rather large uncertainty in the interaction parameters. The uncertainty is even larger in the case of PSP. PSP crystallizes in the β′ modification, and the phase diagrams are therefore mixed β–β′. Consequently the 3-suffix Margules parameters determined from these diagrams that indicate the solubility in β-PSP and in the β′-modification of the other TAG will show a very large error.

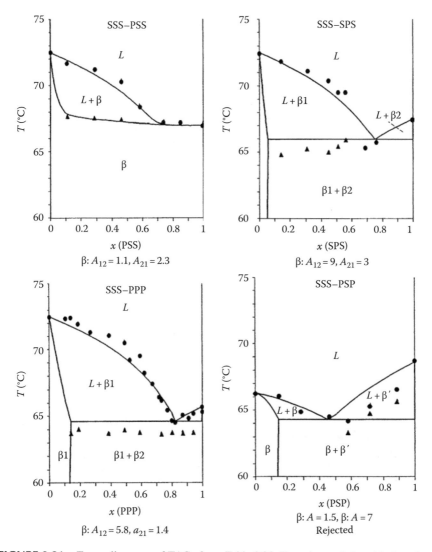

FIGURE 9.21 T, x, y diagrams of TAGs from Table 9.20. Experimental data (circles: clear points; triangles: softening points) and best fit (lines) with the 2- or 3-suffix Margules equation ($a = A/RT$).

The data of Lutton (1955) and de Bruijne and Reckweg for SSS-PPP are in good agreement, while the data of the other authors mentioned by Rossell (Kerridge, Kung, Joglekar) disagree and cannot be fitted at all.

Lutton's and de Bruijne's data for SSS–PSP also agree very well. Both deviate from the data of Perron (Ollivon and Perron, 1979), which show a greater discrepancy between solidus and liquidus. When fitting the G^E models to Perron's data, the residual error is much larger, but similar interaction coefficients are obtained.

The diagrams of SSS–PSS, SSS–PSP, PSS–SPS, PSS–PPS, PSS–PPP, SPS–PSP, and SPS–PPS are perfectly acceptable and are described rather well.

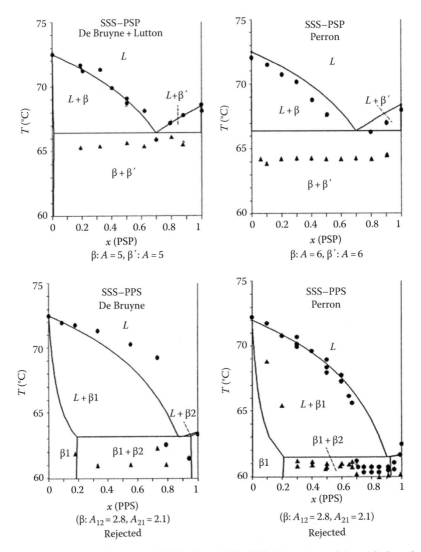

FIGURE 9.22 *T, x, y* diagrams of TAGs from Table 9.20. Experimental data (circles: clear points; triangles: softening points) and best fit (lines) with the 2- or 3-suffix Margules equation (*a = A/RT*).

The 3-suffix Margules parameter ASSS–SPS is inaccurate, due to lack of data on the SPS side of the diagram. The same holds for the PPP side of the PSP–PPP diagram.

The data for SSS–PPS of de Bruijne are poor; part of the data lies above the ideal liquidus, while the data points on the PPS side of the diagram lie below the eutectic liquidus. The solidus points indicate that the 2-suffix Margules interaction parameter should exceed 2.2. The liquidus of Perron for this binary is somewhat better, although it still does not meet the acceptance criteria of Section 9.7.2.3; on average, the liquidus lies more than 2°C below the eutectic liquidus.

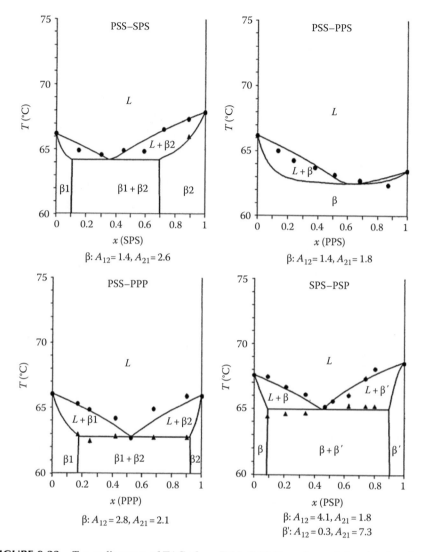

FIGURE 9.23 T, x, y diagrams of TAGs from Table 9.20. Experimental data (circles: clear points; triangles: softening points) and best fit (lines) with the 2- or 3-suffix Margules equation ($a = A/RT$).

At the PSP side of the PSS–PSP diagram, the experimental clear points also lie far below the eutectic liquidus, indicating poor quality of the data. Best fit to the PSS side of the diagram gives an $A(\beta)/RT$ of 1.5 for this system.

The liquidi of Perron and de Bruijne for the system PSP–PPS agree very well, the solidi are in complete disagreement. Best fit to the clear points results in a nearly eutectic solidus, in agreement with de Bruijne's results. The best fit to the complex solidus of Perron, which shows a peritectic and an eutectic, results in a liquidus that fits poorly to the data.

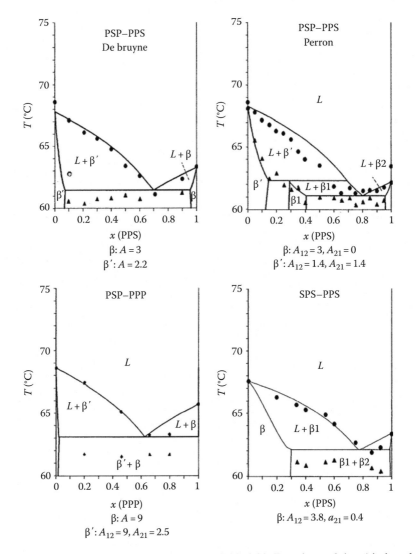

FIGURE 9.24 T, x, y diagrams of TAGs from Table 9.20. Experimental data (circles: clear points; triangles: softening points) and best fit (lines) with the 2- or 3-suffix Margules equation ($a = A/RT$).

In the diagram of PPS–PPP, there are two data points on the PPP side that lie more than 1° below the eutectic liquidus. Those points were given very low weight during fitting. The same occurs in the diagram of SPS–PPP, resulting in a very inaccurate 3-suffix Margules parameter for the solubility of SPS in PPP.

9.7.2.4.2 Saturated Mono-Acid TAGs

The melting points most of these TAGs differ considerably. Therefore no exact values for the interaction parameters can be obtained. In this case fitting is merely

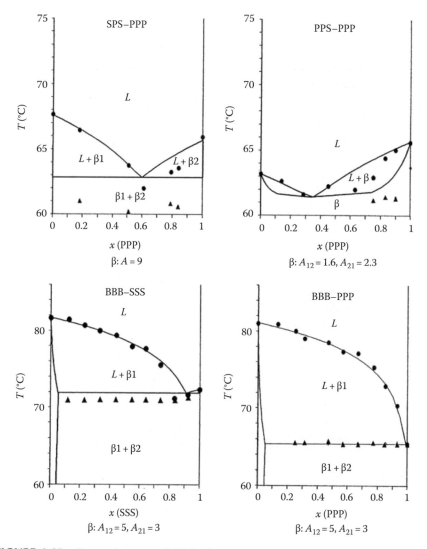

FIGURE 9.25 T, x, y diagrams of TAGs from Tables 9.20 and 9.21. Experimental data (circles: clear points; triangles: softening points) and best fit (lines) with the 2- or 3-suffix Margules equation ($a = A/RT$).

a check whether the reported liquidi agree with expectations (Table 9.21, Figures 9.25 and 9.26).

Indeed all liquidi are described very well. The data of Kerridge and Barbano contain solidus points that lie far above the ideal solidus.

Lutton's data for the systems with LLL confirm that these solidus points are indeed not right. The available solidus points indicate that in all systems solid solubility is only very limited.

TABLE 9.20

Binary Interaction Parameters for the P/S TAGs and Root Mean Square Error between Experimental Points and Fitted Liquidus

Binary	2-Suffix Margules		3-Suffix Margules			Ref
	A/RT	RMSE	A_{12}/RT	A_{21}/RT	RMSE	
SSS–PSS	1.6	0.3	1.1	2.3	0.3	A
SSS–PSP	3–7	0.3	3–7	3–7	0.4	A/B
β'	3–7		3–7	3–7		
SSS–SPS	3–9	0.7	3–9	2–4	0.8	A
SSS–PPS	>2.2?	1.7	—	—	—	A
SSS–PPP	3	0.8	5.8	1.4	0.7	A/B
PSS–PSP	1.5?	1.2	—	—		A
β'	7?		—	—		
PSS–SPS	2.0	0.4	1.4	2.6	0.3	A
PSS–PPS	1.6	0.4	1.4	1.8	0.5	A
PSS–PPP	2.2	0.5	2.8	2.1	0.5	A
PSP–SPS	1.4	0.3	1.8	1–6	0.3	A
β'	1.0		2–7	0.3		
PSP–PPS	3–9	0.9	3–9	1–9	0.9	A
β'	2.2–9		1–9	2.2–9		
PSP–PPP	3–9	0.7	0–9	3–9	0.8	A
β'	0–9		0–9	0–9		
SPS–PPS	2.0	0.6	3–5	0.4	0.4	A
SPS–PPP	5–9	0.8	5–9	5–9	0.8	A
PPS–PPP	1.9	0.7	1.6	2.3	0.7	A
Perron:						
SSS–PSP	5–9	0.7	5–9	5–9	0.8	C
	5–9		5–9	5–9		
SSS–PPS	2.1?	2.4	—	—	2.4	C
PSP–PPS	—	—	3	0	—	C
		1.4	1.4			

Sources: Data from A: de Bruijne, P. and Reckweg, F., *PVD*, 72, 3275; (B): Lutton, E.S., *JAOCS*, 32, 49, 1955; and (C): Perron, R., Petit, J., and Matthieu, A., *Chem. Phys. Lipids*. 6, 58, 1971.

Note: Standard error in the constants is 0.5. When two sets of constants are given, the first are values for the beta and the second set is for the β'-modification.

9.7.2.5 Saturated TAGs + Trans-TAGs

The similarity between elaidic acid and stearic acid is reflected in the nearly ideal miscibility of SSE and SES with SSS (Figures 9.27 and 9.28).

The interaction parameters for PPP–SSE are very similar to that of PPP and SSS. However, the data of Kung for this system look unreliable: there is a very poor

TABLE 9.21

Binary Interaction Parameters and Root Mean Square Error between Experimental Points and Fitted Liquidus for Saturated Mono-Acid TAGs

Binary	2-Suffix Margules		3-Suffix Margules			
	A/RT	RMSE	A_{12}/RT	A_{21}/RT	RMSE	Ref
BBB–SSS	3–9	1.0	4–9	2–9	1	A
BBB–PPP	3–9	0.5	3–9	2–9	0.5	A
SSS–PPP	3	0.8	5.8	1.4	0.7	A/B
SSS–LLL	3–9	0.9	3–9	3–9	0.9	B/C
SSS–888	3–9	2.4	3–9	0–9	2.4	D
PPP–MMM	3–9	0.3	3–9	2–9	0.3	E
PPP–LLL	3–9	0.3	5–9	2–9	0.3	B/C

Sources: Data from (A) de Bruijne, P. and Reckweg, F., *PVD*, 71, 3637; de Bruijne, P., and Reckweg, F., *PVD*, 72, 3275; (B) Lutton, E.S., *JAOCS*, 32, 49, 1955; (C) Kerridge, R., Unpublished work, Private communication to J.B. Rossel, 1952; (D) Barbano, P. and Sherbon, J.W., *JAOCS*, 55, 478, 1977; (E) Krautwurst, J., *Keiler Milchw. Forsch. Ber.,* 22, 255, 1972.

Note: Standard error in the parameters is 0.5, unless otherwise indicated.

correspondence between liquidus and solidus and the liquidus lies well below the eutectic liquidus near the eutectic point. Therefore the interaction parameters cannot be very reliable either.

Clement's data for PPP–EEE had to be disregarded: the liquidus and solidus lie too far above the ideal liquidus and solidus. Simultaneously, Clements measured the system LLL–EEE, using the same technique and batch of EEE. The data are therefore suspect. The fit to these data is with a residual error of 20%, which is rather poor. The interaction coefficients for EEE–LLL should therefore not be taken too seriously (Table 9.22).

9.7.2.6 Saturated TAGs + Mono- and Di-Unsaturated TAGs

Due to the large differences in melting points, it will only be possible to get a coarse indication of the magnitude of the interaction parameters (Figures 9.28 and 9.29).

A complicating factor in the interpretation of these phase diagrams is the fact that the saturated TAGs crystallize in the β-2 modification, while the *cis*-unsaturated TAGs crystallize in the β-3 modification. Complete miscibility of a saturated and a *cis*-unsaturated TAG is impossible: an intermediate β'-2/β'-3 structure is not feasible. In thermodynamic calculations, the two β'-forms must therefore be treated as separate modifications, just like the β'- and β-modifications. This requires β-3 melting points and enthalpies of fusion of β-2 stable TAGs and vice versa. In line with the findings of de Jong (1980) and the observations in Section 15.4, the melting point of the β-form

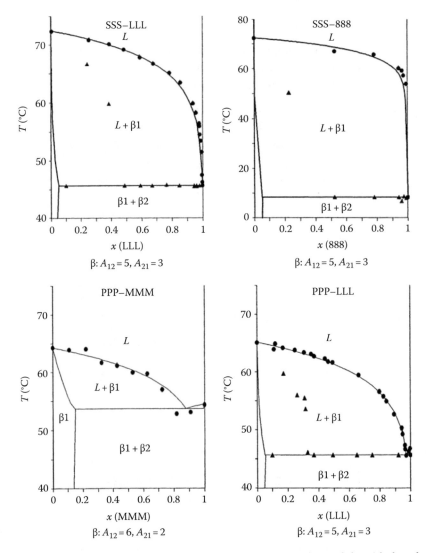

FIGURE 9.26 T, x, y diagrams of TAGs from Table 9.21. Experimental data (circles: clear points; triangles: softening points) and best fit (lines) with the 2- or 3-suffix Margules equation ($a = A/RT$).

that does normally not occur in the pure substance lies 3°C below the normal β melting point, while the enthalpy of fusion is 90% of that of the normal β-enthalpy of fusion.

The liquidi of SSS–SOS (Lutton) and of PPP–POO (Gibon) are described very well. In those cases also, good correspondence with the solidi could be obtained (Table 9.23).

The liquidi of PPP–SOS (Kung, 1950), PPP–POP (Kerridge, 1952), and the data for PPP–POO of Morancelli et al. (1985) are rejected: do not meet the criteria of Section 9.7.2.3: the liquidus lines lie too far below the eutectic liquidus and large parts of the solidus lines are found far above the ideal solidus.

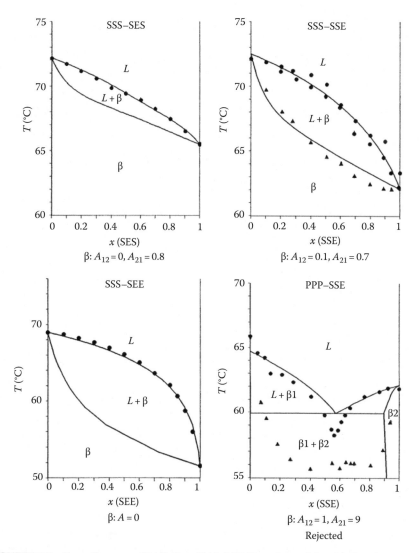

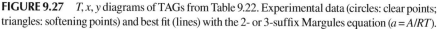

FIGURE 9.27 T, x, y diagrams of TAGs from Table 9.22. Experimental data (circles: clear points; triangles: softening points) and best fit (lines) with the 2- or 3-suffix Margules equation ($a = A/RT$).

9.7.2.7 Unsaturated TAGs

9.7.2.7.1 Mono-Unsaturated TAGs

SSE and SES show ideal miscibility, in line with the good solid solubility of both components in SSS. Elaidic acid behaves like stearic acid (Figures 9.29 and 9.30).

Smith determined solidus and liquidus in independent experiments using NMR. His liquidi of POS–POP, SOS–POP, and SOS–POS, which he obtained after extensive temperature cycling, fit very well. Smith's cycling procedure was less suited for obtaining a reliable solidus, and consequently the correspondence between the solidus and liquidus data is not perfect.

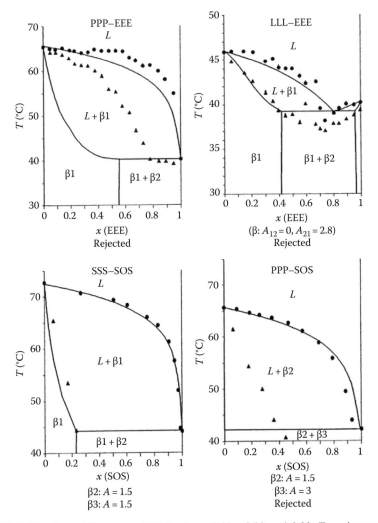

FIGURE 9.28 T, x, y diagrams of TAGs from Tables 9.22 and 9.23. Experimental data (circles: clear points; triangles: softening points) and best fit (lines) with the 2- or 3-suffix Margules equation ($a = A/RT$).

The diagrams of SOS–SSO and POP–PPO could not be fitted. For both systems, there is evidence for formation of a 50–50 compound (Timms, 1984), leading to a maximum in the liquidus lines. The β′-liquid phase diagrams for POP–PPO given by Morancelli et al. (1985) and de Bruijne and Reckweg are in complete disagreement. Morancelli obtained an eutectic at 15% POP, while de Bruijne found an eutectic near 50% POP, in agreement with the eutectic composition that is expected from calculations. Phase diagrams in the unstable β′-modification are suspect: during slow scanning in the DTA or DSC equipment the transition to a more stable modification will occur very easily when a large amount of liquid phase is present. Consequently, not all clear points may be those of the unstable modification.

TABLE 9.22

Binary Interaction Parameters and Root Mean Square Error between Experimental Points and Fitted Liquidus for Saturated + *Trans*-Containing TAGs

Binary	2-Suffix Margules		3-Suffix Margules			Ref
	A/RT	RMSE	A_{12}/RT	A_{21}/RT	RMSE	
SSS–SES	0.4	0.2	0	0.8	0.2	A
SSS–SSE	0.4	0.7	0.1	0.7	0.5	A/B
PPP–SSE	—	—	1?	9?	0.9	B
PPP–EEE	—	—	—	—	—	C
LLL–EEE	—	—	2.8?	0?	—	C

Sources: Data from (A): Grootscholten, P.A.M., Unilever Research Vlaardingen, Private communication, 1987; (B): Kung, H.C., Internal Unilever Report LB 113, 1950; (C): Clements, M.S. and Rossell, J.B., Internal Unilever Report P WN 660044, 1966.

Note: Standard error in the constants is 0.5, unless otherwise indicated.

The binary POP–PEP is another example of a phase diagram of two TAGs, which crystallize in different β-forms: PEP crystallizes into the β-2 form, while POP crystallizes into the β-3 form.

9.7.2.7.2 Other Unsaturated TAGs

All phase diagrams reported for these TAGs seem very unreliable (Figures 9.30 and 9.31). Rossell (1967) gives three diagrams of SOS–SOO, from McGowan, Morancelli, and Rossell. The three diagrams all disagree and have liquidi far below the eutectic liquidus. Therefore they are unreliable. There is one similarity: all diagrams have peritectic point at 24% SOS and about 27°C. This peritectic point can be obtained using the values of the interaction parameters given in Tables 9.24 and 9.25.

The liquidus data of Gibon for PPO–POO are described by the eutectic liquidus. The data were obtained after a stabilization at room temperature for a year. The mixtures are still partially liquid at room temperature and it may be assumed that sufficient recrystallization had occurred to render a reasonable liquidus. The solidus was obtained directly after crash cooling and is therefore unreliable. Moran gives for PPO–POO a liquidus far below the eutectic liquidus. His data were therefore not further used.

Both diagrams of Morancelli for POP–POO also have a liquidus below the eutectic liquidus and were therefore disregarded. In Gibon's DSC curves for POP–POO the final melting peak overlapped with a β′–β transition. Stabilization has been insufficient.

The data of Morancelli for PPO–OPO are very well described by the eutectic liquidus, but the solidus again lies far above the ideal solidus.

Morancelli reported two diagrams for POO-OPO, of which the one given in Figure 9.14 has an eutectic composition where it is expected, at about 50% POO.

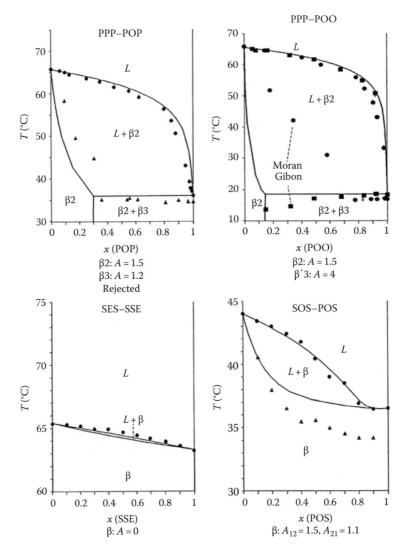

FIGURE 9.29 T, x, y diagrams of TAGs from Tables 9.23 and 9.24. Experimental data (circles: clear points; triangles: softening points) and best fit (lines) with the 2- or 3-suffix Margules equation ($a = A/RT$).

Strangely enough Morancelli judged his other diagram, with a eutectic at 17% POO, more reliable. The correspondence with the solidus is poor in both cases.

Morancelli's diagram for POP–OPO has to be disregarded: the liquidus lies far below the eutectic liquidus. Morancelli explained this by assuming compound formation, but if a compound would have formed, part of the liquidus certainly had to be situated above the eutectic liquidus.

It seems that most of Morancelli's data are not reliable, which arouses doubts about the two diagrams of Morancelli (PPO–OPO and POO–OPO) that did meet the acceptance criteria of Section 9.7.2.3.

TABLE 9.23

Binary Interaction Parameters and Root Mean Square Error between Experimental Points and Fitted Liquidus for the Saturated + Unsaturated TAGs

Binary	2-Suffix Margules		RMSE	Ref
	$A\beta$-2/RT	$A\beta$-3/RT		
SSS–SOS	>1.5	>1	2	A
PPP–SOS	>1	>1	4	B
PPP–POP	>1.3	>1	3.6	C
PPP–POO	1.5	4 (β'-3)	1	D

Sources: Data from (A): Grootscholten, P.A.M., Unilever Research Vlaardingen, Private communication, 1987; (B): Kung, H.C., Internal Unilever Report LB 113, 1950; (C): Kerridge, R., Unpublished work, Private communication to J.B. Rossel, 1952; (D): Gibon, V., Thesis, Universite Notre Dame de la Paix, Namur, Belgium, 1984.

Note: Standard Error in the Constants is 0.5, unless otherwise indicated.

9.7.2.7.3 Triolein

Indeed, all data lie reasonably well on the calculated liquidus, the data of Grootscholten being slightly better than those taken from Rossell (1967; Figures 9.31 and 9.32).

In these diagrams, the difference in melting points is so large that the ideal and eutectic liquidus and solidus almost coincide.

Fitting is in this case merely a test of the quality of the data and does not give much information about the extent of nonideal mixing (Table 9.26).

9.7.2.8 Summarizing

Nearly 120 phase diagrams were considered:

- 84 had to be rejected because they were clearly not correct.
- The liquidus lines were fitted very well for 34 of the remaining 36 phase diagrams. The average deviation from the experimental curves was only 7% of the difference between maximal and minimal liquidus temperature.
- In two cases, no fit could be obtained, which was attributed to compound formation that was not accounted for in the thermodynamic modeling.
- The experimental softening points lay far above the ideal solidus at the side of the higher melting component in five of the 34 remaining diagrams. These five binaries all concerned TAGs with a large difference in melting point. It was made plausible that this deviation was due to an experimental error. Experimental evidence that confirmed this exists for two of the five binaries.

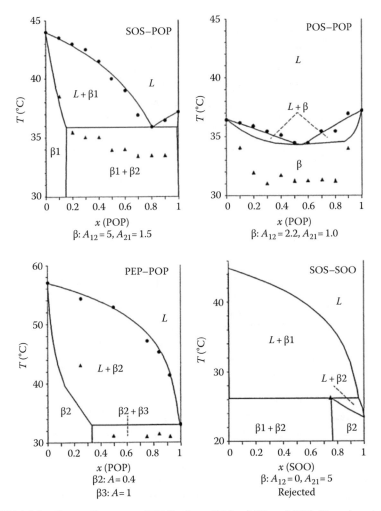

FIGURE 9.30 *T, x, y* diagrams of TAGs from Tables 9.24 and 9.25. Experimental data (circles: clear points; triangles: softening points) and best fit (lines) with the 2- or 3-suffix Margules equation (*a* = *A/RT*).

- The difference between the melting points of two components was less than 15°C in 24 of the 34 binaries that remained. Eleven of these 24 diagrams were almost completely eutectic. This implies that only in 13 diagrams the interaction parameters have a clear influence on the shape of the liquidus. Therefore these 13 diagrams are the most demanding test cases for the performance of the excess Gibbs energy models.

The average root mean square error between measured and fitted liquidus for these 13 remaining diagrams is 0.4°C. This is approximately the experimental error that is claimed. It is impossible to obtain very accurate values of the interaction parameters with an experimental error of this magnitude.

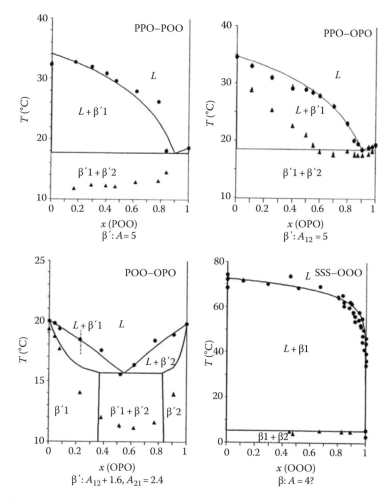

FIGURE 9.31 T, x, y diagrams of TAGs from Tables 9.25 and 9.26. Experimental data (circles: clear points; triangles: softening points) and best fit (lines) with the 2- or 3-suffix Margules equation ($a = A/RT$).

Based on the available data, it is not possible to express a preference for one of the two excess Gibbs energy models used. Hardly any improvement in fit was obtained when the 3-suffix Margules equation was used. Peritectic diagrams that could not be described using the 2-suffix Margules equation, had to be disregarded. Generally the experimental softening points were better described with the 3-suffix Margules equation.

9.7.3   Alternative to Phase Diagram Determination

9.7.3.1   How to Proceed?

The evaluation of all available phase diagrams of TAGs in the previous Section (9.7.2.8) has resulted in interaction coefficients for only 35 binaries. The uncertainty in these interaction parameters is considerable, due to the large experimental error in the diagrams.

TABLE 9.24

Binary Interaction Parameters and Root Mean Square Error between Experimental Points and Fitted Liquidus for Mono-Unsaturated TAGs

Binary	2-Suffix Margules		3-Suffix Margules			Ref
	A/RT	RMSE	A_{12}/RT	A_{21}/RT	RMSE	
SES–SSE	0	0.2	0	0	0.2	A
SOS–SSO	—	—	—	—		B
SOS–POS	1.3	0.2	1.5	1.1	0.2	C
SOS–POP	3.6–6	0.3	5	1.5	0.3	C
POS–POP	1.6	0.5	2.2	1.0	0.4	C
POP–PPO	—		—	—	—	E
POP–PEP	0.4 ± 1	0.9	—	—	—	D
B-3::	1 ± 0.8					

Sources: Data from (A) Grootscholten, P.A.M., *LPVD*, 3074, 84; (B) Freeman, I.P., Internal Unilever Report, Port Sunlight Program, 45, Spring 1957; (C) Smith, K., Unilever Research Colworth, Private communication, 1988; (D) Lovegren, N.V., *JAOCS*, 53, 519, 1976; (E) Morancelli, M., Strauss, H.L., and Snyder, R.G., *J. Phys. Chem.*, 89, 5260, 1985.

Note: Standard Error in the Constants is 0.5, unless otherwise indicated.

TABLE 9.25

Binary Interaction Parameters and Root Mean Square Error between Experimental Points and Fitted Liquidus for Unsaturated TAGs

Binary	2-Suffix Margules		3-Suffix Margules			Ref
	A/RT	RMSE	A_{12}/RT	A_{21}/RT	RMSE	
SOS–SOO	—		0?	5?	—	A
PPO–POO(β′)	1–9?	1.7	1–9?	1–9?	1.7	C
PPO–OPO(β′)	5–9?	0.6	4–9?	2–9?	0.6	B
POO–OPO(β′)	2.0?	0.3	1.6?	2.4?	0.3	B

Sources: Data from (A) Rossell, J.B., *Adv. Lip. Res.*, 5, 353, 1967; (B) Morancelli, M., Strauss, H.L., and Snyder, R.G., *J. Phys. Chem.*, 89, 5260, 1985; (C) Gibon, V., Thesis, Universite Notre Dame de la Paix, Namur, Belgium, 1984.

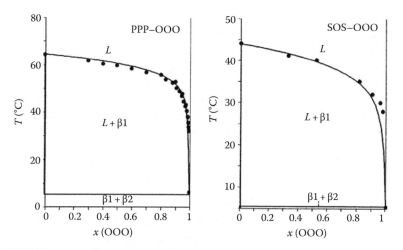

FIGURE 9.32 T, x, y diagrams of TAGs from Table 9.27. Experimental data (circles: clear points, triangles: softening points) and calculated lines.

TABLE 9.26
Binary Interaction Parameters and Root Mean Square Error between Experimental Points and Fitted Liquidus for Triolein

	2-Suffix Margules		3-Suffix Margules			
Binary	**A/RT**	**RMSE**	**A_{12}/RT**	**A_{21}/RT**	**RMSE**	**Ref**
SSS–OOO	0–9	4	—	—		A/B
PPP–OOO	0–9	3	—	—		A/B
SOS–OOO	0–9	2	—	—		A

Sources: Data from (A) Rossell, J.B., *Adv. Lip. Res.*, 5, 353, 1967; (B) Grootscholten, P.A.M., Unilever Research Vlaardingen, Private communication, 1987.

Interaction parameters were almost exclusively obtained for the β-modification. However, it is of much greater importance to know the parameters for the polymorph that is normally found in edible fat products, the β′-modification.

The determination of binary interaction parameters that occur in an excess Gibbs energy model by fitting this model to experimental binary T, x, y solid–liquid phase diagrams has a number of very serious drawbacks:

- It is extremely troublesome, time-consuming, and probably almost impossible to determine accurate phase diagrams. It takes months to years before the samples have been stabilized properly, minor impurities have large effects, and the true start and end melting point of a mixture are very hard to determine.
- Relatively small experimental errors in the position of the liquidus and solidus lead to large uncertainty in the interaction parameters that are

determined from the diagrams. It is questionable whether more accurate β-data can be obtained by determining binary solid–liquid phase diagrams.

- It is not possible to measure reliable phase diagrams of unstable modifications, like a binary β′-phase diagram of β-stable TAGs.
- Determination of binary interaction parameters from phase diagrams is only feasible for TAGs that differless than about 15°C in melting point. At larger melting point differences, the position of liquidus and solidus becomes too insensitive to the value of the interaction parameters.

Determining phase diagrams of TAGs to obtain binary interaction parameters is therefore not practical and another method has to be defined.

9.7.3.2 Formulation of an Alternative Method

The problem of the extremely long stabilization times can probably be solved. In a solid binary system, diffusion rates are extremely low, so that unstable modifications and crystals with a nonequilibrium composition have been shown to persist for years (Gibon, 1984). But if a surplus of a liquid TAG is added to the binary system, things become different. The phase diagram of such a ternary system (PSP/SEE/OOO) is shown in Figure 9.33. When the temperature is sufficiently above the melting point of the liquid component (OOO), the solidus surface in the diagram coincides with the binary side plane of the two crystallizing TAGs (PSP and SEE), even when ideal mixing in the solid phase occurs. The concentration of the liquid component in the solid phase is then negligible. The solid phase in this ternary system is still a binary mixture.

If the fat is crystallized in very small (<1 μm) crystals, to create a large exchange surface with the liquid phase, then the solid fat can recrystallize relatively easily via

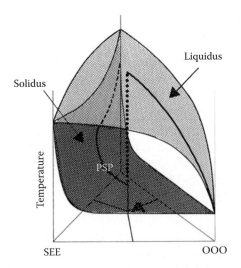

FIGURE 9.33 Schematic ternary phase diagram of the system PSP/SEE/OOO in the β′-modification. The softening point surface is dark shaded, while the clear point surface is light shaded. The lines indicated with solidus and liquidus represent the composition of the solid and liquid phase of a mixture with composition A while it is being melted.

the liquid phase to its equilibrium composition and modification (Zief and Wilcox, 1967; Norton et al., 1985).

Phase diagrams are usually determined by measuring the softening and clear point of a set of mixtures using DSC. Due to impurities and thermal lag that occur both points can only be determined with an accuracy of about 0.2°C–1°C.

In fact, this application of DSC does not use all available information: a DSC curve of a mixture is taken and next the two most undefined points on the curve, its start and end point, are used, while everything in between, that contains lots of information about the phase behavior of the binary system, is wasted. If the complete DSC curve is used for determination of the interaction parameters, rather than only 2 points, it should be possible to increase the accuracy of results with considerably less measurements. In Section 9.3.5, a method to simulate the effect of nonideal mixing in the solid phase on the shape of DSC curves was defined. This method can be applied to fit a complete DSC curve by adapting the binary interaction parameters.

The method for determining interaction parameters that evolve is to fit the complete DSC curve of a binary mixture, to which a surplus of a liquid TAG is added. The liquid TAG assures shorter and better stabilization. The use of the complete curves reduces experimental error reduces the number of measurements and removes the limitation to binary pairs that have a difference in melting point less than 15°C.

9.7.3.3 DSC Curves of Binary Systems Dissolved in a Liquid TAG

Figure 9.34 gives the types of DSC curves that can be obtained for TAGs that differ considerably in melting point by application of the 3-suffix Margules equation. Although in this situation the position of liquidus and solidus in the binary phase diagram is nearly invariant, still considerable differences in curve shape are obtained.

The following aspects feature:

1. All curves contain two melting peaks. The shape of the final melting peak and the start of the first melting peak are hardly influenced by the values of the interaction parameters. This is in line with the invariance of the phase diagram. However, the shape of the first melting peak and the height of the plateau in between the two peaks depend very much on the values of the interaction parameters.
2. The first melting peak ends sharply if demixing in the solid state occurs.
3. If no demixing occurs, the complete shape of the first melting peak is indicative for the value of the interaction parameters, which implies that the interaction parameters can be determined very accurately.
4. If demixing occurs, the height of the plateau just after the first peak is indicative for the magnitude of the interaction parameters. This implies that in these cases, the interaction parameters can be determined less accurately, due to the influence of noise and the uncertainty in the baseline.
5. Above values of $A/RT = 4$ the curve shape does not depend on the magnitude of the interaction parameters any more. However, as at $A/RT = 4$ already nearly complete solid phase immiscibility occurs, this is not a real problem.
6. For the 2-suffix Margules equation peak shape and peak position are closely related: upon increasing magnitude of the interaction parameters the first

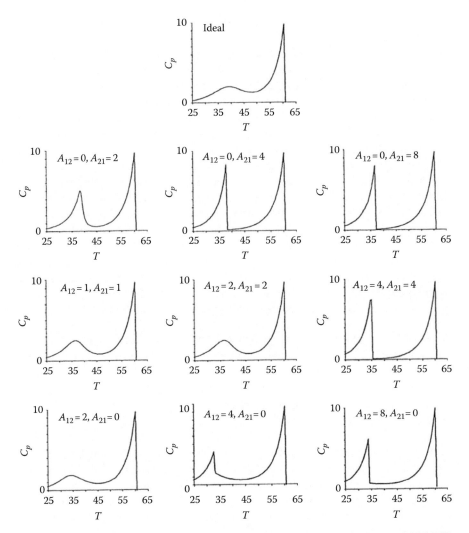

FIGURE 9.34 Theoretically possible DSC melting curves of a ternary mixture of 25% PSP, 25% SOS, and 50% OOO crystallized in the β′-2 modification. The curves are calculated using the 3-suffix Margules equation. The values of the binary interaction parameters are indicated.

melting peak becomes sharper and shifts toward lower temperatures. Therefore 2-suffix Margules parameters can be determined very easily. Bringing in asymmetric behavior with the 3-suffix Margules equation allows shifting of sharp peaks to higher temperatures and of broad humps toward lower temperatures. In fact, if the component with the highest melting point is [1] and the other [2], then the shape of the first melting peak is mainly determined by A_{21} and its position mainly by A_{12}. The 3-suffix Margules parameters determined from the curves will necessarily have a larger error, as the uncertainty in the melting points of 1°C–1.5°C has to be taken into account now.

9.7.3.4   What Experiments?

In most edible fat products, the fat has crystallized into the β′-modification. There are no reliable data available about miscibility in the β′-modification.

For fat fractionation calculations, it is important to know the solubility of mono-unsaturated TAGs in the solid phase of saturated TAGs both in the β′- and β-modifications. It is shown in Section 15.7.3 that the scarce information that exists is unreliable.

In many edible fat products up mostly partially hydrogenated fats where used up to 1995 to provide structure. These are rich in elaidic acid. After 1995 these have been gradually replaced in the industry by palm-oil based fats where structure comes from TAGs of the h2u and h2m type and nowadays elaidic acid rich fats are hardly used (Wesdorp, 1996; Dijkstra et al., 2008).

Of the many systems that can be studied, it seems therefore most relevant to determine interaction parameters of systems where monounsaturated and *trans*-containing TAGs are combined with saturated TAGs.

In order to enable the determination of β′-interaction parameters, assume that during determination of the DSC curve, which must be taken at low scan rates in order to reduce thermal lag, no transition to the β-modification will occur. Near the clear point the solid phase usually almost completely consists of the component with the highest melting point. If this component is β-stable, then a transition is very likely. However, if for this component a β′-stable TAG is taken, the chances of keeping the system sufficiently long into the β′-modification to allow a measurement are much higher. Moreover, even if in this case a solid phase in the β-modification, rich in the crystallizing component with the lower melting point, comes about, the DSC curve will still contain a melting peak of a mixed β′ phase that holds information about the miscibility in the β′-modification.

In view of this, the authors have selected a number of systems, of a β′ stable saturated TAG and a number of mono- and di-unsaturated TAGs, including some that contain elaidic acid:PSP with SOS, SSO, POP, PPO, PEE, EPE, SEE, ESE, and EEE. MPM with SOS, SSO, POP, PPO, PEE, EPE, SEE, ESE, and EEE.

These systems allow the study of the influence of a difference in size as well as the influence of position and nature of the unsaturated chains. The data themselves are relevant for normal edible fat products.

9.7.4   Experimental

9.7.4.1   Principles of DSC

With DSC, two cups, one containing a few milligrams of the sample to be investigated and the other containing an inert reference material, are placed in two identical microfurnaces. Both furnaces are mounted in cavities of a large aluminum block or heat sink. The aluminum block is kept at a constant temperature, well below the temperature range of the experiment. In the base of each furnace are two identical platinum resistance elements. One is used to provide heat to the furnace, the other that is mounted directly under the furnace base, serves as temperature sensor (Figure 9.35).

The equipment is controlled by two control loops: an average temperature control or scan loop that enables simultaneous heating of both microfurnaces at a constant heating rate (in °C/min). The other loop is a differential temperature control loop from

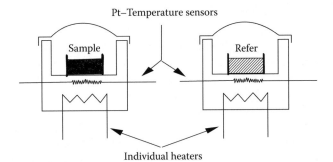

FIGURE 9.35   Schematic view of a DSC apparatus.

which the instrument output signal is obtained. This loop adjusts the heat flow to one of the furnaces if a process in one of the samples takes place that gives out or takes up heat, such that the temperature difference between the two cups always remains zero.

The difference in the amount of energy that must be supplied to the samples to heat both at the same rate plotted as a function of temperature is the DSC curve. It is directly proportional to the apparent heat capacity of the sample as a function of temperature.

DSC is often confused with DTA. In DTA, both samples are heated in the same microfurnace with a constant heat flow. The temperature difference that occurs between the samples, when the sample gives out or takes up heat, is recorded, the DTA curve. The DTA curve contains essentially the same information as the DSC curve. With DTA, fast heating and cooling rates as well as isothermal measurements are not possible and heat effects are derived quantities rather than directly measured quantities. However DTA is technically less complex.

9.7.4.2   Thermal Lag

The main problem in the use of DSC curves is the thermal lag that occurs. During scanning, the temperature of the inner part of the sample lags behind that of the apparatus. This thermal lag can completely disturb the shape of the DSC curve at large sample sizes and scan rates (Figure 9.36).

It is best to check for the presence of thermal lag by obtaining the steepness of the peak at the end melting point. Ideally this should be a straight line perpendicular to the temperature axis. At very low scanning rate or small sample size, thermal lag is negligible, but the DSC signal becomes very weak, leading to noisy DSC curves. In the experimental procedure, the right balance between thermal lag and sensitivity must be found by trial and error.

9.7.4.3   Experimental Procedure

The work was largely carried out using a Perkin Elmer DSC-2, equipped with an IBM-AT for data acquisition and control of the apparatus by means of software developed at the Department for Thermal Analysis of the Unilever Research Vlaardingen. Some work was done on a Perkin Elmer DSC-7, which is mechanically equivalent, but equipped with more advanced Perkin Elmer machinery for data acquisition and control, making it more sensitive. The heat sink was cooled with a solid CO_2/acetone

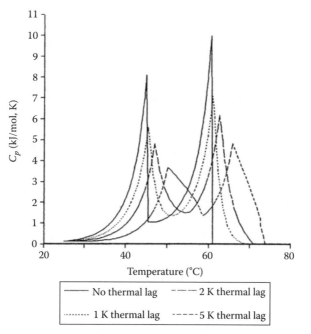

FIGURE 9.36 Influence of thermal lag on the curve shape of the system 25% PSP, 25% SEE, and 50% OOO (simulated curves).

mixture (DSC-2) or liquid nitrogen (DSC-7). The equipment was calibrated using pure indium (T_f = 156.6°C) and gallium (T_f = 29.8°C).

About 10 mg of a sample was weighed into an aluminum cup, which was closed and sealed. The sample was inserted into the apparatus, melted at 80°C and kept for 10 min at that temperature to prevent possible memory effects. Next the sample was cooled at 20°C/min to 5°C below the temperature where the melting of a fully eutectic β′ phase would take place. The sample was stabilized for 1 min at that temperature and subsequently heated at 1°C–2°C/min. The melting curve was recorded. This procedure is repeated several times with varying stabilization times: 1 min stabilization, 5 min, 15 min, 1 h, 3 h, 24 h, 1 week, and 1 month.

Usually stabilization times less than 1 h resulted in melting curves of the β′-modification, sometimes disturbed by a transition to the β-modification. Longer stabilization times mostly lead to a 3-phase β–β′ liquid system.

Stabilization longer than a week usually did not result in changes any more. To verify that indeed crystallization had taken place in the right polymorphic form, x-ray diffractograms were taken after the same cooling and stabilization procedure for some of the mixtures.

The influence of stabilization time on the polymorphic form in which crystallization has taken place is illustrated in Figure 9.37 for the system PSP/SEE/OOO. The shortest stabilization time only shows melting peaks of β′-crystal phases. After somewhat longer stabilization, a β-melting peak starts to appear as a shoulder, while in the completely stabilized sample, a sharp melting peak of a SEE-rich β phase is present next to a melting peak of a PSP-rich β′ phase.

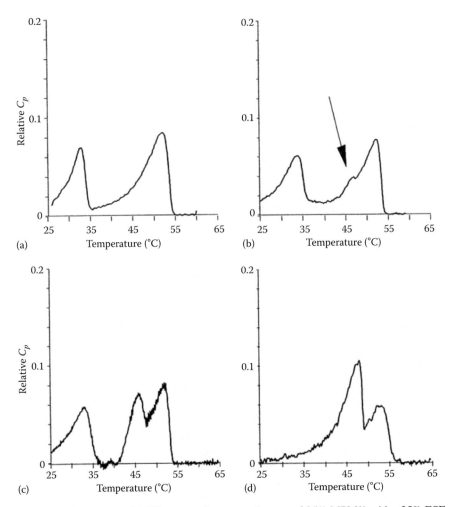

FIGURE 9.37 Measured DSC curves of ternary mixtures of 25% MPM% with ±25% ESE and ±50% OOO. Note the appearance of a melting peak of a separate ESE-rich β phase at 46°C and the disappearance of the melting peak of an ESE-rich β′ phase at 34°C upon increasing stabilization time. (a) = 1 min 20°C. (b) = 15 min 20°C. (c) = 60 min 20°C. (d) = 7 days 20°C.

At the sample size and scan rate used, thermal lag maybe neglected: the curve shape does not change significantly when the scan rate is further decreased, while the noise increases. Normally a scan rate of 2.5°C/min was used, sometimes, if no β′–β transition occurred and the noise level was acceptable a scan rate of 1.25°C/min was used.

The TAGs were obtained from Dr. A. Fröhling of the section Organic Chemistry of Unilever Research Vlaardingen. They were pure on TLC. To remove oxidation products and partial glycerides, the unsaturated TAGs were treated over a silica column. The TAGs were as extra purification recrystallized from hexane. To prevent oxidation during measurements and storage, 0.01% of butylhydroxyanisol (BHA) was added as antioxidant. The GLC and HPLC analysis results of the TAGs are given in Appendix 9.C.

9.7.4.3.1   Determination of Interaction Parameters

The interaction parameters were determined from the measured curves by adapting the parameters using a Simplex procedure until the sum of squared errors between the calculated and measured curve are minimal. Heats of fusion and melting points are obtained from the correlations of Section 9.4. As the melting points from the correlations, as well as experimental melting points, have an accuracy of about 1°C, we allowed the fitting procedure to vary the melting point maximally ±1°C around the value from the correlation. Both calculated as well as measured curves are normalized such that the area under the curves from 25°C up to the clear point equals unity. To reduce calculation time, the number of data points was reduced to one per 0.4°C. One iteration required about 40–60 s on a Compaq 386/25 PC under MS-DOS.

9.7.5   RESULTS

9.7.5.1   PSP and MPM with SEE and ESE

The measurements that were carried out and the results of the fitting procedure are given in Table 9.27 and Figures 9.38 and 9.39.

The fit to the measured curves is most satisfactory. The minor overshoot of the calculated curves at the top of sharp peaks must be ascribed to the inertness of the measuring technique, due to factors like thermal lag and the sampling time of the A/D converter in the DSC-2.

Only the 2-suffix Margules parameters are given, as no improvement of fit could be obtained by application of the 3-suffix Margules equation. While the interaction parameters for the β′-modification are indeed very accurate, the interaction

---

TABLE 9.27

DSC Measurements and 2-Suffix Margules Parameters for the Highest Melting Binary Pair Determined from These Measurements

Modifi-cation	PSP	MPM	SEE	ESE	OOO	Scan Rate (K/min)	Stabili-zation Time (min)	Stabili-zation Temp. (°C)	A/RT
β′	0.26		0.25		0.49	2.5	15–180	20	0 ± 0.2
β′		0.28	0.24		0.48	2.5	1–15	20	3.0 ± 0.2
β′	0.26			0.24		2.5	1–180	20–25	0 ± 0.2
β′						2.5	1–15	20	3.0 ± 0.2
β	0.26		0.25			2.5	14d	20	≥ 2.5
2.5 β						2.5	1d–3d	20	≥ 2.5
β	0.26		0.24			1.25	5d–28d	20–25	≥ 2.0
β						1.25	1m–7d	20	2 ± 1

Note: Composition in mole fractions, temperatures in °C, time in minutes, unless otherwise indicated: d $\triangleq$ days.

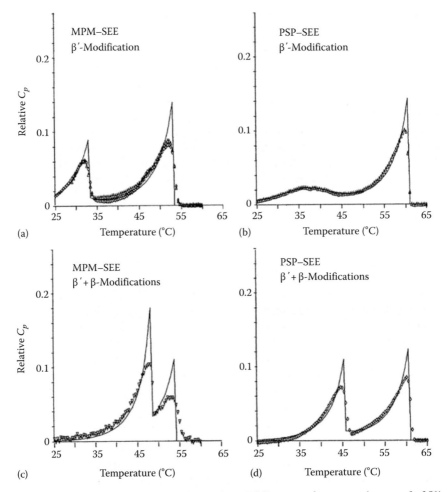

FIGURE 9.38 Measured (dots) and fitted (lines) DSC curves of ternary mixtures of ±25% PSP or MPM% with ±25% SEE and ±50% OOO. The modification(s) in which crystallization has taken place is indicated.

parameters for the β'-modification are only rough estimates. In combination with a β'-stable TAG, a solid phase split, leading to a DSC curve with two sharp peaks, will always occur. Only the position of the peak at the lower temperature end of the curve is indicative of the magnitude of β-interaction parameter, while its shape does not change.

The behavior of ESE and SEE is exactly the same, which is in correspondence to the close similarity of elaidic and stearic acid. Both mix ideally with PSP in the β'-modification, but show a solid solubility less than 15% in the β'-modification. With MPM, the solubility in the β'-modification is reduced to only 7%, showing the influence the increased difference in molecular size.

X-ray diffractograms for PSP-SEE-OOO are in line with these DSC observations: initially a typical β' diffraction pattern is obtained, with two strong maxima

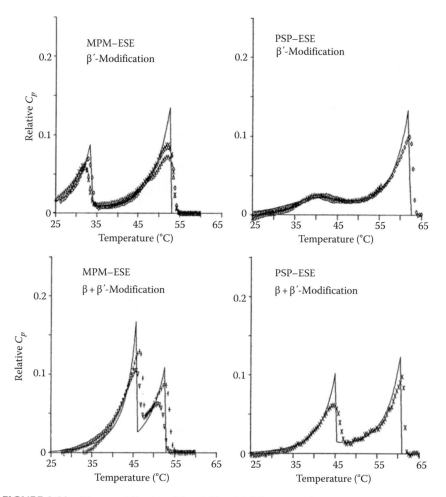

FIGURE 9.39 Measured (dots) and fitted (lines) DSC curves of ternary mixtures of ±25% PSP or MPM% with ±25% ESE and ±50% OOO. The modification(s) in which crystallization has taken place is indicated.

at 3.80 and 4.20 Å, which are characteristic of the β′-modification and a single long spacing, indicating the existence of only one crystalline phase. The diffraction pattern differs from that of pure PSP, indicating that a single mixed β′ phase indeed exists, in agreement with the complete miscibility derived from the DSC results (Figure 9.40).

After one day, the diffraction pattern has changed. The β′-short spacings are still present, but also a short spacing at 4.55 Å has come up, which is characteristic of the β-modification. The long spacing has doubled, indicating the presence of a second crystalline phase. The DSC results are confirmed again: after long stabilization, a β phase coexists with a β′ phase. The positions of the extra β-diffraction peaks correspond closely to those of pure SEE, in agreement with the prediction of a nearly pure β-SEE phase, which can be derived from the DSC curve.

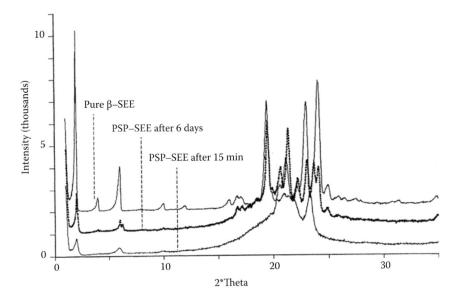

FIGURE 9.40 X-ray diffractograms of the mixture 25% PSP, 25% SEE, and 50% OOO, after 15 min and 6 days of stabilization at 20°C. The diffractogram of pure SEE is also given.

9.7.5.2 PSP and MPM with EPE and PEE

The measurements that were carried out and the results of the fitting procedure are given in Table 9.28 and Figures 9.41 and 9.42.

Naturally palmitic and elaidic acids do not show the similarity in behavior that elaidic and stearic acids do. Consequently PEE and EPE can be expected to show differences

TABLE 9.28

DSC Measurements and 2-Suffix Margules Parameters for the Highest Melting Binary Pair Determined from These Measurements

Modifi-cation	PSP	MPM	PEE	EPE	OOO	Scan Rate (K/min)	Stabili-zation Time (min)	Stabili-zation Temp. (°C)	A/RT
β′	0.26		0.25		0.49	0.62	14d	15	0 ± 0.2
β′		0.28	0.24		0.48	2.5	1–5	5–15	1.8 ± 0.5
β′	0.26			0.25	0.49	2.5	7d	15	2.0 ± 0.5
β′		0.27		0.25	0.48	1.25	3d	20	2.3 ± 0.5
β	0.26		0.25		0.49	0.62	30d	15	noβ
β		0.28	0.24		0.48	2.5	15–10d	15	≥2.5
β	0.26			0.25	0.49	2.5	7d	15	≥2.5
β		0.27		0.25	0.48	1.25	3d	20	≥3

d, days.

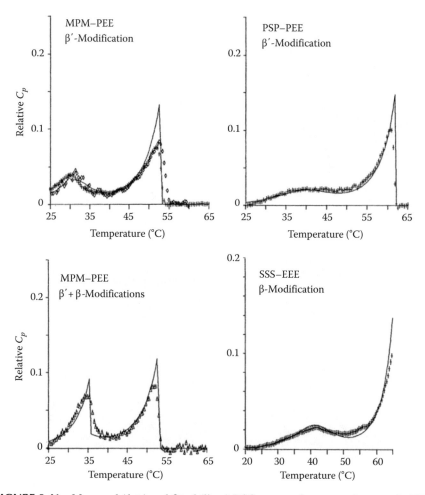

FIGURE 9.41   Measured (dots) and fitted (lines) DSC curves of ternary mixtures of ±25% PSP or MPM% with ±25% PEE and ±50% OOO and of a mixture of 25% SSS, 25% EEE, and 50% OOO. The modification(s) in which crystallization has taken place is indicated.

in mixing behavior. This is indeed observed: while PEE shows a complete miscibility with both PSP and MPM in the β′-modification, EPE shows only limited solubility.

PEE shows a very interesting behavior. The system PSP/ PEE/OOO has remained in the β′-modification, even after a month of stabilization. This is confirmed by the x-ray results for this system. Looking at the results obtained for SEE, ESE, and EPE, a realistic estimate for the PSP–PEE β-interaction parameter is $A/RT = 3$. In that case, calculations show that this system is β′-stable at all temperatures, although PEE itself is β-stable: the extra stability of the β-modification is not enough to compensate for the large excess Gibbs energy of a solid phase in that modification. A nearly pure β-PEE phase will only crystallize when the ratio PSP/PEE in the system drops below 0.5. But even for such mixtures the driving force to β is very low, making a speedy recrystallization unlikely. After 1 week of stabilization of such a mixture, only a

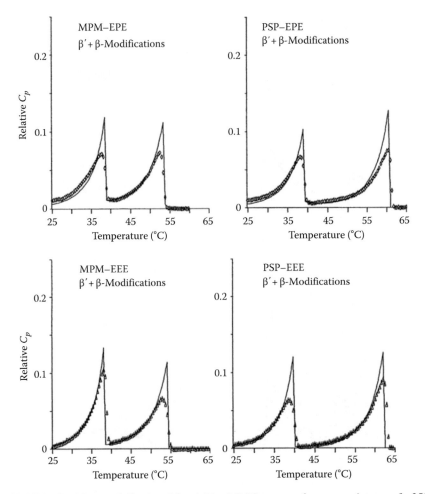

FIGURE 9.42 Measured (dots) and fitted (lines) DSC curves of ternary mixtures of ±25% PSP or MPM% with ±25% EPE or EEE and ±50% OOO. The modification(s) in which crystallization has taken place is indicated.

minor broadening of the first β′-melting peak at the position, where a β-melting peak would occur, was observed.

The system MPM/PEE/OOO does crystallize into the β-modification. Contrary to PSP and PEE, MPM and PEE show nonideal mixing in the β′-modification, and therefore the difference in excess Gibbs energy between the two modifications is not large enough to counterbalance the stability difference. The same holds for the PSP/EPE/OOO system that was studied.

Even without stabilization, the DSC curves of the systems with EPE showed exothermic peaks of the β′–β transition. A complete β′-melting curve can therefore not be obtained. The estimate for the β′ -interaction parameter is based on the shape and height of the plateau just after the first melting peak in the curve of a completely stabilized sample.

9.7.5.3 PSP and MPM with EEE

The measurements that were carried out and the results of the fitting procedure are given in Table 9.29 and Figures 9.41 and 9.42.

These systems all recrystallized quickly into the β-modification. Even without stabilization, the melting curves of the β'-modification were disturbed by exothermic peaks of the β'–β transition. Therefore the values of the β'-interaction parameters had to be estimated from the height of the plateau just after the first melting peak. The plateau height is nearly zero.

For a reliable determination of the β'-interaction parameter, the β'-melting point of EEE is required. The authors were not able to determine a β'-melting point of their EEE sample, only α- and a β-melting points were found. This is in line with the findings of Hagemann (1975) and earlier authors (Vazquez Ladron and Castro Ramos, 1971), who also could not detect a β'-melting point of EEE. A very old report of Malkin and Carter (1947) gives 37°C for the β'-melting point. This value does not compare well with the data of Hagemann (1975) for a number of glycerol tri-*trans*-octadecenoates. Hagemann found a steady decrease in the β'-melting point from 43°C for *trans*-17-octadecenoic acid to 28°C for trans-11-octadecenoic acid. Extrapolating the data of Hagemann to 9-octadecenoic acid (E) results in a β'-melting point of 27°C–28°C for EEE.

If 37°C is used as β'-melting point of EEE, than the 2 suffix Margules interaction parameter for PSP-EEE in the β'-modification must be greater than 2.5 in order to explain a zero plateau height. If a melting point of 27°C is used, A/RT (PSP-EEE) = 0 ± 1.

From binary phase diagrams, we concluded that SSS mixes nearly ideally with SES, SSE, and SEE in the β-modification (Section 9.7.3). In addition to the DSC curve of PSP–EEE–OOO, we also determined the curve of SSS–EEE–OOO in the β-modification (Figure 9.43). The result is not very surprising: SSS and EEE also mix ideally in the β-modification. Clearly the elaidic and stearic acid chains

---

TABLE 9.29

DSC Measurements and 2-Suffix Margules Parameters for the Highest Melting Binary Pair Determined from These Measurements

Modifi-cation	PSP	MPM	PEE	OOO	Scan Rate (K/min)	Stabili-zation Time (min)	Stabili-zation Temp. (°C)	A/RT
β'	0.26		0.25	0.49	2.5	1–7d	15	0 ± 1
β'		0.27	0.25	0.48	2.5	30–70	15–20	≥2?
β	0.26		0.25	0.49	2.5	1–7d	20	≥2.0
β		0.27	0.25	0.48	2.5	30–70	15–20	≥2.0

Note: Composition in mole fractions, temperatures in °C, time in minutes, unless otherwise indicated: d ≙ days.

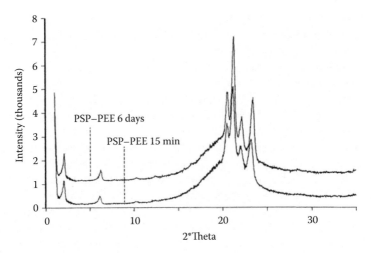

FIGURE 9.43 X-ray diffractograms of the mixture of 25% PSP, 25% PEE, and 50% 000, after 15 min and 6 days of stabilization at 20°C.

behave more or less equivalent, which is confirmed by our findings for PSP-SEE and PSP-ESE for both β′- and β-modifications.

If S and E behave equivalent, then the interaction parameter for the binary PSP–EEE should be equal to those of PSP–SEE and PSP–ESE. As the latter parameters both equal zero, expect that A/RT (PSP–EEE) will also be zero. This implies that the β′-melting point of EEE must be 27°C ± 2°C, and that Malkins' value for the β′-melting point of EEE, 37°C, is not right.

9.7.5.4 PSP and MPM with *cis*-Unsaturated TAGs

The measurements of mixtures of PSP and MPM with SSO, PPO, SOS, and POP resulted in more complicated curves than those of the previous mixtures. Contrary to the previous TAGs, which all crystallized into the β′-2 and β-2 modifications, most *cis*-unsaturated TAGS crystallize into the β′ and β-3 modifications. Only POP crystallizes into the β′-2 form (Table 9.30).

TABLE 9.30
Polymorphic Forms of the
***Cis*-Unsaturated TAGs**

PSP				
PSP	β′-2		MPM	β′-2
SOS	β′-3	β-3	PPO	β′-3
SSO	β′-3		POP	β′-2	β′-3

Sources: Sato, K. et al., *JAOCS*, 66, 664, 1989; Vazquez Ladron, R. and Castro Ramos, R., *Gracas y Aceites*, 22, 401, 1971.

The formation of a continuous solid solution between a component that crystallizes into the β'-2 form and a component that crystallizes into the β'-3 form seems on structural grounds impossible: an intermediate β'-2/β'-3 structure is not feasible. In thermodynamic calculations, the two β'-forms must therefore be treated as separate modifications, just like the β'- and β-modifications.

This has two implications:

1. The DSC curve of the β'-modification of a mixture of a β'-2 and a β'-3 forming TAG must always contain two sharp peaks, because demixing in the solid phase will occur even when the components mix ideally both in the β'-2 as in the β'-3 phase.
2. For calculation of the interaction parameters, the heat of fusion and the melting point of a hypothetical β'-2 form of the β'-3 forming TAG and of the hypothetical β-3 form of the β'-2 forming TAG must be known. These hypothetical pure component properties cannot be measured. They must be estimated, which will cause a considerable uncertainty in the value of the interaction parameters.

Quite surprisingly, after short stabilization of the MPM–SSO–OOO and MPM–PPO–OOO mixtures, DSC curves in which the first melting peak is a broad hump were obtained, rather than the sharp peak that was expected for these β'-2/β'-3 mixtures. After prolonged stabilization, a clear polymorphic transition was observed, and finally the expected DSC curves with two sharp peaks were obtained. The final melting peak remained in place during this transition. The broad hump indicates that initially crystallization must have taken place into a single, mixed solid phase, which has to be in the β'-2 modification. The final, stable, separate PPO- or SSO-rich β'-3 phase is only formed later. This behavior confirms the assumption that the 2-layer and 3-layer forms of the same modification must be treated thermodynamically as separate, independent 'states', that need not always occur in pure components. The order of the polymorphic transitions that is observed is therefore:

$$\alpha \rightarrow \beta'-2 \rightarrow \beta'-3 \qquad (9.93)$$

From the work of de Jong (1980), it can be concluded that the melting point of a hypothetical β-3 modification of PSP and MPM is about 3°C less than the melting point of a hypothetical β-2 form of these TAGS. In analogy to the β-modification, assume that the β'-3 melting points of MPM and PSP also lie 3°C below their experimental β'-2 melting points. Based on the enthalpy of fusion data of Chapter 4, we conclude that the heat of fusion of the β'-3 form is about 90% of that of the β'-2 modification.

Similarly, it is assumed that the heat of fusion of the hypothetical β'-2 form of SSO, PPO and SOS is 90% of that of the β'-3 modification. If the 2-suffix Margules equation is used, the melting points of the β'-2 modification of these TAGs can be calculated from the β'-2 DSC curves of the mixtures with MPM:

SSO: $T_f (\beta'\text{-}2) = 39 \pm 0.5°C$ $(T_f (\beta'\text{-}3) = 42°C)$
SOS: $T_f (\beta'\text{-}2) = 29 \pm 2°C$ $(T_f (\beta'\text{-}3) = 36.5°C)$
PPO: $T_f (13'\text{-}2) = 31 \pm 0.5°C$ $(T_f (\beta'\text{-}3) = 34°C)$

TABLE 9.31

DSC Measurements and 2-Suffix Margules Parameters for the Highest Melting Binary Pair Determined from These Measurements

Modification	PSP	MPM	SOS	SSO	OOO	Scan Rate (K/min)	Stabilization Time (min)	Stabilization Temp. (°C)	A/RT
β′-2	0.32		0.19		0.49	2.5	15	5	0 ± 0.3
β′-2		0.29	0.23		0.48	2.5	1	5	2 ± 0.7
β′-2	0.26			0.24	0.50	2.5	1–7d	10	No fit
β′-2		0.28		0.24	0.48	1.25	1–60	15	2 ± 0.2
β′-3	0.32		0.19		0.49	2.5	15	5	0 ± 0.5
β′-3		0.29	0.23		0.48	1.25	8d	10	1.7 ± 0.7
β′-3		0.28		0.24	0.48	1.25	3d	15	2 ± 0.5
β	0.32		0.19		0.49	1.25	1d	28	1.5 ± 0.5
β		0.29	0.23		0.48	1.25	8d	10	1.0 ± 0.5

d, days.

The measurements that were carried out and the results of the fitting procedure are given in Tables 9.31 and 9.32 and Figures 9.44 through 9.47.

The broad first melting peaks in the β′-2 DSC curves of PSP–POP and PSP–SSO have maxima that lie several degrees above the temperature where these maxima should be situated when $A/RT = 0$. These DSC curves could only be fitted with the 3-suffix Margules equation, using negative values for $A_{PSP-POP}$ and $A_{PSP-SSO}$ (Table 9.33).

TABLE 9.32

DSC Measurements and 2-Suffix Margules Parameters for the Highest Melting Binary Pair Determined from These Measurements

Modification	PSP	MPM	POP	PPO	OOO	Scan Rate (K/min)	Stabilization Time (min)	Stabilization Temp. (°C)	A/RT
β′-2	0.26		0.26		0.48	1.25	15–7d	10	No fit
β′-2	—	0.29	0.23		0.48	2.5	1	5	0.8 ± 0.2
β′-2	0.25			0.28	0.47	1.25	7d	15	−1.5?
β′-2		0.28		0.25	0.47	1.25–2.5	1–60	10	0.8 ± 0.4
β′-3	0.25			0.28	0.47	1.25	26d	5	—
β′-3		0.28		0.25	0.47	1.25–2.5	7d	10	−0.2 ± 0.4
β	0.26		0.26		0.48	—	—	15	No
β		0.29	0.23		0.48	1.25	7d	5	3.5 ± 1

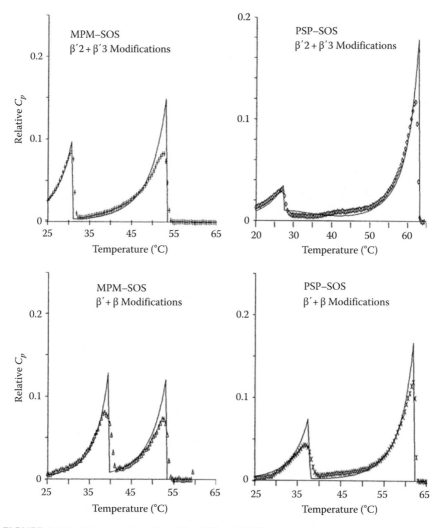

FIGURE 9.44 Measured (dots) and fitted (lines) DSC curves of ternary mixtures of ±25% PSP or MPM% with ±25% SOS and ±50% OOO. The modification(s) in which crystallization has taken place is indicated.

Apparently the β'-2 form of POP and SSO is enormously stabilized by the presence of PSP, their almost exact saturated counterpart. Calculations show that this stabilizing effect makes the mixture β'-2 stable. Indeed, no transition to the β-modification (POP) or the β'-3 modification (SOS) was observed even after long stabilization.

The stabilizing effect seems also present in the curves of PSP–PPO and PSP–SOS. In these cases, an exact determination of the β'-2 state was not possible: a β'-2/β'-3 transition always disturbed the DSC curve. The shape of the PSP–PPO–OOO melting curves above 35°C suggest $A/RT = -1.5$.

The 2-suffix parameters for the β-modification of the mixtures with SOS are very low, compared to those for the β'-modification. It is possible to use 3-suffix

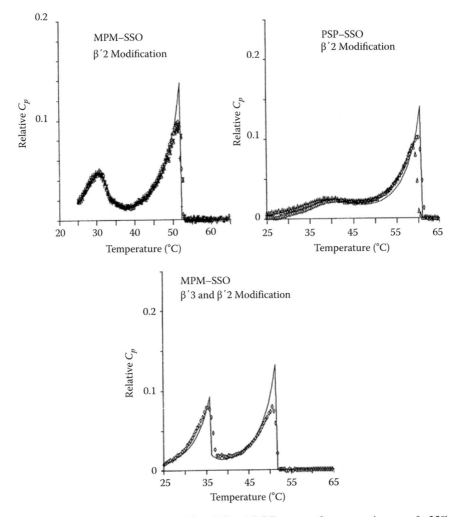

FIGURE 9.45 Measured (dots) and fitted (lines) DSC curves of ternary mixtures of ±25% PSP or MPM% with ±25% SSO and ±50% OOO. The modification(s) in which crystallization has taken place is indicated.

Margules β'-interaction parameters that are more in line with the results for the β'-modification for the description of the curve (Table 9.33). However, they have no statistical significance.

9.7.6   Discussion

9.7.6.1   Use of DSC Melting Curves
The results show that it is possible to use DSC melting curves of ternary mixtures to determine binary interaction parameters. This method is much quicker and more reliable than the determination of a phase diagram. It allows the study of phase

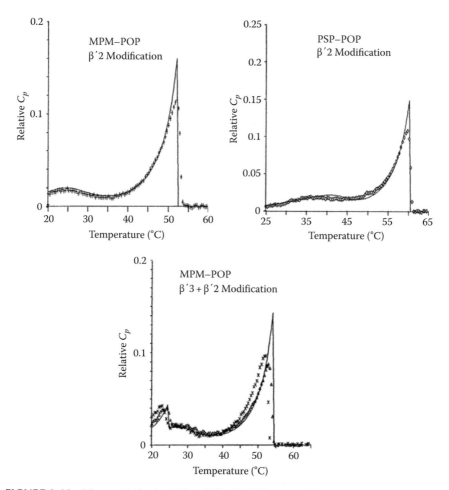

FIGURE 9.46 Measured (dots) and fitted (lines) DSC curves of ternary mixtures of ±25% PSP or MPM% with ±25% POP and ±50% OOO. The modification(s) in which crystallization has taken place is indicated.

behavior in unstable modifications. Thus we have obtained a new, powerful, and versatile method for studying the solid–liquid phase behavior.

9.7.6.1.1 3-Suffix Margules Equation

The 3-suffix Margules interaction parameters can be obtained from the combination of peak shape and peak position in the DSC melting curve. This dependency on exact peak position implies that the uncertainty in the melting points of the pure components of about 1°C will translate itself into an error in the 3-suffix Margules parameters of about 0.5. Due to this uncertainty in the parameters, asymmetric behavior in systems that show only small differences in the two parameters cannot be detected. In those cases, the 2-suffix Margules equation performs equally well. In this work, TAGs were studied that show considerable differences in stereochemical nature.

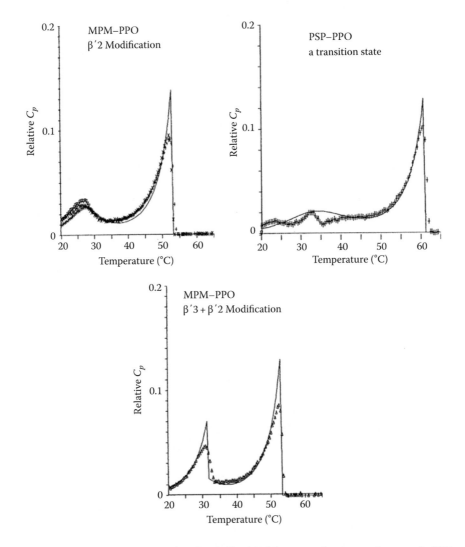

FIGURE 9.47 Measured (dots) and fitted (lines) DSC curves of ternary mixtures of ±25% PSP or MPM% with ±25% PPO and ±50% OOO. The modification(s) in which crystallization has taken place is indicated.

Yet only in the most extreme case, for TAGs with oleic acid, which crystallize in very different lattices, clear asymmetric behavior and the need for using the 3-suffix Margules equation were apparent.

9.7.6.2 Binary Interaction Parameters

9.7.6.2.1 β'-Modification

For the first time, reliable information about mixing behavior in the β'-modification has been obtained. On average, solid solubility in the β'-modification is higher than in the β-modification, but contrary to the α-modification, nonideal mixing can occur.

TABLE 9.33

2- and 3-Suffix Margules Parameters for Some of the Systems of Tables 9.31 and 9.32

System	Modification	A/RT	A_{12}/RT	A_{21}/RT
PSP–POP	β'-2	—	-3 ± 1	0 ± 0.5
PSP–SSO	β'-2	—	-2 ± 1	0 ± 0.5
PSP–SOS	β	1.5 ± 0.5	1.5 ± 0.5	3 ± 1.5
MPM–SOS	β	1.0 ± 0.5	0.5 ± 1	3 ± 2

The influence of size differences is clear: none of the unsaturated TAGs mixes very well with the smaller TAG MPM, while often even ideal miscibility is found with PSP, which is about similar in size. The position of the fatty acid chains on the glycerol influences the mixing behavior with MPM only very slightly, and the size difference dominates. But when the size difference is small, the chain position is of great influence, as can be seen from the data on PSP–PEE vs. PSP–EPE, PSP–SSO vs. PSP–SOS, and PSP–PPO vs. PSP–POP. It is not clear why these relatively small differences can have such large effects on mixing behavior. Simulation of the disturbance of a crystal lattice by insertion of another TAG using molecular mechanics may help to create understanding. This will be attempted in Section 15.8.

9.7.6.2.2 β-Modification

Due to the use of β'-stable TAGs as the highest melting component in the systems that were studied by DSC, β-interaction parameters could only be determined very roughly from the position of the first melting peak. That means that the uncertainty in the melting points already has to be taken into account in the 2-suffix Margules parameters, while the determination of statistically significant 3-suffix Margules parameters is impossible.

However, as illustrated for the system SSS and EEE, determination of accurate interaction parameters for the β-modification from DSC-curves is very well possible if a β-stable TAG is used as highest melting component.

In the survey of the binary phase diagrams, complete solid miscibility was only found for pairs of TAGs that are very similar in size, like SSS and PSS. Our DSC results are in line with this finding. In Section 9.8, the data will be used in an attempt to find a relation between the influence of structural differences of TAGs and the magnitude of the interaction parameters.

9.7.6.3 Kinetics

It is striking how strongly the kinetics of transformation from β' to β depend on the miscibility in the β'-modification: if the components mix very well in the β'-modification, the transformation takes at least several days, while if solid phase immiscibility occurs, it takes only a few seconds to an hour to complete the transformation.

The explanation is twofold:

The combination of poor β-phase miscibility and nearly ideal mixing in the β'-modification reduces the Gibbs energy difference between the two modifications, which is the driving force for recrystallization.

It also leads to a β'-solid phase composition that is completely different from that in the demixed β-modification, which makes the transformation kinetically much more difficult. In the case of poor β'-phase miscibility, the solid phase composition is nearly equal to that in the β-modification, so that the barrier for the transformation is much smaller.

9.7.7 TERNARY SOLIDS

Although the DSC melting curves of Section 15.7.6 were measured using ternary mixtures, the composition of the mixture and the temperature range of the experiment were selected such that only a binary solid phase was present. This enabled the determination of binary interaction parameters in the solid phase.

Crystallized fats are normally multicomponent solid phases. The binary interaction parameters that were determined in this work can only be used for prediction of multicomponent phase behavior if ternary, quaternary, and higher interaction terms can be neglected. A way of checking this is comparing measured ternary phase diagrams with phase diagrams predicted from binary interaction parameters. In the literature, there are only a few ternary phase diagrams available: that of the cocoa butter (CB) TAGs SOS/POS/POP from Smith (1988) and those of the palm oil TAGs PPP/POP/POO and PPP/ PPO/POO from Gibon (1984). The data from Gibon are not very reliable, due to the very poor stabilization procedure and the high DSC scan rates that were used (Section 15.7.2). Calculated and measured ternary clear point curves for the three systems are given in Figures 9.48 and 9.49. For SOS–POS–POP also, a diagram is given with the isotherms for 25% solids.

As can be seen, the agreement between theory and experiment is very good for the SOS/POS/POP ternary, taking into account the error in the measurements of 1°C. Also the temperatures at which a mixture contains 25% solids are predicted within 1°C.

The PPP/POP/POO and PPP/PPO/POO ternaries also show good agreement between measurements and calculations. The agreement near the PPP corner at the PPP–POO side of the diagrams is somewhat less good. However, the clear points that are given by Gibon for this PPP/POO binary are probably not correct as they lie well below the eutectic liquidus.

The results suggest that the use of binary interaction parameters is indeed sufficient for the description of multicomponent TAG systems. Later in this chapter it is shown that also in multicomponent systems excellent results can be obtained using only binary interaction parameters (Section 9.9).

The clear point diagram for the CB-TAGs SOS/POS/POP may be used to define the possible compositions of CBEs (cocoa butter equivalents). CBEs must have the same clear point as CB, and so all possible CBE compositions will be restricted to the isotherm through the CB composition. The 25% isosolids diagram demonstrates

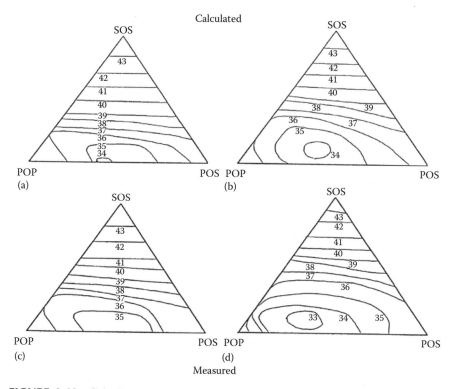

FIGURE 9.48 Calculated and measured ternary isosolids diagrams of SOS–POS–POP. Measured by Smith (1988). (a) and (c): clear point diagrams; (b) and (d): isotherms with 25% solid fat. (From Smith, K., Unilever Research Colworth, Private communication, 1988.)

that even for such difficult systems as SOS–POS–POP solids content can be predicted by application of solid–liquid equilibrium thermodynamics.

9.7.8 CONCLUSION

In the β- and β′-modifications, TAGs show limited solid miscibility that can be described with rather simple models for the excess Gibbs energy: the 2- and 3-suffix Margules equation.

Eighty-four of the 120 binary phase diagrams of TAGs, which are available in the literature, are rejected because they are clearly not correct. The solidus lines of most of the remaining 36 phase diagrams show large inconsistencies. The majority of these phase diagrams were measured after 1971. The 36 phase diagrams can be described within experimental error both by the 2- and by the 3-suffix Margules equations.

The excess Gibbs energy models contain binary interaction parameters. The large experimental errors in the binary phase diagrams lead to a very large inaccuracy in the interaction parameters that are determined from these diagrams.

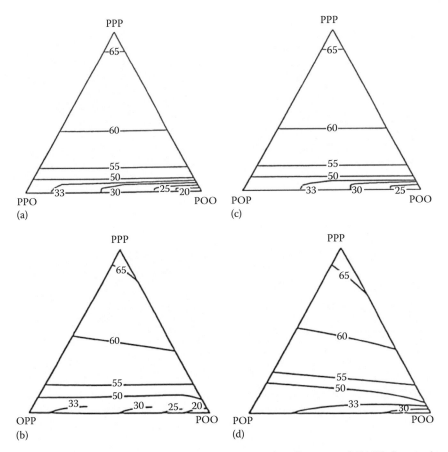

FIGURE 9.49 Calculated and experimental clear point diagrams of TAGS from palm oil. (a) and (c): calculated; (b) and (d): measured by Gibon (1984). (From Gibon, V., Thesis, Universite Notre Dame de la Paix, Namur, Belgium, 1984.)

Due to the serious drawbacks in the use and determination of binary phase diagrams, a new method for the study of solid–liquid phase behavior has been defined. It is possible to obtain quantitative information about mixed crystal formation from a single, complete DSC melting curve of a ternary system, consisting of two crystallizing TAGs and one liquid TAG.

The method is much quicker, more reliable, and more versatile than the traditional study of phase behavior of TAGs via binary phase diagrams.

For the first time, accurate information was obtained about the degree of mixed crystal formation in the β'-modification. Generally solid phase miscibility is higher in the β'-modification than in the β-modification. Large differences in molecular size reduce solid phase miscibility. Also the position of the fatty acid on the glyceryl group has a great influence on solid phase miscibility.

Surprisingly the mixing behavior of TAGs in the solid phase can usually be described with sufficient accuracy by the very simple regular solution model or

2-suffix Margules equation. Only in a few cases, the use of the more complex 3-suffix Margules equation was required.

Ternary systems could be described with binary interaction parameters only.

Kinetics of recrystallization showed a clear relation with the calculated thermodynamic driving force and the degree of rearrangement of the solid phase that is required.

9.8 PREDICTING INTERACTION PARAMETERS

Now that it has been shown that the mixing behavior of TAGs in the α-, β'-, and β-modifications can be described with rather simple excess Gibbs energy models, there remains only one step to be taken: finding a procedure for predicting the binary interaction parameters for any pair of TAGs. Therefore, the relation between geometrical differences and the magnitude of the interaction parameter is studied. Molecular mechanics are used to obtain an impression of the lattice distortions that are brought about by incorporation of guest TAGs in a lattice of a host TAG. Finally a first, empirical, set of rules is given for prediction of the binary interaction parameters.

9.8.1 ARE INTERACTION PARAMETERS RELATED TO STRUCTURAL DIFFERENCES?

In the conclusion of Section 9.2, four steps were mentioned that had to be taken to obtain a description of the liquid–multiple solid phase equilibria in fats. Those four steps were the subject of the preceding sections. It was concluded that the solid–liquid phase equilibrium in a TAG mixture can be described, provided the binary interaction parameters for all possible pairs of TAGs in the mixture are known. The large number of TAGs in a natural oil makes it impossible to determine the parameters experimentally. Therefore a method must be developed for predicting these binary parameters. Fortunately TAGs are chemically very similar, so the degree of nonideal mixing will only be determined by sterical effects. In the next sections, a relation between the binary interaction parameters and structural differences will be discussed.

9.8.1.1 Degree of Isomorphism

In his work on mixed crystals, Kitaigorodskii (1984) investigated the solid miscibility of several hundreds of pairs of organic compounds. Based on this investigation, he formulated his "major rule of substitutional solid solubility" of organic compounds:

1. Solid solubility is determined by geometric factors if
   • There is no electron transfer between the two components to be mixed
   • The components to be mixed have no permanent dipole moments
   • The components to be mixed do not form strong hydrogen bridges
2. If solid solubility is solely determined by geometrical factors, then two components will only mix in the solid state if their "degree of isomorphism" exceeds 0.85.
   The degree of isomorphism $\in$ is defined as follows: superimpose the molecules of the components so as to maximize the intermolecular overlap.

Let the volume of the non-overlapping parts be V_{non} and the volume of the overlapping parts v_0. Then the degree of isomorphism or coefficient of geometrical similarity is given by

$$\epsilon = 1 - \frac{V_{non}}{v_0} \tag{9.94}$$

3. In addition to this second condition, complete solid state miscibility in any proportion is only possible if the molecular packing in the crystal of the pure components is similar, the crystals have the same symmetry and the atoms occupy the same crystallographic positions.

Generally, Kitaigorodskii's rule gives a correct prediction of the occurrence of fully eutectic behavior and the presence of a miscibility gap. However, the reverse is not true: sometimes a miscibility gap is found, even though Kitaigorodskii's rule has predicted a good solid state miscibility. The reason for this is that when a guest molecule is inserted in a host lattice, then, depending on the nature of the guest molecule, its protruding part may occupy a lattice site in an area where packing is very dense. Obviously, solid solubility will be considerably less than in the case of a guest molecule of the same size that protrudes into a loosely packed area. Therefore, generally, Kitaigorodskii's rule can only be used for qualitative statements on solid phase miscibility.

According to Kitaigorodskii, quantitative predictions of solid solubility can be obtained by calculating the lattice distortion:

1. An impurity is placed in an undistorted lattice of the pure component.
2. The difference in interaction energy of the impurity with the lattice and of the host with the lattice is calculated.

    The simplest approach is to leave the host lattice completely undisturbed and calculate the conformation of the impurity that gives minimal interaction energy. A more sophisticated approach is to allow also some conformational changes in the host lattice near the impurity ("crystal elasticity"). Unfortunately, this increases computing time enormously.

The lattice distortion obtained in this way can be used to calculate the excess Gibbs energy and hence the maximum solid solubility. The results are reasonable estimates of the maximum solid solubility. Surprisingly, Kitaigorodskii only calculates the maximal solid solubility and does not use this information to calculate a complete binary phase diagram. Most of these complex computer calculations were carried out for relatively simple atomic crystals. Only one example of molecular crystals is mentioned (diphenyl-dipyridyl).

In mixtures of TAGs, the lattice distortion always occurs on the same lattice sites, the methyl-end-plane region, and is always caused by the same functional group: CH_2–CH_3. Because of this, it may very well be that the degree of isomorphism correlates much better with the solid solubility of TAGs than with that of an arbitrary pair of organic substances.

The condition for complete solid state miscibility of Kitaigorodskii's rule implies that complete solid solubility in the equilibrium state cannot occur when two TAGs differ in their most stable polymorphic form, like PSP and SSS. This was already implicitly assumed in all previous calculations by treating the polymorphic forms as different states of the substance, equivalent to the liquid and gas state. However, this condition also implies that complete miscibility cannot occur in mixtures of two TAGs that crystallize in different submodifications, as is the case in a mixture of a β-3 forming TAG and a β-2 forming TAG.

In the next sections, the following is discussed: the extent to which the simple parameter ∈ correlates with the interaction parameters of TAGs, whether it is possible to calculate the lattice distortion by impurities in TAG crystals, and whether this lattice distortion can be used for prediction of interaction coefficients will be discussed.

9.8.1.2 TAGs and the Degree of Isomorphism ∈

There are two major lattice distortions that occur in TAG mixtures: those caused by differences in chain length and those caused by *cis*-unsaturated double bonds. Assume that the distortion in the lattice of a saturated TAG that is caused by a *trans*-double bond is negligible, in line with the nearly ideal miscibility that was found for SSS, SES, SSE, SEE, and EEE. The two major distortions will be considered separately.

9.8.1.2.1 β'-Modification

In Table 9.34, the binary interaction parameters for the β'-modification of saturated and *trans*-unsaturated TAGs are listed together with the degree of isomorphism. As a matter of convenience, v_{non} is the sum of the absolute differences in carbon number of each of three chains and for v_0 the sum of the carbon numbers of the smallest chain on each glyceryl position.

The correlation between the degree of isomorphism ∈ and the binary interaction parameter is striking. In agreement with the results of Kitaigorodskii, the limit of complete miscibility, corresponding to $A/RT = 2$, is reached at ∈ = 0.85. Moreover, ideal miscibility is found if ∈ > 0.92. The degree of isomorphism explains the large

TABLE 9.34

2-Suffix Margules Parameters for the β'-Modification and the Degree of Isomorphism of Saturated and *Trans*-Unsaturated TAGs (from Section 9.7.5)

Binary Pair	A/RT	∈
PSP–PEE	0 ± 0.2	0.96
PSP–SEE	0 ± 0.2	0.92
PSP–ESE	0 ± 0.2	0.92
PSP–EEE	0 ± 2	0.92
PSP–EPE	2 ± 0.5	0.88
MPM–PEE	1.8 ± 0.5	0.82
MPM–EPE	2.3 ± 0.5	0.82
MPM–SEE	3 ± 0.2	0.77
MPM–ESE	3 ± 0.2	0.77
MPM–EEE	3 ± 2	0.77

TABLE 9.35

2- or 3-Suffix Margules Parameters for the β-Modification and the Degree of Isomorphism of Saturated and *Cis*-Unsaturated TAGs

Binary Pair	Modification	A_{12}/RT	A_{21}/RT	ϵ
PSP–POP	β'-2	-3 ± 1	0 ± 0.5	1-
PSP–PPO	β'-2	-1.5 ± 1.5	-1.5 ± 2	0.92-
PSP–SSO	β'-2	-2 ± 1	0 ± 0.5	0.92-
MPM–POP	β'-2	0.8 ± 0.4	0.8 ± 0.2	0.86-
MPM–PPO	β'-2	0.8 ± 0.5	0.8 ± 0.3	0.86-
MPM–SSO	β'-2	2 ± 0.7	2 ± 0.2	0.77-

Binary Pair	Modification	A/RT		
PSP-SOS	β'-3	0 ± 0.5		0.92-
MPM–PPO	β'-3	-0.2 ± 0.4		0.86-
MPM–SOS	β'-3	1.7 ± 0.7		0.77-
MPM–SSO	β'-3	2 ± 0.5		0.77-

Note: The modification in which the *cis*-unsaturated tag has crystallized is given. A – behind the value of ϵ indicates that the contribution of the *cis*-double bond to $\in$ is not incorporated in the value of $\in$.

difference in miscibility with PSP that has been found for PEE and EPE in Section 15.7. Although the number of data is too small for a decisive statement, it seems possible to use $\in$ for predicting the binary interaction parameters of pairs of these TAGs.

The correlation for *cis*-unsaturated TAGs will be less simple, as the *cis*-double bond disturbs the regular zigzag (*trans* configuration) of the saturated chains in the crystal lattice. From the *cis*-unsaturated TAGs that were studied in Section 9.7.6, only POP crystallizes in the β'-2 modification, while SOS, SSO, and PPO are reported to crystallize in the β'-3 modification (Vazquez Ladron and Castro Ramos, 1971; Gibon, 1984; Sato et al., 1989). In Section 9.7, SSO and PPO initially seem to crystallize in a β'-2 form that does not exist in pure PPO and SSO. The values of the 2- or 3-suffix Margules equation interaction parameters for the β'-2 and β'-3 forms are listed in Table 9.35 together with $\in$. The extra contribution of a *cis*-double bond to ∞ is not incorporated in the number, but indicated by a – (minus).

Here again, the limited amount of data for both β'-forms indicate that the degree of isomorphism is related to the magnitude of the binary interaction parameters. Surprisingly the miscibility of saturated and *cis*-unsaturated TAGs seems to be slightly better than that of saturated and *trans*-unsaturated TAGs with the same degree of isomorphism.

9.8.1.2.2 β-Modification

Most data for the β-modification that are available have a considerable error margin. Yet, as appears from Table 9.36, a correlation between degree of isomorphism and the binary interaction parameters seems present.

TABLE 9.36

2- and 3-Suffix Margules Interaction Parameters and the Degree of Isomorphism for Binary Pairs of -2 or -3 Forming TAGs

Binary Pair	Modification	A/RT	A_{12}/RT	A_{21}/RT	$\in$
SSS–SES	β-2	0.4 ± 0.5	0 ± 0.5	0.8 ± 0.5	1
SSS–SSE	β-2	0.4 ± 0.5	0.1 ± 0.5	0.7 ± 0.5	1
SSS–SEE	β-2	0 ± 1	0 ± 1	0 ± 1	1
SSS–EEE	β-2	0 ± 0.2	0 ± 0.2	0 ± 0.2	1
SES–SSE	β-2	0 ± 0.5	0 ± 0.5	0 ± 0.5	1
SSS–PSS	β-2	1.6 ± 0.5	1.1 ± 0.5	2.3 ± 0.5	0.96
SSS–SPS	β-2	>3	>3	>2	0.96
PSS–PPS	β-2	1.6 ± 0.5	1.4 ± 0.5	1.8 ± 0.5	0.96
SPS–PPS	β-2	2 ± 0.5	3– 5	0.4	0.96
PSP–PPP	β-2	>3	>3	>1	0.96
PPS–PPP	β-2	1.9 ± 0.5	1.6 ± 0.5	2.3 ± 0.5	0.96
SSS–PSP	β-2	>3	—	—	0.92
SSS–PPS	β-2	>2.2?	—	—	0.92
PSS–SPS	β-2	2 ± 0.5	1.4 ± 0.5	2.6 ± 0.5	0.92
PSS–PPP	β-2	2.2 ± 0.5	2.8 ± 0.5	2.1 ± 0.5	0.92
SPS–PPP	β-2	>5	>5	>5	0.92
PSP–PPS	β-2	>3	>3	>1	0.92
PSP–SEE	β-2	>2.5	—	—	0.92
PSP-ESE	β-2	>2	—	—	0.92
PSP–EEE	β-2	>2	—	—	0.92
SSS–PPP	β-2	3 ± 0.5	5.8 ± 1	1.4 ± 1	0.88
SPS–PSP	β-2	1.4 ± 0.5	1	1.8 ± 0.5	0.88
PSP–EPE	β-2	>2.5	—	—	0.88
PPP–MMM	β-2	>3	—	—	0.86
MPM–PEE	β-2	>2.5	—	—	0.82
MPM–EPE	β-2	>3	—	—	0.82
BBB–SSS	β-2	>3	—	—	0.77
MPM–SEE	β-2	>2.5	—	—	0.77
MPM–ESE	β-2	2 ± 1	—	—	0.77
PPP–LLL	β-2	>3	—	—	0.67
BBB–PPP	β-2	3	—	—	0.63
SSS–LLL	β-2	>3	—	—	0.50
SSS–888	β-2	>3	—	—	0
SOS–POS	β-3	1.3 ± 0.5	1.5 ± 0.5	1.1 ± 0.5	0.96
POS–POP	β-3	1.6 ± 0.5	2.2 ± 0.5	1.0 ± 0.5	0.96
SOS–POP	β-3	>3.6	5 ± 1	1.5 ± 1	0.92

In line with expectations, the degree of isomorphism that is required for complete miscibility in the β-modification is higher than for the less densely packed β′-modification: in the β′-modification $A/RT = 2$ is already reached when $\in = 0.94$, while miscibility in the β′-modification at $\in = 0.94$ is still ideal. The few data for the β-3 modification agree with those for the β-2 modification.

There are a few noticeable exceptions in the data set, all concerning SPS and PSP: in combination with each other, the miscibility is better than could be expected from their low degree of isomorphism, while in combination with other saturated TAGs that have a high degree of isomorphism with these two TAGs, miscibility seems much less than expected.

Although SSS and SPS both crystallize in a β-2 lattice, according to de Jong (1980), they crystallize into different β-2 submodifications. SSS, PSS, PPS, and PPP crystallize into the β-2A submodification, while PSP and SPS both crystallize into the β-2B submodification. Although these submodifications are very similar, apparently the condition for complete solid miscibility of Kitaigorodskii's rule is not fulfilled. The β-2 submodifications should have been treated as independent polymorphs. The effect of treating the submodifications as independent polymorphs is illustrated in Figure 9.50 for $A/RT = 1.8$, a value in agreement with the other data of TAGs with a degree of isomorphism $\in = 0.96$. It is clear that the treatment as independent polymorphs gives a much better fit to the data.

Out of the four β-2 submodifications that occur, the β-2B submodification has the lowest packing density in the methyl terrace area. This could explain the "too large" miscibility that was experimentally found for SPS–PSP.

9.8.1.2.3   Conclusions

Kitaigorodskii's major rule of substitutional solubility applies to TAGs. Within one submodification, the degree of isomorphism correlates very well with the binary interaction parameters. Contrary to the general case of organic compounds, the use of the degree of isomorphism for an empirical prediction of interaction parameters between TAGs seems feasible.

It turns out that in the description of phase equilibria in TAGs not only the three basic modifications have to be considered as separate states, but also each subform of these three modifications. A practical problem in doing so is the fact that only the heat of fusion and melting point of the most stable subform can be determined experimentally.

9.8.2   Calculation of Lattice Distortion

When the degree of isomorphism is used for predicting binary interaction parameters, it is implicitly assumed that the two suffix Margules equation gives an adequate description of the phase behavior of TAGs. Although this is quite often true, it is not always the case. Moreover, the validity of the 2-suffix Margules equation is theoretically unlikely. There is no reason why an impurity with a protuberance will lead to a lattice distortion of the same magnitude as an impurity that causes a hole in the lattice, or why a protuberance of a constant size should cause a lattice distortion of the same magnitude regardless whether it is a protuberance of the fatty acid on 1,

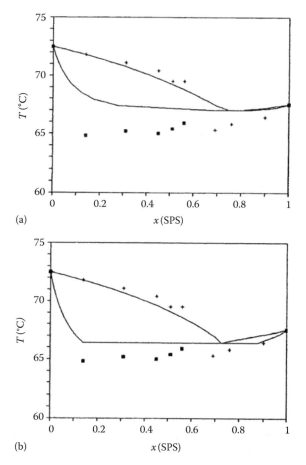

FIGURE 9.50 The influence of discontinuous miscibility in the β-2A and β-2B submodifications on the phase diagram of SSS–SPS. (a) continuous miscibility with $A/RT = 1.8$ and (b) discontinuous miscibility with $A/RT = 1.8$ for both submodifications.

2, or 3 position of the glycerol group. To get a clearer insight into this matter, more sophisticated molecular considerations are required.

9.8.2.1 Equivalent Distortions in the β-2 Modification

All saturated TAGs and *trans*-unsaturated TAGs that were considered in this work crystallize in the β-2 modification. According to de Jong (1980) and using his nomenclature, all mono-acid TAGs—PSS, PPS, SES, SSE, SEE, and EEE—crystallize in the β-2A submodification, while PSP, SPS, and probably also PEP crystallize into the β-2B submodification. These two submodifications have the same angle of tilt, 60° of the fatty acid chains, but differ in the shape of their methyl end plane. The β-2B modification is somewhat less densely packed it this area. Within one submodification the shape of the methyl terrace does not change, but the position of the "steps" in the terrace is shifted (Figure 9.51).

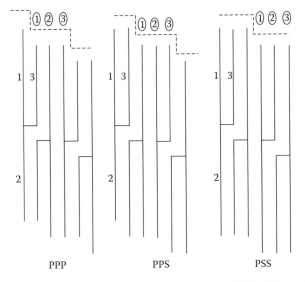

FIGURE 9.51 Equivalent positions on the methyl terraces of PPP, PPS, and PSS (marked by a circled number).

In Figure 9.51, equivalent positions on the methyl terrace are marked by a circled 1, 2, or 3. Exchanging a molecule by a molecule of another TAG will corrupt the methyl end plane. If a PPP molecule is exchanged by PPS, then an ethyl group is added to the first chain of a "step" of the methyl terrace. The distortion of the lattice caused by this addition will be called a (1,0,0) distortion, while that caused by removing an ethyl group from the same chain will be denoted as a (−1,0,0) distortion. A (0,1,0) distortion is in this nomenclature the addition of an ethyl group to the chain in the middle of a "step" in the methyl terrace. It is also possible to bring about a (1,0,0) distortion in a PPS crystal: by exchanging PPS with PSS. As the distortion to the lattice is exactly the same, the activity coefficients at infinite dilution of PPS in PPP and PSS in PPS must be equal. Therefore, according to Equation 9.88, the 3-suffix Margules parameters $A_{PPS-PPP}$ and $A_{PSS-PPS}$ should be identical as well, which was found indeed ($A_{PPS-PPP} = 1.6$, $A_{PSS-PPS} = 1.43$). Table 9.37 lists all 26 different distortions to the lattice of β-2A and β-2B TAGs that can be caused by addition or removal of ethyl groups.

These 26 different distortions occur at only three levels of the degree of isomorphism.

A complicating factor is the asymmetry of the TAGs PPS and PSS: PPS causes a (1,0,0) distortion in PPP, while its mirror image SPP leads to a (0,0,1) disturbance. These two distortions are not necessarily equivalent.

Notwithstanding the fact that sometimes asymmetric TAGs are involved, a good agreement is found between the values of the three suffix Margules parameters of equivalent distortions (Table 9.38).

The 3-suffix Margules parameters for the binary SSS–PPS could not be determined because of poor data quality. If the effects of chirality are negligible, they should be the same as those of PSS–PPP.

TABLE 9.37

TAGs That Cause Equivalent Distortions When Inserted in the Crystal Lattice of the Six TAGs That Can Be Formed from P and S

Disturbance of Methyl Terrace		TAGs Crystallizing in the -2A Modification				TAGs in the -2B Modification		
		PPP	SSS	PPS	PSS	PSP	SPS	∈
A	−1 −1 −1	MMM	PPP	MMP	MPP	MPM	PMP	0.88
B	−1 −1 0	PMM	SPP	MMS	MSP	MPP	PPP	0.92
C	−1 −1 1	SMM	APP	MMA	MAP	MPS	PSP	0.88
D	−1 0 −1	MPM	PSP	PMP	MPS	PPM	PMS	0.92
E	−1 0 0	PPM	SSP	PMS	MSS	PPP	PPS	0.96
F	−1 0 1	SPM	ASP	PMA	MAS	PPS	PSS	0.92
G	−1 1 −1	MSM	PAP	SMP	MPA	SPM	PMA	0.88
H	−1 1 0	PSM	SAP	SMS	MSA	SPP	PPA	0.92
I	−1 1 1	SSM	AAP	SMA	MAA	SPS	PSA	0.88
J	0 −1 −1	MMP	PPS	MPP	PPP	MSM	SMP	0.92
K	0 −1 0	PMP	SPS	MPS	PSP	MSP	SPP	0.96
L	0 −1 1	SMP	APS	MPA	PAP	MSS	SSP	0.92
M	0 0 −1	MPP	PSS	PPP	PPS	PSM	SMS	0.96
N	0 0 1	SPP	ASS	PPA	PAS	PSS	SSS	0.96
O	0 1 −1	MSP	PAS	SPP	PPA	SSM	SMA	0.92
P	0 1 0	PSP	SAS	SPS	PSA	SSP	SPA	0.96
Q	0 1 1	SSP	AAS	SPA	PAA	SSS	SSA	0.92
R	1 −1 −1	MMS	PPA	MSP	SPP	MAM	AMP	0.88
S	1 −1 0	PMS	SPA	MSS	SSP	MAP	APP	0.92
T	1 −1 1	SMS	APA	MSA	SAP	MAS	ASP	0.88
U	1 0 −1	MPS	PSA	PSP	SPS	PAM	AMS	0.92
V	1 0 0	PPS	SSA	PSS	SSS	PAP	APS	0.96
W	1 0 1	SPS	ASA	PSA	SAS	PAS	ASS	0.92
X	1 1 −1	MSS	PAA	SSP	SPA	SAM	AMA	0.88
Y	1 1 0	PSS	SAA	SSS	SSA	SAP	APA	0.92
Z	1 1 1	SSS	AAA	SSA	SAA	SAS	ASA	0.88

The distortions of Table 9.38 are arbitrarily marked from A–Z for the β-2A submodification and from A′–Z′ for the β-2B submodification. Using these markers, the phase diagrams can be classified: for example, SSS–PPP is a Z/A phase diagram. The binary interaction parameters for all Z/A phase diagrams should be the same, regardless of the excess Gibbs energy model used.

If chirality is neglected, the 15 phase diagrams can be classified into nine different groups. The value of the binary interaction parameter within a group seems to be constant.

These structural considerations show clearly that there is no reason at all for the assumption that the interaction parameters for the binary pairs PSP-PPP (type P/E′)

TABLE 9.38
3-Suffix Margules Parameters for Some Equivalent
Distortions ($a \triangleq A/RT$)

Code	Distortion			
P	(0,1,0)	$\alpha_{PSP-PPP} = >3$	$\alpha_{SPS-PPS} = >3$	
U	(1,0,−1)	$\alpha_{PSP-PPS} = >3$	$\alpha_{SPS-PPS} = 2.6$	
J	(0,−1,−1)	$\alpha_{PPS-SSS} = 2.3$	$\alpha_{PPP-PSS} = 2.1$	
M	(0,0,−1)	$\alpha_{PSS-SSS} = 2.3?$	$\alpha_{PPP-PPS} = 2.3$	$\alpha_{PPS-PSS} = 1.8$
V	(1,0,0)	$\alpha_{SSS-PSS} = 1.1$	$\alpha_{PPS-PPP} = 1.6$	$\alpha_{PSS-PPS} = 1.4$
Y	(1,1,0)	$\alpha_{PSS-PPP} = 2.8$	$\alpha_{SSS-PPS} = 2.3?$	

and PPS–PPP (type VN/M) should be the same, in spite of the fact that the degree of isomorphism of both pairs is 0.96. In the previous section, it was found that the binaries with PSP and SPS are exceptions to the empirical "rule" that the degree of isomorphism is related to the magnitude of the 2-suffix Margules parameter. This was explained by assuming that continuous miscibility cannot exist between the β-2A and β-2B modification, in spite of the close resemblance between the two β-modifications. An alternative explanation is simply that the (0,1,0) distortion caused by PSP in PPP is not equivalent to the (1,0,0) distortion of PPS in PPP. This alternative explanation implies that the degree of isomorphism is too coarse a measure to correlate the interaction parameters.

Only calculation of the magnitude of the lattice distortions can decide which of the explanations is most likely (Table 9.39).

9.8.2.2 β-2A Lattice Distortion Calculations

The calculations were carried out using the programs Insight and Discover from Biosym. The program calculates the molecular configuration with minimal energy. The molecular energy is the sum of the internal energy from bond length stretching, bond angle stretching, and torsion, and the nonbond-associated energy due to Coulomb and van der Waals forces.

The magnitude of a (1,0,0), a (0,1,0), and a (0,0,1) distortion to LLL have been calculated by comparison of the lattice energy of a pure LLL crystal with the lattice energy of an LLL crystal in which one molecule of an impurity [I] has been inserted per 54 molecules of LLL.

In the latter calculations, we allowed the impurity to obtain the conformation with minimal energy, while the LLL molecules were fixed to the conformation that they have in a pure crystal. Calculations in which the complete impure crystal was allowed to rearrange went beyond the capacity of the program. The results are given in Table 9.40.

The order of magnitude of the crystal energy corresponds very well with the data of de Jong (1980) for CCC.

The (0,1,0) lattice distortion, which is brought about by the β-2B-forming TAG LML, has the same order of magnitude as those caused by the two β-2A forming TAGs LLM and MLL. The very poor miscibility in the solid phase that was found

TABLE 9.39

The P/S Binaries Classified in the Types Defined in Table 9.37, Together with the 2-Suffix Margules Parameters

Binary	Type	A/RT	Binary	Type	A/RT
SSS–PSS	V/ME	1.6	SSS–PPP	Z/A	3
PSS–PPS	VX/MS	1.6			
PPS–PPP	VN/M	1.9	PSP–SPS	C /I	1.4
SSS–PSP	Q′/D	>3	PSS–SPS	F′L/U	2
			PSP–PPS	U/F′H′	β′
SSS–SPS	N′/K	>3			
PSS–PSP	N′/K		PSP–PPP	P/E′	β′
			SPS–PPS	P/E′	2
SSS–PPS	Y/BJ	>2.3?			
PSS–PPP	QY/J	2.2	SPS–PPP	W/B′	>5

Note: Where asymmetric tags are involved, a three- or four-letter code is used, indicating the distortions caused by both mirror images.

TABLE 9.40

Lattice Distortions U/RT to a β-2A Crystal of LLL Caused by Impurities at Infinite Dilution in LLL

Type	Caused by	Distortion U/RT	U (Distorted Crystal) kJ/mol	U (Pure "Impurity") kJ/mol
(1, 0, 0)	LLM	1.0	258.9	270.1
(0, 1, 0)	LML	1.9	251.9	268.8
(0, 0, 1)	MML	3.6	255.0	270.1
(0, 0, 0)	LLL	0	253.8	253.8

experimentally for TAGs equivalent to LML and LLL can therefore only be explained by assuming that continuous miscibility between the β-2A and β-2B submodification is not possible, in line with the findings in the previous section. The magnitude of the lattice distortion U/RT agrees strikingly well with the magnitude of the interaction parameter $A/RT = 1.8$ that we obtained from the experimental data (Figure 9.50).

The (1,0,0) distortion caused by LLM is much smaller than the (0,0,1) distortion caused by its mirror image MLL. The effects of chirality may clearly not be disregarded. Here again, the average value of the distortions U/RT corresponds reasonably well to the values of the interaction parameter A/RT of 1.6–2.3 that were found experimentally.

The order of magnitude of the calculated distortions corresponds with that expected from the value of the interaction parameters. However, we would expect that the distortions calculated in this way would lead to a systematic overestimate of the interaction parameters. The actual distortion is smaller, because in reality some relaxation of the LLL lattice near the distortion will occur. Moreover, entropy effects are neglected. Apparently both effects are relatively small.

It has become evident that the lattice distortion calculations can be used to obtain good insight into the influence of structural differences on the solid state miscibility. We will continue this approach in future work. In view of the results obtained, the good performance of the 2-suffix Margules equation and the good correlation of the interaction parameter with the degree of isomorphism remain very surprising.

9.8.3 EMPIRICAL METHOD

Although Larsson and Hernqvist (1982) have announced a detailed analysis of the crystal structure of the β′-modification, similar to that of de Jong (1980) for the 13-modification, at present the β′-crystal structure of TAGs is still not exactly known. The instability of the β′-modification and the inability to grow large single crystals are the main obstacles. The crystal structure of *cis*-unsaturated TAGs has not been unambiguously revealed either. This implies that it is not yet possible to study the influence of impurities on the β′-crystal lattice by molecular modeling. The fundamental insight into the relation between nonideal miscibility of TAGs and structural differences can probably best be obtained from these lattice distortion calculations. While this is impossible, a semiempirical approach to the problem of this chapter, finding a method to predict binary interaction parameters, has to be followed.

9.8.3.1 Method

The correlation between the degree of isomorphism, ϵ, and the 2-suffix Margules parameter within one submodification can serve as basis for such a semiempirical predictive method. The following are assumed:

1. Within one submodification, the 2-suffix Margules binary interaction parameter A/RT is
   For a β′-modification:

$$\epsilon > 0.93: \frac{A}{RT} = 0 \tag{9.95}$$

$$\epsilon \leq 0.93: \frac{A}{RT} = -19.5\,\epsilon + 18.2 \tag{9.96}$$

and for a β-modification:

$$\epsilon > 0.98: \frac{A}{RT} = 0 \tag{9.97}$$

$$\epsilon \leq 0.98: \frac{A}{RT} = -35.8\,\epsilon + 35.9 \tag{9.98}$$

2. Both 3-suffix Margules parameters are also given by Equations 9.92 through 9.95, except for the β'-2 modification of a binary pair of which one of the TAGs belongs to the h_2u TAG-group. If the h_2u-type TAG is indicated by 2, then A_{21} is given by Equations 9.92 and 9.93, while A_{12} follows from

$$\frac{A_{12}}{RT} = -21.7 \in +18.7 \qquad (9.99)$$

3. If $A/RT > 8$, $A/RT = 8$ can be used in calculations.
   The relations in Equations 9.95–9.99 were obtained by linear regression through the data of Tables 9.34 through 9.36. The correlation coefficient is 0.9, while the standard error in the estimated interaction parameters is 0.3–0.5. The values of the regression constants clearly illustrate the better solid miscibility in the less densely packed β'- modification.
   Interaction parameters within a submodification are useless if the pure component properties of the TAG, crystallized in that submodification, are not known. Often these properties cannot be measured, because the pure components do not occur in the desired submodification. Therefore the following additional assumptions are made in line with our observations in Section 9.4 and those of de Jong (1980).

4. The melting point of a hypothetical -3 form of a TAG that crystallizes into the β'-2 or β-2 form is 3°C less than the melting point of the corresponding -2 form. The heat of fusion is 90% of that of the corresponding -2 form. Similarly the melting point of a hypothetical -2 forms of a TAG that crystallizes into β'-3 of β-3 form is 3°C less than that of the corresponding -3 form, while the heat of fusion is 90% of that of the -3 form.

5. Submodifications in the β'-modification are, if they exist, continuously miscible.

6. It is only possible to form continuous solid solutions of two TAGs in the β-2 modification when they both crystallize in the same β-2 submodification. Table 9.41 lists the four different β-2 submodifications that occur in

---

TABLE 9.41

Summary of the β-2 Submodifications in Which Different TAG Families (x,y) Crystallize

| | β-2 Submodification | | |
A	B	C	D
(0, 0)	(−2, 0)	(4, 2)	(2, 2)
(0, 2)	(2, 0)	(4, 4)	
(2, 2)	(2, 4)		

---

practice, using de Jong's nomenclature, representing a TAG according to the method defined in Section 9.4 as $(p \cdot p + x \cdot p + y)$ and thus indicating TAG families by (x, y).

7. The melting point and heat of fusion of the hypothetical β-2 forms, which do not occur in the pure TAG, are assumed to lie 3°C below that of the stable β-2 form, while the heat of fusion is 90% of that of the stable form.

    Thermodynamic calculations now become extremely complicated, as the number of independent polymorphs that have to be considered, has increased with each extra assumption from 3 to 8. Because most binary pairs of TAGs that crystallize into different β-2 modifications are not completely miscible even when they would crystallize in the same submodification, often equivalent calculation results are obtained if assumption 7 is replaced by the following assumption.

8. In calculations, only one β-2 polymorph can be used. However, if the two TAGs of a binary crystallize in different submodifications, the two suffix Margules parameter that is obtained from Equations 9.94 and 9.95 is augmented by 1.

    The phase diagram of SSS–SPS (Figure 9.50) that was calculated allowing the occurrence of 2 β-2 modifications and $A/RT = 1.8$ in each submodification is nearly the same as the phase diagram given in Section 9.7.2.4.1 using only one β-modification and A/RT = 3.

The difference of 3°C between the melting point of the stable and hypothetical β'- and β-forms is a best guess, based on de Jong's findings for the melting points of saturated mono-acid TAGs in different submodifications. The value of 90% for the heat of fusion of hypothetical submodifications is a best guess that is based on the data given in Section 9.4.

9.8.3.2 Discussion

It is obvious that the empirical method given is far from perfect, as it is based on a small number of data, which necessitated a number of speculative assumptions. However, the underlying notion that only TAGs with a high degree of geometric similarity will mix well in a densely packed solid phase is most likely correct. Therefore, the main effects of structural differences on solid phase miscibility are covered by the method, although not always very precise. In order to improve the method, more data are required and a number of issues need to be solved:

1. There is still a lack of reliable data on binary interaction parameters of TAGs, both for the β'-modification and for the β-modification. Especially on mixtures with *cis*-mono- and di-unsaturated TAGs more information is required.

2. The crystal structure of the β'-modification and its possible submodifications as well as the structure of the polymorphic forms of *cis*-unsaturated TAGs have to be known.

3. It must be established to what extent the structure of two submodifications must be different before continuous miscibility of the two submodifications cannot occur, so that treatment as independent polymorphs is necessary.

4. If two submodifications need to be treated as independent polymorphs, a method must be developed to estimate the thermodynamic properties of those polymorphs that do not occur in the pure component.
5. To obtain insight in the magnitude of different lattice distortions, more mixtures have to be studied by molecular modeling, not only in the β-modification, but also in the β'-modification.

In spite of its serious limitations, in Section 9.9 this empirical method will be tested to see how it performs in practical situations.

9.8.4 Conclusion

The nonideal mixing behavior of TAGs in the solid phase can be explained by geometrical considerations.

Within one (sub)modification the binary interaction parameters show a clear correlation with the degree of isomorphism, as defined by Kitaigorodskii (1984).

Based on the good correlation between degree of isomorphism and the 2-suffix Margules interaction parameter, a method has been defined that predicts the binary interaction parameters from structural differences.

The work that is required to improve this predictive method was outlined.

9.9 PRACTICAL APPLICATIONS

In the previous sections, all steps that are required to meet the objective of this work (prediction of melting ranges and solid phase composition of fats) have been taken. In this section, the method developed will be used for predictions in a number of practical situations: the prediction of melting ranges of margarine fat blends, the prediction of the composition of fat fractions obtained from fractional crystallization, and the understanding of recrystallization phenomena. Some examples outside the area of edible fats are considered as well. Finally the conclusions of this chapter are summarized.

9.9.1 Prediction of Melting Ranges

The primary objective of this chapter is the development of a general method to predict the melting range of a fat blend from its composition. In this section, we will investigate to what extent this objective has been attained.

In Section 9.6, it was shown that the α-melting ranges, or "α-lines" of the fat blends of nine different commercial fat spreads could be predicted very well.

If the binary interaction parameters for the β'-modification are estimated with the procedure outlined in Section 9.8.3, it is in principle possible to calculate the β'-melting ranges of these fats as well. A complicating factor in the calculations is the large number of components, usually between 30 and 400 in these fats. Even if only four coexisting β' phases are formed, the flash calculation of Section 9.3 must already handle matrices with a 2000×2000 dimension, which will make the calculation procedure too slow for practical use.

To keep the calculations manageable, pseudocomponents are defined as follows: all components that are nearly isomorphous ($\in > 0.95$) and that differ 5°C or less in their β′-2 melting point are taken together as a new pseudocomponent that has the polymorphic behavior and the size of the component that contributes most to the pseudocomponent. Components and pseudocomponents with a concentration less than 0.1% are neglected.

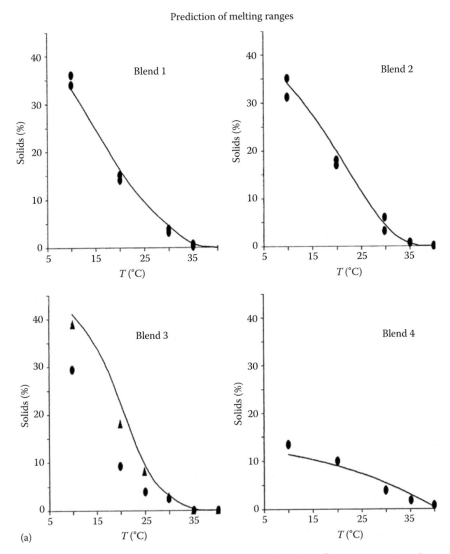

FIGURE 9.52 (a) Predicted (lines) and experimental (points) β′-melting ranges of several commercial fat blends. The numbers refer to the composition of the fat blend, given in Table 9.19. The 2-suffix Margules equation was used for the excess Gibbs energy. dots: solid phase measured with standard procedure triangles (blend 3 only): determined in a week-old margarine.

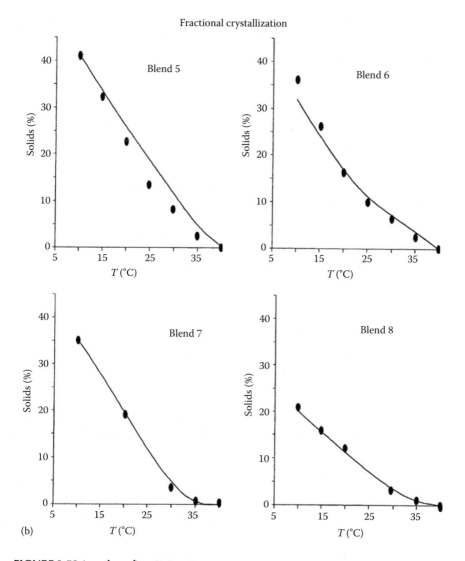

Fractional crystallization

FIGURE 9.52 (continued) (b) Predicted (lines) and experimental (points) β′-melting ranges of several commercial fat blends. The numbers refer to the composition of the fat blend, given in Table 9.19. The 2-suffix Margules equation was used for the excess Gibbs energy.

In this way blend 1–8 from Table 9.19 were reduced to 15–20 component mixtures. Only blend 9 remained too complex. The β′-melting ranges of these 15–20 component mixtures were calculated and plotted in Figure 9.52, together with the experimental values. The experimental values were obtained with wide range NMR of samples that were molten, kept for 1 h at 80°C, rapidly cooled to 0°C, stabilized for 16 h at 0°C and for 30 min at the temperature of measurement. With this procedure, one of the Unilever standard methods, most normal fat blends have crystallized completely.

Although the predictions show a larger deviation than those for the α-melting ranges, the agreement with the experimental data is still very good, in spite of the large simplifications that were made in the composition and in the estimation of the interaction parameters. The standard error of 3% solids compares well with the experimental error of 1% solids and the effects of kinetics of crystallization, that are normally estimated to be in the order of a few percent of solids.

Out of the empirical methods, mentioned in Section 9.1, only the multiple linear regression/linear programming method performs better, but only within the limited range of compositions for which this method is valid. As it is based on interpolations within a finely meshed raster of experimental data points, this is not too surprising.

Only for blend 3, which contains 23% of POP, the experimental points, determined with the standard method, lie far below the calculated line. Blend 3 is a typical example of a fat blend that shows extreme "postcrystallization." Postcrystallization means that in the first week after production still a considerable amount of solid fat crystallizes. The solid fat content that was determined in a margarine sample of a week old is also plotted in Figure 9.52. Now the agreement with the calculated data is very good. Obviously, blend 3 has not crystallized completely when the standard procedure is used.

Thus it has been demonstrated that melting ranges of the practically relevant polymorphic forms of fat blends can be predicted by application of solid–liquid phase equilibrium thermodynamics.

9.9.2   FRACTIONAL CRYSTALLIZATION

The second objective of this chapter is the development of a method that predicts the solid phase composition of a fat at a certain temperature, in order to enable modeling of a fractional crystallization process. At present fractional crystallization is primarily used for fractionation of palm oil into palm olein, a liquid fraction of palm oil and palm stearin, a solid fraction.

Palm oil was heated to 80°C for 1 h and subsequently cooled to the fractionation temperature. Five hours after the appearance of the first turbidity the palm olein was filtered off in a filter press at 12 bar. The separation efficiency (amount of olein that is obtained over the total amount of liquid phase that is present in palm oil) was determined. NMR showed that crystallization had taken place into the β-modification. The TAG compositions of the palm oil, the olein, and the stearin were analyzed by A_gNO_3-HPLC (Kitaigorodskii, 1984). The results are given in Table 9.42, together with the composition that was calculated, using the procedure as outlined earlier and the measured separation efficiency.

The calculated data agree very well with the experimental ones. Van Putte and Bakker (1987) states that palm oil crystallizes into the β'-modification when the crystallization takes place at 27° or less, while above this temperature the β-modification is observed. In line with this observation, calculations show that in palm oil up to 25°C–26°C a stable β'-solid phase coexists with two β-solid phases, while above this temperature only β-solid phases remain.

This example shows clearly that application of solid–liquid phase equilibrium thermodynamics to fat fractionation processes is feasible.

TABLE 9.42

Experimental and Predicted Compositions of Palm Oil Fractions (Calculated for the Most Stable State, the β-Modification)

	Fract. at 29°C		Fract. at 32°C	
	Measured	Calc.	Measured	Calc.
Palm oil				
h_3	8.8			
hOh	33.0			
hhO	7.1			
hlh	9.5			
Rest	41.7			
Stearin				
h_3	45.3	46.2	54.7	53.3
hOh	25.4	29.3	21.5	17.2
hhO	6.1	4.1	5.5	4.3
hlh	5.3	4.9	4.3	4.7
Rest	17.9	15.4	14.0	20.1
Olein				
h_3	2.2	0.1	2.8	0.2
hOh	34.4	33.7	34.7	35.8
hhO	6.1	7.8	5.9	7.6
hlh	10.6	10.5	10.1	10.3
Rest	46.7	47.7	46.4	46.0

9.9.3 Recrystallization Phenomena

9.9.3.1 Influence of Precrystallization and Temperature Cycling

In Section 9.7.4, DSC curves of mixtures of a crystallizing binary pair of TAGs in a surplus of a liquid TAG were determined. The samples were stabilized by rapid cooling to a stabilization temperature. In those experiments, stabilization at temperatures below the onset of the first melting peak in the DSC thermogram in the end always resulted in the same DSC curve, regardless of the stabilization procedure that was followed. However, if a well-stabilized sample was subjected to temperature cyclization by increasing the temperature from the stabilization temperature to a cycling temperature well above the onset of the first melting peak and backward, then the DSC curves that were measured directly after the cycling step often had a completely different shape. The effect even occurred at cycling rates as low as 0.6°C/min. It was not possible to fit such DSC curves with the 2- or the 3-suffix Margules equation.

The shape of the curves of the cycled samples could be obtained using the well-known concept of shell formation (Zief and Wilcox, 1967), which was already mentioned in Section 9.2. Upon heating, the sample part of the solid phase dissolves. It is assumed that upon cooling the remaining crystals are covered with a layer of solid

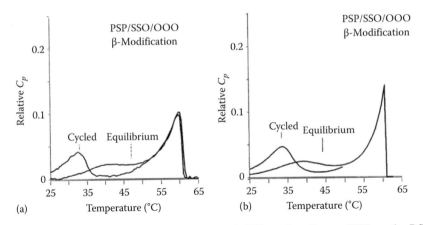

FIGURE 9.53 (a) 25% PSP/25% SSO/50% OOO. Effect of cycling to 50°C on the DSC melting curve. Experimental curves, taken at 2.5°C/min. (b) Simulation of the effect of cycling to 50°C on the DSC melting curve of the system of Figure 9.30.

fat that is in equilibrium with the liquid phase that is present. This shell effectively keeps the inner part of the crystals from gaining the equilibrium composition.

The resulting DSC melting curve will be the sum of that a sample that has an overall composition equal to that of the liquid phase at the cyclization temperature and the curve of the solid phase that remained at the cyclization temperature. This DSC curve will deviate from the equilibrium curve at temperature below the cyclization temperature. Figure 9.53a shows that this is exactly what is observed, while the curves in Figure 9.53b, which are calculated assuming shell formation, confirm that shell formation can describes the observations.

Shell formation is much less likely with the isothermal stabilization procedure that is followed in this work: always a completely liquid sample is very quickly supercooled to the stabilization temperature. Therefore all solids are formed at the stabilization temperature.

Elaborate cyclization procedures were followed in order to obtain the binary phase diagrams given in Section 9.7.2. The cycling effect described here offers another explanation for the large discrepancies between solidus and liquidus that were found: cyclization leads to the formation of a solid phase that starts to melt off at lower temperatures. After initial melting, a solid phase that is enriched in the highest melting component remains.

Shell formation also offers the explanation for the differences in fraction composition that are obtained with fractional crystallization of oils and fats when different cooling rates are used (Keulemans, 1986): slow cooling to the fractionation temperature causes shell formation and so results in a solid phase that is enriched in the higher melting components, while quick cooling results in the equilibrium composition. This work enables the quantification of the magnitude of these effects.

Shell formation also plays a role in the normal margarine "votator" process. Fat spreads are usually prepared by quickly cooling and emulsifying a mixture of a molten fat blend and milk or water in a scraped surface heat exchanger or "A-unit." The supercooled emulsion is subsequently crystallized in a pin stirrer, or "C-unit." In many cases, this is followed by further cooling and crystallization in second series of A and C-units, so that the complete votator process sequence becomes A–C$_1$–A–C$_2$. It is

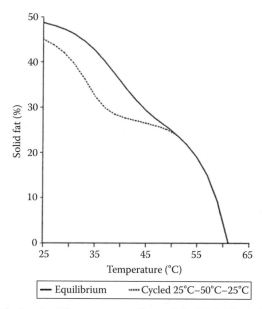

FIGURE 9.54 Calculated solids temperature lines of the TAG blend of Figure 9.53 (25% PSP/25% SSO/50% OOO). Solid line = equilibrium line, dashed line = after precrystallization at 50°C.

often impossible to complete crystallization in this second C-unit to the β′-equilibrium, even when a very long residence time is provided. Upon storage, the equilibrium solids content will eventually be reached, resulting in a considerable posthardening and change of rheological properties. In fact, the two-step crystallization in this votator process is a build in cyclization step: in the first C-unit a solid phase is formed, which is encapsulated in the second C-unit with a solid phase of different composition.

The effect of this precrystallization on solids content can be quite dramatic, as is illustrated in Figure 9.54 for the PSP/SSO/OOO model blend that was used in Section 9.7.5.4: the difference in solid content between precrystallized and equilibrium situation can be as large as 10%–15% solids. After the $A–C_1–A–C_2$ votator process the fat can only crystallize to a point on the lower solids line. But upon storage, it will recrystallize to the equilibrium composition in several weeks, which explains the increase in solids content and the rise in product hardness.

Fat blends in which the TAGs show considerable solid phase miscibility will be more sensitive to precrystallization effects than blends in which the TAGs show no solid phase miscibility.

9.9.3.2 Sandiness

A well-known phenomenon is the development of sandiness in fat spreads with a fat blend containing only liquid oils and hydrogenated low erucic acid rape seed oil (Hernqvist et al., 1981; Hernqvist and Anjou). Sandiness is the development upon storage of large β-fat crystals in a product that initially had crystallized into the β′-modification. Those large crystals give a sensation to the consumer as if sand has been erroneously added to the product.

Hydrogenated rape seed oil consists for 95% out of TAGs from C18 fatty acids (E, O, and S). The degree of isomorphism between these TAGs is 1, which implies ideal solid solubility both in the β'- and β-modifications. In Section 9.7.6.3 it was found that the rate recrystallization from β' to the β-modification depends on the thermodynamic driving force and on the degree of rearrangement of the solid phase that is required. Because the rape seed TAGs mix ideally in both modifications, the thermodynamic driving force is the full stability difference between the two modifications, while virtually no change in solid phase composition has to take place.

In this reasoning, the tendency to develop sandiness can be reduced by additions of TAGs that reduce the thermodynamic driving force and increase the degree of rearrangement in the solid phase that is required. Such TAGs should mix nearly ideally with the TAGs from rape seed oil in the β'-modification, but must demix in the β-modification. TAGs that have a degree of isomorphism with the rape seed TAGs of about 0.92 and a melting point above 30°C fulfill this criterion. Candidates are therefore the TAGs P_2S, P_2E, P_2O, A_2S, A_2S, A_2O, and also BS_2, B_2E, and B_2O. The solubility of these inhibiting TAGs in the β-modification of the rape seed TAGs is according to the findings in Section 8 nearly 10%. Therefore, expect that the β'–β transition in hydrogenated rape seed oil mixtures will be retarded considerably if the crystallizing TAGs in the fat blend consist for more than 10% of inhibiting TAGs. Hydrogenated rape seed oil already contains 5% of those TAGs, so that addition of only 5% of an inhibiting fat is required to retard sandiness.

TAGs with a degree of isomorphism with the rape seed TAGs of less than 0.86 are expected to have much less effect: they do not even cocrystallize with the rape seed TAGs in the β'-modification, so that still hardly any solid phase rearrangement is required for the formation of the β-modification.

Hydrogenated soybean oil has a TAG composition that is very similar to that of hydrogenated rape seed oils, but the concentration of sandiness inhibiting TAGs is higher: it varies from batch to batch around the critical value of 10%, which implies that depending on the batch of oil sandiness can occur. And indeed from time to time, we have observed sandiness in products based on hydrogenated soybean oil.

Hydrogenated high erucic rape seed oil, contains depending on the hydrogenation procedure followed, more than 10% of inhibiting TAGs. Therefore sandiness is not so often observed in products of this rape seed oil breed (Hernqvist et al., 1981; Hernqvist and Anjou).

Hernqvist already mentioned the possibility of addition of inhibiting TAGs, but he followed another reasoning and is not able to specify which TAGs and what amounts are required. By trial and error, Hernqvist found no sandiness in a mixture of TAGs from rape seed and 20% PSP and he suggests that other inhibiting TAGs may exist. Indeed PSP belongs to the inhibiting TAGs mentioned.

9.9.3.3 Conclusion

The examples in this section have made clear that, although quantitative predictions are not always possible, the methods of this work help create understanding of recrystallization phenomena, which can directly be translated into practical methods to influence recrystallization.

9.9.4 APPLICATIONS OUTSIDE EDIBLE OILS AND FATS

9.9.4.1 Solid–Liquid Phase Behavior of *n*-Alkanes

Contrary to TAGs, which have only one thermodynamically stable polymorph ("monotropic polymorphism"), each polymorphic form of medium to long chain odd numbered *n*-alkanes is stable within a certain temperature and pressure region ("enantiotropic polymorphism"). Far below the melting point the β′-form is stable. However, from about 10°C below to the melting point upward the α-modification ("rotator phase") is the stable polymorph. Upon heating a β′-form of an odd numbered *n*-alkane, it will first transform to the α-modification at the so-called transition temperature and finally melt at the α-melting point.

The enantiotropic polymorphism gives rise to most peculiar solid–liquid phase diagrams, of which an example is depicted in Figure 9.55. The phase diagram consists of three single-phase regions, a liquid, an α- and a β-region. The regions are separated by two phase domains, a cigar-shaped α-L and a β′-α region with a minimum transition temperature (Wurflinger, 1972; Morancelli et al., 1985). Although on first sight the phase behavior seems completely different, remarkable agreement exists between the phase behavior of TAGs and these *n*-alkanes.

The cigar-shape of the α-L region indicates that miscibility in the α-modification of these *n*-alkanes must be nearly ideal. This is in agreement with our findings for the α-modification of TAGs.

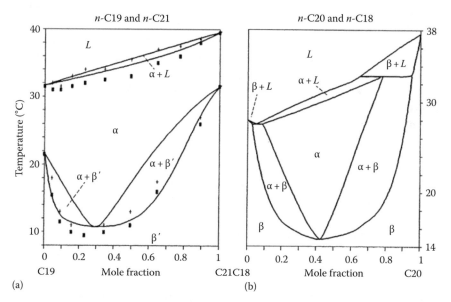

FIGURE 9.55 Phase diagrams of binary *n*-alkane mixtures. (a) *n*-C19 with *n*-C21, points: data from Würflinger (1972) lines: calculated with ideal α and liquid phase miscibility and nonideal mixing in the β′-modification with 3-suffix Margules parameters $A_{19\_21}/RT = 0.4$ and $A_{21\_19}/RT = 1.7$. (b) *n*-C18 with *n*-C20. Calculated assuming ideal miscibility in the α-modification and liquid phase and non ideal miscibility in the β′-modification with regular solution parameter $A/RT = 1.6$. The α-melting points were taken 1.5°C below the β-melting points. (From Wurflinger, A., Thesis, University of Bochum, Bochum, Germany, 1972.)

The shape of the β'-α region and the existence of a single phase β'-region suggest non ideal mixing in the β'-modification, with a value for a regular solution binary interaction parameter in the β'-modification around 1.3–1.8. This is the same order of magnitude of the interaction parameters that have been found for the β'-modification of a binary pair of TAGs with a degree of isomorphism of 0.89.

When these values are used to describe the mixing behavior in the α- and β'-modification, the phase diagram of Figure 9.55 is obtained easily, applying the models and techniques used in this work for TAGs.

The main problem in the calculations is the absence of values for the β'-melting points. The enantiotropic behavior of these n-alkanes makes an experimental determination of β'-melting point impossible. However, they can be calculated from the measured β'-α-transition temperatures using the following expression:

$$T_{f,\beta'} = T_{\text{transition}} \cdot \frac{\Delta H_{\beta'}}{\Delta H_{\beta'} + \Delta H_{\alpha} \left(\dfrac{T_{\text{transition}}}{T_{f,\alpha}} - 1 \right)} \tag{9.100}$$

The expression follows from the fact that at the transition temperature the Gibbs energy of the two polymorphs must be equal. The differences in heat capacity are neglected. Calculated β'-melting points are listed in Table 9.42. Although the transition temperatures lie 10°C or more below the α-melting point, the calculated β'-melting points lie only 1°C or 2°C below the α-melting point.

The table also shows that, in spite of the higher melting point, entropy and heat of fusion of the α-modification are only 70% of the β-values, which is in agreement with the values found for TAGs. In fact, a rise of only 5% in α-melting entropy would change the enantiotropic polymorphism into the normal monotropic behavior that is observed for TAGs.

The phase diagram that is calculated from the values of Table 9.43 and the assumptions about the mixing behavior in the various polymorphic forms is plotted in Figure 9.55a, together with the data of Würflinger. The agreement is striking, especially when taking into account that it must be almost impossible to determine the equilibrium α–β' coexistence region correctly by experiment. It would require large rearrangements in the solid state within the time of measurement, while solid state diffusion is extremely slow.

Some even-chain n-alkanes are just monotropic: the α-melting point lies just below the β-melting point. Binary mixtures of those components are enantiotropic, which gives rise to interesting phase diagrams (Wurflinger, 1972). This type of phase diagrams, too, is obtained without problems, assuming ideal mixing in the α-modification and non ideal mixing in the β'-modification with $A_{12}/RT = 1.3–1.8$ (Figure 9.55b).

Thus it is illustrated that very simple excess Gibbs energy models, when combined with a proper treatment of polymorphism, can describe the main features of the complex solid–liquid phase behavior of n-alkanes.

9.9.4.2 Petroleum Waxes

In parallel with this chapter, a number of articles appeared from Won (1986, 1989) and Hansen (1987) on petroleum waxes. If mineral oils are brought from reservoir

TABLE 9.43
Thermodynamic Data of Some Odd-Chain *n*-Alkanes

Carbon Number	T_f, α	$T_{f\beta'}$	ΔH_α	ΔH_β	ΔS_α	ΔS_β	T_{trans}
9	−53.5	−54.2	16	22	71	100	−56.0
11	−25.5	−28.1	22	29	90	118	−36.5
13	−5.4	−8.1	28	36	106	136	−18.0
15	9.9	7.3	35	44	123	157	−2.3
17	21.9	19.4	41	52	138	176	10.5
19	32.0	29.6	46	59	150	197	22.0
21	40.2	38.3	48	63	152	203	32.5
23	47.5	45.5	54	76	169	238	40.5
25	53.5	51.5	58	84	177	258	47.0
27	58.8	56.9	60	89	182	270	53.0
29	61.2	60.2	66	98	198	293	58.0

Sources: Wurflinger, A., Thesis, University of Bochum, Bochum, Germany, 1972.
Note: T in °C, H in kJ/mol and S in J/K, mol.

temperature and pressure to atmospheric conditions, sometimes a solid wax phase crystallizes, which causes fouling of the equipment and pipelines.

This wax phase is amorphous, so it is in complete rotational disorder. In line with our findings for the α-modification of TAGs and the rotator phase of alkanes, it can be expected that these wax phases behave like ideal solid solutions. Consequently it is possible to calculate the wax appearance point of these oils.

Indeed, Won and Hansen found that the melting range of these complex petroleum waxes was predicted correctly by assuming ideal solid phase miscibility.

9.9.4.3 β-Substituted Naphthalenes

Another class of components that show polymorphism are the β-substituted naphthalenes. Like the odd-chain alkanes, the polymorphism is enantiotropic, with a high-temperature polymorph that shows some rotational disorder and more densely packed lower temperature polymorphs. Consequently, the phase diagrams are very similar to those of the enantiotropic odd-chain *n*-alkanes. An extensive study of binary phase diagrams of these components has been made by Chanh and Haget (Chanh and Haget, 1972, 1975; Haget et al., 1985). Contrary to the *n*-alkanes, the rotational disorder in the high temperature α-polymorph is not very large, so that a fairly high degree of isomorphism is required for ideal α-miscibility. In the more densely packed lower temperature, polymorphs miscibility is even less ideal than in the α-modification, which in mixtures will lead to an extension toward lower temperatures of the region where the α-modification is stable. This is illustrated in Figure 9.56 for β-thionaphthalene and β-bromonaphthalene with a degree of isomorphism of 0.97. These components mix nearly ideally in the α-modification and highly nonideal in the other modifications. β-Thionaphthalene and β-fluoronaphthalene, with a degree isomorphism of only 0.87, already show nonideal mixing in the α-modification.

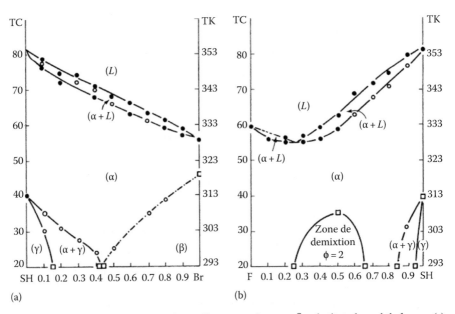

FIGURE 9.56 T, x, y binary phase diagrams of some β-substituted naphthalenes. (a) β-thionaphthalene and β-bromonaphthalene. (b) β-thionaphthalene and β-fluoronaphthalene.

Although Chanh et al. do not report pure component properties for each polymorphic form, so that calculation of the phase diagrams from their data is not possible, it is clear that the approach of this work offers a good starting point for the development of a description of the complex solid phase behavior of this class of compounds.

9.9.5 CONCLUSIONS OF THIS CHAPTER

The objective of this chapter is the development of a predictive method for the melting range and the solid phase composition of edible oils and fats. We have shown that this objective can be attained by application of solid–liquid phase equilibrium thermodynamics to all three basic polymorphic forms in which TAGs crystallize.

In order to develop this thermodynamic description:

- Existing multicomponent multiphase vapor liquid flash calculation algorithms have been adapted to deal with a number of mixed solid phases that show polymorphism.
- The heat of fusion and the melting points of TAGs have been correlated with a number of structural parameters.
- Liquid TAG mixtures and TAG mixtures in the α-modification have been treated as ideal mixtures.
- The deviation from ideal miscibility in the β′- and β-modifications has been described by the 2- and the 3-suffix Margules equations.
- A new method for the determination of the binary interaction parameters that occur in the 2- or 3-suffix Margules equation has been developed. The method

is based upon the interpretation of the complete DSC melting curve of a mixture of the binary pair and a surplus of a liquid TAG. It is quicker, more versatile and more reliable than the use of binary T, x, y solid–liquid phase diagrams.

- The binary interaction parameters that occur in the 2- or the 3-suffix Margules equation have been correlated with the degree of isomorphism of the binary pair, a parameter that indicates geometrical similarity.

Now the thermodynamic framework has been set up, the main scope of future research in this area should be the development of a more refined method to predict binary interaction parameters from structural differences.

9.10 SUMMARY

The properties of many food products, like margarine, reduced fat spreads, bakery products, snacks, cake, ice cream, and chocolate, are to a large extent determined by the melting range of the fat that they contain.

This chapter gives a general method to predict the melting range and solid-phase behavior of fats from their overall composition. It appears that solid fats are complex mixtures of several mixed crystalline solid phases in several polymorphic forms.

The ultimate amount, polymorphic form, and composition of the solid phase in a fat are determined by the position of the thermodynamic equilibrium solely. However, the crystallization process may lead to significant deviations from the equilibrium composition in practical situations. Yet, the starting point for any general predictive method of the solid-phase content and polymorphic behavior is an understanding of the solid–liquid phase equilibrium.

Fats consist of hundreds of different triacylglycerols (TAGs) that, like other long-chain hydrocarbons, can basically crystallize in three main forms:

1. The unstable α-modification
2. The metastable β'-modification
3. The stable β-modification

The thermodynamic framework describing the solid–liquid phase equilibrium for all three polymorphic forms of TAGs is given. The thermodynamic equations that determine these equilibria are worked out. Methods are developed that generate the information needed to solve this set of equations.

1. Calculation methods for gas–liquid systems are adapted to work for complex multicomponent liquid oil–solid fat systems that show polymorphism. Michelsen's method for the initial estimate combined with a Gibbs free energy minimization according to Murray is recommended as the best solution method for the thermodynamic equations.
2. The melting enthalpies and melting points of pure TAGs are obtained from correlations of the TAG structure with existing and new experimental data.
3. In the liquid (oil) phase TAGs are found to exhibit ideal mixing.
4. The degree of mixed crystal formation of TAGs in the solid fat phases can be described by simple excess Gibbs energy models: the 2- and 3-suffix

Margules equations. Binary interaction parameters of the Margules model are determined from existing and newly determined phase diagrams. Miscibility of TAGs in the solid phase is found to be governed by size differences of TAGs only. An empirical correlation is developed that predicts Margules binary parameters from the difference in structure of the two TAGs.

A large part of published binary phase diagrams of TAGs appears to be inaccurate due to experimental errors and stabilization issues.

A simple, fast, and accurate DSC method using a ternary system is given that allows easy determination of binary phase diagrams of two TAGs. It also uniquely allows the determination of phase diagrams of the unstable polymorphic forms.

The experimental data of multicomponent solid-phase behavior of fats and oils show that

- TAGs form ideal mixed crystals in the unstable α-modification. Chain mobility and space in the methyl end plane of the crystals is apparently sufficient to accommodate differences in size between TAGs. α-Melting ranges of a large number of commercial fats are predicted very well assuming complete and ideal mixed crystal formation.
- In the β'-form still a large extent of mixed crystal formation is found; however at chain length differences of 4 or more mixed crystal formation is only limited and TAGs start to crystallize into two separate partially mixed solid phases.
- In the most densely packed β-modification, TAGs only form perfect mixed crystals when there are no chain length differences. Any difference in chain length leads to crystallization in two separate partially mixed solid phases, and at differences in chain length exceeding 4, co-crystallization has virtually ceased.
- Applying the prediction of binary interaction parameters from differences in structure to all binary pairs of TAGs in a commercial fat, it is possible to calculate the β'- and β-melting ranges, the polymorphic behavior, and the solid fat and liquid oil compositions of the commercial fat surprisingly accurately from the composition of the fat blend.

This can be used for optimization of blend composition, fractionation processes, and understanding of product defects upon storage and storage temperature cycling. A better insight is obtained into the occurrence and prevention of undesired recrystallization phenomena that occur in fats. Also applications outside the world of edible fats are shown to be feasible.

LIST OF SYMBOLS

A	Interaction coefficient
A	Interaction parameter
A	Interaction parameter
a	A/RT
B_{11}	Second virial coefficient
c_p	Heat capacity (J/mol, K)

f	A function
G	Gibbs free energy (J)
g	Molar Gibbs free energy (J/mol)
K	Molar distribution coefficient
k, K	A constant
K^{AB}	Distribution constant
L	Long spacing
M	Molar weight (kg/mol)
m	Mass (kg)
N	Number of components
n	Number of moles (mol)
$n_{o,1,e}$	Number of o, 1, or e chains
P	Number of phases
P	Pressure (Pa)
p	Shortest outer TAG chain
$P*$	Vapor pressure of pure component
q	Middle TAG chain
R	Gas constant (J/K, mol)
r	Longest outer TAG chain
T	Temperature (K)
T_f	Normal melting point (K)
T_∞	Hypothetical melting point of polyethylene (K)
V	Volume (m³)
v	Volume of a molecule
v	Liquid molar volume (m³/mol)
x	Interaction parameter
x	Mole fraction (mol/mol)
x	$P - q$
Y	Activity coefficient
y	$P - r$
z	Overall mole fraction
Δ	Number of double bonds
ΔH_f	Heat of fusion (kJ/mol)
ΔH_f	Heat of fusion (kJ/mol) at T_f
ΔS_f	Melting entropy (J/K, mol)
δ, θ	Constants
γ	Activity coefficient
Φ	Phase fraction
Φ_v	Volume flow rate (m³/s)
φ	Volume fraction
μ	Chemical potential (J/mol)
τ	Angle of tilt
$\in$	Degree of isomorphism
$\%S$	Solid phase content
$\times$	Mole fraction

APPENDIX 9.A: PURE COMPONENT DATA

9.A.1 SATURATED TAGs

TABLE 9.A.1
Enthalpy of Fusion for the α-Modification of Saturated TAGs

X	Y	TAG	n	ΔH_f (kJ/mol)	Predicted	Residual	FREQ
−6	0	SLS	48	70.0	85.1	−15.1	1*
−6	2	PCS	44	80.0	71.8	8.2	1*
−4	0	PLP	44	65.0	76.6	−11.6	1*
−4	0	SMS	50	90.0	92.8	−2.8	1*
−4	2	PLS	46	70.0	79.5	−9.5	1*
−4	4	MCS	42	67.0	63.7	3.3	1*
−2	0	PMP	46	79.0	86.5	−7.5	1*
−2	0	SPS	52	103.0	102.7	0.3	
−2	2	PMS	48	93.0	89.4	3.6	1*
0	0	888	24	18.4	31.8	−13.4	3*
0	0	CCC	30	57.3	47.9	9.4	2*
0	0	LLL	36	69.8	64.1	5.7	3*
0	0	MMM	42	81.9	80.3	1.6	3*
0	0	PPP	48	95.8	96.5	−0.7	3*
0	0	SSS	54	108.5	112.7	−4.2	3*
0	0	BBB	66	143.2	145.1	−1.9	1
0	0	24.24.24	72	160.9	161.3	−0.4	1
0	2	88C	26	26.0	34.7	−8.7	1*
0	2	MMP	44	82.0	83.2	−1.2	1*
0	2	PPS	50	100.0	99.4	0.6	1*
0	4	LLP	40	67.0	67.4	−0.4	1*
0	4	MMS	46	87.0	83.6	3.4	1*
2	0	8C8	26	51.0	37.8	13.2	1*
2	0	CLC	32	67.0	54.0	13.0	1*
2	0	LML	38	83.0	70.2	12.8	1*
2	0	MPM	44	93.0	86.4	6.6	1*
2	0	PSP	50	112.2	102.6	9.7	2*
2	2	MPP	46	89.0	89.3	−0.3	1*
2	2	PSS	52	106.0	105.5	0.5	1*
2	4	LMP	42	74.0	73.5	0.5	1*
2	4	MPS	48	86.0	89.7	−3.7	1*
2	6	CLP	38	57.0	59.0	−2.0	1*
2	6	MPA	50	72.0	91.3	−19.3	1
4	0	MSM	46	99.0	87.8	11.2	1*
4	0	SBS	58	128.0	120.2	7.8	1*
4	2	MSP	48	91.0	90.7	0.3	1*
4	4	LPP	44	83.0	75.0	8.0	1*
6	0	LSL	42	66.0	72.2	−6.2	1*
6	0	LSL	42	66.0	72.2	−6.2	1*

TABLE 9.A.1 (continued)
Enthalpy of Fusion for the α-Modification of Saturated TAGs

X	Y	TAG	n	ΔH_f (kJ/mol)	Predicted	Residual	FREQ
6	0	MAM	48	88.0	88.4	−0.4	1*
6	4	LSP	46	76.0	75.5	0.5	1*
6	6	CPP	42	53.0	61.0	−8.0	1*
6	6	LSS	48	70.0	77.2	−7.2	1*

Note: Data Marked by a * include measurements from this chapter. FREQ indicates the number of measurements that has been averaged.

TABLE 9.A.2
Enthalpy of Fusion for the β′-Modification of Saturated TAGs

X	Y	TAG	n	ΔH_f (kJ/mol)	Predicted	Residual	FREQ
0	0	LLL	36	86.0	88.0	−2.0	1
0	0	MMM	42	106.0	111.1	−5.1	3*
0	0	PPP	48	126.5	134.3	−7.8	2*
0	0	SSS	54	156.5	157.5	−1.0	2*
0	2	MMP	44	100.0	106.5	−6.5	1*
0	2	PPS	50	124.0	129.7	−5.7	1*
0	4	LLP	40	90.0	84.4	5.6	1*
0	4	MMS	46	93.0	107.6	−14.6	1*
2	0	8C8	26	62.0	63.5	−1.5	1*
2	0	CLC	32	87.5	86.7	0.8	2*
2	0	LML	38	112.0	109.9	2.1	1*
2	0	MPM	44	127.0	133.0	−6.0	1*
2	0	PSP	50	165.5	156.2	9.3	4*
2	4	LMP	42	94.0	106.3	−12.3	1*
2	4	MPS	48	111.0	129.5	−18.5	1*
4	0	MSM	46	148.0	133.0	15.0	1*
4	0	SBS	58	189.0	179.4	9.6	1*
4	2	MSP	48	143.5	128.4	15.1	2*
4	4	LPP	44	110.0	106.3	3.7	1*
4	4	MSS	50	128.0	129.5	−1.5	1*
6	0	LSL	42	87.0	107.8	−20.8	0*
6	0	MAM	48	117.0	130.9	−13.9	1*
6	4	LSP	46	107.0	104.2	2.8	1*
6	6	LSS	48	104.0	111.6	−7.6	1*

Note: Data marked by a * include measurements from this work. FREQ indicates the number of measurements that has been averaged.

TABLE 9.A.3
Enthalpy of Fusion for the β-Modification
of Saturated TAGs

X	Y	TAG	n	ΔH_f (kJ/mol)	Predicted	Residual	FREQ
−10	0	S8S	44	141.0	131.7	9.3	1
−8	0	P8P	40	122.0	116.2	5.8	3
−8	0	SCS	46	143.0	139.5	3.5	2
−6	0	PCP	42	122.5	124.0	−1.5	2
−6	0	SLS	48	132.0	147.4	−15.4	2*
−6	2	PCS	44	125.0	123.3	1.7	1*
−4	0	PLP	44	121.5	133.1	−11.6	2*
−4	0	SMS	50	146.5	156.5	−10.0	2*
−4	2	PLS	46	123.0	132.4	−9.4	1*
−4	4	MCS	42	88.0	106.3	−18.3	1*
−2	0	PMP	46	137.0	148.3	−11.3	2*
−2	0	SPS	52	170.3	171.7	−1.4	4*
−2	2	PMS	48	152.0	147.6	4.4	1*
0	0	888	24	69.2	74.7	−5.4	4*
0	0	CCC	30	95.0	98.0	−3.0	5*
0	0	LLL	36	122.2	121.3	0.9	9*
0	0	MMM	42	146.8	144.7	2.1	8*
0	0	PPP	48	171.3	168.0	3.3	16*
0	0	SSS	54	194.2	191.3	2.8	20*
0	2	88C	26	59.0	73.9	−14.9	1*
0	2	LLM	38	116.0	120.6	−4.6	1
0	2	MMP	44	131.0	143.9	−12.9	1*
0	2	PPS	50	166.3	167.3	−1.0	4*
0	4	LLP	40	117.0	117.8	−0.8	1*
0	4	MMS	46	145.0	141.2	3.8	1*
2	0	PSP	50	166.0	173.6	−7.6	1
2	2	LMM	40	118.0	126.2	−8.2	1
2	2	MPP	46	140.0	149.6	−9.6	1*
2	2	PSS	52	175.0	172.9	2.1	3*
2	4	LMP	42	125.0	123.5	1.5	1*
2	4	MPS	48	137.0	146.8	−9.8	1*
2	6	CLP	38	95.0	105.0	−10.0	1*
2	6	MPA	50	122.0	151.6	−29.6	0*
4	4	LPP	44	146.0	119.0	27.0	0*
4	4	MSS	50	139.0	142.4	−3.4	1*
6	0	LSL	42	131.0	124.8	6.2	1*
6	0	MAM	48	157.0	148.1	8.9	1*
6	4	LSP	46	124.0	121.3	2.7	1*
6	6	CPP	42	89.0	102.8	−13.8	1*
6	6	LSS	48	123.0	126.1	−3.1	1*

Note: Data marked by a * include measurements from this work. FREQ indicates
the number of measurements that has been averaged.

TABLE 9.A.4

Melting Points of the α-Modification of Saturated TAGs

X	Y	TAG	n	T_f (°C)	Predicted (°C)	Residual	FREQ
−12	0	S6S	42	27.8	23.7	4.1	2
−10	0	P6P	38	14.6	10.2	4.4	2
−8	0	SCS	46	30.0	34.4	−4.4	2
−6	0	PCP	42	20.0	24.3	−4.3	2
−6	0	SLS	48	36.0	39.3	−3.3	2
−4	0	MCM	38	16.0	13.4	2.6	1
−4	0	PLP	44	32.6	31.4	1.2	2
−4	0	SMS	50	43.8	44.3	−0.5	4
−4	0	BSB	62	61.1	61.7	−0.6	1
−2	0	LCL	34	5.0	2.9	2.1	2
−2	0	MLM	40	24.0	23.7	0.3	1
−2	0	PMP	46	39.1	38.5	0.6	5
−2	0	SPS	52	50.7	49.4	1.3	6
−2	2	PMS	48	40.9	41.3	−0.4	2
−1	0	P.15.P	47	43.4	41.7	1.7	1
−1	0	s.17.s	53	53.1	51.8	1.3	1
0	0	888	24	−51.0	−47.8	−3.2	1
0	0	9.9.9	27	−26.0	−26.1	0.1	1
0	0	CCC	30	−11.5	−9.3	−2.2	2
0	0	11.11.11	33	2.0	4.2	−2.2	2
0	0	LLL	36	15.6	15.2	0.4	4
0	0	13.13.13	39	24.5	24.3	0.2	3
0	0	MMM	42	32.6	32.0	0.6	6
0	0	15.15.15	45	39.5	38.7	0.8	4
0	0	PPP	48	44.7	44.4	0.3	5
0	0	17.17.17	51	49.9	49.4	0.5	4
0	0	SSS	54	54.7	53.8	0.9	8
0	0	19.19.19	57	59.1	57.7	1.4	2
0	0	AAA	60	62.9	61.3	1.6	2
0	0	21.21.21	63	65.0	64.4	0.6	1
0	0	BBB	66	69.1	67.3	1.8	2
0	1	11.11.L	34	7.5	7.3	0.2	1
0	2	CCL	32	0.0	−3.2	3.2	1
0	2	LLM	38	19.0	19.4	−0.4	1
0	2	MMP	44	34.5	35.2	−0.7	3
0	2	PPS	50	46.4	46.9	−0.5	5
0	4	CCM	34	3.0	−0.8	3.8	2
0	4	LLP	40	20.0	21.4	−1.4	1
0	4	MMS	46	35.6	37.0	−1.4	3
0	4	SSB	58	56.7	57.2	−0.5	2
0	6	CCP	36	2.0	3.2	−1.2	1

(continued)

TABLE 9.A.4 (continued)
Melting Points of the α-Modification of Saturated TAGs

X	Y	TAG	n	T_f (°C)	Predicted (°C)	Residual	FREQ
0	6	LLS	42	20.5	24.5	−4.0	2
0	8	CCS	38	32.0	9.6	22.4	0
0	10	66P	28	−7.4	−50.2	42.8	0
0	10	88S	34	5.0	−9.0	14.0	0
0	12	66S	30	6.8	−34.1	40.9	0
0	14	88B	38	26.0	9.0	17.0	0
0	16	66B	34	31.0	−9.1	40.1	0
1	0	P.17.P	49	48.2	46.5	1.7	1
1	0	S.19.S	55	55.5	55.4	0.1	1
2	0	CLC	32	6.0	0.8	5.2	1
2	0	LML	38	24.0	22.0	2.0	1
2	0	MPM	44	36.2	36.9	−0.7	3
2	0	PSP	50	47.2	48.0	−0.8	5
2	2	CLL	34	5.0	6.1	−1.1	1
2	2	LMM	40	22.0	25.7	−3.7	1
2	2	MPP	46	36.0	39.8	−3.8	1
2	2	PSS	52	50.1	50.4	−0.3	5
2	4	MPS	48	41.9	41.5	0.4	3
4	0	CMC	34	3.0	4.4	−1.4	1
4	0	LPL	40	19.0	24.7	−5.7	1
4	0	MSM	46	33.0	39.1	−6.1	1
4	0	SBS	58	56.0	58.2	−2.2	1
4	4	CMM	38	15.0	11.7	3.3	1
4	4	LPP	44	32.0	30.3	1.7	1
4	4	MSS	50	44.0	43.6	0.4	1
4	4	SBB	62	61.3	61.4	−0.1	1
6	0	CPC	36	6.0	6.9	−0.9	1
6	0	LSL	42	21.0	26.8	−5.8	1
6	0	PBP	54	47.4	51.4	−4.0	1
6	6	CPP	42	23.0	17.6	5.4	1
6	6	LSS	48	36.0	34.9	1.1	1
6	6	PBB	60	55.9	56.8	−0.9	1
8	0	CSC	38	34.0	11.5	22.5	0
8	8	CSS	46	42.5	26.6	15.9	0
12	0	6S6	30	0.0	−31.6	31.6	0

Note: FREQ indicates the number of measurements that has been averaged.

TABLE 9.A.5
Melting Points of the β′-Modification of Saturated TAGs

X	Y	TAG	n	T_f (°C)	Predicted (°C)	Residual	FREQ
−8	0	SCS	46	53.0	54.3	−1.3	1
−6	0	PCP	42	44.5	47.7	−3.2	2
−6	0	SLS	48	57.3	57.0	0.3	2
−4	0	MCM	38	40.0	39.4	0.6	1
−4	0	PLP	44	49.6	51.2	−1.6	2
−4	0	SMS	50	58.8	59.5	−0.7	3
−4	0	BSB	62	71.8	70.5	1.3	2
−2	2	PMS	48	56.1	53.5	2.6	1
0	0	888	24	−19.5	−16.7	−2.8	2
0	0	9.9.9	27	4.0	3.0	1.0	1
0	0	CCC	30	16.8	17.3	−0.5	2
0	0	11.11.11	33	27.9	28.2	−0.3	4
0	0	LLL	36	35.1	36.7	−1.6	9
0	0	13.13.13	39	41.8	43.6	−1.8	4
0	0	MMM	42	45.9	49.3	−3.4	5
0	0	15.15.15	45	50.8	54.1	−3.3	2
0	0	PPP	48	55.7	58.1	−2.4	6
0	0	17.17.17	51	61.5	61.6	−0.1	4
0	0	SSS	54	64.3	64.6	−0.3	7
0	0	19.19.19	57	65.6	67.3	−1.7	2
0	0	AAA	60	69.5	69.6	−0.1	2
0	0	21.21.21	63	71.0	71.7	−0.7	1
0	0	BBB	66	74.8	73.6	1.2	2
0	1	11.11.L	34	28.5	28.3	0.2	1
0	2	88C	26	5.5	−24.9	30.4	0
0	2	CCL	32	26.0	13.6	12.4	0
0	2	LLM	38	37.8	34.8	3.0	3
0	2	MMP	44	48.5	48.2	0.3	2
0	2	PPS	50	58.7	57.5	1.2	3
0	4	CCM	34	31.0	15.6	15.4	0
0	4	LLP	40	43.0	36.2	6.8	1
0	4	MMS	46	49.3	49.3	0.0	2
0	4	SSB	58	69.7	65.0	4.7	1
0	6	LLS	42	39.8	41.0	−1.2	2
0	8	CCS	38	38.0	30.2	7.8	1
0	10	66P	28	12.0	−22.2	34.2	0
0	10	88S	34	25.0	15.3	9.7	1
0	12	66S	30	17.0	−6.7	23.7	0
1	0	P17P	49	61.2	60.9	0.3	2
1	0	S19S	55	69.8	66.7	3.1	1
2	0	8C8	26	18.5	14.3	4.2	1

(continued)

TABLE 9.A.5 (continued)
Melting Points of the β′-Modification of Saturated TAGs

X	Y	TAG	n	T_f (°C)	Predicted (°C)	Residual	FREQ
2	0	CLC	32	37.7	34.5	3.2	2
2	0	LML	38	49.8	47.5	2.3	3
2	0	MPM	44	59.5	56.6	2.9	5
2	0	PSP	50	67.7	63.4	4.3	10
2	0	11.13.11	35	42.6	41.6	1.0	1
2	2	8CC	28	12.8	10.3	2.5	2
2	2	CLL	34	31.0	32.4	−1.4	1
2	2	LMM	40	42.0	46.4	−4.4	1
2	2	MPP	46	52.0	56.0	−4.0	1
2	2	PSS	52	61.8	63.0	−1.2	3
2	4	CLM	36	36.7	33.9	2.8	1
2	6	LMS	44	45.5	50.8	−5.3	1
2	8	CLS	40	40.0	43.4	−3.4	1
4	0	CMC	34	30.0	35.2	−5.2	1
4	0	LPL	40	46.7	48.2	−1.5	2
4	0	LPL	40	42.5	48.2	−5.7	1
4	0	MSM	46	55.2	57.2	−2.0	3
4	0	SBS	58	64.0	69.0	−5.0	1
4	4	CMM	38	38.0	34.7	3.3	1
4	4	LPP	44	49.5	48.1	1.4	1
4	4	MSS	50	58.3	57.4	0.9	2
4	4	SBB	62	71.5	69.4	2.1	1
4	6	LPS	46	47.0	51.5	−4.5	1
4	8	CMS	42	42.0	44.1	−2.1	1
6	0	CPC	36	36.0	34.8	1.2	1
6	0	LSL	42	43.0	48.0	−5.0	1
6	0	PBP	54	61.5	64.0	−2.5	1
6	6	CPP	42	41.0	39.3	1.7	1
6	6	LSS	48	51.4	51.4	0.0	8
6	6	PBB	60	66.1	66.2	−0.1	1
8	0	CSC	38	40.0	39.4	0.6	1
8	8	CSS	46	46.0	47.6	−1.6	1

Note: FREQ indicates the number of measurements that has been averaged.

TABLE 9.A.6
Melting Points of the β-Modification of Saturated TAGs

X	Y	TAG	n	T_f (°C)	Predicted (°C)	Residual	FREQ
−12	0	S6S	42	53.1	53.6	−0.5	2
−10	0	P6P	38	45.4	45.7	−0.3	3
−10	0	S8S	44	54.0	57.0	−3.0	5
−8	0	P8P	40	48.5	49.9	−1.4	1
−8	0	SCS	46	57.4	60.1	−2.7	4
−6	0	PCP	42	51.8	53.7	−1.9	2
−6	0	SLS	48	60.3	62.9	−2.6	3
−6	2	PCS	44	54.2	51.2	3.0	3
−4	0	MCM	38	43.5	46.0	−2.5	1
−4	0	PLP	44	53.9	57.2	−3.3	4
−4	0	SMS	50	63.3	65.6	−2.3	9
−4	2	PLS	46	57.0	54.9	2.1	2
−4	4	MCS	42	51.7	45.3	6.4	3
−2	0	C8C	28	20.0	19.9	0.1	1
−2	0	LCL	34	37.4	38.5	−1.1	3
−2	0	MLM	40	49.8	51.4	−1.6	2
−2	0	PMP	46	59.9	60.8	−0.9	6
−2	0	SPS	52	68.0	68.1	−0.1	8
−2	2	PMS	48	59.6	58.7	0.9	3
−2	4	MLS	44	54.5	50.9	3.6	2
−2	6	LCS	40	41.8	41.7	0.1	1
−1	0	P.15.P	47	56.5	57.7	−1.2	1
−1	0	S.17.S	53	65.7	64.9	0.8	2
0	0	888	24	9.1	11.4	−2.3	6
0	0	CCC	30	31.6	31.8	−0.2	8
0	0	LLL	36	45.7	45.9	−0.2	9
0	0	13.13.13	39	44.1	45.8	−1.7	5
0	0	MMM	42	57.1	56.2	0.9	10
0	0	15.15.15	45	54.6	55.5	−0.9	4
0	0	PPP	48	65.9	64.0	1.9	9
0	0	17.17.17	51	63.9	63.0	0.9	6
0	0	SSS	54	72.5	70.2	2.3	9
0	0	19.19.19	57	70.6	69.0	1.6	3
0	0	AAA	60	77.8	75.3	2.5	4
0	0	21.21.21	63	77.0	73.9	3.1	2
0	0	BBB	66	81.7	79.4	2.3	7
0	0	24.24.24	72	86.0	82.9	3.1	1
0	1	11.11.L	34	29.2	29.8	−0.6	2
0	2	88C	26	11.5	7.9	3.6	1
0	2	CCL	32	30.0	28.9	1.1	1
0	2	88C	26	11.5	7.9	3.6	1
0	2	CCL	32	30.0	28.9	1.1	1
0	2	LLM	38	42.3	43.4	−1.1	4
0	2	MMP	44	53.3	54.1	−0.8	5

(*continued*)

TABLE 9.A.6 (continued)
Melting Points of the β-Modification of Saturated TAGs

X	Y	TAG	n	T_f (°C)	Predicted (°C)	Residual	FREQ
0	2	PPS	50	62.6	62.2	0.4	10
0	2	15.15.17	47	54.0	53.5	0.5	1
0	4	CCM	34	34.5	30.6	3.9	1
0	4	LLP	40	45.6	45.3	0.3	5
0	4	MMS	46	56.6	55.9	0.7	4
0	4	SSB	58	70.7	70.3	0.4	1
0	6	CCP	36	35.0	35.1	−0.1	1
0	6	LLS	42	45.1	48.7	−3.6	8
0	8	CCS	38	42.5	40.1	2.4	2
0	10	88S	34	31.0	29.3	1.7	2
0	12	66S	30	22.6	15.4	7.2	2
0	14	88B	38	38.0	40.1	−2.1	1
0	16	66B	34	34.0	29.3	4.7	1
2	0	8C8	26	20.5	17.9	2.6	1
2	0	CLC	32	34.0	36.4	−2.4	1
2	0	MPM	44	55.0	58.8	−3.8	1
2	0	PSP	50	65.3	66.2	−0.9	2
2	2	8CC	28	19.0	14.6	4.4	2
2	2	CLL	34	34.1	33.6	0.5	2
2	2	LMM	40	47.5	46.9	0.6	2
2	2	MPP	46	55.8	56.8	−1.0	4
2	2	PSS	52	64.4	64.4	0.0	9
2	4	LMP	42	48.5	48.8	−0.3	3
2	4	MPS	48	58.5	58.7	−0.2	5
2	6	LMS	44	49.0	52.0	−3.0	2
2	8	CLS	40	44.0	44.1	−0.1	1
4	0	CMC	34	34.0	37.7	−3.7	1
4	0	SBS	58	69.9	73.8	−3.9	2
4	4	CMM	38	43.5	36.7	6.8	1
4	4	LPP	44	54.4	50.5	3.9	4
4	4	MSS	50	60.9	60.5	0.4	6
4	4	SBB	62	73.5	73.9	−0.4	1
4	6	LPS	46	52.0	53.8	−1.8	1
4	8	CMS	42	45.0	45.7	−0.7	1
6	0	CPC	36	40.0	41.2	−1.2	1
6	0	LSL	42	49.8	53.8	−4.0	3
6	0	MAM	48	59.0	63.0	−4.0	1
6	0	PBP	54	65.5	70.0	−4.5	2
6	6	CPP	42	45.5	44.5	1.0	1
8	0	CSC	38	44.5	45.7	−1.2	1
8	8	CSS	46	48.3	52.8	−4.5	3
10	0	8S8	34	41.0	35.6	5.4	1
12	0	6S6	30	32.0	22.5	9.5	1

Note: FREQ indicates the number of measurements that has been averaged.

9.A.2 Unsaturated TAGs

TABLE 9.A.7
Enthalpy of Fusion for the α-Modification of Unsaturated TAGs

X	Y	TAG	N	ΔH_f (kJ/mol)	Predicted	Residual	FREQ
−2	0	EPE	52	79	79	0	2*
0	0	SSO	54	71	81	−10	2*
0	0	SOS	54	73	81	−8	1
0	0	OOO	54	37	18	19	1
0	0	SEE	54	89	89	0	1*
0	0	EEE	54	78	78	0	1*
0	0	ESE	54	92	89	3	2*
0	2	PPE	50	118	88	30	0
0	2	PPO	50	53	68	−15	2*
2	0	PEP	50	122	91	31	0
2	0	POP	50	70	71	1	4*
2	2	POS	52	78	74	4	2
2	2	PEE	52	81	82	−1	2*

Note: Data marked by a * include measurements from this chapter. FREQ indicates the number of measurements that has been averaged.

TABLE 9.A.8
Enthalpy of Fusion for the β′-Modification of Unsaturated TAGs

X	Y	TAG	n	ΔH_f (kJ/mol)	Predicted	Residual	FREQ
−2	0	SPO	52	126	117	9	1
0	0	SOS	54	111	129	−18	2
0	0	SOO	54	110	101	9	1
0	0	SSO	54	125	129	−4	2*
0	0	OOO	54	79	73	6	1
0	2	PPO	50	111	101	10	2*
2	0	PEP	50	135	140	−5	1*
2	0	POP	50	104	128	−24	6
2	2	PSO	52	111	123	−12	1
2	2	POS	52	114	123	−9	1
2	2	POO	52	95	95	0	1

Note: Data marked by a * include measurements from this chapter. FREQ indicates the number of measurements that has been averaged.

TABLE 9.A.9

Enthalpy of Fusion for the β-Modification of Unsaturated TAGs

X	Y	TAG	n	ΔH_f (kJ/mol)	Predicted	Residual	FREQ
-2	0	SPO	52	126	117	9	1
-2	0	EPE	52	130	140	10	2
-2	0	OPO	52	126	111	14	1
0	0	EEE	54	148	144	4	11
0	0	ESE	54	155	160	-5	3
0	0	lll	54	84	78	6	1
0	0	SES	54	163	175	-12	2
0	0	SEE	54	155	160	-5	1
0	0	SOS	54	154	161	-7	5
0	0	OOO	54	100	101	1	5
0	2	PPE	50	157	151	6	3
0	2	SOA	56	158	160	-2	1
2	0	POP	50	140	143	-3	4
2	0	PIP	50	100	136	-36	0
2	0	PEP	50	150	158	-8	1
2	2	POS	52	150	143	7	2
2	2	PEE	52	134	141	-7	2

Note: Data marked by a * include measurements from this chapter. FREQ indicates the number of measurements that has been averaged.

TABLE 9.A.10
Melting Points of the α-Modification of Unsaturated TAGs

TAG	T_f (°C)	Predicted (°C)	Residual	FREQ	OMOD
CCE	15	8.9	6.1	1	?
COC	−16.4	−22.8	6.4	1	Alpha
LLl	15.5	17	−1.5	1	?
Lll	−11.5	−15.6	4.1	1	?
MMl	20.5	22.2	−1.7	1	?
Mll	−8.5	−7.6	−0.9	1	?
MOM	11.7	4.8	6.9	2	Alpha
PPO	18.4	16.8	1.6	4	Alpha
PSl	36.5	32.6	3.9	1	?
PSO	25.9	21.3	4.6	2	Alpha
PEP	39.4	35.6	3.8	2	Alpha
PEE	22.8	26.1	−3.3	1	?
PlO	13.2	11.2	2	1	?
POP	16.6	17.6	−1	6	Alpha
POS	19.6	21.3	−1.7	4	Alpha
POl	13.3	11.2	2.1	1	?
SES	41.5	44	−2.5	2	Alpha
SES	46.5	44	2.5	0	
SlS	37.9	37.3	0.6	3	?
SOS	22.9	27.2	−4.3	8	Alpha
88E	3	5.4	−2.4	1	?
EPE	26	27	−1	1	Alpha
EPE	32	27	5	0	
ESS	43	44	−1	1	?
ESE	34	32	2	1	Alpha
EES	28.8	32	−3.2	1	Alpha
EEE	15.8	19.1	−3.3	4	Alpha

(continued)

TABLE 9.A.10 (continued)
Melting Points of the α-Modification of Unsaturated TAGs

TAG	T_f (°C)	Predicted (°C)	Residual	FREQ	OMOD
lPS	34.2	33.1	1.1	1	?
lPO	11.7	11.1	0.6	1	?
lSl	−3	0.2	−3.2	1	?
lSO	16.5	15.8	0.7	1	?
llO	−16.4	−15.6	−0.8	2	?
lleO	−24.7	−22.5	−2.2	1	?
lOO	−2.2	−5.7	3.5	1	?
leSl	−9.2	−7.9	−1.3	1	?
leSO	5.2	6.1	−0.9	1	?
lelS	−8.3	−7.9	−0.4	2	?
lelO	−22.4	−22.5	0.1	2	?
lelele	−44.6	−42.8	−1.8	1	al-
phaleleO	−17.8	−19.9	2.1	2	?
leOS	−2.5	6.1	−8.6	1	?
OPS	17.9	22.1	−4.2	2	Alpha
OSS	30.3	27.2	3.1	6	Alpha
OSO	1	−0.7	1.7	1	Alpha
OleO	−15	−14.3	−0.7	1	?
OOO	−33.7	−28.8	−4.9	4	Alpha

Note: FREQ indicates the number of measurements that has been averaged, OMOD indicates the modification that was originally reported for the melting point.

TABLE 9.A.11
Melting Points of the β′-Modification
of Unsaturated TAGs

TAG	T_f (°C)	Predicted (°C)	Residual	FREQ	OMOD
CCO	4.4	2.7	1.7	2	?
CEE	25	20.1	4.9	1	?
LLO	18	15.7	2.3	2	?
MOM	26.4	26.4	0	2	β
PPl	26.5	24.9	1.6	1	?
PPO	34.6	34.2	0.4	8	β
PSO	40	38	2	6	β
PES	48.5	51.4	−2.9	1	?
Pll	−4.2	−2.9	−1.3	3	?
Plel	−7.5	−6.2	−1.3	1	?
Plele	−10.5	−10.2	−0.3	1	?
POP	33.2	34.8	−1.6	8	β
SOS	37	42.2	−5.2	9	β
SOS	43	42.2	0.8	0	β
ESS	56.7	54.9	1.8	1	?
ESE	43.2	45.1	−1.9	0	β
EEE	37	35.7	1.3	1β	
IPl	−3	−2.3	−0.7	1	?
llS	2.5	3.8	−1.3	2	?
lll	−25.3	−25	−0.3	3	β
lOS	−3.5	−1.2	−2.3	1	?
lOl	−39	−46.5	7.5	1	?
lePl	−4	−4	0	1	?
leSS	27.8	28.1	−0.3	2	?
lelle	−15.5	−15.5	0	1	?
lelel	−16.5	−15.5	−1	1	?
leOl	−28.5	−31.7	3.2	1	?
leOle	−11.1	−14.9	3.8	2	?
OSS	41.9	42.2	−0.3	10	β
OSO	20.5	19.3	1.2	1	β
OlS	−10.4	−1.2	−9.2	2	?
OleS	14.2	7.6	6.6	1	?
OOA	29.2	23	6.2	1	?
OOB	33.3	25.8	7.5	1	?
0.0.24	36.1	27.4	8.7	1	?
OOO	−10	−3.4	−6.6	5	β

Note: FREQ indicates the number of measurements that has been averaged, OMOD indicates the modification that was originally reported for the melting point.

TABLE 9.A.12
Melting Points of the β-Modification
of Unsaturated TAGs

TAG	T_f (°C)	Predicted (°C)	Residual	FREQ	OMOD
CCl	−0.5	−9.7	9.2	1	?
CMO	13.9	16.8	−2.9	1	?
COC	−4.8	1.6	−6.4	0	β
COC	5.9	1.6	4.3	2	β
COO	−0.3	5	−5.3	2	?
LLE	27	31.6	−4.6	1	?
LPO	29.5	24.6	4.9	1	?
LEE	35.5	38	−2.5	1	?
LOL	16.5	15.8	0.7	0	β
LOO	5.1	4.7	0.4	3	?
MME	39.5	40.9	−1.4	1	?
MMO	23.9	22.4	1.5	2	?
MEE	40	39.8	0.2	1	?
MOM	28	28	0	2	β
MOP	27	30	−3	1	?
MOS	27	32.7	−5.7	1	?
MOO	12.8	8.3	4.5	4	?
PPE	50.2	50.1	0.1	3	?
PEP	55.3	53.7	1.6	4	β
PEP	53.9	53.7	0.2	0	β
PEE	44.2	44.1	0.1	3	?
PIP	27.1	27.6	−0.5	2	β
PIS	24.5	28.8	−4.3	1	?
POP	37.2	37.6	−0.4	10	β
POS	31	38	−7	0	β
POS	37.1	38	−0.9	8	β
POO	18.5	14.6	3.9	0	β
SES	60.5	59.3	1.2	3	β
SleS	35.8	31.4	4.4	2	?
SOA	41.5	44	−2.5	1	β
EPE	44.5	46.9	−2.4	1	β
ESS	61.1	59.3	1.8	4	β
ESE	49.7	50.2	−0.5	2	β
EES	49.7	50.2	−0.5	5	β
EEE	42.2	44	−1.8	9	β
ISS	35.8	35.4	0.4	4	?
III	−12.3	−14	1.7	6	β

TABLE 9.A.12 (continued)
Melting Points of the β-Modification of Unsaturated TAGs

TAG	T_f (°C)	Predicted (°C)	Residual	FREQ	OMOD
leleS	−0.5	−0.5	0	1	?
lelele	−24.2	−23.5	−0.7	3	β
leOO	−13.1	−12	−1.1	2	?
OPS	40.2	41	−0.8	0	β
OPO	19.6	17.9	1.7	3	β
OSO	23.9	22.7	1.2	4	β
OlO	−9.5	−7	−2.5	1	?
OOS	23.5	22.7	0.8	0	β
OOO	4.8	6.1	−1.3	9	β
PEP	54	51.2	2.8	0	?

Note: FREQ indicates the number of measurements that has been averaged, OMOD indicates the modification that was originally reported for the melting point.

APPENDIX 9.B: SPECIFIC RETENTION VOLUMES OF SEVERAL PROBES IN STATIONARY PHASES OF LIQUID TAGs

TABLE 9.B.1
Specific Retention Volume in mL/g

T (°C)	82	92	102	112	123	133
n-Pentane	16.856	13.332	10.611	8.567		
n-Hexane	38.053	29.16	22.595	18.307	14.618	11.571
n-Heptane	88.048	64.449	48.319	39.669	29.083	22.295
2-Methyl-pentane	28.852	23.087	18.014	14.722	11.722	9.36
3-Methyl-pentane	33.438	26.465	20.88	16.188	13.428	10.789
Benzene	88.658	65.532	50.591	39.684	31.332	24.767
Toluene	218.379	154.476	114.394	85.939	64.479	50.534
Cyclohexane	73.4	55.932	42.915	33.832	26.582	20.903

Note: Stationary phase: SSS.

TABLE 9.B.2
Specific Retention Volume in mL/g

T (°C)	82	92	102	112	123	133
n-Pentane	16.448	13.525	11.09	8.323		
n-Hexane	39.175	30.356	23.319	18.389	14.449	11.715
n-Heptane	90.897	66.559	50.618	38.246	29.328	23.399
2-Methyl-pentane	30.691	23.922	18.619	14.482	12.019	9.926
3-Methyl-pentane	34.607	27.125	21.358	16.77	13.487	11.41
Benzene	97.106	72.451	54.744	42.351	33.476	26.445
Toluene	235.44	169.84	123.941	91.414	69.741	53.632
Cyclohexane	74.389	57.106	44.007	34.218	26.791	21.699

Stationary phase: MMM.

TABLE 9.B.3
Specific Retention Volume in mL/g

T (°C)	82	92	102	112	123	133
n-Pentane	15.803	12.58	10.341	8.067		
n-Hexane	37.093	28.536	22.103	17.575	14.291	11.052
n-Heptane	84.266	61.824	47.214	36.302	28.01	22.596
2-Methyl-pentane	28.69	22.432	17.473	13.829	11.433	9.33
3-Methyl-pentane	33.008	25.125	19.726	15.558	12.576	10.561
Benzene	105.237	77.315	59	46.098	34.87	27.754
Toluene	250.008	177.156	130.416	97.67	72.598	55.754
Cyclohexane	68.781	52.18	40.569	31.692	24.866	20.14

Note: Stationary phase: 8C8.

TABLE 9.B.4
Specific Retention Volume in mL/g

T (°C)	82	92	102	112	123	133
n-Pentane	15.772	12.92	10.023	8.432		
n-Hexane	38.571	29.43	22.959	18.083	14.644	11.976
n-Heptane	88.55	65.675	49.205	37.647	29.347	23.427
2-Methyl-pentane	29.374	22.84	18.173	14.563	11.875	9.806
3-Methyl-pentane	33.769	26.388	20.544	16.667	13.366	11.047
Benzene	93.42	69.591	53.302	41.705	32.528	25.958
Toluene	226.576	162.709	119.042	89.083	67.384	52.15
Cyclohexane	73.139	56.066	43.033	34.312	26.868	21.905

Note: Stationary phase: 0.4487 SSS, 0.5513 MMM.

TABLE 9.B.5
Specific Retention Volume in mL/g

T (°C)	82	92	102	112	123	133
n-Pentane	15.637	12.695	10.335	8.961		
n-Hexane	37.808	29.054	22.138	18.762	14.216	11.513
n-Heptane	87.566	64.7	48.113	38.308	28.667	22.664
2-Methyl-pentane	29.191	22.873	17.671	14.681	11.656	9.642
3-Methyl-pentane	33.639	26.154	20.204	16.969	13.17	10.806
Benzene	89.447	67.982	51.08	41.244	31.129	24.872
Toluene	216.024	157.377	114.093	88.601	65.358	50.193
Cyclohexane	71.81	55.042	42.012	34.242	26.328	21.026

Note: Stationary phase: 0.7116 SSS, 0.2884 MMM.

TABLE 9.B.6
Specific Retention Volume in mL/g

T (°C)	82	92	102	112	123	133
n-Pentane	16.106	12.78	10.684	7.576		
n-Hexane	38.33	29.045	23.253	18.239	15.18	11.53
n-Heptane	89.874	65.641	49.02	37.321	29.777	22.787
2-Methyl-pentane	29.958	22.945	18.225	14.872	12.261	9.598
3-Methyl-pentane	34.145	26.14	20.739	16.836	13.721	10.981
Benzene	103.725	76.097	58.447	44.336	35.324	27.454
Toluene	249.649	176.011	129.15	95.406	73.275	56.006
Cyclohexane	72.801	54.314	41.793	33.111	26.712	20.865

Note: Stationary phase: 0.1557 SSS, 0.8443 8C8.

TABLE 9.B.7
Specific Retention Volume in mL/g

T (°C)	82	92	102	112	123	133
n-Pentane	15.707	12.973	10.513	8.432		
n-Hexane	38.224	29.198	22.737	18.077	14.426	11.369
n-Heptane	88.634	63.696	48.437	37.146	28.879	22.157
2-Methyl-pentane	29.616	22.65	17.992	14.799	11.769	9.388
3-Methyl-pentane	33.493	25.654	20.25	16.368	12.962	10.658
Benzene	98.094	71.955	54.47	42.629	33.273	26.636
Toluene	235.818	166.864	121.852	91.705	68.986	52.84
Cyclohexane	72.015	53.896	41.858	32.677	26.331	21.08

Note: Stationary phase: 0.3593 SSS, 0.6407 8C8.

TABLE 9.B.8
Specific Retention Volume in mL/g

T (°C)	82	92	102	112	123	133
n-Pentane	16.053	12.907	10.38	8.446		
n-Hexane	38.167	28.82	23.03	18.049	14.425	11.9
n-Heptane	88.696	65.02	49.16	37.255	28.732	23.11
2-Methyl-pentane	29.275	23.16	18.376	14.462	11.607	9.706
3-Methyl-pentane	33.845	26.658	20.643	16.313	13.603	11.093
Benzene	93.481	70.449	53.696	41.189	32.5	26.462
Toluene	226.756	162.489	119.081	88.857	67.427	52.692
Cyclohexane	72.627	55.369	42.955	33.206	26.383	21.492

Note: Stationary phase: 0.6023 SSS, 0.3977 8C8.

APPENDIX 9.C: Purity of the TAGs Used in Section 9.7

TAG	Purity (%)	Main Impurities
SSS	99.6	—
PSP	99.3	—
MPM	>99.6	—
SEE	91.4	SES 8%
ESE	94.3	4.5% SSE, 1% SEO
PEE	91.1	8.6% PEP, PPE, EPE
EPE	89.8	9.1% PEP, PPE, PEE, 1.1% PEO
EEE	96.8	1.7% ESE, SEE
SOS	99.5	—
SSO	90.7	4% SOS, 3% SSE
POP	99.0	1% PEP
PPO	98.5	1.5% PPO

APPENDIX 9.D: BINARY PHASE DIAGRAMS OF TAGs: DATA

	SSS–PSS			SSS–PSP			SSS–SPS	
X	Solidus	Liquidus	X	Solidus	Liquidus	X	Solidus	Liquidus
0.00	72.5	72.5	0.00	72.5	72.7	0.00	72.5	72.5
0.11	67.7	71.8	0.19	65.2	71.6	0.14	64.8	71.8
0.29	67.5	71.2	0.20		71.4	0.31	65.2	71.1
0.46	67.4	70.4	0.32	65.4	71.3	0.45	65.0	70.4
0.58		68.5	0.40		69.9	0.51	65.4	69.5
0.74		67.3	0.50		68.6	0.56	65.9	69.5
0.85		67.1	0.50	65.6	69.0	0.69		65.3
1.00	67.2	67.2	0.62	65.4	68.2	0.76		65.8
			0.70		66.0	0.90		66.4
			0.80	66.1	67.3	1.00	67.5	67.5
			0.88	65.7	67.8			
			1.00	68.4	68.4			
			1.00		68.7			

	SSS–PPS			SSS–PPP			PSS–PSP	
X	Solidus	Liquidus	X	Solidus	Liquidus	X	Solidus	Liquidus
0.00		72.3	0.00	72.5	72.5	0.00	66.2	66.2
0.10	68.8	71.8	0.14	63.8	72.4	0.15		66.1
0.18	61.9	71.8	0.19	64.0	72.0	0.29		64.9
0.20	65.4	70.8	0.39	63.7	71.1	0.46		64.5
0.20		70.9	0.49	64.0	70.6	0.58	63.3	64.2
0.30	61.2	69.9	0.59	63.8	69.5	0.72	64.7	65.4
0.30	60.8	70.7	0.73	63.7	66.3	0.89	65.7	66.6
0.30		70.2	0.80	63.7	64.7	1.00	68.7	68.7
0.33	61.0	71.4	0.87	63.8	65.1			
0.40	61.1	69.6	0.94	63.8	65.2			
0.40	60.9		1.00	65.7	65.7			
0.50	61.0	68.9						

	SSS–PPS			PSS–SPS			SSS–SPS	
X	Solidus	Liquidus	X	Solidus	Liquidus	X	Solidus	Liquidus
0.50	60.8	68.3	0.00	66.3	66.3	0.00	66.1	66.1
0.50	60.6	67.9	0.15		64.9	0.14		65.0
0.55	61.0	70.3	0.30		64.6	0.24		64.3
0.59	60.8	67.3	0.45		64.9	0.38		63.8
0.60	61.0	67.8	0.59		64.9	0.51		63.2
0.65	60.9	66.2	0.72		66.6	0.68		62.8
0.67	60.8	65.7	0.89	66.0	67.4	0.87		62.3
0.67	61.1		1.00	67.9	67.9	1.00	63.5	63.5
0.70	59.5	61.0						
0.70	60.2	61.2						

(continued)

(continued)

SSS–PPS

X	Solidus	Liquidus
0.71	59.6	60.7
0.73	62.3	69.3
0.75	59.7	60.4
0.75	59.7	60.8
0.79	61.0	62.5
0.80	59.8	60.4
0.80	59.6	60.8
0.85		60.7
0.85		60.0
0.85		60.4
0.91		60.8
0.91		60.5
0.91		60.2
0.94		61.6
0.95		60.6
0.95		61.1
0.99		60.2
0.99		61.7
1.00	63.4	63.4

PSS–SPS

X	Solidus	Liquidus

PSS–PPP

X	Solidus	Liquidus
0.00	66.1	66.1
0.17	63.0	65.3
0.25	62.4	64.9
0.42	62.9	64.2
0.53		62.7
0.68	62.8	65.0
0.90	62.8	66.0
1.00	66.0	66.0

SSS–SPS

X	Solidus	Liquidus

PSP–SPS

X	Solidus	Liquidus
0.00	68.5	68.5
0.20	65.2	68.1
0.26	65.2	67.4
0.37	65.3	66.1
0.48		65.6
0.53		65.2
0.66	64.6	66.1
0.79	64.6	66.7
0.91	64.4	67.5
1.00	67.6	67.6

PSP–PPS (De Bruijne)

X	Solidus	Liquidus
0.00	68.6	68.6
0.10	60.6	67.1
0.21	60.4	66.2
0.30	60.8	65.7
0.40	60.8	64.8
0.50	61.1	63.5
0.60	60.7	62.7
0.71	61.2	61.2
0.9	61.3	62.4
1.0	63.4	63.4

PSP–PPP

X	Solidus	Liquidus
0.00	68.6	68.6
0.20	61.7	67.4
0.46	61.5	65.1
0.66	61.7	63.2
0.80	61.7	63.3
1.00	65.8	65.8

SPS–PPS

X	Solidus	Liquidus
0.00	67.6	67.6
0.20		66.3
0.34	61.1	65.7
0.40	60.8	65.3
0.54	61.1	64.9
0.60	61.3	64.2
0.75		62.7
0.86	60.6	61.9
0.92	60.3	62.2
1.00	63.3	63.3

PSP–PPS (Perron)

X	Solidus	Liquidus
0.00		68.2
0.05	65.6	67.8
0.10	64.1	67.5
0.15	62.5	66.9
0.20	62.9	66.4
0.25	61.9	66.2
0.30	61.6	65.6

SPS–PPP

X	Solidus	Liquidus
0.00	67.7	67.7
0.18	61.0	66.5
0.51	60.2	63.8
0.61		62.0
0.79	60.8	63.3
0.84	60.6	63.6
1.00	66.0	66.0

PPS–PPP

X	Solidus	Liquidus
0.00	63.2	63.2
0.14		62.7
0.29		61.6
0.45		62.3
0.63		62.0
0.75	61.1	63.0
0.83	61.4	64.4

(continued)

PSP–PPS (Perron)		
X	Solidus	Liquidus
0.35	61.8	64.7
0.40	60.6	64.1
0.50	60.9	63.5
0.60	60.9	61.9
0.71	60.9	61.8
0.75	60.4	61.3
0.80	60.6	61.5
0.85	60.9	61.6
0.90	60.4	61.5
0.95	60.7	61.8
1.00	62.2	62.2

SPS–PPP		
X	Solidus	Liquidus

BBB–SSS		
X	Solidus	Liquidus
0.00	81.7	81.7
0.12	71.0	81.5
0.23	71.0	80.7
0.33	71.0	80.0
0.44	71.0	79.5
0.55	71.0	78.0
0.64	71.0	77.7
0.74	71.0	75.6
0.83	71.0	73.9
0.92	71.5	71.5
1.00	72.4	72.4

PPS–PPP		
X	Solidus	Liquidus
0.90	61.4	65.1
1.0	63.7	65.7

BBB–PPP		
X	Solidus	Liquidus
0.00	81.1	81.1
0.13	67.0	80.9
0.25	65.4	80.0
0.31	65.4	79.0
0.47	65.4	78.5
0.57	65.4	77.4
0.67	65.4	77.2
0.77	65.4	75.4
0.85	65.4	73.0
0.93	65.4	70.4
1.00	65.4	

SSS–LLL		
X	Solidus	Liquidus
0.00	72.6	72.6
0.15	45.7	
0.25	66.7	71.0
0.39	59.9	70.1
0.48	45.7	69.2
0.59	45.7	68.0
0.68	45.7	66.9
0.79	45.7	65.1
0.85	45.7	63.6
0.937	45.7	60.0
0.955	45.7	58.4
0.975		56.4
0.980	45.7	56.1
0.983		54.6
0.988	53.4	53.4
0.993	51.6	51.6
0.995	47.6	47.6
1.000	46.4	46.4

SSS–888		
X	Solidus	Liquidus
0.00	72.5	72.5
0.23	50.2	
0.52	8.4	67.0
0.78	8.4	65.9
0.94	8.4	60.6
0.96	7.0	59.1
0.97	8.4	57.7
0.99	8.4	54.1
1.00	8.4	8.4

PPP–MMM		
X	Solidus	Liquidus
0.00	64.3	64.3
0.11	64.0	64.0
0.22	64.0	64.0
0.32	61.8	61.8
0.43	61.4	61.4
0.53	60.1	60.1
0.63	59.8	59.8
0.72	57.2	57.2
0.82	52.9	52.9
0.91	53.2	53.2
1.00	54.5	54.5

(*continued*)

(continued)

PPP–LLL			SSS–SES			SSS–SSE		
X	Solidus	Liquidus	X	Solidus	Liquidus	X	Solidus	Liquidus
0.00	65.1	65.1	0.00		72.2	0.00	72.2	72.2
0.12	63.9	64.9	0.10		71.7	0.10	69.8	71.9
0.17	59.7	64.2	0.20		71.2	0.20		71.2
0.25	56.0	63.8	0.30		70.6	0.2	68.0	71.5
0.31	55.6	63.5	0.40	69.9	0.3	0.3	67.3	71.2
0.32	53.6		0.50		69.5	0.30		70.6
0.33	46.1		0.60		69.0	0.40		70.0
0.36		63.1	0.70		68.3	0.4	65.6	70.9
0.37	45.7	62.9	0.80		67.5	0.50		69.2
0.44		62.4	0.90		66.6	0.5	64.6	70.2
0.46		61.9	1.0		65.5	0.60		68.4
0.49	45.7	61.7				0.6	64.1	68.6
0.66	45.7	59.6				0.69	63.1	66.5
0.75	45.7					0.70		67.4
0.80		56.5				0.80	62.5	65.5
0.84		55.1	SSS–SEE			0.89	62.1	64.6
0.89		52.8	X Solidus	Liquidus		0.90		65.8
0.92	45.7		0.20	71.2		0.94	62.1	63.4
0.94		50.1	0.30	70.6		1.00	62.1	62.1
0.95		49.6	0.40	70.0				
0.967		47.1	0.50	69.2				
0.970		46.7	0.60	68.4				
0.973	45.7	45.7	0.70	67.4				
0.989		46.3	0.80	66.2				
0.995		46.4	0.90	65.8				
1.000	45.7	45.7	1.00	63.4				

PPP–SSE			PPP–EEE			LLL–EEE		
X	Solidus	Liquidus	X	Solidus	Liquidus	X	Solidus	Liquidus
0.00	65.8	65.8	0.00	65.4	65.4	0.00	46.0	46.0
0.05	60.8	64.5	0.05	64.2	65.1	0.05	44.8	46.0
0.10	59.5	64.2	0.09	64.2	65.0	0.15	43.5	46.0
0.14		63.0	0.14	63.7	65.1	0.21	42.6	45.9
0.21	57.6	62.9	0.19	62.9	64.7	0.30	41.5	44.8
0.29	56.4	62.3	0.23	61.6	64.5	0.36	40.4	45.1
0.41	55.6	61.3	0.28	61.4	64.2	0.40	39.3	44.3
0.50		59.8	0.33	61.0	64.3	0.45	38.9	44.1
0.54	56.2	58.7	0.38	60.0	65.0	0.50	38.9	44.1
0.56		58.3	0.43	59.0	64.6	0.55	38.1	43.3
0.59	55.7	58.6	0.48	55.9	64.6	0.60	40.0	42.5
0.61		59.3	0.53	55.4	64.7	0.67	37.7	42.6
0.64	56.2	59.8	0.58	52.2	64.4	0.70	37.3	39.6
0.67	56.1	60.4	0.63	50.6	63.0	0.74	37.1	38.2

(continued)

PPP–SSE

X	Solidus	Liquidus
0.70	56.0	
0.76		61.3
0.81	56.0	
0.87	57.1	61.6
0.92	59.3	61.9
1.00	61.9	61.9

PPP–EEE

X	Solidus	Liquidus
0.68	46.9	62.5
0.73	43.9	61.3
0.78	40.3	61.2
0.84	40.0	60.0
0.89	39.8	58.0
0.95	39.3	55.0
1.00	40.3	40.3

LLL–EEE

X	Solidus	Liquidus
0.80	37.8	39.1
0.84	37.8	39.7
0.89	38.4	40.2
0.95	38.9	40.0
1.00	39.5	40.3

SSS–SOS

X	Solidus	Liquidus
0.00	72.8	72.8
0.06	65.5	
0.17	53.6	
0.23	44.3	
0.27		70.9
0.51	44.3	69.3
0.61	44.3	68.3
0.75	44.3	66.1
0.83	44.3	64.5
0.91	44.3	61.3
0.95	44.3	57.8
0.97	44.3	52.0
0.99	44.3	44.8
1.00	44.3	44.3

PPP–SOS

X	Solidus	Liquidus
0.00	65.8	65.8
0.09	61.6	65.5
0.19	54.5	64.7
0.26	50.0	64.2
0.35	44.0	63.7
0.47	40.6	62.6
0.57	41.0	61.1
0.69	41.2	59.0
0.79	41.0	56.1
0.88	41.6	49.5
0.93	41.0	44.0
0.94	41.6	41.3
0.97	41.8	41.8
1.00	42.3	42.3

PPP–POP

X	Solidus	Liquidus
0.00	65.9	65.9
0.06		65.4
0.10	58.4	65.0
0.13	49.6	64.6
0.25	44.8	63.6
0.35	35.2	62.7
0.45		61.6
0.56	35.2	60.5
0.63	35.2	59.5
0.80	34.5	56.4
0.85		53.6
0.88	34.7	50.7
0.95		43.0
0.97		39.5
0.98		37.9
0.99		37.1
1.00	34.6	34.6

PPP–POO

X	Solidus	Liquidus
0.00	65.7	65.7
0.06	65.3	65.3
0.08	64.5	64.5
0.15	13.6	64.7
0.32	14.7	63.3
0.49	17.1	61.3
0.68	17.7	58.1
0.83	17.9	54.5
0.92	18.5	50.0
1.00	18.6	18.6

SES–SSE

X	Solidus	Liquidus
0.00	65.4	65.4
0.10	65.3	65.3
0.20	65.2	65.2
0.30	65.0	65.0
0.40	64.9	64.9
0.50	64.7	64.7
0.60	64.5	64.5
0.70	64.2	64.2
0.80	64.0	64.0
0.90	63.6	63.6
1.00	63.3	63.3

SOS–POS

X	Solidus	Liquidus
0.00	36.5	36.5
0.10	34.2	36.5
0.20	34.2	37.0
0.30	34.5	38.5
0.40	35.0	39.0
0.50	35.5	40.5
0.60	35.5	41.8
0.70	36.5	42.5
0.80	38.0	43.0
0.90	40.5	43.5
1.00	44.0	44.0

(*continued*)

(continued)

	SOS–POP			POS–POP			POP–PEP	
X	Solidus	Liquidus	X	Solidus	Liquidus	X	Solidus	Liquidus
0.00	44.0	44.0	0.00	36.5	36.5	0.00	33.0	33.0
0.10	38.5	43.5	0.10	34.0	36.2	0.08	31.1	41.7
0.20	35.5	43.0	0.20	32.0	36.0	0.16	31.5	45.3
0.30	35.0	42.5	0.30	31.0	35.5	0.25	31.1	47.1
0.40	35.0	41.5	0.40	31.8	35.2	0.50	31.1	52.9
0.50	34.0	40.0	0.50	31.5	34.0	0.75	36.0	54.4
0.60	34.0	39.0	0.60	31.5	34.5	1.00	57.0	57.0
0.70	33.5	37.0	0.70	31.5	35.5			
0.80	33.5	36.0	0.80	31.5	35.5			
0.90	33.5	36.5	0.90	34.0	37.0			
1.00	37.2	37.2	1.00	37.2	37.2			

	PPO–POO			PPO–OPO			POO–OPO	
X	Solidus	Liquidus	X	Solidus	Liquidus	X	Solidus	Liquidus
0.00	32.6	32.6	0.00	34.7	34.7	0.00	19.3	20.0
0.17	11.8	32.7	0.11	28.9	33.1	0.04	18.7	19.8
0.29	12.3	32.0	0.25	25.3	31.1	0.08	17.6	19.4
0.40	12.2	31.0	0.40	22.7	29.6	0.23	14.0	18.4
0.47	12.1	29.7	0.50	21.1	29.1	0.38	12.0	17.6
0.62	12.8	28.0	0.55	19.3	28.7	0.52	11.2	15.6
0.78	13.1	26.3	0.60	18.2	28.0	0.62	11.1	16.3
0.84	14.6	18.3	0.70	17.8	26.2	0.77	11.6	18.4
1.00	18.6	18.6	0.80	18.2	23.1	0.91	14.0	18.9
			0.85	17.8	20.2	1.00	19.8	19.8
			0.87	17.6	19.8			
			0.90	17.6	18.7			
			0.95	17.8	19.1			
			0.98	18.2	19.2			
			1.00	19.3	19.3			

	SSS–OOO			PPP–OOO			SOS–OOO	
X	Solidus	Liquidus	X	Solidus	Liquidus	X	Solidus	Liquidus
0.00	73.9	73.9	0.00	64.5	64.5	0.00	42.8	42.8
0.01		72.5	0.10			0.12	42.4	42.4
0.11		71.9	0.20			0.33	41.3	41.3
0.29		70.3	0.30			0.52	40.3	40.3
0.33	5.4	68.9	0.40			0.81	35.3	35.3
0.51		68.4	0.50			0.90	32.1	32.1
0.54		64.5	0.60			0.96	30.0	30.0
0.58		64.2	0.70			0.98	28.2	28.2
0.62		65.4	0.80			1.00	5.4	5.4
0.71		64.9	0.83		54.3			
0.80		64.1	0.88		52.3			

(continued)

SSS–OOO			PPP–OOO			SOS–OOO		
X	Solidus	Liquidus	X	Solidus	Liquidus	X	Solidus	Liquidus
0.83	5.4	65.4	0.90					
0.84		62.7	0.91		50.3			
0.84		60.7	0.93		49.0			
0.85		66.1	0.942		47.3			
0.88		60.0	0.956		44.7			
0.89	4.9	62.4	0.971		42.7			
0.89		66.6	0.978		41.0			
0.899		61.9	0.989		38.0			
0.907		57.3	0.991		35.7			
0.920		56.0	0.993		34.3			
0.920		54.0	0.996		32.7			
0.947		54.7	1.000	6.7	6.7			
0.949		57.8						
0.960		52.0						
0.960		51.3						
0.974		47.3						
0.974		44.7						
0.975		55.0						
0.980		50.0						
0.987		41.3						
1.000	5.4	5.4						

Note: x in mole fraction, solidus and liquidus in °C.

REFERENCES

Asselineau, L. and J. Jacq. 1989. Fifth Annual Conference on Fluid Properties and Phase Equilibrai. Banff. Canada.

Bailey, A.N. 1950. *Melting and Solidification of Fats*, New York: Interscience.

Barbano, P. and J.W. Sherbon. 1977. *JAOCS*. 55: 478.

Billmeyer, F.W. 1975. *J. Appl. Phys.* 28: 115.

Birker, P.J.M.W.L., S. de Jong, E.C. Roijers, and T.C. van Soest. 1991. *JOACS* 68: 895.

Bommel, M. 1986. Thesis. Rijksuniversiteit Utrecht.

Broadhurst, M.G. 1962. *J. Res. NBS-A.* 66: 241.

de Bruijne, P. and J. Eedenburg. 1983. Unilever Research Report PVD 83 3062.

de Bruijne, O. and F. Reckweg. 1972. *Phase transition behaviour of mixtures of tri-behenate with tristearate and tripalmitate.* Unilever R&D internal report P VD 72 3275.

de Bruijne, P. and F. Reckweg. *P VD.* 71: 3637.

de Bruijne, P. and F. Reckweg. *P VD.* 72: 3275.

de Bruijne, P., M. van den Tempel, and M. Knoester. 1972. *Chem. Phys. Lipids.* 9: 309.

Brumbaugh, E.E., M.L. Johnson, and C. Huang. 1990. *Chem. Phys. Lipids.* 52: 69.

Busfield, W.K. et al. 1990. *JAOCS.* 67: 171.

Cebula, D.J., D.J. McClements, and M.J.W. Povey. 1990. *JAOCS.* 67: 76.

Chanh, N.B. and Y. Haget. 1972. *Acta Cryst.* 28: 3400.

Chanh, N.B. and Y. Haget. 1975. *J. Chem. Phys.* 72: 760.

Chapman, D. 1962. *Chem. Rev.* 62: 433.

Christophersen, M.J.N. 1986. Study of some fat blends with DCS. *Unilever Research Report.* 86: 3274.

Clements, M.S. and J.B. Rossell. 1966. Internal Unilever Report P WN 660044.

Crowe, C.M. and M. Nishio. 1975. *AIChE J.* 21: 528.

De Smedt, A. and V. Gibon. 1990. Private communication.

Desphande, D.D., D. Patterson, H.P. Schreiber, and C.S. Su. 1974. *Macromolecules* 7: 530.

Desty, D.H. et al. 1962. *Gas Chromatography*, London: Butterswoth.

Dijkstra, A.J., R.J. Hamilton, and W. Hamm. 2008. *Trans Fatty Acids*. Wiley-Blackwell.

Dollhopf, W., H.P. Grossmann, and U. Leute. 1981. *Colloid and Polymer Sci.* 259: 267.

Duffy, P. 1853. *J. Chem. Soc.* 5: 197.

Fifth International Conference on Phase Equilibria for Chemical Process Design, 1989.

Fischer, J. and D. Moeller. 1985. 1: 246.

Fischer, J. and D. Möller. 1985. *Proc. of the 2nd CODATA symposium on critical evaluation and prediction of phase equilibria in multicomponent systems, Paris, France 11–13 september 1985*. Elsevier Amsterdam, Vol I, 246.

Flory, P.J. and A. Vrij. 1963. *JAOCS.* 85: 3548.

Fredenslund, A., A. Gmehling, and P. Rasmussen. 1977. *Vapor-Liquid Equilibria Using UNIFAC*, Amsterdam: Elsevier.

Freeman, I.P. 1957. Internal Unilever Report, Port Sunlight Program. Spring 1957: 45.

Gandasasmita, I. 1987. Thesis. Technische Universiteit Delft.

Garti, N., J. Schlichter, and S. Sarig. 1988. *Fat Sci. Technol.* 90: 295.

Gautam, R. and W.D. Seider. 1979. *AIChE J.* 25: 991, 999.

Gibon, V. 1984. Thesis. Universite Notre Dame de la Paix, Namur.

Gibon, V. 1986. Thesis. Universite de Namur.

Gray, M.W. and N.V. Lovegren. 1978. *JAOCS.* 55: 310, 601.

Grob, R.L. 1985. *Modern Practices of Gas Chromatography*, New York: J Wiley.

Grootscholten, P.A.M. 1984. *Calculation of solid liquid equilibria in fats*. Unilever R&D internal report LPVD 84 3074.

Grootscholten, P.A.M and S.J. Jančić. 1984. *Industrial Crystallization*. Springer.

Grootscholten, P.A.M. 1986. Jancic: Industrial crystallization.

Grootscholten, P.A.M. 1987. Unilever Research Vlaardingen, Private communication.

Grootscholten, P.A.M. Calculation of solid liquid equilibria in fats. LPVD 3074; 84.

Hagemann, H.W. 1975. *JAOCS.* 52: 204.

Hagemann, H.W. 1975. *JAOCS.* 60: 1123.

Hagemann, H.W. 1988. Crystallization and polymorphism of fats. In: *Surfactant Science*, Garti, N., K. Sato, Eds, New York: Marcel Dekker.

Hagemann, J.W. and J.A. Rothfus. 1988. *JAOCS.* 65: 1493.

Hagemann, J.W. and J.A. Rothfuss. 1983. *JAOCS.* 60: 1123.

Haget, Y. et al. 1985. Second codata symposium on prediction of phase equilibria in mulitcomponent systems.

Haget, Y. et al. 1985. *Proc. of the 2nd CODATA symposium on critical evaluation and prediction of phase equilibria in multicomponent systems, Paris, France 11–13 september 1985*. Elsevier Amsterdam, vol II, 636.

Hannewijk, J., A.J. Haighton, and P.W. Hendrikse. 1964. *Analysis and Characterization of Oils, Fats, and Fat Products*, London: Interscience.

Hansen, H.A. and A. Fredenslund. et al. 1987. A thermodynamic model for predicting wax formation in crude oils. Danmark: Tekniske Hojskole.

Hendriske, P.W. Dilatation of fats. Unilever Research.

Hernqvist, L. 1984. Thesis. University of Lund.

Hernqvist, L. and K. Anjou. *Fette Seife Anstrichmittel.* 64: 85.

Hernqvist, L. and K. Larsson. 1982. *Anstrichmittel*. 84: 349.

Hernqvist, L. et al. 1981. *J. Sci. Food Agric*. 32: 1197.

Human, H., J. Ende, and L. Alderliesten. 1989. Poster at AOCS meeting Masstricht, the Netherlands.

Janelli, L. 1985. 1: 246.

Janelli, L. 1985. *Proc. of the 2nd CODATA symposium on critical evaluation and prediction of phase equilibria in multicomponent systems, Paris, France 11–13 september 1985.* Elsevier Amsterdam, Vol II, 631.

de Jong, S. 1980. Thesis. Rijksuniversiteit Utrecht.

de Jong, S. and T.C. van Soest. 1978. *Acta Cryst*. B34: 1570.

Juriaanse, A.C. 1985. Butterlike margarines. Unilever Research Report VD 85 6003.

Kellens, M., W. Meeussen, C. Riekel, and H. Reynaers. 1990. *Chem. Phys. Lipids*. 52: 79.

Kerridge, R. 1952. Unpublished work. Private communication to J.B. Rossel.

Keulemans, K. 1986. Unilever Research Vlaardinger: Private communication.

Kitaigorodskii, A.I. 1973. *Molecular Crystals and Molecules*, London: Academic Press.

Kitaigorodskii, A.I. 1984. *Mixed Crystals: Solid State Sciences 33*, Berlin: Springer Verlag.

Knoester, M., P. de Bruijne, and M. van den Tempel. 1968. *J. Crystal. Growth*. 3: 776.

Knoester, M., P. de Bruijne, and M. van den Tempel. 1972. *Chem. Phys. Lipids*. 9: 309.

Krautwurst, J. 1972. *Keiler Milchw. Forsch. Ber*. 22: 255.

Kung, H.C. 1950. Internal Unilever Report LB 113.

Larsson, K. 1964. *Arkiv. Kemi*. 23: 35.

Larsson, K. 1986. In: *The Lipids Handbook*, Gunstone, F.D., J.L. Harwood, F.B. Padley, Eds, London: Chapman and Hall.

Lovegren, N.V. 1976. *JAOCS*. 53: 519.

Lutton, E.S. 1955. *JAOCS*. 32: 49.

Luzatti, V., H. Mustacchi, A. Skoulios, and F. Husson. 1960. *Acta Cryst*. 13: 660.

Malkin, T. and M.G.R. Carter. 1947. *J. Chem. Soc*. 1947: 554.

Michelsen, M. 1982, 1982, 1989. *Fluid Phase Eq*. 9: 1, 9: 21, 1989: 16.

Morancelli, M., H.L. Strauss, and R.G. Snyder. 1985. *J. Phys. Chem*. 89: 5260.

Muller, A. 1932. *Proc. Roy Soc*. 138: 514.

Murray, W., P.E. Gill, and M. Wright. 1981. *Practical Optimization*, London: Academic Press.

Norton, I.T. 1984. Unilever Research Report PCW 84 1065.

Norton, I.T. et al. 1985. *JAOCS*. 62: 1237.

Norton, I.T., C.D. Lee-Tuffnell, S. Ablett, and S.M. Bociek. 1985a. *JAOCS*. 62: 1237, 1261.

Norton, I.T., C.D. Lee-Tuffnell, and D.W. Rowlands. 1985b. *Unilever Research Colworth*. 85: 1175.

Null, H.R. 1980. *Phase Equilibria in Process Design*, New York: Krieger.

Nyvlt, J. 1967. *J. Chem. Prum*. 18: 260.

Ollivon, M. and R. Perron. 1979. *Chem. Phys. Lipids*. 25: 395.

Ollivon, M. and R. Perron. 1982. *Thermochimicia Acta*. 53: 183.

Pamplin, B.R. 1980. *Crystal Growth*, New York: Pergamon Press, pp. 526–530.

Perron, R. 1984. *Rev. Fr. Corps. Gras*. 31: 171.

Perron, R. 1986. *Rev. Fr. Corps. Gras*. 33: 195.

Perron, R., J. Petit, and A. Matthieu. 1971. *Chem. Phys. Lipids*. 6: 58.

Poot, C.G. and G. Biernorth. 1986. In: *The Lipids Handbook*, Gunstone, F.D., J.L. Harwood, F.B. Padley, Eds, London: Chapman and Hall.

Prausnitz, J.M. 1986. *Molecular Thermodynamics of Fluid Phase Equilibria*, New York: Prentice Hall.

van Putte, K. and B. Bakker. 1987. *JAOCS*. 64: 1138.

Renon, H. (Editor). 1989. *Proceedings of the Fifth International Conference of Phase Equilibria for Chemical Process Design, Banff, Canada*. Elsevier Amsterdam.

Rossell, J.B. 1967. *Adv. Lip. Res*. 5: 353.

Sato, K. 1987. *Food Microstructure.* 6: 151.

Sato, K. et al. 1989. *JAOCS.* 66: 664, 1614.

Shah, V.B. 1980. Thesis. University of Toledo.

Simpson, T.D. and J.W. Hagemann. 1982. *JAOCS.* 84: 349.

Small, D. et al. 1984. *The Physical Chemistry of Lipids: Handbook of Lipid Research 4,* New York: Plenum Press.

Smith, K. 1988. Unilever Research Colworth. Private communication.

Smith, J.M. and H.C. van Ness. 1987. *Introduction to Chemical Engineering Thermodynamics,* 4th edn. New York: McGraw-Hill.

Swank, D.H. and J.C. Mullins. 1986. *Fluid Phase Eq.* 30: 101.

van den Tempel, M. 1979. Physicochimie des Composes Amphiphiles. *Colloques Nat* 938: 261.

Timms, R.E. 1978. *Chem. Phys. Lipids.* 21: 113.

Timms, R.E. 1984. *Prog. Lip. Res.* 23: 1.

Vazquez Ladron, R. and R. Castro Ramos. 1971. *Gracas y Aceites.* 22: 401.

Vetterling, W., S. Teukolsky, W. Press, and B. Flannery. 1990. *Numerical Recipes,* Cambridge: Cambridge University Press.

Wesdorp, L.H. and J.A. van Meeteren. 1989. *The influence of TAG composition on crystallization rates, part II.* Unilever R&D internal report LP VD 89 3140.

Wesdorp, L.H. and M. Struik. 1990. European Patent EP 0264149.

Wesdorp, L.H. 1996. *Lipid Techn.* 8/6: 129.

Wieske, T. 1970. Internal Unilever report BAH D7019.

Wieske, T. and K. Brown. 1986. Private communication.

Won, K.W. 1986. Fourth international conference on fluid properties and phase equilibria for chemical process design, Helsingor, Denmark.

Won, K.W. 1989. Fifth international conference on fluid properties and phase equilibria for chemical profess design, Banff, Canada.

Wurflinger, A. 1972. Thesis. University of Bochum.

Yap, P.H., J.M. de Man, and L. de Man. 1989. *JAOCS.* 66: 693.

Zacharis, H.M. 1975. *Ad X-ray Anal.* 18: 535.

Zacharis, H.M. 1977. *Chem. Phys. Lipids.* 18: 221.

Zief, M. and W.R. Wilcox. 1967. *Fractional Solidification,* New York: Marcel Dekker.

10 Experimental Methodology

Rodrigo Campos

10.1 INTRODUCTION

This chapter is a step-by-step description of the most common analytical methods used in the study of physicochemical properties of fat crystal networks. The methodologies covered include those used in the investigation of nucleation and crystallization kinetics, thermal properties and polymorphism, as well as micro (microstructure and rheology) and macroscopic (texture and oil migration) properties. A summary of these techniques, along with the parameters obtained and the instrumentation used, is listed in Table 10.1. In aiming to provide the reader with a complete overview of the experimental methodologies, a brief theoretical background is included which covers the fundamental concepts related to the phenomena and instrumentation used. This is followed by a comprehensive description of experimental procedures, from sample preparation to data analysis, including illustrative examples. This is by no means an exhaustive review of all the analytical methods available in the area of fat material science research, but rather the most commonly used methods in academic and industrial research laboratories.

10.2 CRYSTALLIZATION

By studying the development of solid crystalline material as a function of time, different kinetic and thermodynamic parameters which provide insight into the nature of the crystallization process can be obtained. Three different analytical techniques can be used to investigate crystallization: light scattering, microscopy, and pulsed nuclear magnetic resonance (pNMR).

For all crystallization work, completely melt fat samples. In order to get valid results, it is essential that the fat's thermal history and crystal memory are erased. A temperature of 80°C for 30 min is recommended for most edible fats. High melting point fats will require a higher temperature to erase all crystal memory; for example, palm oil needs to be heated at 120°C for 10 min. A general guideline is to melt the fat at least 10°C above the temperature at which it completely melts.

TABLE 10.1

Summary List of Variables Measured, Parameters Obtained, and Instrumentation Used by the Experimental Methodologies Covered in This Chapter

Area of Study		Variable Measured	Calculated Parameters	Instrumentation
Crystallization	Crystallization kinetics	Solid fat content (SFC) as a function of time	Crystallization rate (k) Crystallization index (n) Equilibrium SFC (SFC_∞) Half-time of crystallization ($t_{1/2}$)	Pulsed nuclear magnetic resonance (NMR) minispec
	Crystallization induction times	Scattered light signal as a function of time	Crystallization induction times ($\tau_{crystallization}$) Nucleation rate (J) Free energy of nucleation (ΔG_c)	Cloud point analyzer (light scattering)
	Early crystallization growth	Number of features as a function of time	Crystal growth rate (J_c)	Polarized light microscopy (PLM)
Thermal properties	Melt profiles	SFC as a function of temperature	Melting curves	Pulsed nuclear magnetic resonance (NMR) minispec
	Solution behavior	SFC as a function of temperature	Iso-solid diagrams	Pulsed nuclear magnetic resonance (NMR) minispec
	Thermal behaviors	Heat flow differential between sample and reference	Peak temperature of crystallization (T_C) and melting (T_M) Onset temperature of crystallization (T_{oC}) and melting (T_{oM}) Enthalpy of crystallization (ΔH_C) and fusion (ΔH_m)	Differential scanning calorimeter (DSC)

Polymorphism	Crystal structure	Diffraction of x-rays	Short spacings (subcell spacings); Long spacings (longitudinal packing of lamellae)	X-ray diffractometer (XRD)
Microstructure	Microstructure	Crystal features	Thickness of crystal domain; Number of particles (N_p); Crystal size; Fractal dimension (D_f and D_b)	Polarized light microscope
Mechanical properties	Small deformation rheology	Strain as a function of applied stress	Complex modulus (G^*); Shear storage modulus (G'); Shear loss modulus (G''); Phase angle (tan δ); Fractal dimension (D_f)	Rheometer
	Large deformation testing	Deformation under constant force rate	Yield force; Yield point	Texture analyzer
Fractal dimension		Distribution of mass of the fat crystal network at different scales	Particle counting (D_f); Box counting (D_b); Physical counting (D_f)	Polarized light microscope/rheometer—NMR
Oil migration	Oil loss assay	Oil transferred from sample to filter paper	% Oil loss	Analytical balance
	Scanning imaging	Pixel intensity changes in time	Movement of oil migration front	Flatbed scanner

10.2.1   Nucleation Events

The crystallization process of a fat begins when it is supercooled or undercooled (ΔT). Supercooling of a melt system (e.g., fat) is equivalent to the supersaturation of a solution system (e.g., sugar solution). A melted fat is supercooled when it is taken to a crystallization temperature (T_C) below its melting temperature (T_M); thus, it is defined as $\Delta T = T_C - T_M$. In the melt, fat molecules are in random thermal motion; upon supercooling, solid nuclei embryos form. The newly formed embryos (very small regions of the new solid phase) will continue to form and dissolve until they reach a critical size. The Gibbs free energy of the embryos will reach a maximum at this critical size. In order to keep the free energy to a minimum, these embryos will either redissolve or continue to grow, forming stable nuclei. These nuclei will grow continuously as other lipid species in the melt diffuse onto its growing surface. Crystal growth continues, followed by aggregation into clusters and formation of bridges with each other, eventually resulting in the creation of a crystal network (Hartel, 2001; Timms, 2003; Walstra, 2003; Marangoni, 2005a).

Nucleation is influenced by the degree of supercooling (ΔT) and the activation free energy of nucleation (ΔG_c). When a fat system is highly supercooled, the energy barrier for nucleation is lower; this translates into a lower induction time (τ), which in turn is inversely proportional to the nucleation rate (J), namely,

$$ J \approx \frac{1}{\tau} \tag{10.1} $$

The study of nucleation events is based on the formation of solid crystal nuclei and their quantification. The time required for detectable evidence of stable nuclei is τ. Experimentally obtained values of τ combine the time for stable nuclei to form ($\tau_{nucleation}$) and the time required for these nuclei to grow sufficiently and be detected by the analytical tools used (i.e., cloud point analyzer, microscopy, NMR). Although NMR is widely used to measure the SFC of fats, its sensitivity does not allow the accurate measurement of less than 1% crystalline material. Crystals can be visually observed in the sample tubes before the NMR detects any solids. For this reason, NMR is considered to be insensitive to the initial nucleation events that take place in the melt. Other analytical methods are used to study early stages of crystallization (i.e., determination of τ and ΔG_c). Such techniques include monitoring the changes in turbidity of a crystallizing sample through measurements of light scattering intensity or visually monitoring formation of crystals through microscopy. For example, τ values determined by light scattering are going to be shorter than those determined by visualization of crystal features by microscopy. It is critical to always keep in mind instrument's sensitivity and thus the phenomena are being analyzed.

10.2.1.1   Measurement of Inductions Times by Light Scattering

An alternative approach to measuring nucleation induction times is light scattering. A cloud point analyzer such as the phase transition analyzer (Phase Technology, Richmond, British Columbia, Canada) shown in Figure 10.1a is used to quantify the

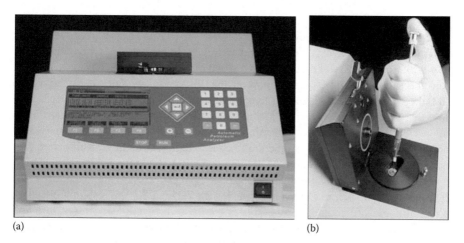

(a)   (b)

FIGURE 10.1 Phase transition analyzer (Phase Technology) used in nucleation studies of fats (a). Samples of melted fats are loaded into the sample chamber using a positive displacement pipette (b). A beam of light is impinged on the sample. When nucleation and early crystallization events take place, the incident beam is scattered by the solid–liquid phase boundaries onto detectors in the chamber, and the detected signal is automatically recorded.

light scattered by growing crystalline mass in a sample as it is being cooled from the melt. A relatively small sample is placed in the sample chamber (Figure 10.1b). A beam of light is impinged upon the sample while detectors in the chamber record the signal being scattered from the growing crystals. Prior to any nucleation events, the recorded signal strength is constant. As nuclei start to appear in the sample, the incident beam is scattered by the solid–liquid phase boundaries onto the detectors. As more and more nuclei develop, the signal output increases. In time, nuclei grow into crystals and the signal increases proportionally, until it reaches saturation and plateaus. The user is cautioned not to use the plateau signal to quantify the total solids at equilibrium. The signal units are arbitrary and do not correspond to absolute values of crystallized solid fat. For this purpose, pNMR, described further in this chapter, is recommended. The sample chamber is maintained under dry conditions to prevent vapor condensation and the temperature of the fat is precisely controlled using a Peltier element. This technique is reliable, reproducible, and allows for the accurate monitoring of the early stages of nucleation.

10.2.1.1.1 Procedure

10.2.1.1.1.1 Sample Preparation Melt the sample in an oven at a temperature that ensures that all crystal memory is erased. Pipette 150 μL of melted sample into the sample chamber of the phase transition analyzer (Figure 10.1b). A displacement pipette is used in order to guarantee that the entire sample is ejected from the pipette.

10.2.1.1.1.2 Experimental Procedure Prior to running any sample, ensure that the coolant supply (clean tap water free of deposits) is on and the desiccant materials are in good condition. This is critical to the temperature control of the sample and to avoid condensation in the sample chamber. Using the instrument's software set the following experimental parameters: holding temperatures, heating or cooling rates, and soak or holding times.

The first step in the experimental procedure is to hold the sample at a temperature that will erase all crystal memory. Even when the sample is melted prior to loading into the sample chamber, it is important that all crystal memory is erased as the initial step of the run. A holding temperature of 80°C or the maximum temperature that the instrument reaches, at a rate of 20°C/min or as high as the instrument allows, and soak (hold) time of 30 min is recommended. The second step, for nucleation studies, is to cool the sample to the chosen temperature of crystallization (T_C) at a fast cooling rate followed by a holding time. Rapid cooling is recommended to avoid nucleation events at temperatures above the desired T_C. If there are any crystallization events before the sample reaches the defined T_C, the results will reflect a dynamic crystallization process (crystallization as a function of temperature and time) rather than crystallization in time under isothermal conditions. The holding time needs to be long enough to include nucleation and early crystal growth events. If the chosen holding time is too short, it may not allow sufficient time to monitor the formation of nuclei. Trial and error will dictate the optimum values to be used for each sample at particular T_C's. Once the experimental parameters are entered, the run can be started.

10.2.1.1.1.3 Data Analysis   Throughout the entire experimental run, the instrument saves signal strength, time, and temperature data. Data is saved using the instrument's software format. Export data from the instrument's software into a spreadsheet for further analysis. Only the data from the isothermal crystallization step (not the initial melting of the sample) is used for analysis. Construct a plot of signal strength as a function of time (Figure 10.2). The first point in the time axis corresponds to the time point when the observed sample temperature reaches T_C; this will be time zero. τ can be defined either as the point in time when the signal deviates from the baseline by 1% or by extrapolating the linear portion of crystal growth to

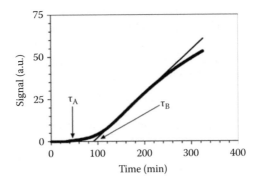

FIGURE 10.2 Light scattering signal as a function of time from which the induction time for nucleation (τ) is obtained either when the signal deviates from the baseline by 1% (τ_A), or as the onset of linear turbidity development (τ_B) obtained from the intersection of the crystallization's curve linear growth portion with the time axis. Results may vary depending on how the induction time is calculated (45 vs. 92 min).

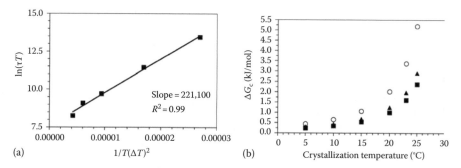

FIGURE 10.3 Plot of $\ln(\tau T_C)$ as a function of $1/T_C(\Delta T)^2$ of the retentate obtained from the fractionation of milk fat by short path distillation fitted to a straight line (a). The slope of the line is used to calculate free energies of nucleation (ΔG_C) at each crystallization temperature (T_C) using the Fisher–Turnbull equation (b). Shown are calculated values of ΔG_C for milk fat (o) and retentates obtained from fractionation of milk fat by short path distillation at 175 (■) and 200°C (▲).

the time axis (Wright et al., 2001). To determine this, extrapolate the linear portion of the plot to the x axis. In either case, the induction time is the point at which recognizable nuclei are detected by the scattered light. An example of τ determination by the two approaches herein described is shown in Figure 10.2.

The Fisher–Turnbull equation is typically used in fats systems to quantify the activation free energy of nucleation for a crystallization process (Marangoni, 2005a). The variation in the crystallization free Gibbs energy (ΔG_c) can be calculated from τ values obtained over a range of at least five T_C's. The range of temperatures at which to calculate τ is found by trial and error. Using experimentally determined τ's, T_C's, and the melting point (I_M) of the sample, construct a plot of $\ln(\tau T_C)$ versus $1/T_C(\Delta T)^2$ like the one shown in Figure 10.3a. Use the slope of the resulting line (m), along with the Boltzmann constant ($k = 1.38 \times 10^{-23}$ J/K), the sample's I_M and each I_C in the Fisher–Turnbull equation shown below to calculate ΔG_c at each T_C.

$$\Delta G_c = \frac{mk}{\left(T_M - T_C\right)^2} \tag{10.2}$$

Figure 10.3b illustrates the effect of ΔT on the ΔG_c of native milk fat and the retentates obtained under different fractionation temperatures by short path distillation.

10.2.1.2 Monitoring Early Crystal Growth by Polarized Light Microscopy

Polarized light microscopy (PLM) allows the birefringent solid microstructural elements of the fat crystal network to be directly observed as sharp bright features against a dark background. PLM is not only used in the quantification of the microstructural elements of a crystal network, but can also be useful in imaging early crystal growth. Early crystal growth is referred here as the instance when the microscope can detect the appearance of a crystal (at the magnification scaled used) under controlled crystallization conditions. A light microscope allows the user to determine the times required for crystals to reach the critical size that allows them to be visible. The sensitivity of a light microscopy is lower than the cloud point analyzer. Thus, τ values obtained by PLM are considered to reflect early crystallization events, rather than nucleation events. Along with τ, this methodology is useful in determining crystal growth rates. This section will only briefly describe the experimental procedure to be followed in the estimation of early crystal growth using PLM and will focus on the analysis of data. The topic of microscopy is covered later in this chapter. For more detailed information regarding calibration of the microscope, image acquisition, image processing, and basic concepts of microscopy, refer to Section 5.1.

10.2.1.2.1 *Procedure*

10.2.1.2.1.1 Sample Preparation   Melt the sample in an oven at a temperature that ensures that all crystal memory is erased. Using a preheated capillary tube, place a small droplet (about 10 μL) of melted fat on a preheated (80°C) glass slide. Carefully place a preheated glass cover slip over the sample to produce a film of uniform thickness.

10.2.1.2.1.2 Experimental Procedure   Immediately transfer the sample slide to a temperature-controlled stage set at the temperature of study. Capture digital grayscale images under polarized light using 90° crossed polarizer every 15 s for 30 min to monitor the crystallization process. Polarization at 90° allows for the removal of most of the nonbirefringent background signal. The time intervals provided here are a suggested starting point. When performing the experiment, the user will have to determine the length of the experiment as well as appropriate time intervals. Longer time intervals for an extended period of time may be used for slower crystallization processes or low degrees of undercooling.

10.2.1.2.1.3 Data Analysis   The first step in the calculation of crystal induction time and crystal growth rate is to threshold the acquired series of images using an image analysis software (e.g., Photoshop [Adobe Systems Incorporated, San Jose, CA]). It is important to ensure that the threshold value used accurately represents the microstructure imaged in the original grayscale images, and that the same threshold value is applied to all the micrographs in the series. The resulting images will have white features representing crystalline mass surrounded by black liquid oil. Invert the images so that the crystalline matter is represented by black features on a white background. Figure 10.4a through f shows a series of images acquired at different time intervals during the crystallization of a mixture of high and low melting milk fat fractions at 5°C. The result of processing these images (i.e., threshold and inversion) is shown in Figure 10.4g through l. Quantify the amount of black in each inverted thresholded image using an image analysis software (e.g., Photoshop). Plot the % black, which is proportional to the solid fraction of the network, as a function of crystallization time. Obtain τ by extrapolating from the linearly increasing portion of the crystallization curve to the time axis as shown in Figure 10.5a.

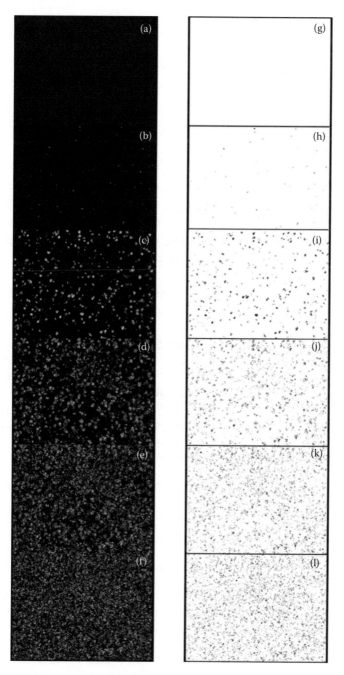

FIGURE 10.4 Polarized light micrographs taken at time 0 (a), 12.5 (b), 13 (c), 13.5 (d), 14 (e), and 15 (f) seconds of static isothermal crystallization at 5°C of a mixture of high and low melting fractions of milk fat (30:70) used in the analysis of early crystal nucleation events. Images (g through l) are thresholded and inverted images of (g through i) are used for quantitative image analysis.

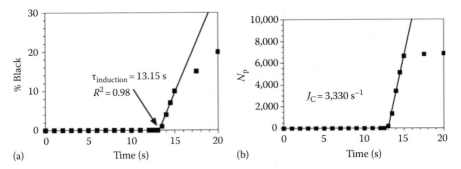

FIGURE 10.5 Estimation of the crystallization induction times (τ_C) (a) and the crystal growth rate (J_C) (b) from polarized light micrographs of a mixture of high and low melting fractions of milk fat (70:30) crystallized at 5°C.

Another useful parameter is the rate of crystal growth (J_C), which can be estimated from the obtained images using the following relationship:

$$J_C \approx \frac{dN_p}{dt} \tag{10.3}$$

where N_p is the number of particles at a time t in the region of linear increase of crystalline material. For this, compute the number of particles or black features of the thresholded inverted images with a suitable image analysis software package (i.e., Photoshop). Plot the number of particles (N_p) as a function of time. Calculate the first derivate of the portion of the curve that reflects linear crystal growth, as shown in Figure 10.5b. The first derivate corresponds to the slope of the linear portion of the curve and is obtained via a linear regression. This value is an estimate of J_C, which is inversely proportional to τ.

10.2.2 Crystallization Kinetics by Nuclear Magnetic Resonance

The solid fat content (SFC) of a fat sample can be measured by nuclear magnetic resonance (NMR). Figure 10.6 shows a benchtop NMR instrument, the Bruker Minispec PC/20 series NMR Analyzer (Bruker, Milton, Ontario, Canada). In this technique, the response of hydrogen protons aligned at equilibrium in a magnetic field to a short intense pulse of radio frequency (RF) energy is measured. The RF pulse is applied at 90° to the direction of the magnetic field. When this RF pulse is applied, the protons tip away from the magnetic field and become excited. The NMR signal observed right after the application of the RF pulse is proportional to the total number of hydrogen protons present in the sample. Once the RF is removed, protons relax from their excited state back to their equilibrium state (in line with the magnetic field). The NMR signal consequently decays as a result of this relaxation. Free induction decay (FID) is the name given to the time-domain signal obtained during this relaxation process. The FID is different in the solid and the liquid phases of the fat sample. For protons in the solid state, the signal decays relatively quickly; while for protons in the liquid state, the signal slowly decays. Distinct decay times

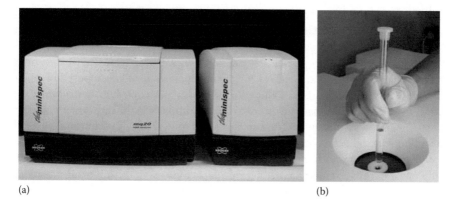

(a)   (b)

FIGURE 10.6   NMR bench top instrument (Bruker PC/20 series NMR Analyzer, Bruker) (a) and glass NMR tube containing a sample of melted fat to be analyzed for solid fat content (b).

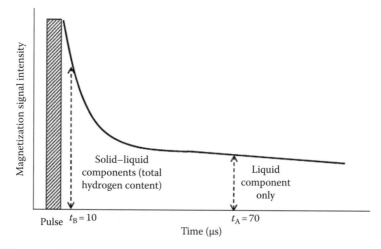

FIGURE 10.7   FID for a typical fat sample showing the magnetization signal as a function of time from which the signal for hydrogen protons present only in the liquid component and signal for all hydrogen protons present in the sample (solid and liquid components) determined at t_A and t_B. The SFC is calculated as the difference between the signals associated with t_A and t_B.

are observed in Figure 10.7. The initial decay in the magnetization signal intensity is characteristic of the solid fraction. The remaining magnetization displays a slower decay, arising from the liquid fraction.

There are two approaches to calculating the SFC: direct and indirect method. In the direct method, SFC is quantified directly by the minispec by sampling the decaying NMR signal at two suitable time points, indicated in Figure 10.7 as t_A and t_B. These correspond to the liquid signal and the total signal (solid and liquid signal) respectively. The percent solid is calculated as the ratio between the NMR response obtained from the protons in the solid phase and the NMR response obtained from the protons in both the solid and the liquid phases of the sample. When determining SFC with this method, the instrument calculates the ratio and reports the percent SFC. For its simplicity, it is the most commonly used approach to measure SFC.

The indirect method is particularly useful when working with fat-based samples which contain nonfat solid particles (e.g., confectionery blends composed of sugar, nuts, milk, and cocoa solids in addition to fat) which also contribute to the NMR signal. If the SFC of these materials were to be determined via the direct method, inaccurate SFC values would result from an erroneous solid signal. To determine the SFC, the indirect method compares the FID signal obtained from the sample's liquid phase at a given time/temperature of interest with the FID signal of the same sample when fully melted. Thus the indirect method calculates SFC using the following equation:

$$\text{SFC} = 100 - \left[\left(\frac{S^T}{S^{60°C}} \right) \left(\frac{R_{oil}^{60°C}}{R_{oil}^T} \right) \right] \times 100 \tag{10.4}$$

where

S^T is the signal intensity of the sample at the measuring point of interest (i.e., temperature or time of storage)

$S^{60°C}$ is the signal intensity at 60°C (when the sample is completely melted)

$R_{oil}^{60°C}$ is the signal intensity of a reference oil at 60°C (olive oil)

R_{oil}^T is the signal intensity of the reference oil at the measuring point of interest (i.e., temperature or time of storage) (Bruker, 1989)

The first ratio term of Equation 10.4 corresponds to the NMR response from the protons in the liquid phase at the time/temperature of interest and the response from the protons in the fully melted sample. The second ratio term of Equation 10.4 corresponds to a temperature correction, as the NMR signal is also a function of temperature in addition to proton response. For example, olive oil that is fully melted between 5°C and 60°C will have a higher NMR signal at 60°C compared to at 5°C. Olive oil is typically used as a reference oil, but any oil that does not crystallize at the temperature range of interest can be used. In Equation 10.4, 60°C is used as the temperature at which the fat is completely melted. This is true for most edible fats. However, when working with samples containing high melting point fats, it is advisable to use a temperature 10°C above the melting point of the fat.

10.2.2.1 Procedure

10.2.2.1.1 Sample Preparation

Melt the sample in an oven at a temperature that ensures that all crystal memory is erased. Transfer the molten fat to preheated (80°C) glass NMR tubes (10 mm diameter, 1 mm thickness, and 180 mm height) to a height fill of 3 cm (Figure 10.6b). Wipe the outside of the tube with tissue paper, making sure that the exterior is clean and dry. It is critical that the tubes are dry prior to making any measurements, as the hardware can easily be damaged by water droplets. To prevent the entry of dust, solvent vapors, or moisture, close the tubes with a cap or rubber stopper.

10.2.2.1.2 Experimental Procedure

Once the sample is prepared, immediately transfer the tubes to a water bath that has been preset to the desired temperature of study. It is important to make sure that the sample is completely molten before readings are taken. If the sample is cooled during preparation, heat the prepared pNMR tube to melt the sample and ensure that all crystal memory is erased. Take SFC readings at appropriate time intervals with a pNMR analyzer (Figure 10.6a). The time intervals chosen and duration of the experiment will depend on the fat and temperature of crystallization. For samples exhibiting rapid crystallization (such as in situations of high degrees of undercooling), measurements may be taken every 30 s, whereas for slower crystallization processes (low degrees of undercooling), measurements taken every 5 min may suffice. Continue to take measurements until the SFC readings reach a plateau and the system reaches equilibrium.

10.2.2.1.3 Data Analysis

Construct crystallization curves with the data obtained by plotting the SFC(%) as a function of time, as shown in Figure 10.8. Fit the crystallization curves to the Avrami equation using nonlinear regression with the assistance of data-fitting software (i.e., Graph Pad, SigmaPlot, Origin, etc.). The Avrami model describes crystallization processes in which there is an initial lag period, followed by a rapid increase in crystal mass (Avrami, 1939). This model has been used to describe the liquid to solid phase transition (i.e., crystallization) of fats (Metin and Hartel, 1998;

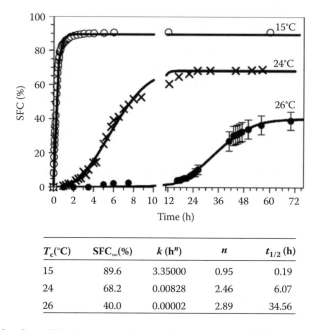

$T_c(°C)$	$SFC_\infty(\%)$	k (h^n)	n	$t_{1/2}$ (h)
15	89.6	3.35000	0.95	0.19
24	68.2	0.00828	2.46	6.07
26	40.0	0.00002	2.89	34.56

FIGURE 10.8 Crystallization curves of cocoa butter cooled at 5°C/min and crystallized at 15°C, 24°C, and 26°C. Data are collected using NMR (benchtop Bruker PC/20 series NMR). Crystallization curves were fitted to the Avrami equation. The resulting Avrami parameters describe the crystallization kinetics of cocoa butter under these crystallization conditions.

Wright et al., 2000). Figure 10.8 shows an example of crystallization data fitted to the Avrami equation. The Avrami model (Avrami, 1939) as applied to fats has the following form:

$$\frac{SFC(t)}{SFC_\infty} = 1 - e^{-kt^n} \tag{10.5}$$

where

SFC(t) is the % SFC as a function of time
SFC$_\infty$ is the limiting SFC as time approaches infinity (equilibrium SFC)
k the Avrami constant which represents the crystallization rate
n the Avrami exponent or index of crystallization which gives information on the crystal growth mechanism

The Avrami index is a combined function of the time dependence of nucleation and the number of dimensions in which growth takes place. Nucleation can be either instantaneous, with nuclei appearing all at once early on in the process; or sporadic, with the number of nuclei increasing with time (Sharples, 1996; Wright et al., 2000). Possible values for the Avrami exponent for different types of nucleation and crystal growth are shown in Table 10.2.

Half-times of crystallization, which reflect the magnitudes of the rate constants, represent the time required for 50% conversion of the material and are calculated using the following relationship, derived from the Avrami model:

$$t_{1/2} = \left(\frac{0.693}{k}\right)^{1/n} \tag{10.6}$$

TABLE 10.2
Avrami Exponent (n) Values for Different Types of Nucleation and Dimensionality of Growth

Avrami Exponent (n)	Various Types of Growth and Nucleation
$1 + 0 = 1$	Rod-like growth from instantaneous nuclei
$1 + 1 = 2$	Rod-like growth from sporadic nuclei
$2 + 0 = 2$	Disc-like growth from instantaneous nuclei
$2 + 1 = 3$	Disc-like growth from sporadic nuclei
$3 + 0 = 3$	Spherulitic growth from instantaneous nuclei
$3 + 1 + 4$	Spherulitic growth from sporadic nuclei

Source: Sharples, A., Overall kinetics of crystallization, In: *Introduction to Polymer Crystallization*, Sharples, A., Ed. Edward Arnold Publishers, Ltd., London, U.K., pp. 44–59.

10.3 THERMAL PROPERTIES

10.3.1 MELT PROFILES BY SOLID FAT CONTENT

Melt profiles are be constructed by measuring the SFC as a function of temperature. SFC is defined as the quantity of solid glycerides in a sample, determined as the percentage of hydrogen nuclei (protons) present in the solid phase measured by NMR (AOCS Official Method Cb-16b-93 or IUPAC Standard Method 2.150). For a brief description regarding NMR methodology, refer to Section 10.2.2.

SFC values at a determined temperature depend on the thermal history of the crystallized fat. In some cases there is the need for polymorphic stabilization of the sample. For this, the construction of melting profiles by SFC makes a distinction between stabilizing and non-stabilizing fats. Stabilizing fats are those that require polymorphic transformation and thus require a tempering period (the convention is to use 40 h at 26°C) for the crystal structure to reach its most stable form before any SFC is read. Examples of stabilizing fats are cocoa butter and confectionery fats with 2-oleo-disaturated triacylglycerides. Non-stabilizing fats do not undergo polymorphic transformation, and thus do not require a tempering period. Examples of non-tempering fats include lard, palm oil, coconut oil, and palm kernel oil.

When performing a melting profile experiment, samples may be subjected to the different temperatures of interest in either a parallel or a serial fashion. When done in parallel, one sample tube is prepared for each temperature to be measured; hence, there will be as many tubes as temperatures at which SFC levels are determined. Alternatively, in the serial procedure, only one sample tube is prepared which is measured sequentially at all measuring temperatures starting with the lowest. Both methods yield good results. However, for some fats, differences in SFC reading may result from the two procedures (Bruker, 1989) due to polymorphic transformations and recrystallization events taking place during the progressive heating of the fat samples. As such, these differences should be noted and quantified, especially when the analysis affects the commercial value of the analyzed sample. One advantage of the parallel method over the serial method is the considerably shorter period of time required to carry out the experiment. The drawback is the requirement of a number of water baths equal to the temperatures at which SFC is going to be determined. In the following section, both approaches (serial and parallel) are described for the analysis of stabilizing and non-stabilizing fat samples.

10.3.1.1 Procedure

10.3.1.1.1 Sample Preparation

Melt the sample in an oven at a temperature that ensures that all crystal memory is erased. Homogenize the sample and transfer the molten fat to preheated at 80°C glass pNMR tubes (10 mm diameter, 1 mm thickness, and 180 mm height) to a height fill of 3 cm. Wipe the outside of the tube with a tissue, making sure that the exterior is clean and dry. To prevent the entry of dust, solvent vapors, or moisture, close the tubes with a cap or rubber stopper. Depending on the type of fat (i.e., stabilizing or non-stabilizing), the samples are then subjected to a tempering procedures as per the prescribed cooling regimes outlined in Figure 10.9.

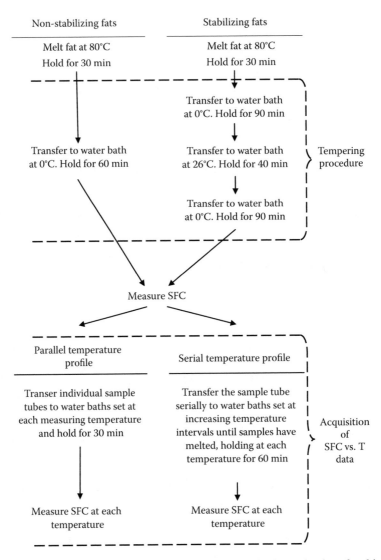

FIGURE 10.9 Experimental temperature profiles used in the determination of melting profiles of stabilizing and non-stabilizing fats by serial and parallel methods.

10.3.1.1.2 Experimental Procedure

After the samples are prepared and tempered, they are placed in water baths set at the different measuring temperatures. The measuring temperatures will depend upon the method being used. Temperatures of 10°C, 21.1°C, 26.7°C, 33.3°C, and 37.8°C are used in the AOCS Official Method Cd 16b-93, while the IUPAC Standard Method 2.150 indicates to measure at 10°C, 20°C, 25°C, 30°C, 35°C, and 40°C. A systematic description of this methodology can be found in the following official methods: AOCS Cd 16b-93, UPAC 2.150, and ISO 8292.

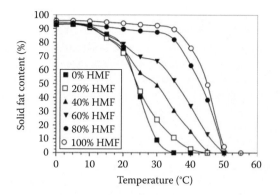

FIGURE 10.10 Melting profiles of (○) 100% HMF, (●) 80% HMF and 20% cocoa butter, (▼) 60% HMF and 40% cocoa butter, (▲) 40% HMF and 60% cocoa butter, (□) 20% HMF and 80% cocoa butter, and (■) 100% cocoa butter.

If a serial method is to be used, measure the SFC after the sample has been cooled to 0°C, and transfer to a water bath set at the lowest measuring temperature. Hold for 30 min in the case of non-stabilizing fats and 60 min in case of stabilizing fats. This will quench cool fat samples, providing a significant degree of undercooling and allowing for crystallization of fat samples. Measure the SFC and transfer to a water bath set at the next lowest measuring temperature, hold tubes for the appropriate time, and measure the SFC. Repeat this procedure until measurement at the highest temperature is performed. At this temperature the fat should be completely melted. In the case of a parallel method, measure the SFC after the sample has been cooled at 0°C. Transfer each sample tube to a water bath set at each measuring temperature, hold for 30 min in the case of non-tempering fats and 60 min in case of tempering fats, and measure the SFC.

10.3.1.1.3   Data Analysis

Construct the melting profile of each sample by plotting the SFC as a function of temperature and join the points with a line. An example of this analysis for mixtures of high melting milk fat fraction (HMF) and cocoa butter is shown in Figure 10.10. From the melting profiles, it is observed that HMF has a different melting behavior relative to cocoa butter. HMF melts between 35°C and 50°C, while cocoa butter melts between 20°C and 35°C. In blends of the two fats, the melting range of the resulting mixture increases as the proportion of HMF is increased. In other words, at any given temperature, the SFC increases as the fraction of HMF is increased.

10.3.2   Iso-Solid Phase Diagram Construction

When two distinct fats are combined, the physicochemical properties of the mixture are not necessarily the same as those of the original components. To better understand and predict the physicochemical properties of a mixture of two independent fat systems, it is necessary to study their phase behavior through the construction of phase diagrams. Phase diagrams are used to examine the conditions at which thermodynamically different fats come to equilibrium between the solid and liquid phase. In this context, iso-solid phase diagrams are used to study the solution behavior of a binary system. These diagrams have been extensively used in the study of phase behavior of mixtures of confectionery fats with milk fat and milk fat fractions (Hartel, 1996; Marangoni and Lenki, 1998; Marangoni, 2002; Timms, 2003).

10.3.2.1 Procedure

10.3.2.1.1 Sample Preparation

Completely melt the two components of which their phase behavior will be studied. Prepare mixtures (w/w) of the two components in 10% increments from 0% to 100%. Mix thoroughly while the samples are still melted to ensure homogenized composition. Transfer 3 g of each mixture to glass pNMR tubes (10 mm diameter, 1 mm thickness, and 180 mm height). Treat sample tubes (i.e., melting and tempering) as described previously in the melt profile determination by pNMR section.

10.3.2.1.2 Experimental Procedure

After the mixtures are tempered, measure the SFC at determined temperature intervals. For more information regarding the generation of SFC data points, refer to the experimental procedure section on melting profiles by SFC.

10.3.2.1.3 Data Analysis

Construct a melting profile of each mixture by plotting the SFC as a function of temperature as shown in Figure 10.10. Fit the points to a cubic spline curve with the assistance of curve fitting software (i.e., Graph Pad, Sigma Plot, Origin, etc.). A cubic spline fitting produces a curve that passes through every data point, bending and twisting as needed. Based on the fitting results, obtain temperatures corresponding to SFC values in the range of interest in 2%–5% SFC increments. Construct a plot of temperature as a function of composition (% of component A) for each SFC value for all blends. Perform an additional cubic spline curve fit to smooth the final iso-solid data lines. Figure 10.11 illustrates examples of iso-solid diagrams for mixtures of HMF with middle melting milk fat fraction (MMF) (a), HMF with cocoa butter (b), and MMF with cocoa butter (c). In the construction of the diagram, temperature data were generated for SFC levels in 5% increments from 0% to 100%. Each line observed corresponds to the same SFC level.

When two distinct fats are mixed, three different solution behaviors can potentially take place: compatibility (given by a monotectic diagram), partial compatibility, and incompatibility (shown by a eutectic diagram). Iso-solid diagrams are useful in the study of solution behavior as they describe the storage temperature requirements as a function of the blend composition to keep SFC constant. The different solution behaviors between components of a binary fat mixture are shown in the iso-solid diagrams in Figure 10.11. Figure 10.11a illustrates the compatibility between the TAG molecules of MMF and HMF resulting in the formation of mixed crystals (solid solution) due to the dissolution of one component into the other. When HMF is mixed with cocoa butter, partial compatibility is observed (Figure 10.11b).

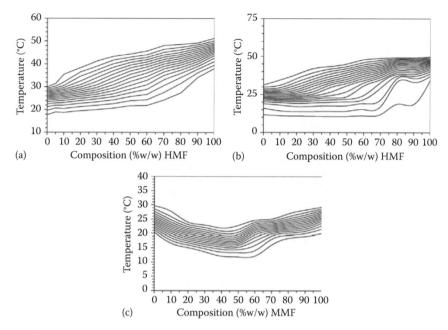

FIGURE 10.11 Iso-solid phase diagram for MMF–HMF (a), HMF—cocoa butter (b), and MMF—cocoa butter (c) mixtures constructed with data obtained from melt profiles (SFC as a function of temperature).

There is partial solid solution formation, both a monotectic and a slight amount of thermodynamic incompatibility of the two components. Incompatibility between MMF and cocoa butter is observed in Figure 10.11c. A eutectic between the two components is seen as a valley in the iso-solid lines. The relevance of this valley or eutectic is that between 25 and 75% MMF, a decreased in temperature is required to keep the solid fraction constant. In other words, the melting point of blends between 25 and 75% MMF in cocoa butter is lower than either pure MMF or cocoa butter.

10.3.3 Thermal Behavior by Differential Scanning Calorimetry

The thermal behavior of fat samples can be studied using differential scanning calorimetry (DSC). A DSC instrument, shown in Figure 10.12, measures the difference between the temperature or heat input between a sample pan and a reference pan. This differential is used to characterize changes associated with phase transitions (i.e., crystal formation or melting). Thermograms associated with sample phase changes are obtained as the temperature is increased or decreased at a controlled rate.

10.3.3.1 Procedure

10.3.3.1.1 Sample Preparation

Melt the sample in an oven at a temperature that ensures that all crystal memory is erased. The sample preparation involves the encapsulation of the fat in pans designed for calorimetric analysis. Examples of DSC pans are shown in Figure 10.13. At all

FIGURE 10.12 Differential scanning calorimeter DSC Q 1000 (TA Instruments, Mississauga, Ontario, Canada) used in the thermal analysis of edible fats and oils. The unit on the right houses the sample cell and the instrument controls. The unit on the left is an external refrigerating cooling system (RCS).

times, use clean tweezers to handle the sample pans. Touching them as well as using dirty tweezers and placing sample pans on dirty surfaces can potentially leave a residue that may affect the results. Weigh the empty sample pan and lid (Figure 10.13a). Deposit 1–10 mg of fat sample into the sample pan (Figure 10.13b). Care should be taken not to cover the inside lip of the pan with the sample. This would result in sample spill during sealing of the pan, and thus inaccurate sample weight (Figure 10.34b). Place the lid on the sample pan as illustrated in Figure 10.13c and seal the pan with

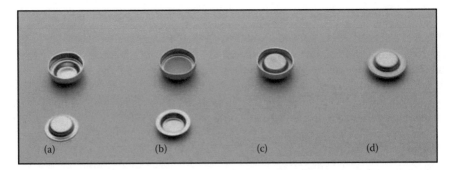

FIGURE 10.13 Illustration of the encapsulation process of a DSC pan: initially sample pan and lid are weighted (a), 1–10 mg of melted fat are place in the sample pan (b), the pan lid is placed over the sample pan being careful not to spill any fat onto the lip of the pan (c), and finally, the pan is sealed with an encapsulation press (shown in Figure 10.14).

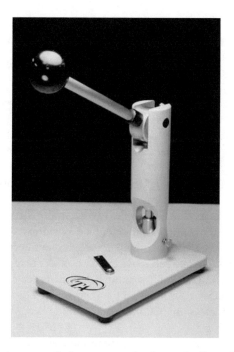

FIGURE 10.14 Sample encapsulating press used to hermetically seal sample pans used in DSC (TA Instruments, Mississauga, Ontario, Canada).

the sample encapsulation press (Figure 10.14). Always inspect the pan after sealing, making sure that the bottom and the seal are smooth and that no fat spilt out while closing the pan. Additionally, prepare one empty sealed pan. This will be used as the "reference" pan. Record the weights of the sample, sample pan and reference pan.

10.3.3.1.2 Instrument Calibration

When operating a DSC, an inert purge gas flow must be used at all times. The purpose of the purge gas is to provide a smooth thermal "blanket" medium, eliminating hot spots and allowing a better cooling of the cell. Additionally, it removes any moisture and oxygen from the cell that may have accumulated over time. Nitrogen is most commonly used as a purge gas as it is inexpensive, inert, and easily available; additionally, it does not interfere with heat measurements due to its low thermal conductivity. Helium may also be used as a purge gas if faster cooling rates are required. The external refrigerating cooling system must be turned on. It provides the necessary controlled cooling and heating rates during the experimental run.

A series of calibrations have to be performed prior to any experimental run to obtain adequate results. Typically, a calibration is carried out once in the course of all experiments, as calibration results are saved and reused each time an experiment is run. The calibration only needs to be redone, when either the purge gas is changed, the heating or cooling rate are changed, an experimental run is going to go over the calibrated temperature range or after the burning of the cell to clean any leaks and residues.

The experimental procedures that need to be carried out will depend on the instrument model. Typically, it involves a baseline calibration and a temperature and heat of fusion calibration.

The baseline calibration compensates for subtle differences between the reference and sample thermocouples. Most instruments perform only one run for this calibration, which consists of heating the empty cell through the instrument's temperature range. Theoretically, with an empty cell, the heat flow signal should be zero and the baseline should have a minimum slope. The output of this calibration is the calculation of a baseline slope and offset values which are used to flatten the baseline and zero the heat flow signal.

TA Instruments' (New Castle, DE) TA Q series DSCs use a second calibration run to calculate two additional terms for the baseline calibration. The second calibration run is performed using two sapphire disks of known weights (no pans) placed on the sample and reference positions. Both calibrations use the same method (i.e., same initial temperature, temperature ramp rate, and final temperature). The sample and reference sapphire disks heating rates are assumed to be equal to the heating rates of the sample and reference calorimeters. Sapphire is used as it has no phase transitions in the instrument's temperature range. Once these other two parameters are obtained, the baseline is constructed by the instrument calibration software.

The temperature and heat of fusion calibration ensures that the sample thermocouple reading is correct under the experimental conditions chosen. It involves the use of standard materials (i.e., high purity metals such as indium and gallium) which are melted at the same heating rate used in the analysis of samples. This calibration gives three different outputs: the cell constant, onset slope, and the temperature calibration (Thermal Advantage Manual, 2000). The cell constant is the ratio between the theoretical and experimentally measured heat of fusion of the standard. Onset slope, or thermal resistance, is a measure of the temperature drop that occurs when melting a sample in relation to the thermocouple. Theoretically, a standard sample should melt at a constant temperature. As it melts and draws heat—melting is an endothermic process—a temperature difference develops between the sample and the sample thermocouple. The thermal resistance between these two points is calculated as the onset slope of the heat flow vs. temperature curve on the front side of the melting peak. The temperature calibration corrects between the observed and theoretical transition temperature (melting point) of the standard material used. If only one standard is used, then the calibration shifts the sample temperature by a constant amount. A two-point calibration shifts the temperature with a linear correction (straight line) and projects this correction to temperatures above and below the two calibration points.

10.3.3.1.3 Experimental Procedure

Method segments are used to describe the thermal treatment to which the sample will be subjected. These contain commands that define target temperatures, heating and cooling rates, and isothermal holding times. An example of a procedure used to characterize the thermal behavior of a fat sample is shown in Table 10.3. In this

TABLE 10.3

Method Log for the Thermal Analysis of Cocoa Butter by DSC

Method Log:

1. Equilibrate at 60.0°C   } Melting to erase
2. Isothermal for 30.0 min   } all crystal memory

3. Data storage: on
4. Jump to 25.0°C
5. Equilibrate at 25°C   } Crystallization
6. Isothermal for 15.0 min
7. Ramp 5°C/min to 60°C

8. Equilibrate at 60°C
9. Isothermal for 2 min   } Melting
10. Data storage: off
11. End of method

Note: In this method, the sample is initially crystallized at 25°C for 15 min, followed by melting at a heating rate of 5°C/min.

method, the sample in initially melted at 60°C to ensure that all crystal memory is erased, after which it is cooled at the highest rate attainable by the instrument to a temperature of 20°C. The samples is allowed to crystallize at 20°C for 60 minutes and subsequently melted by heating to 60°C at a rate of 5°C/min. The purpose of this method is to examine the initial crystalline structure formed. If the objective of the user is to analyze the thermal behavior of the sample after extended storage times, samples are crystallized in the DSC cell, following steps 1 through 6 in the method shown in Table 10.3, after which they are transferred to temperature controlled incubators. After the defined storage period, sample pans are transferred to the DSC for melting, taking care that the temperature of the cell is equal to the storage temperature of study. It is recommended to include a step for isothermal equilibrium at the defined storage temperature for 15 min prior to melting of the sample (steps 7 through 10 in Table 10.3). This will allow the DSC cell, along with the reference and sample pans, to stabilize at the defined storage temperature prior to the melting of the sample.

Sample weights as well as sample and reference pans weights are entered into the software. The external refrigerating cooling system and purge gas must be turned on prior to running any sample. Depending on the instrument model, sample loading can be done manually or automatically. If loading of sample is done manually, care should be taken to center the pans on the pan platforms inside the cell. In the sample cell, there are positions specific for both the reference and the sample pan. Each instrument maker will have a different configuration, thus the user needs verify to positions for both reference and sample pans. Figure 10.15 illustrates the sample cell of a TA Q1000 (TA Instruments, New Castle, DE) instrument.

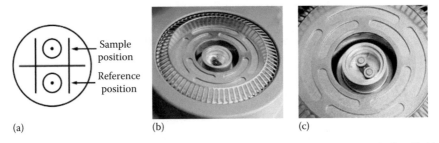

(a)   (b)   (c)

FIGURE 10.15 Positioning of reference and sample pans in a DSC standard cell (a) (Adapted from *TA Instruments Advantage Manuals*, TA Instruments, Mississauga, Ontario, Canada.), view of an empty DSC cell (b), and view of a DSC with the sample and reference pans in position (c).

10.3.3.1.4 Data Analysis

Results from an experimental DSC run are represented as a thermogram whereby heat flow is plotted as a function of time or temperature. Data analysis can be obtained using the instrument's analysis software (e.g., Universal Analysis 2000, TA Instruments, New Castle, DE; Pyris, Perkin Elmer, Waltham, MA). A spreadsheet containing the data acquired during the experiment may also be created by exporting the numerical data points from the instrument's software. Further analysis of the data may be performed using suitable software applications (e.g., Microsoft Excel, Graph Pad Prism, etc.).

Figure 10.16 shows an example of the thermogram resulting from the crystallization and subsequent melting of cocoa butter. In the thermogram both phase transition

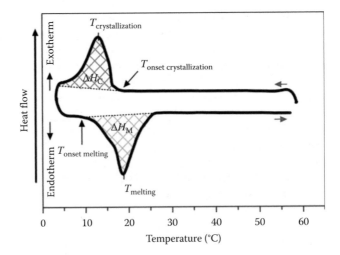

FIGURE 10.16 Illustration of a thermogram (plot of heat flow as a function of temperature) of obtained by DSC. The example shown is cocoa butter crystallized by cooling from 60°C to 5°C at a rate of 5°C/min, followed by melting to 60°C at a rate of 5°C/min. Crystallization corresponds to the upper peak, while melting corresponds to the lower peak. Indicated are parameters typically used to thermally characterize fat samples.

phenomena (i.e., crystallization and melting) are observed. Crystallization is an exothermic event, as energy is released during the process, which is observed in the thermogram as an upward peak (↑). On the other hand, melting is an endothermic event in which energy is absorbed, resulting in a decrease of the energy flow of the system, and is observed in the thermogram as a downward peak (↓). The user must keep in mind that different instruments' software's have the option to display exothermic events not only as positive (↑) but also negative (↓) deviations in heat flow from the baseline. Consequently, endothermic events are displayed as positive (↑) events.

Several parameters can be calculated to describe the thermal behavior of the sample. These include the peak temperature of crystallization (T_C) and melting (T_M), onset temperature of crystallization (T_{oC}), onset of melting (T_{oM}), and enthalpy of crystallization (ΔH_C) and fusion (ΔH_m). When analyzing pure samples, peaks are very narrow as all species crystallize or melt at a distinct temperature. When working with multicomponent systems, such as natural edible fats and oils which are a blend of TAG molecules with a range of different crystallization and melting points, peaks are broader. I_C and I_M are calculated as the peak's point of maximum deviation from a linear baseline drawn between two points on the curve. It corresponds to the temperature at which the sample has the highest heat flow. This temperature is related to the temperature at which the largest proportion of lipid species crystallize or melt, depending on the phase transition that is being analyzed, and is considered an average of the crystallizing or melting point of the sample. T_{oC} is the temperature at which the first crystallites are formed, while T_{oM} is the temperature at which the fat sample begins to melt. Both parameters are defined as the temperature at which the heat flow begins to deviate from the base line. The peak area corresponds to the enthalpy of the phase transition, which is calculated as the area under the thermogram curve. When integrating the area under the peak, two limit points are established which correspond to the start and end temperatures at which the thermal event is observed. In the analysis, a baseline is drawn between these upper and lower limits. This baseline can be linear, sigmoidal, or extrapolated. A linear baseline is a straight line drawn between the upper and lower limits and is used when the baseline changes linearly with time. A sigmoidal baseline is an s-shaped line that changes in level or slope, which compensates for changes in baseline that take place during the transition. It is also possible to extrapolate the baseline. To do this, four points are selected and tangent lines drawn between pairs of consecutive points, so that the extrapolated baseline intersects the range of data bracketed by the other two points. This is carried out by the instrument's analysis software as it evaluates potential tangent lines between consecutive points, and uses the tangent line that will intersect the data between the remaining points. The area under the peak is proportional to the total amount of crystalline material formed during crystallization or the amount of solid phase which is converted to liquid phase during melting.

Crystallization and melting are the two most commonly studied phase transitions by DSC. The following sections provide the user important considerations for the study of crystallization and melting of fats along with examples.

10.3.3.1.5 Crystallization Studies

For crystallization experiments, the sample is melted to erase all crystal memory, cooled at a controlled rate to the desired temperature of crystallization, and maintained isothermally at the temperature of study to allow for its crystallization. The cooling rate used is critical for two reasons. First, if slow cooling rates (typically, below 5°C) are used, nucleation can potentially commence during the cooling stage. If nucleation and early crystallization events take place above the defined crystallization temperature, the results will reflect a dynamic crystallization process (in which crystallization is a function of time and temperature) rather than isothermal crystallization. For isothermal crystallization experiments, high cooling rates (above 10°C/min) are recommended so that nucleation takes place at the desired crystallization temperature and not at any other temperatures. Secondly, the cooling rate used will directly affect the resulting crystalline form of the fat. Under rapid cooling conditions, TAG molecules usually crystallize in metastable polymorphic forms, which subsequently transform into polymorphs of higher stability. On the other hand, at slow cooling rates, TAG molecules of similar chain lengths have time to associate with each other in more stable geometrical arrangements, resulting in the formation of a more stable polymorphic form. Due to the dependence of fat crystallization on the degree of undercooling and the cooling rate used, different crystallization events (i.e., crystallization of different crystalline forms) can be observed when using different cooling rates.

In the characterization of fat samples, it is possible to monitor crystallization as a function of time or temperature. Figure 10.17 shows the crystallization of cocoa butter as a function of temperature while cooling (a) and as a function of time under isothermal conditions (20°C) (b). The type of method to run will depend on the user's objective. By running a method similar to that outlined in Table 10.3, where a fat sample is cooled from a temperature above its melting point to an undercooled state, it is possible to determine temperatures at which phase transitions take place (Figure 10.17a). The user must keep in mind that these runs will study dynamic crystallization processes in which both temperature and time will affect the phase transition. On the other hand, if the user is interested in understanding how the fat behaves at a given temperature, an isothermal run is required (Figure 10.17b). In such run, it is important

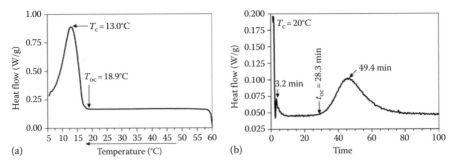

FIGURE 10.17 Crystallization of cocoa butter studied by DSC as a function of temperature (a) by cooling from 60°C to 5°C at a rate of 5°C/min, and time (b) by cooling from 60°C to 20°C at a rate of 5°C/min and holding isothermally at 20°C for 100 min.

to cool the sample at high cooling rates to the temperature of crystallization of interest so that the observed crystallization events will be a function only of time.

10.3.3.1.6   Melting Studies

For melting experiments, the sample is heated at a controlled rate until it is completely liquid. The heating rate used is also critical as different rates will yield slightly different results. At very low melting rates (1°C/min or lower), the heat transfer is going to be slow allowing for molecular rearrangements to occur. Some thermal events, such as melting of existing polymorphic forms, polymorphic transformations, recrystallization, and melting of the newly formed polymorphic forms, may be observed when a sample is melted slowly. Fast heating rates (5°C/min or higher) result in sample melting before any structural changes can take place. It is recommended, from experience in our laboratory, that fat samples be melted at a rate of 5°C/min.

Beyond the rates at which a sample is run, different crystallization conditions (i.e., T_C, temperature cycling, cooling rate, and storage time) yield distinct crystalline forms of the same fat and thus result in unique melting curves. For example, melting of three different cocoa butter samples with different thermal history (crystallization at 5°C for 2 min, 5°C for 5 days, and 22°C for 28 days) using the same DSC melting methodology can yield three distinct melt curves, as shown in Figure 10.18. Upon analysis of the samples by x-ray diffraction (XRD) (XRD is covered later in this chapter), it was confirmed that each peak corresponds to a different polymorph. In this way, DSC is frequently used to characterize the polymorphic forms present in the fat, relating experimental melting temperatures with reported values. The reader should be cautioned that although DSC melting points are used as an indication of the polymorphic state of fat samples, they should always be confirmed by XRD.

Specific thermal melting behavior will be observed in fat samples that have been crystallized under identical conditions, but stored for different periods of time.

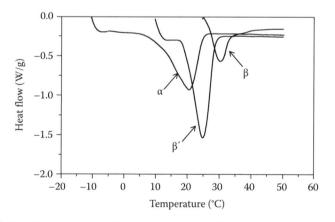

FIGURE 10.18   Characteristic melting thermograms of different polymorphs of cocoa butter. Prior to melting in the DSC, samples were crystallized at 5°C for 2 min for the α form, 5°C for 5 days for the β′ form, and 22°C for 28 days for the β form. (Adapted from McGauley, S. The relationship between polymorphism, crystallization kinetics, and microstructure of statically crystallizes cocoa butter. Master in Science Thesis. University of Guelph: Guelph ON, Canada, 2001.)

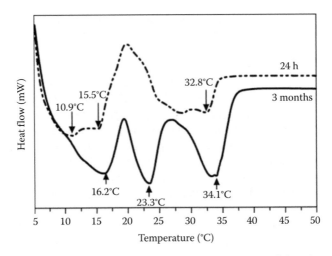

FIGURE 10.19 Melting thermograms of milk fat crystallized at 5°C and stored for 24 h (dotted line) and 3 months (continuous line) under isothermal conditions at 5°C.

As such, there are two different approaches that may be taken. A fat can be melted either immediately after crystallization has taken place, or after it has been crystallized and stored for a predetermined period of time. Melting a sample shortly after solidification allows the examination of the initial metastable crystalline structures, and the polymorphic transformations which the fat underwent during the crystallization process. On the other hand, when a sample is crystallized and stored for a specific period of time, it is possible to characterize the stable crystalline structures present in the sample, and any polymorphic transformations that may have taken place during storage. It is critical to keep the storage time consistent when comparing melting data from different samples. Otherwise, the user will be doing irrelevant comparisons.

The effect of storage time on the thermal properties of a crystallized sample is shown in Figure 10.19 for milk fat samples melted after being crystallized at 5°C and stored for either 24 h or 3 months. The sample melted after only 24 h of storage displays two major melting peaks, corresponding to the melting of two different fractions. At roughly 20°C, there is what seems to be an exothermal event, believed to be the polymorphic transformation of a metastable crystal structure to one of higher stability as the sample is being melted. Upon melting of the same material after 3 months of storage, three distinct fractions are observed with slightly higher peak melting temperatures relative to the 24 h sample. No exotherm event is observed indicating that no further polymorphic changes take place while the sample is melted. The data on Figure 10.19 suggest that polymorphic transformations took place during storage leading crystal structures of higher stability.

10.4 POLYMORPHISM

Triacylglycerol (TAG) molecules have the ability to pack in different geometric conformations during crystallization. Different polymorphs exhibit distinct crystalline structures with identical chemical composition but with dissimilar

Characteristic	α Form	β′ Form	β Form
	Hexagonal	Orthorhombic	Triclinic
Chain packing	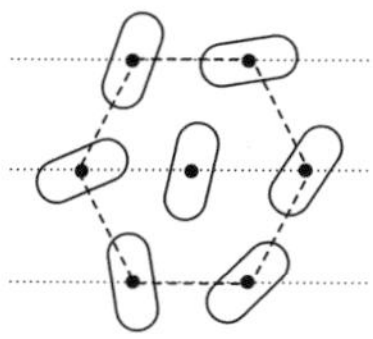	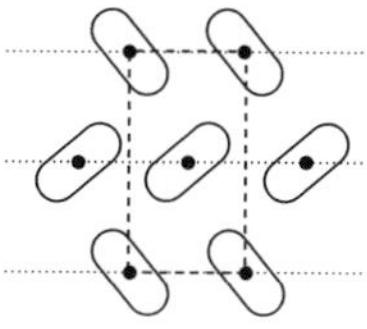	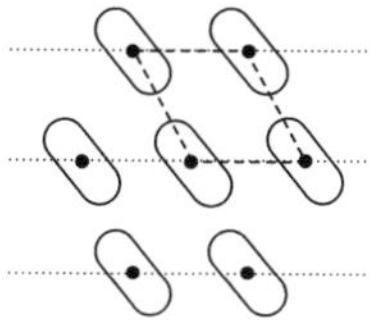
Short spacing(s)	0.415 nm	0.38 and 0.42 nm	0.37, 0.39, and 0.46 nm
Density	Least dense	Intermediate	Most dense
Melting point	Lowest	Medium	Highest
Stability	Least stable	Intermediate	Most stable

FIGURE 10.20 Characteristics of three different polymorphs found in fats. (Adapted from Sato, K. et al. Progress in Lipid Research, 38, 91, 1999; Marangoni, A.G. Crystallography. In: Fat Crystal Networks, Marangoni, A.G. (Ed.), Marcel Dekker, New York, pp. 1–20, 2005c.)

physical properties (i.e., density, stability, and melting point). In crystallized fats, three different polymorphic forms have been reported which are designated by the Greek letters α, β′, and β (Sato, 2001). The characteristics of these different polymorphic forms in TAGs are summarized in Figure 10.20. During crystallization the initial polymorphic form found is that of least stability—the α form. The α form has the least constraints regarding incorporation of TAG molecules from the melt onto growing crystals and hence has the lowest energy barrier to form. With time, molecules rearrange into more stable forms (i.e., β′ and β) as they possess lower energy states. The latter phenomenon is defined as polymorphic transformation, and is always toward forms of higher stability. Polymorphic transformations are exothermic phase transitions, and take place until molecules adopt an ideal conformation and arrangement, achieving an efficient close packing (Larsson, 1994).

10.4.1 X-Ray Diffraction

Powder XRD is used in the study of fat polymorphism. Through characteristic Bragg peak positions, different polymorphs are identified, based on the fact that different polymorphic geometries scatter x-rays in different ways. This technique also allows the determination of the molecular packing that form the lamella as well as the calculation of the thickness of the primary crystal (smallest unit in a structure that when joined together, fills the entire volume of the crystal).

X-rays striking crystalline fats are scattered into two major regions, the wide angle and the small angle region. Wide diffraction angles correspond to the characteristic small spacings of the subcell within the crystal lattice, providing information about polymorphism. Small diffraction angles correspond to the long spacings of the unit cell, which provide insight on how the TAG molecules stack together to form lamellar structures (Small, 1966; de Man, 1992; Larsson, 1994). The thickness of the crystal is calculated from the full width at half maximum of a peak in the small angle region, using the Scherrer analysis, discuss later in this section.

Bragg (1933) found that the peak position of the diffracted x-ray beam can be understood using a notion of mirror reflections, where the incoming x-ray gets diffracted from various atomic planes in the crystal lattice. Peak formation results from the constructive interference of the diffracted x-ray beam. The constructive interference can only happen when the distance travelled by different parallel diffracted x-rays differ by an integer number (n) of the incident wavelength (λ). The crystal lattice is considered as a series of atomic planes parallel to each other and separated a distance d apart. The path difference for rays reflected from adjacent planes at different reflection angles (θ) is described by Bragg's law as:

$$n\lambda = 2d\sin\theta \tag{10.7}$$

The integer "n" refers to the order of reflection, which makes reference to the reflections from successive planes. A value of one for "n" corresponds to the first-order reflection and its peaks are of the highest intensity relative to higher order reflections. At higher orders of reflection, the reflecting plane is a distant d/n from the first reflection plane and are of lower intensities.

The peak position angle (θ) on Figure 10.21 corresponds to the separation between the atomic planes (d), and is related to the wavelength of the incoming x-ray (λ) and the particular position of the diffracted plane of atoms (denoted by the integer n). Values of θ are converted into d spacings by using Equation 10.7. When doing calculations, consider the first order of reflection ($n = 1$), as it is the peak of highest intensity.

When experiments are done using synchrotron radiation, results are plotted as intensity versus "q," where

$$q = \frac{4\pi\sin\theta}{\lambda} \tag{10.8}$$

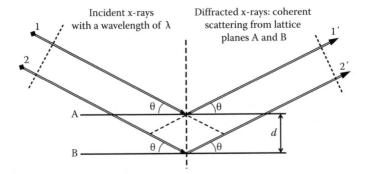

FIGURE 10.21 Geometry of the reflection of x-rays from crystal planes used in the derivation of Bragg's law.

Experiments performed in synchrotron have higher intensity which means more x-rays can be focused onto smaller, laser-like spots, allowing researchers to gather more data in greater detail in less time.

10.4.1.1 X-Ray Diffractometer

X-ray diffractometers measure d spacing values and diffraction intensities. Typical bench top x-ray instruments have an x-ray source, a sample holder, a monochromator, and a detector. Modern x-ray powder diffractometers' essential characteristic is that the relation between the incident and the scattered angle is maintained constant through the analyses. Figure 10.22 illustrates a Rigaku Automated Powder X-ray Diffractomer (Rigaku/MSC, The Woodlands, TX). Shown is the path that x-rays follow from the source (cooper tube that gives a wavelength of 1.54 Å) to the detector. The goniometer is the heart of the equipment. It maintains the desired angle position between the x-ray tube and the sample and between the scattered x-ray and the detector. The x-ray tube is mounted on one of the two arms of the goniometer. The other arm contains the monochromator (optical device that transmits a mechanically selectable narrow band of wavelengths to work with) and the detector. Each arm moves independently and together they follow the measuring diffractometer circle to maintain either a θ/θ or a θ/2θ relation at all times. These two relations refer to the angles between the incident x-rays, the sample, and the diffracted x-rays. The first term corresponds to the angle between the direction of the incident x-rays and the plane of the sample. The second term is the angle between the direction of the incoming x-rays and the direction of the detector from the sample. In the instrument shown in Figure 10.22, the sample holder is in a horizontal permanent position. The distance between the x-ray source to the sample and between the sample and the detector is constant at all times. Instruments will also have a cooling system for the x-ray tube. Some diffractometers allow mounting a temperature control system, like a Peltier temperature control system right underneath the sample holder.

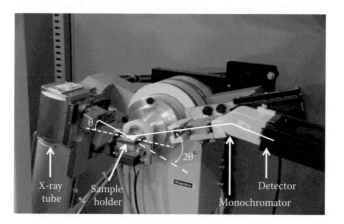

FIGURE 10.22 Goniometer and sample holder of a Rigaku Automated Powder X-Ray Diffractomer, multiflex theta/theta, with a cooper x-ray tube that gives a wavelength of 1.54 Å (Rigaku/MSC). Shown on the figure is the path that x-rays follow from the x-ray to the detector as they are diffracted by the sample.

10.4.1.2 Procedure

10.4.1.2.1 Sample Preparation

The sample is placed in a sample holder which then is situated inside the instrument in the goniometer sample holder. For the Rigaku instrument shown in Figure 10.22, the sample holder is a rectangle piece of glass of 5 cm × 3.5 cm × 1.7 mm, with an indent of 2 cm × 2 cm × 0.4 mm where the material is placed. For semisolid crystallized fat samples, a small flat spatula is used to spread the material onto the sample holder indent. Care should be taken not to compress or orient the sample in any one direction. It is also possible to crystallize the sample directly onto the sample holder, thus preventing any changes in structure and orientation due to sample manipulation.

Working with molten samples also allows studying the material as it crystallizes. Prior to sample preparation, melt the sample in an oven at a temperature that ensures that all crystal memory is erased. The liquid fat sample is then transferred to preheated sample holders with a preheated glass pipettes. Prepared samples are then transferred to temperature-controlled incubators or the instrument's Peltier sample holder for controlled crystallization.

10.4.1.2.2 Experimental Procedure

Prepared sample holders are placed in the goniometer. Care should be taken when opening and closing the door of the XRD instrument, as the x-ray tube might shut down automatically as a safety precaution. Based on the instrument's software, the user must select the kind of scan to perform: fixed time or counts, where the x-ray tube moves a specific interval of degrees, stops and collects data for a specified period of time or amount of counts, before moving to the next position; or continuous, where data are collected continuously as the x-ray tube moves. Method parameters that specify experimental conditions for the analysis such as starting angle, final angle, step size, scan speed, and slits sizes are defined. The slits help maximize the intensity as a function of θ, making sure that the incoming and scattered x-rays are focused onto the sample and onto the detector. For crystallized fat experiments the use of 1/2° slits are recommended. The starting and final angles depend on the objective of the test (i.e., wide or small angle regions). The step size or sampling width represents the step that the detector is moved before collecting data. It is generally maintained at 0.02°. The scan speed determines how long the detector will collect at each position. The faster the detector moves, the less resolution is achieved. A typical full scan for fats, covering both wide and small angle regions, will start at 1° and finish at 30°, with a step size of 0.02° and a scan speed of 1°/min.

10.4.1.2.3 Data Analysis

The instrument software constructs a histogram of the signal intensity as a function of 2θ. The software can also present the signal intensity as a function of the d spacings. Equation 10.7 is used to manually convert results from 2θ to d spacings. Figure 10.23, illustrates a histogram of signal intensity as a function of 2θ for glyceryl tristerate. The sample was first melted to erase all history and then cooled from 80°C to 20°C at a rate of 5°C/min directly on the sample holder on top of the Peltier. The sample was scanned once it has reached 20°C from $2\theta = 1$–30° at a scan speed of 1°/min in continuous mode with a step size of 0.02°.

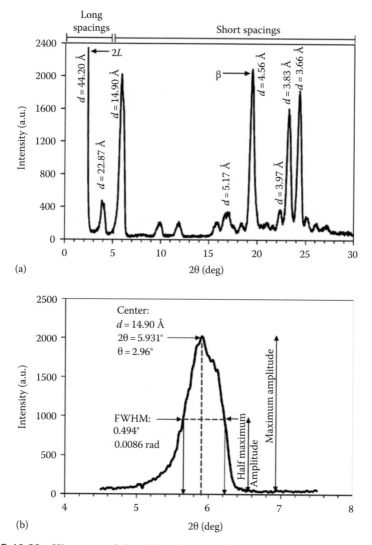

FIGURE 10.23 Histogram of signal intensity as a function of diffraction angle (2θ) for glyceryl tristerate cooled from 80°C to 20°C at a rate of 5°C/min directly on the sample holder on top of the Peltier and scanned when it had reached 20°C from $2\theta = 1°-30°$, at a scan speed of 1°/min in continuous mode with a step size of 0.02°. Indicated are the d spacings in the small and wide angle region that allow determination of longitudinal packing and polymorphism, respectively (a), as well as a close up section in the small angle peak (b) with d spacing of 5.83° that allows calculation of crystal domain thickness.

Small spacings or wide angles ($2\theta = 5°-30°$) are used for the characterization of the polymorphic form present in the sample. The d spacing obtained from the Bragg peaks defines the subcell within the crystal lattice (Small, 1966). Characteristic d values for each polymorph are illustrated in Figure 10.19. The α form has a peak at 4.10 Å, the β' form has two peaks at 3.8 and 4.2 Å, while β form has three peaks at 3.7,

3.9, and 4.6 Å (Larsson, 1994). In the analysis of the histogram shown in Figure 10.23, the d values obtained are 5.167, 4.562, 3.975, 3.8279, and 3.659 Å. The presence of peak at 3.65, 3.82, and 4.562 Å confirms that the predominant form is β.

Long spacings or small angles ($2\theta = 1°–5°$) are used to determine the longitudinal packing of the lamella. The d spacing obtained from the Bragg peak position corresponds to the distances between the lipid bilayers and yields information regarding the unit cell or lamella size. For the histogram shown in Figure 10.23, Bragg peaks positions in the small angle region correspond to d spacings of 44.20, 22.87, and 14.90 Å. The strong first peak, with the highest intensity, is the first-order reflection ($n = 1$ in Bragg's equation). The second peak at position 22.87 Å appears to be the second-order reflection ($n = 2$ in Bragg's equation). Notice at 44.20 Å that first-order reflection is roughly twice the 22.87 Å of the second reflection. Likewise, the third peak at 14.90 Å is the third-order reflection ($n = 3$ in Bragg's equation).

The longitudinal packing of the TAG molecules in the lamella is based on its constituent fatty acid (FA) alignment. The d spacing of the first-order reflection in the small angle region is used to resolve the packing of FA in the lamella by dividing the d spacing by the average FA chain length. As illustrated in Figure 10.24, two different longitudinal packing of TAG molecules conformations are possible. If the length of the longitudinal packing is equivalent to two or three FA chain lengths, it is possible to have two chain length (2L or chair) or three chain length (3L or tune fork) configurations, respectively. For example, when analyzing glyceryl tristerate, shown in Figure 10.23, one chain length (18 carbons) is equivalent to 21.59 Å. To calculate the length of a chain of 18 carbons, one needs to take into account that the carbon atoms in the acyl chain are joined in a zigzag form. Using a value of 2.54 Å (Small, 1966) as the distance between alternate carbons (the sequential or intermediate carbon is away from the line that joins alternate carbons to maintain the zigzag), the length of a chain of 18 carbons is equivalent this distance multiplied by the number of spaces between alternate carbons (e.g., 8.5). Dividing the d spacing

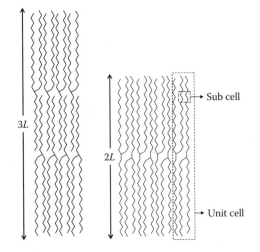

FIGURE 10.24   Schematic representation of a two (2L or chair) and three chain length (3L or tune fork) conformations of TAG longitudinal packing in a fat crystal network.

of the first-order reflection (44.20 Å) by the chain length of steric acid (21.59 Å), one determines that tristearin is in a $2L$ or chain conformation.

The thickness of a crystal or crystal domain, defined by many lamellae stacked together, can be estimated by performing the Scherrer analysis (Cullity and Stock, 2001). This analysis is only valid if the crystal thickness is less that 1000 Å. It should only be performed on a well-defined and symmetric peak in the small-angle region rather than peaks from the wide area region as peak broadening increases with the diffraction angle. The following equation is used:

$$TH = Thickness = \frac{0.9\lambda}{FWHM\cos\theta} \tag{10.9}$$

where
   FWHM is the full width half maximum of the peak of interest (i.e., the width of peak at 50% of the total peak amplitude) (in rad)
   θ is the angle of the center of the peak

The FWHM can be calculated with the instruments operating software or by with any peak analysis software. For glyceryl tristerate, shown in Figure 10.23, we chose to work with Bragg peak at $2\theta = 5.93°$, which makes $\theta = 2.96°$. The FWHM is $0.494°$ or 0.0086 rad. Since $\lambda = 1.54$ Å for a cooper tube, the thickness of the crystal is computed to be thickness is equal to 0.9×1.54 Å/$(0.0086$ rad$) \times \cos(2.96) = 160.96$ Å.

10.5 MICROSTRUCTURE

10.5.1 POLARIZED LIGHT MICROSCOPY

PLM is a well-established analytical technique used to image the microstructure of fat crystal networks by using polarized light rather than bright field illumination. PLM exploits the high contrast between the birefringent solid microstructural elements and the non-birefringent liquid fraction. This contrast is possible as PLM allows the distinction between isotropic and anisotropic materials. Isotropic materials, such as the liquid phase of a fat crystal network, have a single refractive index, hence the same optical properties in all directions. On the other hand, anisotropic materials, such as the solid phase of a fat crystal network, have optical properties that vary with the orientation of incident light (Robinson and Davidson).

The model of light describes electromagnetic waves vibrating at right angles to the direction of propagation with all vibration directions being equally probable. Polarizers are filters of polymer molecules oriented in a single direction, which restrict the propagation of light waves to a single plane, producing what is referred to as plane polarized light. The light waves that vibrate in the same plane as the polymer-oriented molecules are absorbed, while light waves vibrating at perpendicular angles to the direction of the polymer molecules pass through the polarizer (Dutch; Murphy et al.). This phenomenon is represented in Figure 10.25.

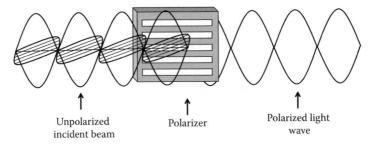

FIGURE 10.25 Schematic diagram of the path that a beam of light follows when it encounters a polarizer. (Adapted from Murphy, D. et al., Polarization of light, http://www.microscopyu.com/articles/polarized/polarizedlightintro.html)

A polarized light microscope is equipped with two polarizers. A first polarizer is positioned in the light path before the specimen to be imaged; a second polarizer, called the analyzer, is placed in the optical pathway between the objective and the observation tube. Anisotropic materials act as beam splitters, such that when light passes through an anisotropic crystal it is refracted into two rays, each polarized and refracted at right angles to each other and traveling at different speeds. This phenomenon of double refraction is called birefringence. The analyzer recombines the two beams traveling in the same direction and vibrating in the same plane. The polarizer ensures that the two beams are in phase at the time of recombination for maximum contrast (Abramowitz et al.).

10.5.1.1 Procedure

10.5.1.1.1 Microscope Alignment

Prior to any imaging or recording performed with the polarized light microscope (shown in Figure 10.26), all its optical components must be aligned. For optimum performance of the microscope, and hence good quality imaging, the light source has to be centered, and the condenser and field diaphragm should be properly adjusted. The following is a general procedure for the alignment of the optical components of the microscope.

To begin, switch the light source on and adjust the illumination intensity. Care should be taken not to switch on the light source at a high intensity as this reduces the life of the light bulb. A microscope with direct observation and recording capabilities (i.e., with a camera) will be equipped with a light path selector knob. The knob position selects the amount of light that will be directed to the observation tube or the phototube. For direct observation, the knob is positioned such that all of the light is directed to the binocular tube. To record images with a camera, the knob should be set at the position that directs light to the phototube, and hence the camera. For calibration and alignment of the microscope, the knob should allow all light to go to the binocular tube. Place a sample slide similar to that which will be imaged on the microscope stage and bring it into focus with the 10× objective. Close the iris diaphragm completely with the field iris diaphragm ring. A blurred image will be seen through the eyepiece. This is the image of the field, which is observed as a blurred polygon as depicted in Figure 10.27a. Move the condenser upward or downward with the condenser height adjustment knob to focus the image of the field diaphragm as

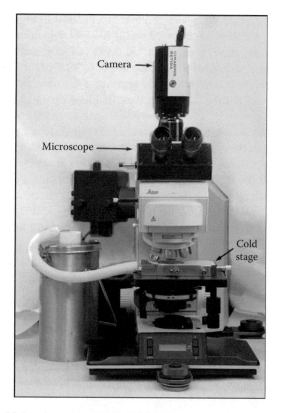

FIGURE 10.26   Light microscope (DMR XA2 Microscope, Leica Microsystems, Wetzlar, Germany) equipped with a polarizer, a camera for image acquisition (Qimaging Retiga, Burnaby, British Columbia, Canada), and a heating–freezing stage (Linkam Scientific Instruments, Surrey, United Kingdom).

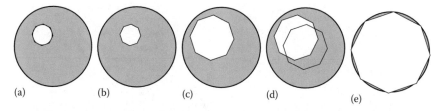

(a)   (b)   (c)   (d)   (e)

FIGURE 10.27   Schematic representation of the field of view in a microscope during illumination alignment. Initially the iris diaphragm is closed completely, a blurry polygon indicates that the condenser has to be adjusted (a) while a focused polygon indicates that the condenser is properly set (b). The diagraph is progressively centered and opened until it covers all the field of view (c–e).

observed in Figure 10.27b. The field iris diaphragm controls the diameter of the ray bundle impinging on the specimen surface increasing the image definition. Open the field iris diaphragm to increase the diameter of the field progressively until it is outside the field of view as observed in Figure 10.27c through e. Center the condenser while opening the diaphragm using the condenser centering screws to bring the diaphragm image into the center of view, as observed in Figure 10.27d. Adjust the aperture of the iris diaphragm based on the imaging conditions such as numerical aperture of the objective, image contrast, depth of focus, and flatness of field. It is suggested to adjust the aperture iris diaphragm to 70% or 90% of the numerical aperture of the objective. Remove an eyepiece from the observation tube and look through the observation tube. Open the aperture iris diaphragm ring until 70% or 90% of the field is covered.

Remove the sample slide from the microscope stage so that a transparent field of view is observed. Set the polarizer and analyzer at position "0," which is the crossed filter position. Rotate the polarizer ring until the desired extinction is obtained. Figure 10.28 demonstrates the difference between an image acquired with no polarization and under full polarization of light in which the microstructural features of the network appear clear and distinct from the liquid phase. Position the light path knob such that light is directed to the photometric tube, allowing the acquisition of images. Cover

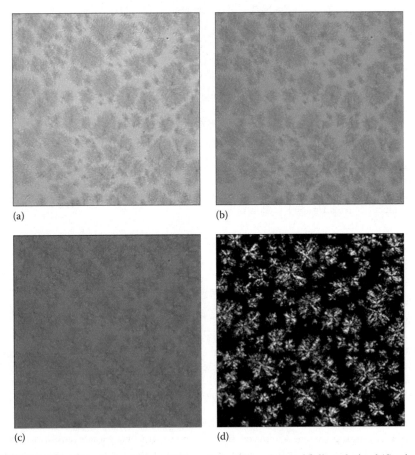

(a) (b)

(c) (d)

FIGURE 10.28 Unpolarized (a), partially polarized (b and c), and fully polarized (d) micrographs of cocoa butter crystallized directly onto the microscope slide by cooling from 80°C to 24°C at 5°C/min and isothermally storing at 24°C for 12 h prior to imaging.

the eyepieces so that no light will go through the binocular tube as this will adversely affect the illumination conditions. Finally, fine focus the sample. The aforementioned procedure is to be performed each time the microscope is turned on and for each objective used. Refer to the microscope manual of operation for further information.

When imaging is done at temperatures different from room temperature, when there is the need for minimum fluctuations in temperature or when a thermal treatment is to be applied on the sample, a heating/freezing stage as the one observed in Figure 10.29 can be used. The temperature-controlled stage is placed over the microscope stage and should be secured with the use of clamps to prevent movement during image acquisition. This should be done prior to the calibration of the microscope so that calibration is performed with the stage in place.

After the microscope optics and illumination have been aligned, it is strongly suggested to take a blank image (with a microscope slide and cover slip yet with no sample) to correct for any issues with uneven background. This blank image is used to remove by subtraction any imbalanced lighting or noisy background from every sample image subsequently taken. The user is advised to refer to the camera software manual for detailed instructions on how to subtract the blank image from each image taken.

10.5.1.1.2 Sample Preparation

In PLM, light passes through the sample and lenses, allowing a visualization of the its magnified structures. It is then essential that samples are not too dense or opaque. There are two approaches when preparing sample slides of fats. Fats can be sampled in liquid form and crystallized directly on the microscope slide, or they can be sampled in solid form after they have been crystallized in bulk. Both sample preparations are described.

To crystallize samples directly on a microscope slide, place a small droplet (about $10\,\mu L$) of melted fat on a clean preheated glass slide, using a preheated capillary tube or a glass pipette. Prior to sample preparation, melt the sample in an oven at a temperature that ensures that all crystal memory is erased. Special care should be taken to avoid any dust or debris on the microscope slide, as this may induce nucleation. Place a preheated glass cover slip over the sample to produce a film of uniform thickness, avoiding the introduction of air bubbles into the sample. The previously described procedure is illustrated in Figure 10.30. Transfer sample slides to temperature-controlled incubators set at the T_C of interest for the desired length of time.

Alternatively, when working with previously crystallized fats, with a small spatula, deposit a small amount of the fat onto the glass slide and place the cover slip on top. Gently press down the cover slip applying uniform pressure, until a semi-translucent film is obtained. During sample preparation, microscope slides and cover slips are pre-tempered at the same temperature at which the bulk crystallized fat was stored. After samples are prepared, transfer slides to temperature-controlled incubators set at the storage temperature of interest for the desired length of time.

10.5.1.1.3 Experimental Procedure

Prior to imaging, visually inspect the slides as only those which appear to have a uniform thickness of sample (i.e., no obvious directionality of the crystallized fat exists)

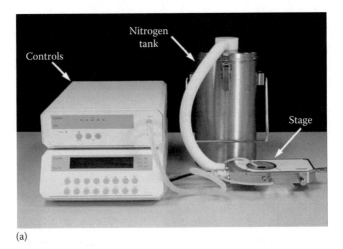

(a)

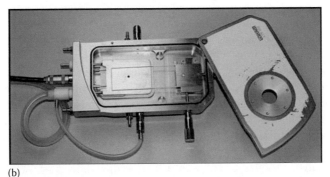

(b)

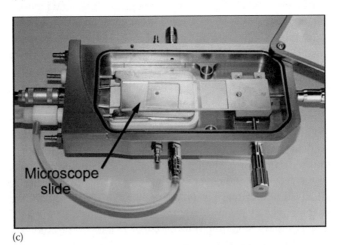

(c)

FIGURE 10.29 Heating–freezing temperature-controlled microscope stage (Linkam Scientific Instruments) used for precise temperature control of prepared microscope slides. Illustrated are components of the Linkam system: (a) electronic controls, a small tank for liquid nitrogen, and the stage. Close ups of the stage (b) (c) where the sample slide is place are shown.

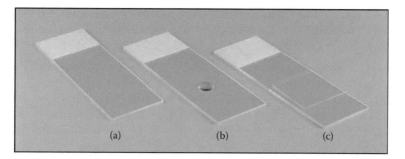

FIGURE 10.30 Illustration of sample preparation from melted fat for imaging of crystal networks by PLM. A clean microscope slide is preheated (a), a small droplet (about $10\,\mu L$) of fat is placed over the microscope slide (b), and a preheated glass cover slip is place over the fat to produce a film of uniform thickness (c).

should be analyzed. Transfer the prepared slide to a programmed heating freezing temperature control stage positioned on the microscope stage (Figure 10.29), which has been equilibrated to the temperature of study. Fine focus the sample and acquire the desired images according to the camera and software setup available. From this setup, grayscale images of the microstructure of the network will be obtained.

Care should be taken with regard to the gain of the camera. The gain of the camera can be automatic or manual. In automatic gain, the camera will automatically amplify the image signal to an appropriate level, whereas in the manual gain, no such adjustments are made. For further information regarding the operation of the camera and software refer to the appropriate operation manuals.

10.5.1.1.4 Data Analysis

Imaging is only the first step in the study of the microstructure of fat crystal networks. The acquired micrographs are nothing but raw data from which qualitative and quantitative information can be obtained through image analysis. When attempting to analyze the microstructure, the first thing to be done is to define what information is of interest. Image processing is then used to improve the visual appearance of images, enhance visibility of structures, and to prepare them for quantitative measurements (Russ, 1999), depending on the features of interest. The correction of any defects in the images during acquisition and the enhancement of visibility of structures are performed on the acquired images. Qualitative analysis can be made when there are obvious and dramatic differences between the micrographs, such as the presence or absence of certain features, as well as large differences in morphology and size. For quantitative analysis, descriptive numeric parameters are obtained that concisely characterize the information of interest in the image. In this section, basic image processing tools such as correction for nonuniform illumination and thresholding will be described, as well as quantitative image analysis techniques such as those used to determine the size and number of features, percent black which is related to the solid fraction of the network, and the calculation of the fractal dimension. For both image processing and analysis, several software packages are available. Unless otherwise noted, this text will describe treatment and analysis performed with Adobe Photoshop 6.0 (Adobe System Incorporated) and the Fovea Pro Image Processing Tool Kit 4.0 plug-ins (Reindeer Games, Inc., Asheville, NC).

10.5.1.1.4.1   Automatic Background Leveling   When the background of a micrograph, which is either darker or lighter than the microstructural features, varies gradually due to nonuniform illumination at the time of image acquisition, it can be corrected by automatic background leveling. Using the function Autolevel Bright in either the Filter > IP • Adjust, the background is modeled by a polynomial function. The image is divided up into a 9 × 9 grid of rectangles, the histogram of each region is examined to find the brightest or darkest values present, and uses the 81 values and their locations to construct a polynomial for brightness as a function of position, which is subtracted from the original image. When the background is the darkest phase present in the image, the function Autolevel Dark is used. In cases when the image contrast varies across the image as well as the background brightness, the function Auto Contrast is used which fits functions to both the brightest and darkest pixels in each region of the image, maximizing the contrast range in all areas of the image (Russ, 1999). In Figure 10.31a, an example of an image acquired under uneven illumination conditions is shown. The result of image processing using the automatic background leveling tool in the defective image is observed in Figure 10.31b. Regardless of this function, extreme care should be taken in the calibration of the instrument to avoid an uneven illumination when imaging.

10.5.1.1.4.2   Thresholding   The purpose of thresholding is to convert the acquired grayscale images (i.e., images which contain a wide array of gray tones that go from pure white to pure black) to binary images (i.e., images merely in black and white). Thresholding is necessary for further image analysis as the algorithms used rely on the discrimination between features and background. In the case of polarized light micrographs of fat crystal networks, the background is set to black and the fat crystals is set to white. The resulting binary image is used for subsequent image analysis.

The function BiLevel in the menu Filter > IP • Adjust is used to threshold grayscale images. When used, a dialog box like the one shown in Figure 10.32 appears. In the

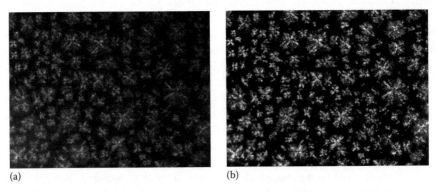

(a)   (b)

FIGURE 10.31   Micrograph of a cocoa butter samples crystallized at 24°C with uneven illumination (a), which is processed using Photoshop (Adobe Systems Incorporated) with the automatic background level function to obtain an image with a uniform background (b).

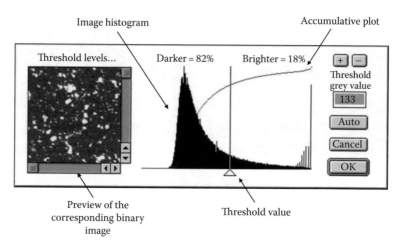

FIGURE 10.32 Dialogue box used to threshold a grayscale image. It contains an image histogram of the number of pixels at each brightness level, a preview of what will result in a binary image, and controls used in the determination of threshold values. (Adapted from Photoshop 6.0, Adobe System Incorporated, San Jose, CA.)

center of the dialog box, the image histogram is presented, which is the number of pixels at each brightness level, along with an accumulative plot. Brightness values of the histogram range from 0 representing pure white to 255 representing pure black. The cursor beneath the histogram can be moved to manually select a particular brightness value within the range of 0–255. On the left side of the dialog box, there is a preview window, in which any part of the corresponding binary image can be screened using the scroll bars. This is used to ensure that the chosen thresholded value represents the original image. The dialog window also displays the area fraction of each phase and the threshold value. An image can be thresholded manually by moving the cursor in the histogram or by entering a threshold value. It is important to ensure that all the solid mass present in the grayscale image is represented in its correspondent thresholded image. This process relies on the human eye to decide what features should be included and measured and thus is inherently subjective and inconsistent. To overcome this problem, a threshold value can be set automatically by the software, by choosing an "Auto" threshold in the dialog box. On each portion of the histogram, the significant differences are evaluated statistically (t-test) by the software. The automatic threshold value is the point of brightness which has the highest probability of significant difference. This automatic threshold value is different for each image and independent of the user. In this way, the subjectivity from the user's input is decreased.

Typically, measurement routines will count or identify black features on a white background. This binary color scheme is the opposite to that obtained when thresholding a fat crystal network imaged by PLM. For this reason the image has to be inverted by using the invert function in the Image > Adjust menu. It is a straightforward function, which interchanges black and white. Figure 10.33 shows an example of a grayscale image (Figure 10.33a) which is thresholded to produce a binary image (Figure 10.33b) and inverted (Figure 10.33c) as to obtain an image of black features on a white background.

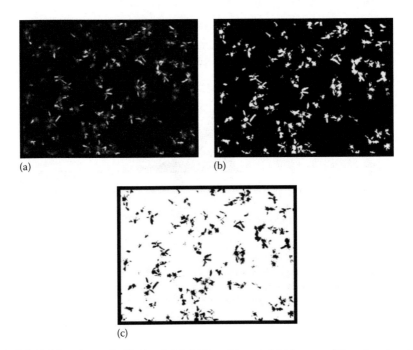

(a)   (b)

(c)

FIGURE 10.33 Grayscale (a), thresholded (b), and inverted (c) images of the microstructure of a crystallized cocoa butter.

10.5.1.1.4.3 Numeric Parameters That Characterize Microstructure Parameters of crystal microstructures such as size (area, length, breadth, and perimeter) or shape (topology, dimensionless ratios, and fractal dimension) are calculated from thresholded and inverted images using image analysis software. Prior to doing any quantitative image analysis, it is necessary to calibrate the magnification scale used by the software. For this, take an image of a microscope scale, usually a 0.1 mm segment divided in 100 divisions. The imaged microscope scale needs to be taken with the same optic alignment as the rest of the images. This scale will be used by the analysis software to measure different microstructural features present in the sample. Using the function IP • Measure Global → Calibrate Magnification, select two points or divisions on the microscope scale image, enter the distance between them and select the units. The software will then count the number of pixels between the two selected points on the microscope scale and equate that to the entered distance. An image of a microscope scale is used as it contains precise linear distances and their corresponding units. Calibrations are saved and recalled as needed. The magnification calibration is not tied to any specific image. It is rather a system constant that stays in effect for all measurements performed.

After the magnification has been calibrated, open the sample image files. Select the function IP • Measure Features > Measure All Features from the filers menu. The function "Measure All Features" returns a data file with many columns of information for each crystal that was accounted for (i.e., area, filled area, length, breadth, equivalent diameter, circumference radius, perimeter, aspect ratio, fractal dimension, etc.).

This function measures all parameters for all crystal features imaged. The data file is a tab-delimited ascii text that can be read by most spreadsheet or data analysis programs. Through simple statistical analysis, the user can characterize the microstructure using descriptive statistics (i.e., average, mean, 25th and 50th percent quartiles, confidence intervals, standard deviation and standard error) of parameters such as equivalent diameter, radius, and area. The number of particles (N_p) is determined as the total number of rows, which corresponds to the total number of crystals or features counted. Distributions on any parameter measured (e.g., crystal size distribution) can be constructed by plotting the number of features of predefined ranges as a function of such ranges. For example, a histogram of the crystal size distribution results from the plot of the number of particles with equivalent diameter up to 5 µm, between 5 and 10 µm, between 10 and 15 µm, between 15 and 20 µm, and so forth until the biggest crystal size is covered. The function IP • Measure Features → Plot (distribution) can also be used to determine distributions.

An example of image analysis is shown in Figure 10.34 where for cocoa butter cooled under static and dynamic (average shear rate of 120 s^{-1}) conditions. In addition to the qualitative observations that can be made from the micrographs, quantitative information is obtained from the average particle size and particle size distribution holograms. Additional qualitative analysis includes the number of particles, average area of particles, and average distance between neighbouring particles.

10.6 MECHANICAL PROPERTIES

The mechanical properties of a fat crystal network are determined by various factors, which include different levels of structure, such as chemical composition, SFC, and crystal habit. Through mechanical testing, the user can obtain experimental evidence related to different levels of structure and their effect on functionality of fat crystal networks. The analysis of the mechanical properties examines how the crystallized material responds to either an applied force (stress) or to deformations (strain) (Daubert and Foegeding, 1998).

10.6.1 SMALL DEFORMATION RHEOLOGY

Fats are known to behave like rigid solids until a deforming stress exceeds the yield value at which point the product starts to flow like a viscous liquid (de Man and Beers, 1987). Small deformation rheology refers to testing procedures that do not cause structural damage to the sample. Typically, a sinusoidal stress is applied to the sample causing some level of strain to be transmitted through the material. The magnitude and time lag of the transmission depends on the viscoelastic nature of the test substance. If the material is more viscous or liquid-like, much of the stress is dissipated in frictional losses. In more elastic or solid-like materials the stress is mostly transmitted.

At small applied stresses, fat crystal networks behave like Hookean solids, in which the stress (σ) is directly proportional to the strain (γ) as indicated by the following equation:

$$\sigma \propto \gamma \qquad\qquad (10.10)$$

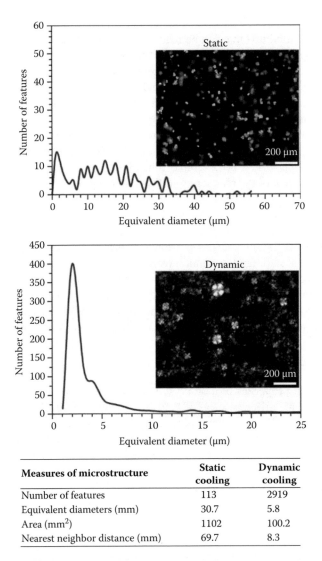

Measures of microstructure	Static cooling	Dynamic cooling
Number of features	113	2919
Equivalent diameters (mm)	30.7	5.8
Area (mm²)	1102	100.2
Nearest neighbor distance (mm)	69.7	8.3

FIGURE 10.34 Example of quantitative image analysis for two samples of cocoa butter that were cooled either statically or dynamically (average shear rate of $120\,s^{-1}$) from 60°C to 24°C/min at a rate of 1°C/min and held isothermally at 24°C. The microstructure was imaged under polarized light 1 min after the cocoa butter has reached 1.5% SFC (determined by pNMR).

It is possible to see the linear relation between the strain and stress on Figure 10.35. The linear viscoelastic region (LVR) goes from the origin to the yield point A. It is in the LVR that a material has the property to deform and recover to its original state, and thus, small deformation experiments are carried out in this region. It is at the yield point A where the relation between the stress and strain stops being linear. This point is called strain at the limit of linearity or stress yield point. The elastic

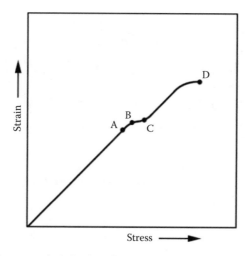

FIGURE 10.35 Stress–strain behavior of a typical elastic system. Indicated are the yield point (A), elastic limit (B), irreversible deformation (C), and fracture (D).

limit quickly follows (point B), where permanent deformation (point C) and sample facture occurs (point D) (Rohm and Weidinger, 1993). Outside the LVR, parameters like the yield stress and yield strain are calculated.

Oscillatory tests are used to examine viscoelastic materials. They are also referred to as small amplitude oscillatory testing because small deformations must be employed to maintain linear viscoelastic behavior. Depending on what stress is applied and how the strain is measured, different moduli (ratios of stress to strain) can be determined. For example, when shear stress is applied, the shear modulus (G) is defined as the ratio of shear stress (σ) to shear strain (γ). Useful parameters that are obtained through small deformation rheology to describe a material include the storage modulus (G'), loss modulus (G''), phase angle (tan δ), and complex modulus (G^*). The method consists of the application of a stress or strain in a oscillatory fashion, where the material is subjected to a sinusoidal perturbation, either in the form of deformation (for controlled rate instruments) or stress (for controlled stress equipment).

Figure 10.36 shows typical stress–strain sinusoidal relationships, in which the resultant strain can be in phase or out of phase with the applied stress. When the sample is an ideal (or purely elastic) solid, the maximum strain occurs when the maximum stress is applied. The stress and strain are said to be in phase. If the material is purely viscous, the stress and strain are out of phase by 90°. δ is the phase lag or shift between the applied stress and the resulting strain curves. δ is also known as loss angle, given in degrees (°). Viscoelastic materials exhibit a behavior that lays somewhere in between purely viscous and purely elastic extremes, where δ is always between 0° and 90° (Steffe, 1996).

Both storage (G') and loss (G'') moduli depend on the phase angle based on the following relationships:

$$G' = \left(\frac{\sigma_o}{\gamma_o}\right)\cos(\delta) \qquad (10.11)$$

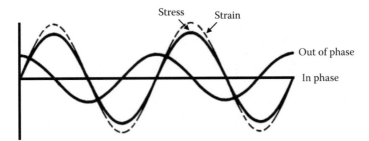

FIGURE 10.36 Applied oscillatory sinusoidal stress wave (dotted line) and the sample's corresponding resulting sinusoidal strain wave (continuous line) as they apply to small deformation rheological testing.

$$G'' = \left(\frac{\sigma_o}{\gamma_o} \right) \sin(\delta) \qquad (10.12)$$

where
  σ_o is the applied stress
  γ_o is the resulting strain

Both moduli have units of Pa. G' is a measure of the deformation energy stored in the sample during the shear process. It is the energy available to act as driving force to compensate for the applied deformation (Mezger, 2002). G' represents the elastic behavior of a sample. G'' is a measure of deformation energy used up in the sample during the shear process. This energy is lost or dissipated during the deformation and hence the material displays irreversible deformation behavior (Mezger, 2002). G'' represents the viscous behavior of a sample. In fat crystal networks, G' has been shown to be related to the hardness and strength of the fat crystal network while G'' has been shown to be related to the spreadability of the fat crystal network.

Both G' and G'' are functions of frequency that can be expressed in terms of the phase shift and the amplitude ratio between the shear stress and strain. The ratio of these two moduli is a common material function used to describe viscoelastic behavior:

$$\tan(\delta) = \frac{G''}{G'} \qquad (10.13)$$

$\tan(\delta)$ is the quotient of the energy lost per cycle divided by the energy stored per cycle, in other words, the ratio of the viscous to the elastic portion of the deformation behavior.

The vectorial resolution shown in Figure 10.37 of the stress–strain ratio illustrates the sum of the viscous and elastic responses in viscoelastic behavior, from which G^* is calculated using G' and G'' from the following equation:

$$G^* = \sqrt{G'^2 + G''^2} \qquad (10.14)$$

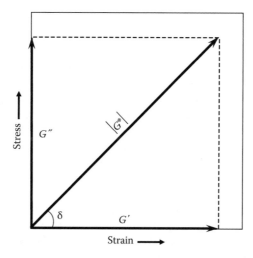

FIGURE 10.37 Vectorial resolution of the complex modulus components. Indicated are the complex modulus (G^*), storage modulus (G'), loss modulus (G''), and phase angle (δ).

10.6.1.1 Procedure

10.6.1.1.1 Sample Preparation

For mechanical testing of fat crystal networks, molten samples are crystallized into disks of uniform dimensions. In out laboratory, disks of solid fat are prepared by crystallizing the samples directly in PVC molds, shown in Figure 10.38. The molds are comprised of a lower, middle, and upper plate. The lower plate contains six screws, which connect the other two sections forming a tight seal. The middle plate serves as

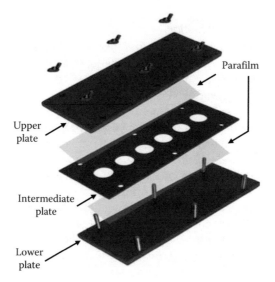

FIGURE 10.38 PVC molds used to prepare fat sample discs of uniform dimensions used in rheological studies.

spacer, containing six screw holes as well as six perforations to be filled with molten sample. The upper plate contains six holes for the screws and fasteners to seal the three sections together. For small deformation rheology, the perforations have a diameter of 20 mm and a height of 3 mm. Between each of the PVC sections, a layer of Parafilm "M" (Fisher Scientific, Pittsburg, PA) is used to create a tight seal, preventing leakage of the molten fat. Wing nuts are used to tightly hold the three sections together.

Prior to sample disk preparation, equilibrate the three parts of the mold and strips of Parafilm to the temperature of study. Assemble the PVC mold as follows: directly place a strip of Parafilm over the lower plate; then, position the middle plate over the Parafilm, aligning the holes along the edge of the middle plate with the corresponding screws on the lower plate; next, carefully lower the middle plate onto the Parafilm, as illustrated in Figure 10.38. Melt the sample in an oven at a temperature that ensures that all crystal memory is erased. Allow sample to cool to approximately 60°C, as temperatures above 60°C will melt the Parafilm. Using a pre-tempered pipette, fill the mold wells until the liquid fat forms a convex surface above the level of the spacer. Place a second strip of Parafilm over the convex surface of the samples. It is critical that no air bubbles are formed between the liquid fat and the strip of Parafilm. Presence of air bubbles on the crystallized samples will cause uneven contact between the sample and instrument's probe surfaces, leading to inaccurate results. To prevent the trapping of air bubbles between the sample surface and the Parafilm, it is recommended to carefully lay the strip of Parafilm, starting slowly at one edge and leading to the opposite edge. If air bubbles form, remove the Parafilm strip, replenish the mold wells with melted sample and cover with a new strip of Parafilm. Finally, place the upper plate on top of the Parafilm and use wing nuts to tightly hold the three sections of the mold together. Transfer molds to temperature-controlled incubators set at storage conditions of interest for samples to crystallize.

10.6.1.1.2 Instrument Calibration

A rotational controlled stress rheometer, like the AR 2000 Shear Dynamic Rheometer (TA Instruments, Mississauga, Ontario, Canada) shown in Figure 10.39, applies an oscillatory sinusoidal stress and records the strain. It consists of a main unit mounted on a cast metal stand, with the electronic control circuitry contained within a separate electronics control box. The main unit has an arm that moves up and down, containing the motor, an optical encoder, and the geometry. The sample holder is on the base of the main unit and includes a Peltier temperature-controlled system connected to a water bath.

At the beginning of any experimental run, turn on the water bath attached to the Peltier system, initiate air flow to the bearing, remove the bearing lock to allow free rotation of the spindle, initiate the computer controller that will trigger a number of system checks, and finally open the control/operation software, ensuring that communication is established between the computer and the instrument.

Before proceeding with the experiment, the user must perform a series of daily checks covering the following: instrument inertia, geometry inertia, bearing friction correction, temperature system selection, mapping of rotational torque, gap zeroing, and gap compensation. These are not to calibrate the instrument, but rather to reset and define

FIGURE 10.39 AR 2000 Dynamic Shear Dynamic Rheometer (TA Instruments, Mississauga, Ontario, Canada).

experimental parameters. The relevance of each of these checks for the TA Instruments' AR 2000 rheometer is described further in this section. The user should review specifics on how to proceed with each check on its own instrument controlling software.

Strictly speaking, the user cannot calibrate the rheometer. Calibration of the instrument is done by a qualified technician from the instrument's maker during maintenance service. On a monthly frequency, it is recommended to verify the instrument's readings by measuring the viscosity of a certified standard Newtonian oil (PTB 1000A) with a nominal viscosity of 1 Pa·s and a yield stress of 0 Pa at 20°C. A 4 min flow test, continuous ramp, controlled stress range from 0 to 88 Pa at 20°C, using a 60 mm 2° stainless steel cone and plate (CP) geometry is suggested. If the resulting Newtonian viscosity is more than 4% of the reference standard value, service maintenance needs to be scheduled.

10.6.1.1.2.1 Instrument Inertia In a rheometer the torque output of the motor is composed of the torque required to overcome the instrument inertia and the torque required to deform the sample. Instrument inertia refers to acceleration or deceleration of the instrument's motor shaft, which can lead to corruption of oscillation readings. The instrument inertia check addresses the torque associated with the instrument, not with the sample. It is critical that when performing this check the air flow to the bearing is optimal (i.e., adequate flow and unblocked). In the case of the TA Instruments' AR 2000 Rheometer, the acceptable range is in the range of $14–16\,\mu Nms^2$.

10.6.1.1.2.2 Geometry Inertia The rheometer's measuring system consists of a static (e.g., Peltier plate) and a moving component. The moving component is the geometry, which is attached to the driving motor spindle. When geometry is in place, it will have a defined inertia. The inertia for each geometry (i.e., measuring system) is determined by the instrument and corrected. It is critical that when performing this check, the specific geometry to be used during experimentation is in place.

10.6.1.1.2.3 Bearing Friction Correction An air bearing is used to provide virtually free friction application of torque to the sample. However, there will be some residual friction. The bearing friction correction compensates for any residual friction.

10.6.1.1.2.4 Mapping of Rotational Torque The air bearing will have small variations in behaviour around one revolution of the shaft. Mapping of such rotational torque stores variations and uses this for real-time corrections when performing measurements. To create the map, the software rotates the drive shaft at a fixed speed, monitoring the torque required to maintain this speed through a full 360° rotation.

10.6.1.1.2.5 Zeroing the Gap Zeroing the gap defines that the distance gap between the geometry and the Peltier plate is zero when the geometry touches the plate in the absence of a sample. Due to thermal expansion and contraction of the geometry's material, it is essential that zeroing the gap is done at the testing temperature. The gap needs to be zeroed after each sample is measured. There are two approaches to zeroing the gap: normal force (set to 1 N) or deceleration. Using normal force, the instrument will define the gap as zero when it reads a normal force of 1 N unto the geometry. When done by deceleration, the gap is set to zero when the rotating geometry comes to a stop upon contact with the base.

10.6.1.1.2.6 Temperature Gap Compensation When running an experiment over a temperature ramp range, small variations in the gap are possible due to thermal expansion of the geometry. Such variations need to be corrected. Refer to specifics on how to ensure that the gap is compensated as a function of temperature in the instrument's controlling software.

10.6.1.1.3 Experimental Procedure

An experimental run is comprised of three steps: sample conditioning (pretest), test, and posttest. Parameters for these steps are entered in the instrument's operation software. The conditioning step allows the sample to equilibrate to the testing temperature and to a normal force after loading onto the instrument. A normal force of 4 N is recommended for crystallized fats. For the calculation of G', G'', and $\tan(\delta)$, oscillation testing is done in the form of a frequency or stress sweep. The test involves ramping the frequency or stress between a minimum and maximum value. The exact minimum and maximum limits are found by trial and error. Following are recommended starting points. For frequency sweeps, a range of 1–1000 Hz is recommended, keeping one of the following parameters constant: torque (μN·m), oscillation stress (Pa), displacement (rad), or %strain or strain. For stress sweeps, it is recommended to run the sample from 0 to 30 kPa at a constant frequency (1 Hz is recommended). All oscillation runs need to be run within the LVR, in which stress is proportional to strain (Equation 10.10). To determine the LVR, a stress sweep is run. The region at which G' and G'' are linearly constant is considered as the LVR. After the sample is run, the sample is kept at a given temperature in the posttest step.

10.6.1.1.3.1 Measuring System As mentioned previously, the complete measuring system consists of the static Peltier base and a geometry that moves. Geometries are constructed of stainless steel, aluminum, or acrylic materials in flat plates or cone shapes. The complete measuring system is then referred to as cone and plate (CP) or parallel plates (PP), where the fixed component is the base plate and the moving component is the cone or plate geometry, respectively. The user must choose the best measuring system to use based on the material to test and the analysis of interest.

Following are general recommendations on what geometry to use for different sample systems. For low-viscosity materials, it is common to use larger diameter geometries (40–60 mm), while smaller diameters (10–20 mm) are used for low-viscosity materials. CP are recommended for liquid and dispersions with particles size less than 5 μm, and PP geometries are used for gels, pastes, soft solids, and polymer melts. In the study of crystallized fat disk samples, a 20 mm diameter stainless steel plate geometry is used. Given that crystallized fats can easily slip during oscillation, sand paper is typically glued to both the geometry and Peltier surfaces to create friction and prevent slippage. A 60-grit aluminum oxide sand paper is commonly attached with epoxy glue (such as Instant Krazy® Glue). A circle of sand paper of identical dimensions as the geometry plate is used for the geometry. For the Peltier plate, it is recommended to cut a circle that is slightly larger than the diameter of the geometry to ensure that it covers the entire sample disk area. After measurements are done, acetone is used to remove sand paper from the measuring system.

10.6.1.1.3.2 Sample Loading After the measurement system is ready (i.e., geometry in place and sand paper glued to both Peltier plate and geometry), all instrument checks are done, and all experimental parameters are entered, place a disk of crystallized sample over the Peltier plate. It is recommended that the Peltier plate is preset at the same temperature at which samples where crystallized and stored. It is critical that the disk is centered so that the geometry will exactly coincide with the sample disk. Lower the geometry until there is mechanical contact with the sample disk. This is done through the instrument's controls. Mechanical contact is standardized either by establishing a normal force or a gap size. Work in our laboratory suggests that using a normal force of 4–5 N provides more accurate results. If using a normalized gap size, compression of not more than 10% of the sample dimension is recommended. For sample disks that are 3 mm high, chose a gap of no more than 2700 μm. Figure 10.40 illustrates how the sample is placed between the Peltier plate, with the sand paper already attached and under the PP geometry.

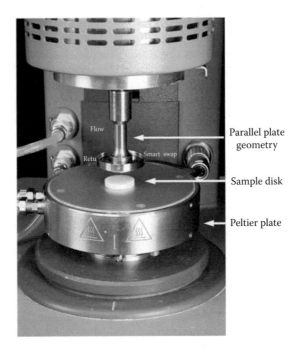

FIGURE 10.40 Position of the precrystallized fat sample disk between the geometry (20 mm diameter parallel place) and the Peltier.

After each sample is run, the sand paper on both the geometry and the Peltier need to be replaced. Generally, six repetitions are recommended for each sample to ensure the most accurate results possible. A minimum of three repetitions is advised.

10.6.1.1.4   Data Analysis

The instrument's operating software collects the data and calculates G', G'', and $\tan(\delta)$ as a function of frequency or stress. Figure 10.41 shows a typical result of G', G'' and $\tan(\delta)$ as a function of the applied shear oscillatory stress for cocoa butter. The constant value for G' and G'' is observed for up to roughly 3000 Pa, which corresponds to the LVR. The vertical dashed line indicates the strain at the limit of linearity and corresponds to the point where G' deviate 1% from the previous constant value. Past the yield point, G' starts to decrease while G'' increases. Throughout the studied range of stress, $G' > G''$ indicating that the elastic behavior dominates the viscous as a result of rigidity in the sample.

10.6.2   LARGE DEFORMATION TESTING

Large deformation testing is based on the deformation of a sample under a constant force rate, to the point where the force exceeds the structural capacity of the sample causing it to permanently deform and break (Wright et al., 2001). By measuring the amount of applied force required to induce a change in the sample, a representative measurement of hardness, spreadability, cutting force, or yield force is obtained

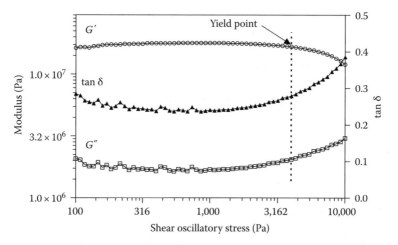

FIGURE 10.41 Storage modulus (G'), loss modulus (G''), and tan(δ) as a function of shear stress for cocoa butter. Indicated is the yield point, which corresponds to the strain at the limit of linearity.

(Wright et al., 2001). One such method involves the compression, by the application of uniaxial parallel force, of a sample between two PPs to determine relative hardness values via the measurement of yield force. The yield force will depend at least on two parameters, the rate of the applied stress and the load. Figure 10.42 shows a typical load–deformation curve, which can be used to derive values for yield stress,

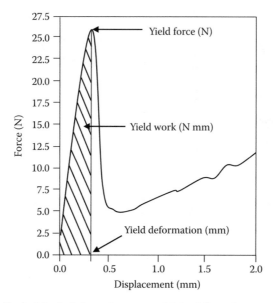

FIGURE 10.42 Typical load–deformation curve obtained from the compression of a sample between two PPs used to characterize its mechanical properties through the following parameters: yield force, displacement at yield force, compressive yield work, and compression modulus.

yield strain, compressive yield work, and depending on the linearity of the onset of compression a compressive modulus may be obtained (Wright et al., 2001). These measurements can be used to provide an index of hardness for fats, which have been successfully correlated to the textural attributes of hardness and spreadability obtained through sensory evaluation (Wright et al., 2001). Unfortunately, these tests are destructive in nature and yield little information about the native microstructure of the system.

10.6.2.1 Procedure

10.6.2.1.1 Sample Preparation

Prepare samples disks by crystallizing the fat sample in PVC molds to ensure uniform dimensions, in the same manner as for small deformation rheology. In the case of large deformation studies, the dimensions of the disks are of 10 mm diameter and 6 mm height. As with small deformation rheology testing, samples are transferred to temperature-controlled incubators set at the desired temperature of study for the period of time on interest for the analysis. Storage of the crystallized samples at 5°C for 24 h is recommended; however, the time–temperature profile may be changed by the user according to the objectives of each study.

10.6.2.1.2 Instrument Calibration

Figure 10.43 shows an example of a Stable Micro System Materials Tester Model MT-LQ (Stable Micro Systems, Surrey, England). It is equipped with a 50 kg load cell calibrated using a 10 kg standard weight. A 34 mm flat plate geometry is used to measure

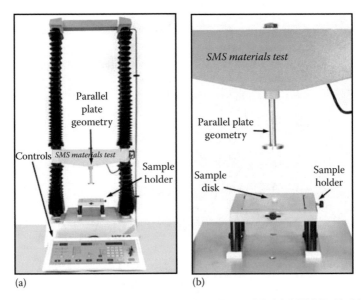

(a)   (b)

FIGURE 10.43 Stable Micro System Materials Tester Model MT-LQ (Stable Micro Systems, Surrey, England), with a 34 mm flat plate geometry used for large deformation rheological analyses (two-plate compression) (a). A close up of the sample holder, and geometry is shown (b).

the apparent yield force. The sample holder is a fixed plate attached to a recirculation water bath in order to maintain the surface of the plate at a constant temperature. The instrument is controlled, and data are compiled by the appropriate software package.

Two calibrations are performed by the instrument: force calibration and probe position calibration (Stable Micro Systems, 1997). For force calibration, a standard weight is used. The instrument will initially analyze the force signal present without any weight exerting force on the cross arm's load cell. Consecutively, the weight is placed over or hung from the instrument cross arm (depending on the model) and the force signal is set to the corresponding weight. Care should be taken that nothing is interfering with the force signal that the load cell in the cross arm is recording.

The probe calibration procedure is used to calibrate the probe position. This will allow the user to control the position of the probe accurately through the instrument operational software, and will enable the measurement of product height to make strain (displacement) calculations. To perform the probe calibration, attach the selected probe geometry to be used (i.e., 34 mm flat plate geometry) and make sure that the probe is clear of any obstruction. Bring the probe close (roughly 10 mm) to the lower plate. Set a value for probe return displacement (e.g., 20 mm) and run the calibration. The instrument will calibrate the probe position by making the tip of the probe touch the lower plate so it can reset the travel displacement to zero, followed by the probe return to the previously established distance. All movement of the probe is tracked. This calibration is required every time the geometry is changed.

10.6.2.1.3   Experimental Procedure

Make sure that both the lower plate and the flat plate geometry are clean prior to usage. Between measurements, it is recommended to clean with ethanol to ensure that all fat residues are eliminated. Set the recirculation water bath at the temperature of study and allow it to equilibrate. Establish the test parameters through the instrument's software. Material testers can be operated in two different modes: compression and extension. The later is used when testing the resistance of a sample upon elongation or extension. Compression mode is used in the study of fats and oils. Experimental conditions include geometry speed (pretest, test, and posttest) and target (how much deformation will be applied onto the sample). The target deformation can be as distance traveled by the geometry (in mm), strain applied (in %), or as a given force (in g). The last key parameter to enter when setting up a method is the trigger; this is the point when the instrument will start to gather data. The trigger can be set to when the method begins (button), after the geometry has traveled a specified distance (pre-travel in mm), or to the point where the geometry touches the sample and reads a specific force (called Auto, in grams). The latter is recommended and typically set to 5 g. A list of recommended parameters based on experimental work performed in our laboratory is shown in Table 10.4.

Once the experimental method is defined and loaded, transfer a sample disk to the lower plate of the material tester and begin the run. The instrument will perform the experiment based on the preestablished parameters and will collect data accordingly.

10.6.2.1.4   Data Analysis

During the experiment the instrument compiles the data, and makes them available in the form of a load–deformation curve. Analysis functions are included in the

TABLE 10.4

Test Parameters Used in Large Deformation Testing

Parameters (Units)	Recommendations
Test mode	Compression
Pretest speed	5 mm/s
Test speed	10 mm/s
Posttest speed	5 mm/s
Target	Distance
Displacement	5 mm
Trigger	Auto
Trigger force	5 g

Note: Included are recommendations of the experimental conditions based on experimental work done in our laboratory, yet they have to be determined by the user based on the analysis objectives and sample characteristics.

instrument's software. Yet, if the user wants to perform further analysis using a plotting software (i.e., Graph Pad, Excel Microsoft, MatLab, SigaPlot, Origin, etc), the data are also available in a spreadsheet form.

Several measurements characterize large deformation rheology and are obtained from the load–deformation curve (Figure 10.42). The yield force or yield stress is measured as the force measurement at the apparent breaking point or yield point of the sample discs. Furthermore, the yield work corresponds to the area under the curve to the yield point, and the yield strain corresponds to the ratio of geometry displacement at the yield point to sample height.

10.7   FRACTAL DIMENSION

Fractal objects have an infinite sequence of structural elements that remain unchanged upon assessment on different scales (Manderlbrot, 1989; Avnir et al., 1998). In other words, they are self-similar, as their characteristic features are repeated at different magnifications. The whole of a self-similar object has the same property as one or more of its parts. In addition to being self-similar, fractals are objects for which it is not possible to obtain a congruent measurement of a particular structural feature (i.e., volume, surface, length, mechanical strength) upon changing the magnitude of the measuring technique (Vicsek, 1992). A classical example being the measurement of the length of England's coastline yielding different values when rulers of different sizes are used. Fractal geometry is used to describe the irregular, almost chaotic, shapes of fractal objects, which cannot be described through Euclidian geometry.

Fractal structures are created by a consecutive aggregation process to form larger objects in a random, iterative fashion. Fat crystal networks are formed by the aggregation of primary crystallites forming clusters which further aggregate until they form a 3D network. The microstructure of fat crystal networks has then been described in terms of fractal geometry (Marangoni, 2005b; Narine and Marangoni, 2005). The characteristic that a specific property (e.g., length, mass, area, force) of fractal objects scales in a power-law fashion within a length scale is used to calculate the fractal dimension of fat crystal networks. Fractal theory predicts that the mass of a fractal aggregate (or distribution of mass within the system) (M) is related to the size of the object (or region of interest [ROI]) (R) in the following power-law fashion:

$$M \sim R^D \tag{10.15}$$

where D is the fractal dimension of the object or the distribution of mass within a region of the network. Three methods to determine the fractal dimension of fat crystal networks are described: particle counting (D_f), box counting (D_b), and physical (rheological) (D_r).

10.7.1 PARTICLE COUNTING METHOD TO DETERMINE FRACTAL DIMENSION

The determination of the fractal dimension by the particle counting method (D_f) uses PLM images and is based on the following relationship:

$$N_p \sim R^{D_f} \tag{10.16}$$

where N_p is the number of reflections or "particles" in a ROI of R size. D_f is calculated by counting the number of crystal reflections or "particles" (N_p) observed within boxes of increasing length size (R) placed over a properly thresholded and inverted image. The value of D_f is obtained from plotting the logarithm of N_p as a function of the logarithm of R, where the slope gives the fractal dimension (Narine and Marangoni, 1999; Marangoni, 2002). The particle counting method is sensitive to the degree of order and packing of mass distributed in a fat crystal network.

The first step in the calculation of D_f by particle counting is to convert the grayscale images to binary images through proper thresholding, making sure all the solid features are represented. An inversion of the thresholded images is necessary, as black features over a white background are counted. An algorithm in Object Image 2.01 software (http://simon.bio.uva.nl/object-image.html) is used in our laboratory to calculate the number of individual features (N_p) which correspond to crystal reflections in a defined ROI. Initially, the particles in the entire image are counted. Sequentially, the ROI is reduced by 5% from all the edges resulting in a smaller ROI from which N_p values are obtained. It is important to note that the values of ROI correspond to area values. The square root of the ROIs, which correspond to R, are used in the calculations of D_f. Once the numbers of particles in each box are obtained, a log-log plot of N_p vs. R is constructed using a standard plotting software package (e.g., Excel, Graph Pad Prism, Sigma Plot, etc.). The slope of the resulting line corresponds to the fractal dimension of the network. Figure 10.44 illustrates the particle counting procedure used to determine D_f.

When calculating the fractal dimension with the particle counting method, there are two important things to consider: the range of sizes of the ROI and the inclusion or

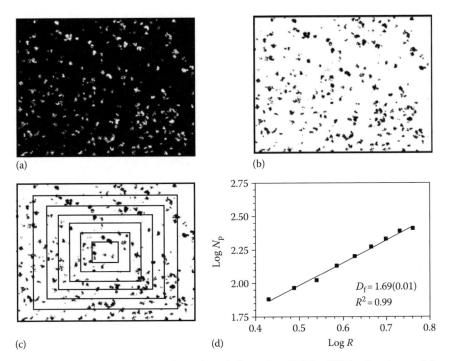

FIGURE 10.44 Determination of the fractal dimension (D_f) by PLM using the particle-counting method. A grayscale image (a) of the fat crystal network is thresholded and inverted (b). The number of crystal particles (N_p) within boxes or ROIs of increasing size placed over the thresholded image are counted (c). The fractal dimension is the slope of the log-log plot (d) of the number of particles (N_p) as a function of the square root of ROI (R).

exclusion of the particles that are touching the edges of each ROI. When using small-sized boxes, the numbers of particles present drops dramatically leading to changes in the estimation of the fractal dimension. It is recommended that boxes which represent 30% of the original image size and below should not be used, thus avoiding any errors in the fractal analysis. It is our experience that the value of D_f changes significantly when the particles that touch the edges of the ROI are included or excluded from the calculation. It is then advised that in the fractal analysis, an average be made between the value of D_f obtained both including and excluding all particles that touch the perimeter of the ROI (Marangoni, 2002). Algorithms developed in our laboratory for the calculation of N_p with object image are reported by Litwinenko (2005).

10.7.2 Box Counting Method to Determine Fractal Dimension

The calculation of the fractal dimension with the box dimension (D_b) method is sensitive to the degree of fill. Like the particle counting method, it uses PLM images and is based on the following relationship:

$$N \sim d^{-D_b} \tag{10.17}$$

where N is the number of boxes of a linear size d which covers the crystal features or reflections present in a 2D plane. D_b is calculated by placing a grid formed by boxes of decreasing size d over a properly thresholded image followed by the quantification of the number of boxes (N) necessary to cover the observed crystal reflections. At a given value of d, the number of boxes needed to cover an array of reflections should be minimized, hence the grid is progressively rotated 90° at set angular increments. The number of occupied boxes is counted and only the minimum value of N is used to calculate D_b. The log-log plot of N versus d is plotted. If the microstructure is of a fractal nature, the plot will yield a straight line, where the slope corresponds to D_b. The box sizes used follow a geometric progression (i.e., 1, 2, 4, 8, ...) in order to obtain an evenly spaced plot (TruSoft Int'l Inc., 1999).

Benoit 1.3 software package (TruSoft Int'l Inc., St. Petersburg, FL) includes a box counting method to determine the fractal dimension and therefore is suggested for this calculation. Although Benoit 1.3 software includes an automatic thresholding function, it is recommended that the threshold be performed using Adobe Photoshop, to ensure that all solid mass depicted in the original grayscale images is represented in the processed image. Benoit 1.3 software quantifies white features in a black background; hence, no inversion of the images is necessary. All image files need to be converted to BMP files, as Benoit 1.3 software analyzes bitmap format images.

To perform the analysis, images are open using Benoit 1.3 software. The software will immediately begin the analysis. During the analysis, grids formed by different size boxes are laid over the binary image at different orientations. This, along with the analysis considerations, is observed on the right side of the screen. On the left side of the screen, the resulting log-log plot is progressively constructed. In the plot, the equation of the line obtained from the log-log plot is shown. Once the analysis has finished, a results window appears which includes the resulting D_b and the standard deviation of the plot. Inconsistent results can be observed for very small or very large d values, which may be eliminated from the analysis by selecting them manually. Figure 10.45 illustrates the particle counting procedure used to determine D_b.

10.7.3 RHEOLOGICAL METHOD TO DETERMINE FRACTAL DIMENSION

The calculation of the fractal dimension by means of rheology (D_r) is based on the fact that the elasticity of fat crystal networks is dependent on microstructure, including the spatial distribution of mass. The shear modulus (G) scales with the volume fraction of solids in a power-law fashion, from which the following relationship is obtained:

$$G \sim \lambda \Phi^{1/(3-D_r)} \tag{10.18}$$

where λ is a constant independent of the volume fraction, but dependent on several primary particle structural parameters as well as intermolecular forces, and Φ is the solid volume fraction or SFC/100 (Marangoni, 2000; Marangoni, 2002; Marangoni

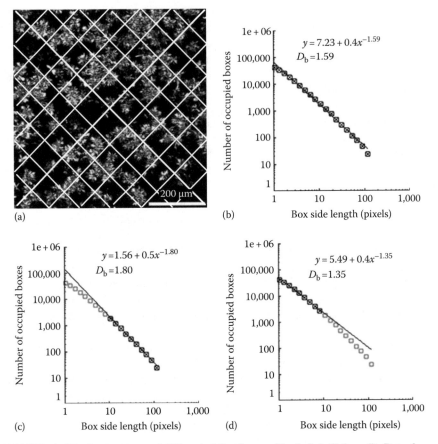

FIGURE 10.45 Screen capture of Benoit 1.3 software (TruSoft Int'l Inc., St. Petersburg, FL, 1999) while the analysis of the fractal dimension is performed. Fractal analysis of cocoa butter crystallized at 24°C by imaging its microstructure (a) followed by its analysis with Benoit 1.3 software (b). In the Figure we observe that the value of D_b changes from 1.59 when using all d values, to 1.8 when manually selecting d values of 10–100 pixels (c) or 1.35 when manually selecting d values of 1–10 pixels (d).

and Rogers, 2003). Based on this relationship, an estimate of D_r of a fat system can be obtained using rheological data by constructing a log-log plot of G' versus Φ values of the network material. Dilution of fat samples with oil allows for a controlled variation of the SFC of the fat crystal network studied. It is critical that the oil used only dilutes without altering the crystal network. The slope of a plot of log G' versus log Φ is used to determine D_r.

10.7.3.1 Procedure

10.7.3.1.1 Sample Preparation and Data Gathering

Prepare a series of dilutions with the sample fat and a vegetable oil (i.e., canola oil) in 2%–5% increments. The oil is used to dilute the solid fraction of the sample fat.

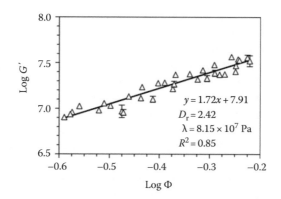

FIGURE 10.46 Log-log plot of the storage modulus (G') as a function of the solid volume fraction (Φ) used in the determination of the fractal dimension (D_r) of a fat crystal network.

Work in our laboratory has shown that canola oil is a good diluent for D_r determinations. Canola oil does not crystallize above 0°C, nor does it appear to solubilize or cocrystallize with a wide range fats. A volume of 20 g of each dilution is recommended, as it is enough for small and large deformation rheology studies. Melt and mix the fat to ensure that it is homogenized and all crystal memory is erased. Transfer sample to either rheology molds (for crystallization of sample discs) or to NMR tubes, as directed in the previous sections. Immediately transfer the molds and NMR tubes to a temperature-controlled incubator set at the temperature of study for 24 h before measurements are taken. Most fractal work entails a temperature regime of 5°C for 24 h, however, the time-temperature profile particular to the studied fat may be decided upon by the user. After the appropriate time of crystallization at the temperature of study has elapsed, proceed to measure the rheological parameters (i.e., G' and G'') and SFC of each dilution. For details how to measure these parameters, refer to previous sections.

10.7.3.1.2 Data Analysis

Construct a plot of the logarithm of G' as a function of the logarithm of Φ (SFC divided by 100), as shown in Figure 10.46. If the microstructure is of fractal nature, the plot will follow a straight line. Fit the data to a linear regression and obtain the slope and the x-intercept. The slope (m) is used to determine D_r based on the following equation:

$$m = \frac{1}{3 - D_r} \quad \text{therefore } D_r = 3 - \frac{1}{m} \tag{10.19}$$

10.8 OIL MIGRATION

Migration of oil in products that contain two or more fat-based components is of critical importance for the food industry, in particular for confectionery products. For example, a chocolate truffle—a curved shape center of chocolate mixed with cream covered in a hard shell of chocolate—has great potential for quality issues through shelf life due to migration of the low melting point molecules in the soft creamy center toward the harder higher melting point chocolate shell. Migration is believed to result from concentration gradients of different melting point lipid species through capillary or diffusion mechanisms (Ziegleder et al., 1996a,b; Miquel et al., 2001; Aguilera et al., 2004; Khan and Rousseau, 2006). Quality issues that result from the migration of fat systems in a complex food product matrix include softening of the hard component, along with some degree of hardening of the soft component. Quality issues are not limited to changes in texture. As one phase migrates into the other, stability, color, and crystal structures also change. Due to the fact that many finished products have extended shelf lives (in the range of months), the understanding and quantification of oil migration between two different fat systems is essential.

Physical and chemical methodologies have been used to approach the study of oil migration. These include time studies to monitor differences in FA or TAG composition by chromatographic methods; variations in the SFC by pNMR; changes in the thermal properties (melting points and melting behaviors) by DSC; and macroscopic measurements of texture through both sensory and instrumental testing. Most of these methods examine real-time oil migration in finished products and are thus timely and expensive. Two alternative approaches using model systems are described in the following sections which aim to gain fundamental understanding of the phenomenon of oil migration. These include gravimetric quantification of oil lost from a sample and a visual method by which the migration of a stained soft phase through a hard phase is quantified.

10.8.1   Oil Loss Assay

The oil loss assay quantifies the amount of oil that is lost by a sample onto filter paper as a function of time. The amount of oil loss is determined gravimetrically. The results from this assay are used to compare samples which have been crystallized and tested identically. The resulting oil loss, reported in %, does not correspond to the mass of oil that a given sample will lose to migration in a multicomponent matrix. It does however provide comparisons between different fat samples or formulations.

10.8.1.1   Procedure

10.8.1.1.1   Sample Preparation

Crystallize the sample fat or fat-based matrix into disks of uniform dimensions. The molds used for rheological testing previously described in this chapter (22 mm diameter by 3 mm height) are recommended. Although it is not strictly necessary to work with sample disks, it is critical to crystallize samples in a format that has the same dimensions, so that the sample mass and area of contact between the filter paper and the sample are maintained constant between replicates and variables tested. For each sampling time point, at least three replicates are recommended. Prepare a different sample disk for each replicate at each time point.

10.8.1.1.2   Experimental Procedure

Place crystallized samples on pre-weighed circular filter paper sheets (Whatman No. 4, 150 mm diameter, GE Healthcare Bio-Sciences Corp., Piscataway NJ, USA) and store in temperature-controlled incubators for the duration of the study.

The storage conditions will depend on the user's research objectives. As a starting point, storage at 20°C a total of 4 weeks is recommended, with readings after 1, 3, 7, 14, and 28 days. At each sampling point, weigh the filter paper with and without the sample. Samples are discarded after weighing. Reusing samples for further time points is not recommended as the contact between the sample and the filter paper will be different.

Alongside the sample filter papers, weigh three filter papers and store them without any sample. These blank filter papers will be used to monitor and correct for changes in weight due to environmental differences in humidity.

10.8.1.1.3 Data Analysis

Calculate the % oil loss using the following equation:

$$\% \text{ Oil loss} = \frac{w_a - w_b}{S_d} \tag{10.20}$$

where
  w_a is the weight of the filter paper after oil has migrated from the sample disk
  w_b is the weight of the filter paper at the beginning of the study before any oil has migrated
  S_d is the weight of the sample disk

With the calculated values of % oil loss, make a plot as a function of storage time. The % oil loss for different blends of a palm mid fraction (PMF) and palm olein are shown in Figure 10.47. This type of analysis reveals that most of the oil loss takes place within the first 3 days. In the formulation of a confectionery cream with olein, the data from Figure 10.47 point toward the value of adding PMF. With only 20% addition of PMF, the oil loss is reduced by almost half. In other words, the stability toward oil migration of olein is greatly increased when blending with PMF.

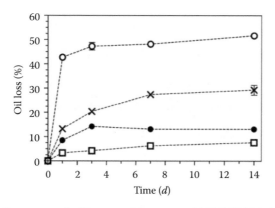

FIGURE 10.47 Percentage of oil lost onto filter paper of 100% PMF (□), 80% PMF: 20% olein (•), 20% PMF: 80% olein (×), and 100% olein (○) stored at 20°C.

10.8.2   FLATBED SCANNER IMAGING TECHNIQUE

This methodology is based on tracking the migration of low-melting point lipid species from a soft fat through a hard fat. Nile red is used to stain the soft fat. In time, Nile red will migrate along with the low-melting point oil phase. The movement of Nile red is captured by imaging with a common flatbed scanner. Through image analysis migration of oil is quantified. The advantage of the methodology is that it allows real-time visualization of migration along with its quantification.

10.8.2.1   Procedure

10.8.2.1.1   Sample Preparation

Assemble molds consisting of two coverslips each adhered using double-sided tape along a glass microscope slide. These coverslips serve a 170 μm height spacer to support an overlying coverslip, as illustrated in Figure 10.48. Melt the hard fat to erase all crystal memory and pour into the indent formed on the microscope slide by the two adhered coverslips. The hard fat used may require specific tempering procedures. For example, when working with cocoa butter, progressive and controlled cooling to 28°C is recommended to crystallize it in the stable β polymorph (Marty et al., 2005). Place a third coverslip on top of the hard fat and cool to 5°C for 40 min. Prepare sample slides in triplicates.

Prepare the soft fat by blending the melted fat with 0.05% Nile red dye (Sigma-Aldrich Co. LLC, St. Louis MO, USA), vortex for three minutes, and store at 5°C

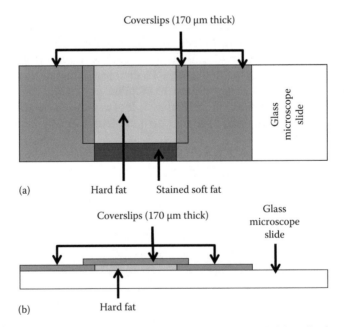

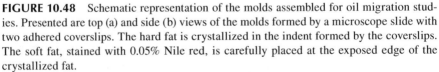

FIGURE 10.48   Schematic representation of the molds assembled for oil migration studies. Presented are top (a) and side (b) views of the molds formed by a microscope slide with two adhered coverslips. The hard fat is crystallized in the indent formed by the coverslips. The soft fat, stained with 0.05% Nile red, is carefully placed at the exposed edge of the crystallized fat.

in the dark. This is done in advance. After the hard fat has been crystallized onto the molds, melt the stained soft fat at 80°C for 30 min. Cool the melted soft fat to just below 30°C to avoid melting the hard fat's crystal structure. Carefully place 25 µL of the stained soft fat at the exposed edge of the crystallized hard fat with the aid of a preheated micropipette to produce a uniform interface between both fats, as shown in Figure 10.48. To prevent light damage throughout sample preparation, keep samples in a light and tight cardboard box. After solidification of the stained phase, transfer slides to temperature-controlled incubators set at the temperature of interest for the duration of the study. Storage conditions and time point at which images are acquired depend on the user's research objectives.

10.8.2.1.2 *Experimental Procedure*

Acquire images at different time points using a flatbed scanner in reflected light mode. In our laboratory, we use a Hewlett Packard Scanjet 6100C DeskScan II (Hewlett Packard Company, Palo Alto, CA). Set image resolution to 24 bit RGB, and 75 pixels per inch with initial document size of 29 by 11 in. In RGB images, colors are reproduced by the combination of red, green, and blue. The known dimensions of glass microscope slides (25 by 75 mm) are used for spatial calibration of the acquired images. A stage micrometer is also imaged for calibration of all images.

The initial image, when time is 0 days is taken directly after the stained soft phase in contact with the hard phase has solidified. Assuming no migration occurs during day 0 of storage; all subsequent image intensity measurements are corrected by subtraction of day 0 value.

10.8.2.1.3 *Data Analysis*

RGB color space scanner images are transformed into hue saturation lightness (HSL) color space with an Adobe Photoshop® 8.0 (Adobe Systems Incorporated) automation using Fovea Pro plug-ins (Reindeer Graphics Inc., Ashville, NC). The HSL color space is a different representation of the color space, taking into consideration hue, saturation, and light, not just primary colors red, green, and blue. It is intended to be more intuitive and perceptually relevant. The saturation and hue channels are individually used to extract monochrome gray scale images. Each gray scale image is blurred using Gaussian smoothing with SD of 2.3 pixels radius (Russ, 2004). The intensity of each smoothed gray scale image is analyzed using ImageJ software (Rasband, W., National Institute of Mental Health, Bethesda, MD; http://rsb.info. nih.gov/ij/), an openly available and license-free software.

Pixel intensity variation along the microscope slide is measured to quantify the migrating front and evaluate the amount of stain throughout the hard fat phase matrix. It is assumed that the amount of stain is equal to the amount of low-melting liquid TAG molecules migrating from the soft fat into the hard fat. Although saturation images represent the amount of stain and hue images represent the stain color (Russ, 2004), work in our laboratory has compared oil migration and amount of oil migrating using images from these two channels.

An invariable central region (10 mm × 18 mm) is defined and all measurements are done within this area to minimize the effects of light scattering at coverslip edges and potential edge effects on the localized migration rates. The uncalibrated pixel intensity or gray-scale value within this central region is plotted as a function of the distance in millimeters from the side that contains the soft fat toward the opposite side containing the hard fat. In the gray scale, white areas have a value of 255 while black areas have a value of 0.

The interface between the stained soft fat and the hard fat is systematically defined as the position corresponding to the maximum brightness value ($\approx$230) in hue component images. The pixel intensity within the stained phase in hue images is lower than at the interface and never reached the maximum value 255. In saturation images, the migration stating point is chosen arbitrarily. Work in our laboratory has not been able to clearly distinguish between the intensity of the stained soft fat and the interface with the hard fat.

To illustrate the migration phenomenon of the soft phase into the hard fat, construct a plot of the measured average pixel intensity as a function of the distance within the microscope slide. The distance plotted starts at the origin of the soft fat and increases toward the opposite plane containing the hard fat. Plots are analyzed using data fitting software, such as GraphPad Prism 4 (GraphPad Software, Inc., San Diego, CA). Curves are fitted using a sigmoidal dose-response (variable slope) model described by the following equation:

$$y = I_h + \left[\frac{I_o - I_h}{1 + 10^{(LogEC50-X)(Hillslope)}} \right] \tag{10.21}$$

where
   I_o is the intensity of the 100% soft fat containing the dye
   I_h is the background intensity of the hard fat without dye

The intensity of I_o is constrained to 255, corresponding to white. The migration distance corresponding to 10% of I_o can then be determined from a transformation of Equation 10.21, namely,

$$X = \log_{10}(EC_{50}) - \frac{1}{\text{Hillslope}} \log_{10} \left(\frac{I_o - I_h}{0.1I_o - I_h} - 1 \right) \tag{10.22}$$

Figure 10.49 shows an example of a grayscale image from which average pixel intensity is measured as a function of distance from the origin of the soft phase toward the hard fat phase within the chosen rectangular central region. Also shown in Figure 10.49 are plots of the raw (b) and normalized (c and d) pixel intensity as a function of distance. The pixel intensity is normalized by dividing all values over the maximum pixel intensity measured. Oil migration kinetics can be assessed using the movement of the dye front based on the distance corresponding to 10% (Figure 10.49c) or 50% (Figure 10.49d) of the maximum intensity. We determined that a 50% value is too imprecise to distinguish between similar profiles such as those illustrated in Figure 10.49. Because oil migration is relatively small within tempered cocoa butter matrices ($\sim$2 mm during the first 10 days) shown in Figure 10.49, it is essential to define a finer threshold value than 50%, and thus, a measurement relying on the 10% value is recommended.

Work in our laboratory has demonstrated that the Nile red stain migrates with the soft fat phase. In a system of cocoa butter and a blend of peanut oil and palm fat stained

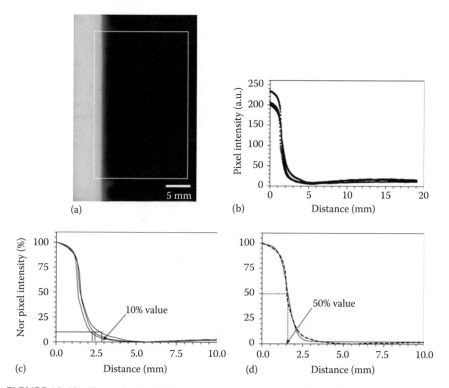

FIGURE 10.49 Example of pixel intensity measurements within the ROI (gray rectangle) of a grayscale image (saturation channel), the stained filling (peanut oil and palm blend) model and the cocoa butter phase appearing light gray and black, respectively (a). Raw pixel intensity profiles of triplicates after 24 h of storage at 18°C (b). Moving front (10% and 50% values) and mass transfer (area under the curve) were determined using normalized pixel intensity values (c and d). A sigmoidal fitting model (solid line in [d]) gave the 50% values while the 10% values were obtained from the normalized data set (c).

with Nile red, the FA profile at different distances from the soft phase toward the cocoa butter were determined. The concentration gradient of linoleic acid into tempered cocoa butter was compared with the pixel intensity decay obtained after 7 days.

ACKNOWLEDGMENTS

Photography courtesy of Arturo Gonzalez de Cosio.

REFERENCES

Abramowitz, M., S. Bradbury, P.C. Robinson, K.R. Spring, B.O. Flynn, J.C. Long, M.J. Parry-Hill, K.I. Tchourioukanov, and M.W. Davidson. Polarized light microscopy. http://www.microscopy.fsu.edu/primer/techniques/polarized/polarizedhome.html

Aguilera, J.M., M. Michel, and G. Mayor. 2004. Fat migration in chocolate: Diffusion or capillary flow in a particulate solid? A hypothesis paper. *J. Food Sci.* 69(7): R167–R174.

AOCS. *Official Methods and Recommended Practices of the American Oil Chemists' Society.* 1999. Champaign, IL: American Oil Chemists' Society. Solid fat content (SFC) by low-resolution nuclear magnetic resonance, official method Cb-16b-93.

Avnir, D., O. Biham, D. Lidar, and O. Malcai. 1998. Is the geometry of nature fractal? *Science.* 279(5347): 39–40.

Avrami, M. 1939. Kinetics of phase change I. General theory. *J. Chem. Phys.* 7(12): 1103–1112.

Bragg, W.L. 1933. *The Crystalline State*, Vol. I. London, U.K.: G. Bell and Sons, Ltd.

Bruker. 1989. *Bruker PC/20 Series Nuclear Magnetic Resonance Analyzer, Instrument's Manual.* Milton, Ontario, Canada: Bruker.

Cullity, B.D. and S.R. Stock. 2001. *Elements of X-Ray Diffraction.* Upper Saddle River, NJ: Prentice Hall.

Daubert, C.R. and E.A. Foegeding. 1998. Rheological principles for food analysis. In: *Food Analysis*, 2nd edn., Nielsen, S.S., Ed., Gaithersburg, MD: Aspen Publishers, Inc., pp. 551–569.

Dutch, S. Light and polarization. http://www.uwgb.edu/dutchs/petrolgy/genlight.htm

Hartel, R.W. 1996. Applications of milk-fat fractions in confectionery products. *J. Am. Oil Chem. Soc.* 73(8): 945–953.

Hartel, R.W. 2001. *Crystallization in Foods.* Gaithersburg, MD: Aspen Publishers, pp. 145–191.

IUPAC. 1992. *Standard Methods for the Analysis of Oils, Fats and Derivatives.* International Union of Pure and Applied Chemistry. Oxford, U.K.: Blackwell Scientific Publications, Solid content determination in fats by NMR, standard method 2.150.

Khan, R.S. and D. Rousseau. 2006. Hazelnut oil migration in dark chocolate—Kinetic, thermodynamic and structural considerations. *Eur. J. Lipid Sci. Technol.* 108: 434–443.

Larsson, K. 1994. Lipids-molecular organization, physical functions and technical applications. The Oily Press LTD, Sweden.

Litwinenko, J.W. 2005. Fat crystal networks: Microstructure—DVD image archive. In: *Fat Crystal Networks*, Marangoni, A.G., Ed. New York: Marcel Dekker.

de Man, J.M. 1992. X-ray diffraction spectroscopy in the study of fat polymorphism. *Food Res. Int.* 25: 471–476.

de Man, J.M. and A.M. Beers. 1987. Fat crystal networks: Structure and rheological properties. *J. Text. Stud.* 18: 303–318.

Manderlbrot, B.B. 1989. Fractal geometry: What is it, and what does it do? *Proc. R. Soc. Lond.* 423: 3–16.

Marangoni, A.G. 2000. Elasticity of high volume fraction fractal aggregate networks: A thermodynamic approach. *Phys. Rev. B.* 62(21): 13951–13955.

Marangoni, A.G. 2002. The nature of fractality in fat crystal networks. *Trends Food Sci. Technol.* 13: 37–47.

Marangoni, A.G. 2005a. Crystallization kinetics. In: *Fat Crystal Networks*, Marangoni, A.G., Ed. New York: Marcel Dekker, pp. 21–82.

Marangoni, A.G. 2005b. The nature of fractality in fat crystal networks. In: *Fat Crystal Networks*, Marangoni, A.G., Ed. New York: Marcel Dekker, pp. 413–440.

Marangoni, A.G. 2005c. Crystallography. In: *Fat Crystal Networks*, Marangoni, A.G., Ed. New York: Marcel Dekker, pp. 1–20.

Marangoni, A.G. and R.W. Lenki. 1998. Ternary phase behavior of milk fat fractions. *J. Agric. Food Chem.* 46: 3879–3884.

Marangoni, A.G. and M.A. Rogers. 2003. Structural basis for the yield stress in plastic disperse systems. *Appl. Phys. Lett.* 82(19): 3239–3241.

Marty, S., K. Baker, E. Dibildox-Alvarado, J. Neves Rodriguez, and A.G. Marangoni. 2005. Monitoring and quantifying oil migration in cocoa butter using a flatbed scanner and fluorescence microscopy. *Food Res. Int.* 38: 1189–1197.

McGauley, S. 2001. The relationship between polymorphism, crystallization kinetics, and microstructure of statically crystallizes cocoa butter. Master in Science Thesis. University of Guelph: Guelph ON, Canada.

Metin, S. and Hartel, R.W. 1998. Thermal analysis of isothermal crystallization kinetics in blends of cocoa butter with milk fat or milk fat fractions. *J. Am. Oil Chem. Soc.* 75: 1617–1624.

Mezger, T.G. 2002. Oscillatory tests. In: *The Rheology Handbook*. Hannover, Germany: Hannoprint, pp. 112–162.

Miquel, M.E., S. Carli, P.J. Couzens, H.J. Wille, and L.D. Hall. 2001. Kinetics of the migration of lipids in composite chocolate measured by magnetic resonance imaging. *Food Res. Int.* 34(9): 773–781.

Murphy, D., K.R. Spring, and M.W. Davidson. Polarization of light. http://www.microscopyu.com/articles/polarized/polarizedlightintro.html

Narine, S.S. and A.G. Marangoni. 1999a. Fractal nature of fat crystal networks. *Phys. Rev. E.* 59(2): 1908–1920.

Narine, S.A. and A.G. Marangoni. 2005. Microstructure. In: *Fat Crystal Networks*, Marangoni, A.G., Ed. New York: Marcel Dekker.

Robinson, P. and M.W. Davidson. Introduction to polarized light microscopy. http://www.microscopyu.com/articles/polarizedintro.html

Rohm, H. and K.H. Weidinger. 1993. Rheological behavior at small deformations. *J. Text. Stud.* 24: 157–172.

Russ, J. 1999. *The Image Processing and Analysis Cookbook*. Ashville, NC: Reindeer Graphics, Inc.

Sato, K., S. Ueno, and J. Yano. 1999. Molecular interactions and kinetic properties of fats. *Progress in Lipid Research*. 38: 91–116.

Sato, K. 2001. Crystallization behavior of fats and lipids—A review. *Chem. Eng. Sci.* 56: 2255–2265.

Sharples, A. 1996. Overall kinetics of crystallization. In: *Introduction to Polymer Crystallization*, Sharples, A., Ed. London, U.K.: Edward Arnold Publishers, Ltd., pp. 44–59.

Small, D.M. 1966. *Handbook of Lipid Research*. New York: Plenum Press.

Stable Micro Systems. 1997. Stable Micro System Materials Tester Model MT-LQ instruction manual, Surrey England.

Steffe, J.F. (Ed) 1996. Viscoelasticity. In: *Rheological Methods in Food Process Engineering*, 2nd edn. East Lansing, MI: Freeman Press, pp. 294–349.

Thermal Advantage User Reference Guide © 1999, 2000, New Castle, DE: TA Instruments.

Timms, R.E. 2003. *Confectionery Fats Handbook: Properties, Production and Application*. Bridgwater, U.K.: The Oily Press, pp. 25–36.

TruSoft Int'l Inc. 1999. Benoit 1.3 software help manual, TruSoft Int'l Inc, St Petersburg Fl.

Vicsek, T. 1992. Fractal Growth Phenomena. 2nd Ed. World Scientific Publishing Company. Pte. Ltd. Singapore, pp. 9–47.

Walstra, P. 2003. *Physical Chemistry of Foods*. New York: Marcel Dekker.

Wright, A.J., R.W. Hartel, S.S. Narine, and A.G. Marangoni. 2000. The effect of minor components on milk fat crystallization. *J. Am. Oil Chem. Soc.* 77: 463–475.

Wright, A.J., M.G. Scanlon, R.W. Hartel, and A.G. Marangoni. 2001. Rheological properties of milk fat and butter. *J. Food Sci.* 66(8): 1056–1071.

Ziegleder, G., C. Moser, and J. Geier-Greguska. 1996a. Kinetics of fat migration within chocolate products. 1. Principles and analytics. *Fett–Lipid*. 98(6): 196–199.

Ziegleder, G., C. Moser, and J. Geier-Greguska, J. 1996b. Kinetics of fat migration within chocolate products. 2. Influence of storage temperature, diffusion coefficient, solid fat content. *Fett–Lipid*. 98(7–8): 253–256.

Index

Tribehenate (BBB) phase diagrams, 317
Tricaprin
    molecular outline, 10, 12
    single crystal, 10, 11
Tripalmitate (PPP) phase diagrams, 317
Tristearate (SSS) phase diagrams, 317

V

Van der Waals interactions
    fluctuating dipoles, 103
    materials, 102–103
    6–12 potential method, 104
    rheological characteristics, 117–118
    spheres, 104
Viscoelastic properties, fats
    creep and recovery/stress relaxation
        Burger model, 152–154
        butter and margarine, 157
        curve, milkfat, 155
        effect, shear work, 158
        effect, temperature, 157
        idealized Newtonian solid, 148, 149
        Kelvin–Voigt solid, 149–150
        loading force and time, 156
        LVR, 148–149
        Maxwell fluid, 150–152
    Deborah numbers, 147
    description, 147
    Weissenberg effect, 147–148

W

Wide angle x-ray scattering (WAXS) region, 15, 20

X

X-ray diffraction (XRD)
    angles, 447–448
    Bragg's law, 448
    data analysis
        full width half maximum (FWHM), 453
        histogram, signal intensity, 450, 451
        long spacings/small angles, 452
        packing, fatty acid (FA), 452–453
        small spacings/wide angles, 451–452
        thickness (TH), crystal, 453
    description, 447
    diffractometer, 449
    experimental procedure, 450
    powder, 64, 65, 88
    sample preparation, 450
XRD, *see* X-ray diffraction (XRD)

Y

Yield stress and elastic modulus
    description, 233
    model
        changes, 238, 239
        critical strain, 236–237
        elastic energy, 233–234
        flocs, 235
        free energy, 233
        macroscopic strain terms, 236
        scaling relationships, 240
        shear modulus, 238
        simulations, 238, 239
        van der Waals' force, 237–238
        volume, system, 234
        weak-link rheological regime, 235–236
        Young's modulus, 237